# Ocean Remote Sensing with Synthetic Aperture Radar

Special Issue Editors

**Xiaofeng Yang**
**Xiaofeng Li**
**Ferdinando Nunziata**
**Alexis Mouche**

MDPI • Basel • Beijing • Wuhan • Barcelona • Belgrade

*Special Issue Editors*
Xiaofeng Yang
Chinese Academy of Sciences
China

Xiaofeng Li
National Oceanic and Atmospheric Administration
USA

Ferdinando Nunziata
Università degli Studi di Napoli Parthenope
Italy

Alexis Mouche
Laboratoire dOcéanographie Physique et Spatiale, Ifremer
France

*Editorial Office*
MDPI AG
St. Alban-Anlage 66
Basel, Switzerland

This edition is a reprint of the Special Issue published online in the open access journal *Remote Sensing* (ISSN 2072-4292) in 2017 (available at: http://www.mdpi.com/journal/remotesensing/special issues/ocean rs SAR).

For citation purposes, cite each article independently as indicated on the article page online and as indicated below:

| Lastname, F.M.; Lastname, F.M. Article title. *Journal Name*. **Year**. *Article number*, page range. |
|---|

**First Edition 2018**

Image courtesy of Xiaofeng Yang, Xiaofeng Li, Ferdinando Nunziata and Alexis Mouche

**ISBN 978-3-03842-720-9 (Pbk)**
**ISBN 978-3-03842-719-3 (PDF)**

# Table of Contents

# About the Special Issue Editors

**Xiaofeng Yang** received his B.S. degree in environmental science from Sichuan University, Chengdu, China, in 2005, and his Ph.D. degree in cartography and geographic information systems from the Institute of Remote Sensing Applications (IRSA), Chinese Academy of Sciences (CAS), Beijing, China, in 2010. From 2010 to 2015, he was an Associate Professor with the Institute of Remote Sensing and Digital Earth, CAS. He is currently a Full Professor with the State Key Laboratory of Remote Sensing Science, RADI, CAS. His research interests include satellite oceanography, synthetic aperture radar image processing, and marine atmospheric boundary layer process studies. Dr. Yang serves as an Editor for MDPIs Remote Sensing. He is an IEEE Senior Member, and the Secretary of Technical Committee on Space Earth Science, Chinese Society of Space Research. He has also served as the Assistant Chief Scientist of the Chinese Water Circle Observation Mission.

**Xiaofeng Li**, Scientist at NOAA, received his B.S. degree in optical engineering from Zhejiang University, China in 1985 and his Ph.D. degree in physical oceanography from the North Carolina State University, USA in 1997. He is the author of more than 100 peer-reviewed publications covering the topics in remote sensing observation and theoretical/numerical model studies of various types of oceanic and atmospheric phenomena, satellite image processing, ocean surface oil spill and target detection/classification with multi-polarization SAR, and development of sea surface temperature algorithms. Dr. Li currently serves as an Associate Editor of IEEE Transactions on Geoscience and Remote Sensing, is an Associate Editor of the International Journal of Remote Sensing, and the Ocean Section Editor-in-Chief of Remote Sensing. He is an Editorial Board Member of the International Journal of Digital Earth, Big Earth Data, and CAAI Transactions on Intelligence Technology.

**Ferdinando Nunziata**, PhD, was born in Italy in 1982. He received his B.Sc. and M.Sc. degrees (summa cum laude) in telecommunications engineering and his Ph.D. degree (curriculum electromagnetic fields) from the Università degli Studi di Napoli Parthenope, Napoli, Italy, in 2003, 2005, and 2008, respectively. Since 2010, he has been an Assistant Professor of electromagnetic fields with the Università degli Studi di Napoli Parthenope. He authored/coauthored more than 60 peer-reviewed journal papers that deal with applied electromagnetics.

**Alexis Mouche**, received his Ph.D. degree in ocean remote sensing from the Universite de Versailles Saint-Quentin, Versailles, France, in 2005. He is currently a Senior Research Scientist with the Laboratoire dOcéanographie Physique et Spatiale, Institut Francais de Recherche pour lExploitation de la Mer, Plouzane, France. His research interests include the interaction of electromagnetic and oceanic waves for ocean remote sensing applications.

# Preface to "Ocean Remote Sensing with Synthetic Aperture Radar"

The oceans covers approximately 71% of the Earth's surface, 90% of the biosphere and contains 97% of Earth's water. In 1978, NASA launched the first SeaSat satellite, primilary aiming at ocean observations and the microwave synthetic aperture radar (SAR) was one of four instruments. Since then, the global oceans have been observed on SAR images, which has a high resolution (<100 m) and a large swath (450 km for ScanSAR mode images). The microwave SAR can image the ocean surface in all weather conditions and day or night. An increasing number of SAR satellites have become available since the early 1990s, such as the ERS-1/-2 and Envisat satellites, the Radarsat-1/-2 satellites, the COSMO-SkyMed constellation, TerraSAR-X and TanDEM-X, the Gaofen-3, among others. Recently, the European Space Agency lauched a new generation of SAR satellites (Sentinel-1A in 2014 and Sentinel-1B in 2016). This operational SAR mission, for the first time, provides researchers with free and open SAR images necessary to carry out broader and deeper investigation of the global oceans.

SAR remote sensing of ocean and coastal monitoring has become a research hotspot in geoscience and remote sensing. This book—Progress in SAR Oceanography—provides an update of the state-of-the-science research on ocean remote sensing with SAR. Overall, the book presents a variety of marine applications in ocean research topics such as, oceanic surface and internal waves characteristics studies, high-resolution sea surface wind retrieval, shallow-water bathymetry mapping, oil spill detection, coastline and inter-tidal zone classification, ship and other man-made objects detection, as well as remotely sensed data assimilation. The book is aimed at a wide audience, ranging from graduate students, university faculty members, scientists to policy makers and managers.

**Xiaofeng Yang, Xiaofeng Li, Ferdinando Nunziata, Alexis Mouche**
*Special Issue Editors*

*Article*

# Directional Spreading Function of the Gravity-Capillary Wave Spectrum Derived from Radar Observations

**Xuan Zhou [1,*], Jinsong Chong [1], Haibo Bi [2,3], Xiangzhen Yu [4], Yingni Shi [5] and Xiaomin Ye [6]**

1. National Key Laboratory of Microwave Imaging Technology, Institute of Electronics, Chinese Academy of Sciences, Beijing 100190, China; chongjinsong@sina.com
2. Key Laboratory of Marine Geology and Environment, Institute of Oceanology, Chinese Academy of Sciences, Qingdao 266061, China; Bhb@qdio.ac.cn
3. Laboratory for Marine Geology, Qingdao National Laboratory for Marine Science and Technology, Qingdao 266200, China
4. Shanghai Radio Equipment Research Institute, Shanghai 200090, China; yxz8302@163.com
5. College of Information Science & Engineering, Ocean University of China, Qingdao 266100, China; nini0303@163.com
6. National Satellite Ocean Application Service, Beijing 100081, China; yxm@mail.nsoas.org.cn

* Correspondence: zhouxuan@radi.ac.cn; Tel.: +86-134-666-60511

Academic Editors: Xiaofeng Li, Ferdinando Nunziata, Alexis Mouche, Raphael M. Kudela and Prasad S. Thenkabail
Received: 3 December 2016; Accepted: 1 April 2017; Published: 12 April 2017

**Abstract:** Directional spreading function of the gravity-capillary wave spectrum can provide the high-wavenumber wave energy distribution among different directions on the sea surface. The existing directional spreading functions have been mainly developed for the low-wavenumber gravity wave with buoy data. In this paper, we use radar observations to derive the directional spreading function of the gravity-capillary wave spectrum, which is expressed as the second-order Fourier series expansion. So far the standard form of the second-order harmonic coefficient has not been proposed to correctly unify the gravity and gravity-capillary wave. Our strategy is to introduce a correcting term to replace the inaccurate gravity-capillary spectral component in Elfouhaily's directional spreading function. The second-order harmonic coefficient at L, C and Ku band calculated by the radar observation is used to fit the correcting term to obtain one at the full gravity-capillary wave region. According to our proposed the directional spreading function, there is a spectral region between the gravity and gravity-capillary range where it signifies the negative upwind–crosswind asymmetry at low and moderate speed range. And this is not reflected by the previous models, but has been confirmed by radar observations. The Root Mean Square Difference (RMSD) of the proposed second-order harmonic coefficient versus the radar-observed one at L, C band Ku band is 0.0438, 0.0263 and 0.0382, respectively. The overall bias and RMSD are −0.0029 and 0.0433 for the whole second-order harmonic coefficient range, respectively. The result verifies the accuracy of the proposed directional spreading function at L, C band Ku band.

**Keywords:** directional spreading function; gravity-capillary wave; radar observations

---

## 1. Introduction

The gravity-capillary wave plays an important role in air-sea interaction because it affects the mass, momentum and energy flux through the air-sea interface. The wind-induced turbulence transfers wind energy from the atmosphere to the gravity-capillary wave by the friction at the interface, and then because the phase speed of the gravity-capillary wave is less than one of the gravity waves, its energy is pillaged by the gravity wave and wind wave will grow [1,2]. The energy propagation is colinear with gravity crests propagation in absence of currents and internal gravity waves. The directional spectrum is used to describe the gravity-capillary waves, which can give the wave energy distribution among different directions on the sea surface. There are two main ways to observe the azimuthal behavior of the directional spectrum: in situ measurement and remote sensing measurement.

Buoy and its array provided two main types of directional spreading functions of ocean wave spectrum. The cosine-shape spreading function was first proposed by Longuet-Higgines et al. [3] according to the motion of a flotation buoy. Mitsuyasu et al. [4] estimated the spreading parameter of the cosine-shape function using a cloverleaf buoy data. Hasselmann et al. [5] improved the cosine-shape functions using wave data collected by pitch-and-roll buoys. The sech-shape spreading function was advanced by Donelan et al. [6] using data collected from a 14-element wave gauge array because they found that the distribution of wave energy in the direction transverse to the main wave direction behaved like a hyperbolic secant. The above-mentioned directional spreading functions measured by buoy are suitable for the gravity wave spectrum. However, the wavelength of the gravity-capillary wave is too short for buoy and its array to measure the azimuthal behavior of the gravity-capillary wave.

Radar is the important method to measure the gravity-capillary wave spectrum due to the Bragg resonance scattering between electromagnetic waves and gravity-capillary waves [7–9]. Apel [10], Caudal et al. [11] and Liu et al. [12] extended the sech-shape spreading function to the gravity-capillary domain using radar data. However, the sech-shape spreading function cannot explain the angular scattering behavior of radar data due to its noncentrosymmetric property. Guissard [13] pointed out that the directional spreading function of gravity-capillary wave spectrum should contain only even harmonics if expressed as a Fourier series. And then it was also applied to a directional spectrum by Elfouhaily et al. [14] and Hwang et al. [15]. Unfortunately, when these directional spreading functions are used to demonstrate scattering properties of the sea surface, there are obvious differences between theoretical calculations and radar observations. For example, the L-band negative upwind-crosswind asymmetry [16–18] of backscatter at low wind speeds is not explained by the existing directional spreading functions.

The radar observation is related to the directional spectrum by sea surface backscatter model, such as the Two-Scale Model (TSM). However, the double integrals in TSM are very inconvenient to calculate the directional spreading function. For the VV polarization, the solutions of Small-Perturbation Method (SPM) are approximately equal to the TSM solutions in intermediate incidence angles. Therefore, the SPM is used to relate the directional spectrum to the GMFs at VV polarization in intermediate incidence angles. The new directional spreading function at the full gravity-capillary wave region is derived from the L-, C- and Ku-band one calculated by the SPM. This paper is organized as follows: Section 2 describes the L-, C- and Ku-band GMFs. The detail of methodology is given in Section 3. Section 4 validate the directional spreading function by radar observation from SMAP SAR, METOP-A ASCAT and QuikSCAT SeaWinds-1. The whole paper is discussed and concluded in Sections 5 and 6, respectively.

## 2. Data Description

In this paper, the L-, C- and Ku-band geophysical model functions (GMFs), which empirically describe the backscattering properties of sea surface, are chosen to serve as a proxy of radar observations to derive a new directional spreading function of the gravity-capillary wave spectrum. The combination of the L, C and Ku bands provide a good coverage of the gravity-capillary wave spectrum for the wavenumber ranging from 25 to 500 rad/m. The following derivation is based on L-band GMF [18], C-band CMOD5 GMF [19] and Ku-band NSCAT2 GMF [20]. The L-band GMF and CMOD5 GMF express the NRCSs as second-order cosine harmonic functions of the radar-observed azimuthal angle with the analytical functions [21]. The NSCAT2 GMF is given as the lookup table with respect to the NRCS, the polarization, the 10-m-height wind speed, the relative wind direction and the incidence angle.

Figure 1 shows the contour plots of SMAP SAR GMF, CMOD5 GMF and NSCAT2 GMF in 40° incidence angles. The contour line of each wind speed is symmetric around the wind direction. At C and Ku band, the maxima of the contour line occur in the upwind (0°) and downwind (180°) directions and minima in the crosswind (90° or 270°) directions. However, the contour lines of low wind speeds overlap ones of moderate and high wind speeds at L band, which is obviously different from C- and Ku-band pattern. This is not explained by the existing directional spreading functions.

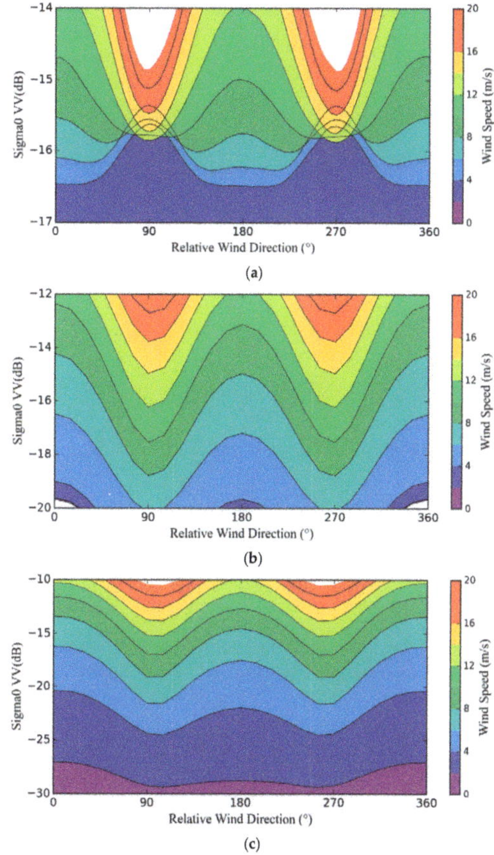


**Figure 1.** The SMAP SAR GMF (**a**); CMOD5 GMF (**b**) and NSCAT2 GMF (**c**) in 40° incidence angles.

## 3. Methodology

The directional spreading function of the gravity-capillary wave spectrum is related to the radar observation through sea surface backscatter model. We first introduce the directional wave spectrum and propose a basic form of directional spreading function. And then TSM and SPM, which are two basic kinds of sea surface backscatter model, are described and compared. It is generally known that TSM is more suitable for a realistic sea surface due to introduce the double integrals to describe the sea surface tilting effect. However, the double integrals in TSM are very inconvenient to be used to calculate the directional spreading function with radar observations. Fortunately, the solutions of SPM are approximately equal to the TSM solutions in intermediate incidence angles at VV polarization.

Finally, we derive the directional spreading function from SPM and calculate its parameters with L-, C- and Ku-band GMFs.

### 3.1. Basic Form of Directional Spreading Function

The directional wave spectrum can provide the directional distribution of ocean wave energy on the sea surface. With the increase of the quality and quantity of available data, more and more directional wave spectrums have been proposed [22,23]. In most cases the directional wave spectrum $\Psi(k,\phi)$ can be described as a function of both the wave wavenumber and the wave direction relative to the wind as follows:

$$\Psi(k,\phi) = \varphi(k)\cdot \mathrm{D}(k,\phi) \tag{1}$$

where $k$ is the wave wavenumber, $\phi$ is the wave direction relative to the wind, $\varphi(k)$ is the omnidirectional wave spectrum and $\mathrm{D}(k,\phi)$ is the directional spreading function defined as:

$$\mathrm{D}(k,\phi) = \frac{\Psi(k,\phi)}{\int_0^{2\pi} \Psi(k,\phi)d\phi} \tag{2}$$

If the directional wave spectrum is expressed as a Fourier series, the directional spreading function should contain only even harmonics:

$$\mathrm{D}(k,\phi) = \frac{1}{2\pi}[1 + \sum_{n=1}^{\infty} \mathrm{a}_{2n}\cos(2n\phi)] \tag{3}$$

where $\mathrm{a}_{2n}$ is the coefficient of even harmonics. In fact the Fourier series expansion is usually truncated to second order [10]:

$$\mathrm{D}(k,\phi) = \frac{1}{2\pi}[1 + \Delta(k)\cos(2\phi)] \tag{4}$$

where $\Delta(k)$ is the second-order harmonic coefficient and the function of both the wavenumber and the wind speed. Unfortunately, up till now the shape of the directional spreading function has been a controversial issue, and the standard form of $\Delta(k)$ has not been given to correctly unify the gravity and gravity-capillary wave [11]. Here, the form of the directional spreading function from Elfouhaily's spectrum is used and $\Delta(k)$ is expressed as:

$$\Delta(k) = \tanh\{a_0 + a_p(c/c_p)^{2.5} + a_m(c_m/c)^{2.5}\} \tag{5}$$

where $a_0$ and $a_p$ are both constants, $a_m$ is the function of $u^*/c_m$, $c$ is the phase speed, $c_p$ is the phase speed of the dominant long wave and $u^*$ is the friction velocity at the sea surface. $a_p(c/c_p)^{2.5}$ and $a_m(c_m/c)^{2.5}$ in Equation (5) are related to the directionality of the gravity wave and the gravity-capillary wave, respectively. However, Elfouhaily's spectrum cannot correctly reflect the directionality of the gravity-capillary wave because radar data is excluded from its development. A correcting term is introduced to replace the gravity-capillary spectral component in the directional spreading function of Elfouhaily's spectrum:

$$\Delta(k) = \tanh\{a_p(c/c_p)^{2.5} + \delta(k,\ U_{10})\} \tag{6}$$

where $a_p$ is equal to 4, $U_{10}$ is the 10-m-height wind speed and $\delta$ is a correction factor which is a function of wavenumber and wind speed.

### 3.2. Comparison and Selection of Sea Surface Backscatter Model

Sea surface backscatter model can describe the relation between radar observation and directional wave spectrum. The SPM and TSM, which are two basic approaches to calculate ocean-surface scattering, are suitable for the small-scale surface and the tilted small-scale surface, respectively.

### 3.2.1. Small-Perturbation Method

According to electromagnetic scattering perturbation theory, the Normalized Radar Cross Section (NRCS) of the gravity-capillary wave surface without regard to the tilting effect can be calculated by the first-order SPM [24]:

$$\sigma_{0pq}(\theta) = 16\pi k_R^4 \cos^4\theta \left|g_{pq}(\theta)\right|^2 \Psi(2k_R \sin\theta, \phi) \tag{7}$$

where $\sigma_0$ is the NRCS, the indices $p$ and $q$ represent transmitting and receiving polarizations, respectively; $k_R$ is the radar wavenumber, $k_R = 2\pi/\lambda$, $\lambda$ is the radar wavelength; $\theta$ is the incidence angle; $g_{pq}(\theta)$ is the first-order scattering coefficient.

### 3.2.2. Two-Scale Model

In fact, the gravity-capillary waves are tilted by the gravity waves of sea surface. The tilting effect modifies the incidence angle $\theta$ referenced to a horizontal surface as the local angle $\theta_i$. Accounting for the sea surface tilting effect, the NRCS is calculated by TSM [24]:

$$\sigma_{0pq}(\theta) = \int_{-\infty}^{\infty}\int_{-\cot\theta}^{\infty} \sigma_{0pq}(\theta_i) P_\theta\left(Z'_x, Z'_y\right) dZ_x dZ_y \tag{8}$$

where $P_\theta\left(Z'_x, Z'_y\right)$ is the slope probability density of the gravity wave as viewed at an incidence angle $\theta$. $Z_x$ and $Z_y$ are the slope components for upwind and crosswind, respectively. $Z'_x$ and $Z'_y$ are expressed as:

$$Z'_x = Z_x \cos\phi + Z_y \sin\phi \tag{9}$$

$$Z'_y = Z_y \cos\phi - Z_x \sin\phi \tag{10}$$

The relation between the slope probability density function $P_\theta\left(Z'_x, Z'_y\right)$ and the function $P\left(Z'_x, Z'_y\right)$ defined by Cox and Munk [25] is:

$$P_\theta\left(Z'_x, Z'_y\right) = (1 + Z_x \tan\theta) P\left(Z'_x, Z'_y\right) \tag{11}$$

The form of $P\left(Z'_x, Z'_y\right)$ is a Gram-Charlier series [25,26]:

$$P\left(Z'_x, Z'_y\right) = \frac{F\left(Z'_x, Z'_y\right)}{2\pi S_u S_c} \exp\left[-\frac{{Z'_x}^2}{2{S_u}^2} - \frac{{Z'_y}^2}{2{S_c}^2}\right] \tag{12}$$

where

$$\begin{aligned} F\left(Z'_x, Z'_y\right) = 1 &- \frac{C_{21}}{2}\left(\frac{{Z'_y}^2}{{S_c}^2} - 1\right)\frac{Z'_x}{S_u} - \frac{C_{03}}{6}\left(\frac{{Z'_x}^2}{{S_u}^3} - \frac{3Z'_x}{S_u}\right) + \frac{C_{40}}{24}\left(\frac{{Z'_y}^4}{{S_c}^4} - 6\frac{{Z'_y}^2}{{S_c}^2} + 3\right) \\ &+ \frac{C_{22}}{4}\left(\frac{{Z'_y}^2}{{S_c}^2} - 1\right)\left(\frac{{Z'_x}^2}{{S_u}^2} - 1\right) + \frac{C_{04}}{24}\left(\frac{{Z'_x}^4}{{S_u}^4} - 6\frac{{Z'_x}^2}{{S_u}^2} + 3\right) \end{aligned} \tag{13}$$

where $C_{40} = 0.4$, $C_{22} = 0.12$, $C_{04} = 0.23$, $C_{21} = 0.01 - 0.0086U$, ${S_u}^2 = 0.005 + 0.78 \times 10^{-3}U$, $C_{03} = 0.04 - 0.033U$, ${S_c}^2 = 0.003 + 0.84 \times 10^{-3}U$.

TSM is more suitable for a realistic sea surface than SPM because accurately expressing the sea surface tilting effect with the double integrals. But the double integrals bring the difficulty for directly calculating the directional spreading function from TSM. In order to simplify the derivation and calculation of directional spreading function, we find in which case the solutions of SPM are approximately equal to ones of TSM by comparing radar backscatters calculated by SPM and TSM in the following section.

### 3.2.3. Comparison of SPM and TSM

Figure 2 shows VV- and HH-polarization NRCS at 5, 12 and 20 m/s wind speeds, which is calculated by SPM and TSM at L-, C- and Ku-band radar frequencies using Elfouhaily's omnidirectional spectrum. For HH polarization, there is an evident disagreement of NRCS calculated by SPM and TSM, especially for high wind speed. This is because the gravity-capillary waves are riding on the gravity waves and are thus tilted with respect to the horizontal. For VV polarization, within the range of about 35°–40° incidence angles, there are a very good agreement between the SPM and TSM solutions. It means that the tilting effect from the gravity waves cannot significantly modify the VV-polarization NRCS and the SPM solutions are approximately equal to the TSM solutions within 35°–40° incidence angles.

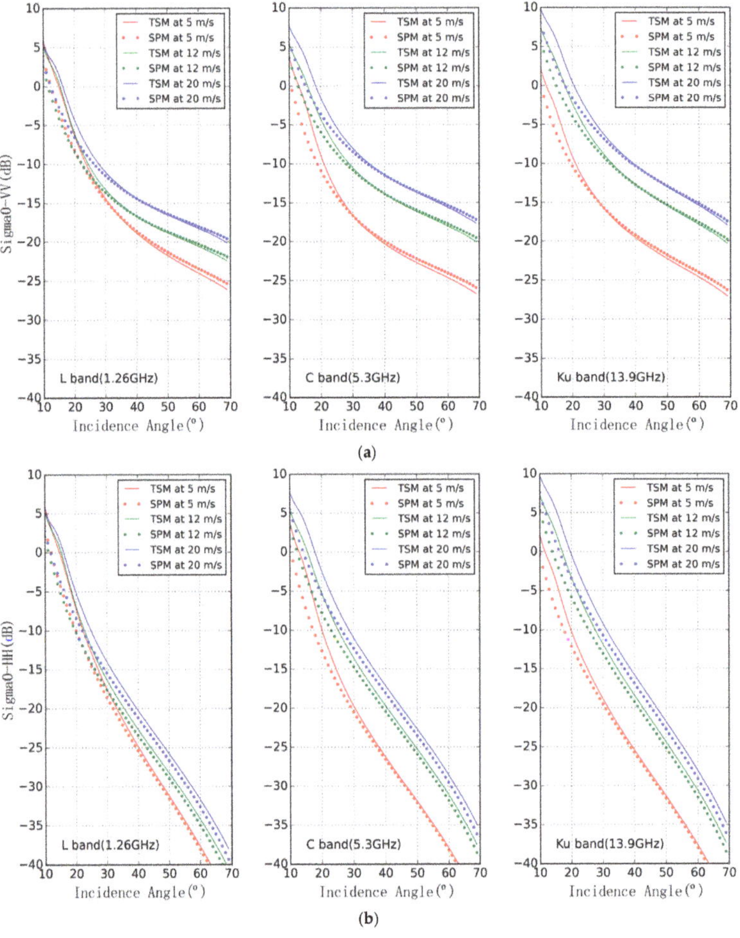


**Figure 2.** Comparison of VV- (**a**) and HH- (**b**) polarization NRCS calculated by SPM and TSM at L band (1.26 GHz), C band (5.3 GHz) and Ku band (13.9 GHz).

### 3.3. Derivation and Calculation of Directional Spreading Function

Wright [27] demonstrated that NRCS calculated by TSM compare favorably with measurements. However, double integrals in TSM (Equation (6)) are very inconvenient to derive the directional spreading function of the gravity-capillary wave spectrum. Fortunately, within 35°–40° incidence angles, the SPM solutions are very close to the TSM solutions at VV polarization. That means the VV-polarization NRCS calculated by Equation (7) is equal to Equation (8). Therefore Equation (7) can be used to retrieve the directional spreading function of the gravity-capillary wave spectrum at VV polarization within 35°–40° incidence angles.

According to Equation (7), the directional spreading function is written as:

$$D(k_B,\phi) = \frac{\sigma_{0vv}\tan^4\theta}{\pi k_B^4 |g_{vv}(\theta)|^2 \varphi(k_B)} \tag{14}$$

where $k_B$ is the wavenumber of the Bragg resonance ocean wave component and related to the radar wavenumber by $k_B = 2k_R \sin\theta$, $\sigma_{0vv}$ represents VV-polarization NRCS and is measured by radar. An empirical functional relationship between the VV-polarization NRCS $\sigma_{0vv}$, the 10-m-height wind speed $U_{10}$, the relative wind direction $\phi$ (the radar azimuth angle with respect to the wind direction) and the incidence angle $\theta$ is generally expressed as:

$$\sigma_{0vv} = A_0(U_{10},\theta)(1 + A_1(U_{10},\theta)\cos\phi + A_2(U_{10},\theta)\cos 2\phi) \tag{15}$$

where the $A_1$ term describes the upwind-downwind difference of NRCS. The difference is weak and cannot be attributed to the contribution of ocean wave spectrum [13]. We do not discuss the upwind-downwind difference in this paper. The $A_2$ term describes the upwind-crosswind asymmetry of NRCS and is calculated by:

$$A_2(U_{10},\theta) = \frac{\sigma_{0vv}^{upwind} + \sigma_{0vv}^{downwind} - 2\sigma_{0vv}^{crosswind}}{\sigma_{0vv}^{upwind} + \sigma_{0vv}^{downwind} + 2\sigma_{0vv}^{crosswind}} \tag{16}$$

where $\sigma_{0vv}^{upwind}$, $\sigma_{0vv}^{downwind}$ and $\sigma_{0vv}^{crosswind}$ are the VV-polarization NRCS along the upwind (0°), downwind (180°) and crosswind (90° or 270°) directions, respectively. Because the radar-observed NRCS is proportional to the directional spreading function in Equation (14), the $A_2$ term can be analogous to the second-order harmonic coefficient $\Delta(k)$ in the directional spreading function of the gravity-capillary wave. Therefore, the second-order harmonic coefficient $\Delta(k)$ is expressed as:

$$\Delta(k) = \frac{\sigma_{0vv}^{upwind} + \sigma_{0vv}^{downwind} - 2\sigma_{0vv}^{crosswind}}{\sigma_{0vv}^{upwind} + \sigma_{0vv}^{downwind} + 2\sigma_{0vv}^{crosswind}} \tag{17}$$

Presently, the L-, C- and Ku-band GMFs, which empirically relate the NRCS, the 10-m-height wind speed, the relative wind direction and the incidence angle, are better developed than other frequencies with radar observation. The combination of the L, C and Ku bands provide a good coverage of the gravity-capillary wave spectrum for the wavenumber ranging from 25 to 500 rad/m. These GMFs can provide the $\sigma_{0vv}^{upwind}$, $\sigma_{0vv}^{downwind}$ and $\sigma_{0vv}^{crosswind}$ at the three frequency bands and are used to derive a directional spreading function of the gravity-capillary wave.

The following calculations are based on the L-band GMF, CMOD5 GMF, and NSCAT2 GMF. Figure 3 shows that the second-order harmonic coefficient $\Delta(k)$ from L-, C- and Ku-band GMF in 35°–40° incidence angles vary with the wavenumber. The L-band $\Delta(k)$ is obviously less than the C- and Ku-band ones at all wind speeds, and even is negative at low wind speeds. That indicates that the upwind-crosswind asymmetry of NRCS at L band is weaker than ones at C and Ku band. However, the Elfouhaily's $\Delta(k)$ has very little variation in the wavenumber ranging from 10 to 1000 rad/m (contain L, C and Ku band) and is positive at all wind speeds. That cannot explain the obvious variation in L-, C- and Ku-band $\Delta(k)$ from radar observations, and is inconsistent with the L-band negative value of radar observations at low wind speeds [15]. Therefore a new directional spreading function should be developed to explain these azimuthal behaviors.

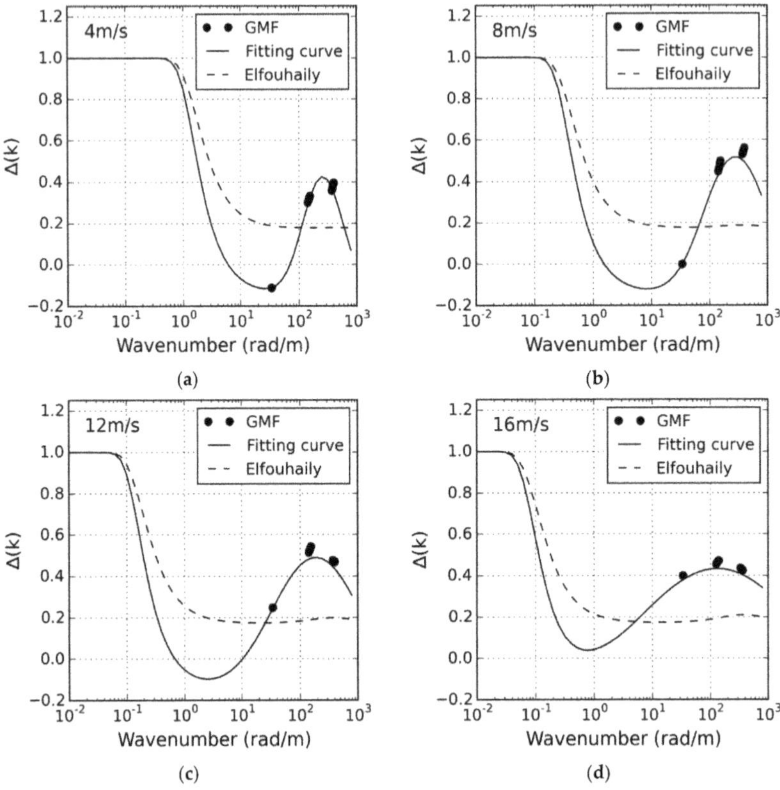


**Figure 3.** The second-order harmonic coefficient $\Delta(k)$ inferred from the GMFs in 35°–40° incidence angles, Elfouhaily's spectrum and fitting curves plots as a function of wavenumber at wind speeds of 4 (**a**); 8 (**b**); 12 (**c**); 16 (**d**) m/s.

According to Equation (17), we use the NRCSs from L-, C- and Ku-band GMF at 35°–40° incidence angle and 2–20 m/s wind speed range to calculate the second-order harmonic coefficient $\Delta(k)$ at wavenumbers of 30–33, 127–142, 333–374 rad/m. And then the second-order harmonic coefficient $\Delta(k)$ at the full gravity-capillary wave region is derived by fitting the L-, C- and Ku-band $\Delta(k)$ to Equation (6) with the Least-Squares-Fitting (LSF) method. $\delta(k, U_{10})$ in Equation (6) is written as:

$$\delta(k, U_{10}) = \delta_0 + 10^{B_1 K^2 + B_2 K + B_3} \tag{18}$$

where $\delta_0$ is a constant and equal to $-0.1467$; $K = log_{10}(k)$, $k$ is expressed in radian per meter; $B_1$, $B_2$ and $B_3$ are the regression coefficients and can be derived in each wind-speed bin. The cubic functions of wind speed are used to model $B_1$, $B_2$ and $B_3$ by the LSF method.

$$B_1 = p_{13}U_{10}^3 + p_{12}U_{10}^2 + p_{11}U_{10} + p_{10} \tag{19}$$

$$B_2 = p_{23}U_{10}^3 + p_{22}U_{10}^2 + p_{21}U_{10} + p_{20} \quad (20)$$

$$B_3 = p_{33}U_{10}^3 + p_{32}U_{10}^2 + p_{31}U_{10} + p_{30} \quad (21)$$

where $p_{1i}$, $p_{2i}$ and $p_{3i}$ are the coefficients of the cubic functions and given in Table 1.

**Table 1.** The regression coefficients for $B_1$, $B_2$ and $B_3$.

| Coefficients | $p_{i3}$ | $p_{i2}$ | $p_{i1}$ | $p_{i0}$ |
|---|---|---|---|---|
| $B_i$ $(i = 1)$ | $3.6924 \times 10^{-3}$ | $-2.1047 \times 10^{-1}$ | 3.9774 | $-2.5721 \times 10^{1}$ |
| $B_i$ $(i = 2)$ | $-3.2332 \times 10^{-3}$ | $1.8138 \times 10^{-1}$ | −3.3790 | $2.1479 \times 10^{1}$ |
| $B_i$ $(i = 3)$ | $7.1639 \times 10^{-4}$ | $-3.9639 \times 10^{-2}$ | $7.2625 \times 10^{-1}$ | −4.5533 |

According to Equations (6) and (18)–(21), we plots the proposed second-order harmonic coefficient $\Delta(k)$ as a function of wavenumber for wind speeds from 2 m/s to 20 m/s with a 4 m/s step in Figure 4. The proposed $\Delta(k)$ is 1 in the gravity wave region and then decreases with the increasing wavenumber. When the wavenumber is close to the gravity-capillary wave region, the proposed $\Delta(k)$ drops to the nadir. The nadir is even negative at low and moderate speed range (2–14 m/s). This feature is confirmed by radar observation but is not reflected by the previous models, such as directional spreading functions of Apel [10], Caudal et al. [11] and Elfouhaily et al. [14]. When the wavenumber is in the gravity-capillary wave region, there exists obviously the peak, which will move toward the low wavenumber under the conditions of high wind speeds. The value of peak varies with the wind speed. Its maximum is about 0.4759 and occurs at the wind speed of about 10 m/s and the wavenumber of about 260 rad/m where the gravity-capillary wave spectrum shows the strongest dependence on the direction.

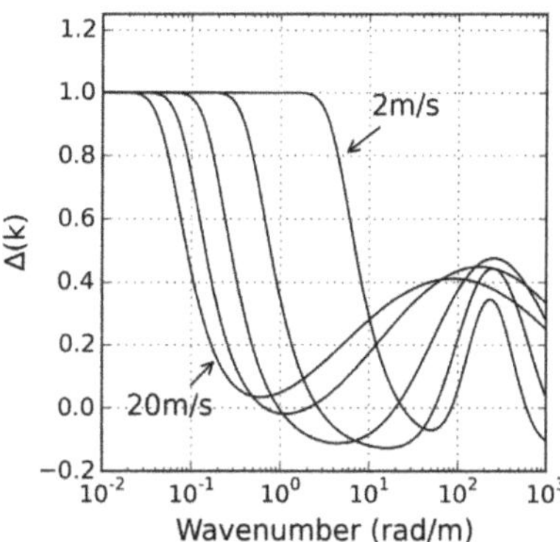


**Figure 4.** The proposed second-order harmonic coefficient $\Delta(k)$ plots as a function of wavenumber for wind speeds from 2 m/s to 20 m/s with a 4 m/s step.

Figure 5 shows the proposed directional spreading function of L, C and Ku band at wind speeds of 4, 8, 12, 16 m/s in polar coordinate. The C- and Ku-band amplitudes at all wind speeds (4, 8, 12 and 16 m/s) and the L-band amplitude at high wind speeds (12 and 16 m/s) along upwind (0°) or downwind (180°) directions are evidently greater than one along crosswind (90° or 270°) directions. In contrast, the L-band amplitude at low and moderate wind speeds (4 and 8 m/s) along upwind (0°) or downwind (180°) directions is less than one along crosswind (90° or 270°) directions, which signifies the negative upwind–crosswind asymmetry. It is consistent with the directional feature observed by Yueh et al. [17], Zhou et al. [18] and Isoguchi et al. [28]. In addition, the difference of the directional spreading function between L, C and Ku band decreases with the increase of wind speed. When the wind speed increases to 16 m/s, the maximum difference is less than 0.02, which means that the directional spreading function of the gravity-capillary wave spectrum has very little variation with the frequency (wavenumber) at high wind speeds.

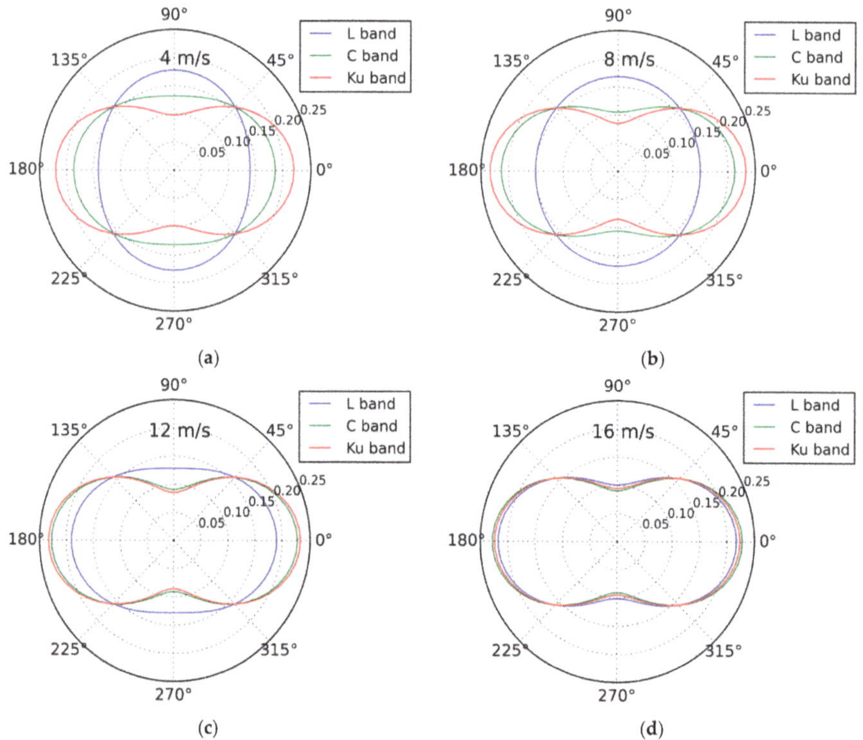


**Figure 5.** The L-, C- and Ku-band directional spreading function plots as a function of the wave direction relative to the wind at wind speeds of 4 (**a**); 8 (**b**); 12 (**c**); 16 (**d**) m/s.

## 4. Verification of Directional Spreading Function

The gravity-capillary wave spectrum is not obtained with traditional wave measuring techniques, and therefore it is not feasible to directly verify the proposed directional spreading function of the gravity-capillary wave spectrum with field data at present. Fortunately, the radar backscatter carries the information of the directional wave spectrum due to the Bragg resonance, thus the proposed directional spreading function can be verified by radar observations from the L-band SAR on the SMAP satellite, the C-band ASCAT scatterometer on the METOP-A satellite and the Ku-band SeaWinds-1 scatterometer on the QuikSCAT satellite. SMAP SAR NRCS, simultaneous DMSP F17 SSMI/S wind speed and NCEP wind direction are used to act as the L-band validation data, and its time range is from 18 to 28 April 2015. ASCAT NRCS and wind field are used to act as the C-band validation data, and its time range from 1 to 10 February 2010. SeaWinds-1 NRCS and wind field are used to act as the Ku-band validation data, and its time range is from 1 to 10 January 2008.

According to Equation (4), the accuracy of the directional spreading function is closely related to the second-order harmonic coefficient $\Delta(k)$, which can be calculated by the VV-polarization NRCS from radar observation along the upwind, downwind and crosswind directions. Therefore, we validate the accuracy of the directional spreading function by comparing the proposed second-order harmonic coefficients and the radar-observed second-order harmonic coefficients.

Figure 6 shows the comparisons of the second-order harmonic coefficient $\Delta(k)$ from the proposed direction spreading function, radar observation and Elfouhaily's spectrum at L, C and Ku band at 2–20 m/s wind speed range. The incidence angles of L-band data from SMAP SAR and C-band data from ASCAT are both 40°, and one of Ku-band data from SeaWinds-1 is 55°. The proposed $\Delta(k)$ varies with wind speeds and is basically consistent with one from radar observation. Other than the above two ones, the Elfouhaily's $\Delta(k)$, which is about 0.2 and has very little variation with wind speeds especially at C and Ku band, is inconsistent with radar observation. The comparisons between the three second-order harmonic coefficients indicates that the proposed direction spreading function is more consistent with radar observation than Elfouhaily's spectrum, which is also reflected by the statistics of the comparisons in Table 2.

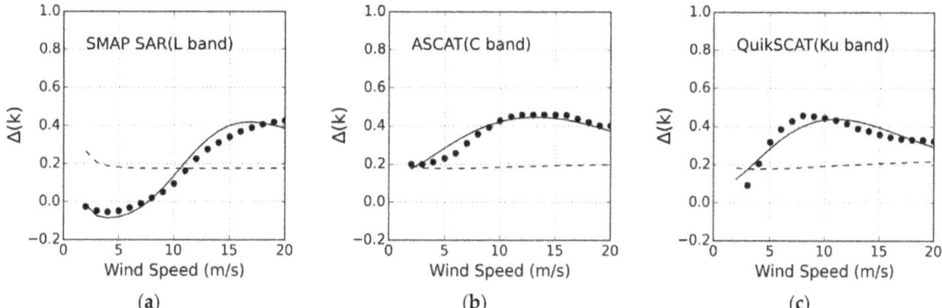


**Figure 6.** Comparisons of the $\Delta(k)$ from the proposed direction spreading function(solid lines), radar observations (dotted lines) and Elfouhaily's spectrum (dashed lines) at L (**a**), C (**b**) and Ku (**c**) band at 2–20 m/s wind speed range.

Table 2 shows the RMSD and CC of the proposed $\Delta(k)$ and Elfouhaily's $\Delta(k)$ versus the radar-observed $\Delta(k)$ at L, C and Ku band at 2–20 m/s wind speed range. The radar-observed $\Delta(k)$ acts as sea truth data. The L-band RMSD and CC of the proposed $\Delta(k)$ is 0.0438 and 0.9745, respectively. The C- and Ku-band RMSDs are reduced to 0.0263 and 0.0382, respectively, and their CCs are reduced to 0.9656 and 0.9009, respectively. Overall, the proposed $\Delta(k)$ has the high accuracy and is remarkably consistent with radar observation. In addition, it is obvious that the RMSD of Elfouhaily's $\Delta(k)$ is greater than one of the proposed $\Delta(k)$, and its CC is less than one of the proposed $\Delta(k)$. In other words, the accuracy of the proposed $\Delta(k)$ is much higher than Elfouhaily's $\Delta(k)$ because the development of Elfouhaily's spectrum does not introduce radar data, which contain the information of the gravity-capillary waves.

**Table 2.** Statistics of the proposed $\Delta(k)$ and Elfouhaily's $\Delta(k)$ versus the radar-observed $\Delta(k)$ at 2–20 m/s wind speed range.

| $\Delta(k)$ | L Band | | C Band | | Ku Band | |
|---|---|---|---|---|---|---|
| | RMSD | CC | RMSD | CC | RMSD | CC |
| the proposed $\Delta(k)$ | 0.0438 | 0.9745 | 0.0263 | 0.9656 | 0.0382 | 0.9009 |
| Elfouhaily's $\Delta(k)$ | 0.2005 | −0.3965 | 0.0909 | 0.5840 | 0.0887 | −0.0257 |

Figure 7 shows the scatterplots of the proposed $\Delta(k)$ and Elfouhaily's $\Delta(k)$ versus the radar-observed $\Delta(k)$ at L, C and Ku band at 2–20 m/s wind speed range. The incidence angles of L-band data from SMAP SAR and Ku-band data from SeaWinds-1 is 40° and 55°, respectively, and ones of C-band data from ASCAT is from 35° to 45°. The radar-observed $\Delta(k)$ acts as sea truth data. Figure 7a shows that the proposed $\Delta(k)$ is basically consistent with radar observation. The bias and

RMSD of the proposed versus the radar-observed $\Delta(k)$ are $-0.0029$ and 0.0433, respectively. Figure 7b shows that Elfouhaily's $\Delta(k)$ is a serious deviation from radar observation and its bias and RMSD are 0.1580 and 0.1426, respectively. Obviously, the proposed $\Delta(k)$ is more consistent with radar observation than Elfouhaily's $\Delta(k)$.

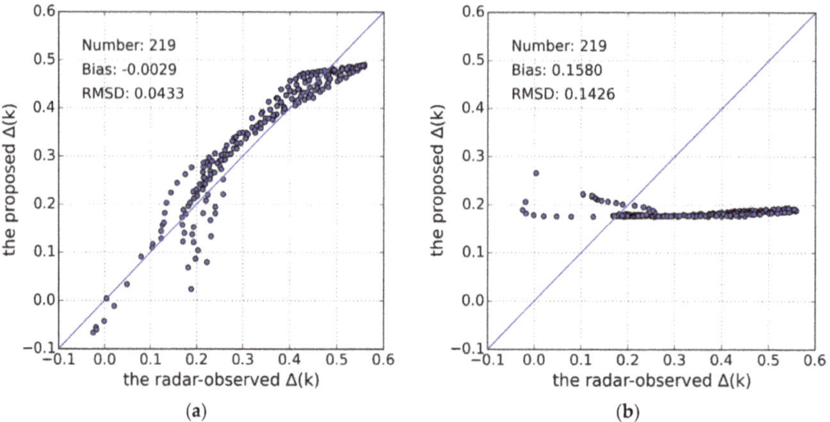


**Figure 7.** The radar-observed $\Delta(k)$ versus the proposed $\Delta(k)$ (**a**) and Elfouhaily's $\Delta(k)$ (**b**).

In conclusion, the proposed direction spreading function has the high accuracy and is basically consistent with radar observation on the basis of comparisons and statistics of the second-order harmonic coefficient.

## 5. Discussion

At present, there is no standard form of the directional spreading function to correctly unify the gravity and gravity-capillary wave. If the directional wave spectrum is expressed as a Fourier series, the cosine-shape spreading function proposed by Longuet-Higgines et al. [3], Mitsuyasu et al. [4], Hasselmann et al. [5], and the sech-shape spreading function proposed by Donelan et al. [6] are all transformed into the hyperbolic tangent form. Due to the natural involvement of the hyperbolic tangent function, the proposed directional spreading function in this paper, which introduces a correcting term to replace the gravity-capillary spectral component in the directional spreading function of Elfouhaily's spectrum, is a good choice to unify the gravity and gravity-capillary wave. The correcting term is a function of wavenumber and wind speed with twelve adjusted parameters, which are derived from L-, C- and Ku-band radar backscatter. It is noted that the X-band GMF has been developed by Li et al. [29] and Ren et al. [30]. But the upwind-crosswind asymmetry in their GMFs needs to be verified by large amounts of data. Therefore, the better developed L-, C- and Ku-band GMFs are only used to derive and verify the directional spreading function of the gravity-capillary wave spectrum.

## 6. Conclusions

In this paper, the directional spreading function of the gravity-capillary wave spectrum is expressed as the second-order Fourier series expansion. It is worthwhile to note that no standard form of the second-order harmonic coefficient has been given to correctly unify the gravity and gravity-capillary wave at the present. Our strategy is to introduce a correcting term to replace the inaccurate gravity-capillary spectral component in Elfouhaily's directional spreading function. And then we derive the correcting term from radar observations.

The Two-Scale Model (TSM) is widely applied to describe the relation between radar observation and directional wave spectrum, but has double integrals to be very inconvenient to study the directional spectrum. Fortunately, the comparison of radar backscatters calculated by Small-Perturbation Method (SPM) and TSM shows that the SPM solutions are approximately equal to the TSM solutions for VV polarization within 35°–40° incidence angles. So we use the SPM to relate radar observations at VV polarization within intermediate incidence angles to the directional wave spectrum.

The radar-observed Normalized Radar Cross Section (NRCS) is proportional to the directional spreading function in radar backscatter model of SPM. So the upwind-crosswind asymmetry of NRCS is analogous to the second-order harmonic coefficient in the directional spreading function. The second-order harmonic coefficient at wavenumbers of 30–33, 127–142, 333–374 rad/m calculated by the NRCS is used to fit the correcting term to obtain one at the full gravity-capillary wave region by the Least-Squares-Fitting (LSF) method. According to the proposed second-order harmonic coefficient, we find that there is the obvious peak at the gravity-capillary wave region, which varies with the wind speed. In addition, there exist the negative values at low and moderate wind speeds in proposed second-order harmonic coefficient, which is different from the previous model but has been confirmed by the L-band radar observation. The Root Mean Square Difference (RMSD) of the proposed second-order harmonic coefficient versus the L-, C- and Ku-band radar-observed one from SMAP SAR, ASCAT and SeaWinds-1 are 0.0438, 0.0263 and 0.0382, respectively. The L-, C- and Ku-band Correlation Coefficient (CC) is 0.9745, 0.9656 and 0.9009, respectively. The overall bias and RMSD are −0.0029 and 0.0433 for the whole second-order harmonic coefficient range, respectively. This means that the proposed second-order harmonic coefficient in the paper has the high accuracy and is consistent with radar observation at L, C and Ku band.

This paper verifies the proposed second-order harmonic coefficient with L-, C- and Ku-band NRCS. It is worthwhile to note that the proposed second-order harmonic coefficient is derived at the full gravity-capillary wave region and the accuracy at other microwave frequency bands needs to be quantitatively verified. With the increase of the quality and quantity of available data at other microwave frequency bands, the future work will be able to further verify and improve it.

**Acknowledgments:** This study was supported by the Natural Science Foundation of China (41276185, 41406215).

**Author Contributions:** Xuan Zhou and Jinsong Chong contributed the main idea and wrote the manuscript; Haibo Bi and Xiangzhen Yu derived and verified the directional spreading function; Xiaomin Ye compared radar backscatter calculated by SPM and TSM. All authors have read and approved the submitted manuscript.

**Conflicts of Interest:** The authors declare no conflict of interest.

## References

1. Sullivan, P.P.; McWilliams, J.C. Dynamics of winds and currents coupled to surface waves. *Annu. Rev. Fluid Mech.* **2010**, *42*, 19–42. [CrossRef]
2. Long, S. Wind-generated water waves in a wind tunnel: Free surface statistics, wind friction and mean air flow properties. *Coast. Eng.* **2012**, *61*, 27–41. [CrossRef]
3. Longuet-Higgins, M.S.; Cartwright, D.E.; Smith, N.D. Observations of the directional spectrum of sea waves using the motions of a floating buoy. In *Ocean Wave Spectra*; Prentice-Hall: Upper Saddle River, NJ, USA, 1963; pp. 111–136.
4. Mitsuyasu, H.; Tasai, F.; Suhara, T.; Mizuno, S.; Ohkusu, M.; Honda, T.; Rikiishi, K. Observation of the directional wave spectra of ocean waves using a cloverleaf buoy. *J. Phys. Oceanogr.* **1975**, *5*, 750–760. [CrossRef]
5. Hasselmann, D.E.; Dunckel, M.; Ewing, J.A. Directional wave spectra observed during JONSWAP 1973. *J. Phys. Oceanogr.* **1980**, *10*, 1264–1280. [CrossRef]
6. Donelan, M.A.; Hamilton, J.; Hui, W.H. Directional spectra of wind-generated waves. *Philos. Trans. R. Soc. Lond.* **1985**, *315*, 509–562. [CrossRef]
7. Shao, W.; Zhang, Z.; Li, X.; Li, H. Ocean wave parameters retrieval from Sentinel-1 SAR imagery. *Remote Sens.* **2016**, *8*. [CrossRef]
8. Shao, W.; Li, X.; Sun, J. Ocean Wave Parameters Retrieval from TerraSAR-X Images Validated against Buoy Measurements and Model Results. *Remote Sens.* **2015**, *7*, 12815–12828. [CrossRef]

9. Li, X.; Pichel, W.; He, M.; Wu, S.; Friedman, K.; Clemente-Colon, P.; Zhao, C. Observation of Hurricane-Generated Ocean Swell Refraction at the Gulf Stream North Wall with the RADARSAT-1 Synthetic Aperture Radar. *IEEE Trans. Geosci. Remote Sens.* **2002**, *40*, 2131–2142.
10. Apel, J.R. An improved model of the ocean surface wave vector spectrum and its effects on radar backscatter. *J. Geophys. Res.* **1994**, *99*, 16269–16291. [CrossRef]
11. Caudal, G.; Hauser, D. Directional spreading function of the sea wave spectrum at short scale, inferred from multifrequency radar observations. *J. Geophys. Res.* **1996**, *101*, 16601–16613. [CrossRef]
12. Liu, Y.G.; Yan, X.H. The wind-induced wave growth rate and the spectrum of the gravity-capillary waves. *J. Phys. Oceanogr.* **1995**, *25*, 3196–3218. [CrossRef]
13. Gutssard, A. Directional spectrum of the sea surface and wind scatterometry. *Int. J. Remote Sens.* **1993**, *14*, 1615–1633. [CrossRef]
14. Elfouhaily, T.; Chapron, B.; Katsaros, K.; Vandemark, D. A unified directional spectrum for long and short wind-driven waves. *J. Geophys. Res.* **1997**, *102*, 15781–15796. [CrossRef]
15. Hwang, P.; Wang, D.W.; Walsh, E.J.; Krabill, W.B.; Swift, R.N. Airborne measurements of the wavenumber spectra of ocean surface waves. Part II: Directional distribution. *J. Phys. Oceanogr.* **2010**, *30*, 2768–2787. [CrossRef]
16. Yueh, S.H.; Tang, W.A.; Fore, G.; Neumann, G.; Hayashi, A.; Freedman, A.; Chaubell, J.; Lagerloef, G.S.E. L-band passive and active microwave geophysical model functions of ocean surface winds and applications to aquarius retrieval. *IEEE Trans. Geosci. Remote Sens.* **2013**, *51*, 4619–4632. [CrossRef]
17. Yueh, S.H.; Tang, W.A.; Fore, G.; Hayashi, A.K.; Song, Y.T.; Lagerloef, G. Aquarius geophysical model function and combined active passive algorithm for ocean surface salinity and wind retrieval. *J. Geophys. Res. Oceans* **2014**, *119*, 5360–5379. [CrossRef]
18. Zhou, X.; Chong, J.S.; Yang, X.F.; Li, W.; Guo, X.X. Ocean Surface Wind Retrieval using SMAP L-Band SAR. *IEEE J. Sel. Top. Appl. Earth Obs. Remote Sens.* **2017**, *10*, 65–74. [CrossRef]
19. Hersbach, H.; Stoffelen, A.; de Haan, S. An improved C-band scatterometer ocean geophysical model function: CMOD5. *J. Geophys. Res.* **2007**, *112*. [CrossRef]
20. Wentz, F.J.; Smith, D.K. A model function for the ocean-normalized radar cross section at 14 GHz derived from NSCAT observations. *J. Geophys. Res.* **1999**, *104*, 11499–11514. [CrossRef]
21. Hwang, P. Ocean Surface Roughness Spectrum in High Wind Condition for Microwave Backscatter and Emission Computations. *J. Atmos. Ocean. Technol.* **2013**, *30*, 2168–2188. [CrossRef]
22. Yurovskaya, M.V.; Dulov, V.A.; Chapron, B.; Kudryavtsev, V.N. Directional short wind wave spectra derived from the sea surface photography. *J. Geophys. Res. Oceans* **2013**, *118*, 4380–4394. [CrossRef]
23. Hwang, P. A note on the ocean surface roughness spectrum. *J. Atmos. Ocean. Technol.* **2011**, *28*, 436–443. [CrossRef]
24. Valenzuela, G.R. Theories for the interaction of electromagnetic and oceanic waves—A review. *Bound. Layer Meteorol.* **1978**, *13*, 61–85. [CrossRef]
25. Cox, C.; Munk, W. Measurement of the roughness of the sea surface from photographs of the sun glitter. *J. Opt. Soc. Am.* **1954**, *44*, 838–850. [CrossRef]
26. Fung, A.K.; Lee, K.K. A Semi-Empirical Sea-Spectrum Model for Scattering Coefficient Estimation. *IEEE J. Ocean. Eng.* **1982**, *7*, 166–176. [CrossRef]
27. Wright, J.W. A new model for sea clutter. *IEEE Trans. Antennas Propag.* **1968**, *16*, 217–223. [CrossRef]
28. Isoguchi, O.; Shimada, M. An L-band ocean geophysical model function derived from PALSAR. *IEEE Trans. Geosci. Remote Sens.* **2009**, *47*, 1925–1936. [CrossRef]
29. Li, X.M.; Lehner, S. Algorithm for Sea Surface Wind Retrieval from TerraSAR-X and TanDEM-X Data. *IEEE Trans. Geosci. Remote Sens.* **2014**, *52*, 2928–2939. [CrossRef]
30. Ren, Y.Z.; Li, X.M.; Zhou, G.Q. Sea Surface Wind Retrievals from SIR-C/X-SAR Data: A Revisit. *Remote Sens.* **2015**, *7*, 3548–3564. [CrossRef]

*Article*

# Doppler Spectrum-Based NRCS Estimation Method for Low-Scattering Areas in Ocean SAR Images

**Hui Meng [1,2,3], Xiaoqing Wang [4,*], Jinsong Chong [1,2,*], Xiangfei Wei [1,2,3] and Weiya Kong [1,2,3]**

1 National Key Laboratory of Science and Technology on Microwave Imaging, Beijing 100190, China; dream_kevin@126.com (H.M.); weixiangfei14@mails.ucas.ac.cn (X.W.); kongweiya13@mails.ucas.ac.cn (W.K.)
2 Institute of Electronics, Chinese Academy of Sciences, Beijing 100190, China
3 School of Electronics, Electrical and Communication Engineering, University of Chinese Academy of Sciences, Beijing 100190, China
4 Institute of Microelectronics of Chinese Academy of Sciences, Beijing 100029, China
* Correspondence: huadaqq@126.com (X.W.); iecas_chong@163.com (J.C.); Tel.: +86-10-8299-5990 (X.W.); +86-10-5888-7125 (J.C.)

Academic Editors: Xiaofeng Yang, Xiaofeng Li, Ferdinando Nunziata, Alexis Mouche and Prasad S. Thenkabail
Received: 6 December 2016; Accepted: 25 February 2017; Published: 28 February 2017

**Abstract:** The image intensities of low-backscattering areas in synthetic aperture radar (SAR) images are often seriously contaminated by the system noise floor and azimuthal ambiguity signal from adjacent high-backscattering areas. Hence, the image intensity of low-backscattering areas does not correctly reflect the backscattering intensity, which causes confusion in subsequent image processing or interpretation. In this paper, a method is proposed to estimate the normalized radar cross-section (NRCS) of low-backscattering area by utilizing the differences between noise, azimuthal ambiguity, and signal in the Doppler frequency domain of single-look SAR images; the aim is to eliminate the effect of system noise and azimuthal ambiguity. Analysis shows that, for a spaceborne SAR with a noise equivalent sigma zero (NESZ) of −25 dB and a single-look pixel of 8 m × 5 m, the NRCS-estimation precision of this method can reach −38 dB at a resolution of 96 m × 100 m. Three examples are given to validate the advantages of this method in estimating the low NRCS and the filtering of the azimuthal ambiguity.

**Keywords:** SAR; ocean; NRCS; Doppler spectrum; azimuthal ambiguity

---

## 1. Introduction

Areas with low normalized radar cross-section (NRCS) appear dark in synthetic aperture radar (SAR) images. It is frequently seen in ocean SAR images, such as those of oil spills, organic films, low wind areas, fronts, upwelling, current shear zones, and dark strips of internal waves and swells [1,2]. Among land targets, the backside of mountains and flat ground such as airport runways are also typical low-backscattering targets. The signal intensities of low-backscattering areas in SAR images are often close to or even less than the noise floor of the SAR system. Taking the ocean surface as an example; the mean NRCS of the ocean surface for the L, C, and X bands ranges from −15 dB to −25 dB at moderate wind speeds and incident angles. However, the NRCS of low-backscattering areas on the ocean surface is much lower than the mean NRCS of the ocean surface. The NRCS of low-backscattering areas of the ocean surface is often less than −30 dB, whereas the noise equivalent sigma zero (NESZ) of most spaceborne SAR systems ranges from −20 dB to −30 dB. Hence, the backscattering signal intensities of low-backscattering areas in ocean SAR images are often less than the noise floor of SAR systems. The NESZ values of typical spaceborne SAR systems are listed in Table 1 [3,4].

**Table 1.** Noise equivalent sigma zero (NESZ) values of typical spaceborne synthetic aperture radar (SAR) systems.

| Satellite | Mode | NESZ (dB) |
|---|---|---|
| ERS-1/2 | | −21 to −24 |
| Envisat | Image | −20 to −22 |
| | Wave | −20 to −22 |
| | Alternating Polarization | −19 to −22 |
| | Wide Swath | −21 to −26 |
| | Global Monitoring | −32 to −35 |
| Radarsat-1 | | <−21 |
| Radarsat-2 | Standard | −31 |
| | Fine | −28 |
| | Wide | −23 |
| | ScanSAR narrow | −23 |
| | ScanSAR wide | −23 |
| JERS-1 | | <−20.5 |
| ALOS/PALSAR | Fine (Swath Width 70 km) | <−23 |
| | Fine (Swath Width 60 km) | <−25 |
| | Polarimetry | <−29 |
| | ScanSAR | <−25 |
| TerraSAR-X | Spotlight (HS) | −23 |
| | Spotlight (LS) | −23 |
| | Stripmap | −22 |
| | ScanSAR | −21 |

Another factor that can affect the image intensities of low-backscattering areas is the azimuthal ambiguity effect of high-backscattering areas. It occurs because the Doppler frequency of the signal reflected from the area illuminated by the azimuthal sidelobe of the antenna exceeds the pulse repetition frequency (PRF). The azimuthal ambiguity signal of a target is located at a position with a certain displacement relative to its real position. This displacement depends on the PRF, the velocity of the platform and the Doppler centroid frequency of the SAR system. A typical value of the azimuthal ambiguity of a spaceborne SAR is about −15 dB to −20 dB. If the NRCS of a high-backscattering area is 15 dB to 20 dB higher than that of a low-backscattering area located at the position where the azimuthal ambiguity signal from the high-backscattering area is present, the azimuthal ambiguity signal could significantly affect the image intensity of the dark area. Azimuthal ambiguities are especially frequent in land–water junctions, because the NRCSs of land targets are much higher than that of the water surface.

Two analyses above indicate that, in order to estimate the certain value of the low areas' NRCS, the effect of the azimuthal ambiguity must be taken into consideration. Nevertheless, the standard radiometric calibration algorithm for SAR images only takes the system noise into account and ignores the azimuthal ambiguity effect, which is expressed in the following equation [5,6].

$$\sigma_{cal} = 10\lg(I - N_0) - K + 10\lg\left[\left(\frac{R}{R_{ref}}\right)^3 \frac{\sin\alpha}{\sin\alpha_{ref}} \frac{G_{ref}}{g(\alpha)G}\right] \quad (1)$$

where $I$, $R$, $\alpha$, and $G$ are the image intensity, slant range, elevation angle and system gain of a certain image pixel respectively; $g(\alpha)$ is the two-way antenna gain at elevation angle $\alpha$, $N_0$ is the system noise, $K$ is the calibration constant, and $R_{ref}$, $\alpha_{ref}$, and $G_{ref}$ are the slant range, elevation angle, and system gain of the reference target, respectively.

However, an accurate system noise $N_0$ is seldom provided in standard commercial spaceborne SAR data products. Moreover, even if a sufficiently accurate $N_0$ is provided, it is possible to obtain a meaningless NRCS of less than or equal to zero because the image intensity is a stochastic variable that may be less than the system noise $N_0$, especially when the NRCS of the target is relatively low. Hence, in most practical NRCS calibration applications, the system noise is also ignored and Equation (1) is simplified as follows [5–10].

$$\sigma_{cal} = 10\lg(I) - K + 10\lg\left[\left(\frac{R}{R_{ref}}\right)^3 \frac{\sin\alpha}{\sin\alpha_{ref}} \frac{G_{ref}}{g(\alpha)G}\right] \tag{2}$$

When calibrated using Equation (2), the NRCS of low-backscattering areas inevitably includes a significant contribution from the system noise and azimuthal ambiguity, which can cause confusion in subsequent image processing and interpretation.

In this paper, a method of NRCS estimation for low-backscattering areas based on a spectrum is proposed. This method requires to know the noise floor $N_0$ and antenna patterns firstly. If $N_0$ and antenna patterns are unknown, we also can estimate them from a single-look complex SAR image. This method can also eliminate the azimuthal ambiguity effect according to the shape of the Doppler spectrum; avoiding a meaningless NRCS estimation by using the maximum likelihood (ML) estimation method and the modified Newton's iteration method.

The rest of this paper is organized as follows. Section 2 gives the details of the algorithms and principles used in this method. In Section 3, three examples are presented to validate the advantages of this method. In Section 4, an analysis of the estimation precision and simulations are stated. Finally, some conclusions are presented in Section 5.

## 2. Spectral-Based NRCS Estimation Method for Low-Scattering Areas in SAR Images

### 2.1. The Principle of the Proposed Method

#### 2.1.1. Analysis of Doppler Spectrum Composition

From the SAR imaging theory [4–11], it is well known that the shape of system noise, azimuthal ambiguity, and the backscattering signal present different patterns in the Doppler spectrum of the SAR raw signal (here, it is supposed that the range match filtering and range cell migration correction have been done), i.e., the system noise power density is a certain constant in the Doppler spectrum, whereas the shape of the Doppler spectrum of the backscattering signal and azimuthal ambiguity depend on the antenna pattern: the backscattering signal and azimuthal ambiguity correspond to the main lobe and side lobe respectively. The Doppler spectrum of the SAR raw signal can be expressed as:

$$E[p_r(f, x_0, y_0)] = \sum_{n=-\infty}^{n=\infty} \overline{\sigma}(x_0 + nD_x, y_0 + nD_y) P_a(f - f_0 + nF_r) + \frac{N_0}{F_r} \tag{3}$$

In Equation (3) $\begin{bmatrix} x_0 & y_0 \end{bmatrix}$ are the center positions of the area where the Fourier transformation apply, and $x_0$ and $y_0$ are the coordinates in the flight and look directions, respectively, $E[\,]$ refers to the mathematic expectation, $f$ denotes the Doppler frequency, and $p_r(f)$ denotes the azimuthal power spectrum of the SAR raw signal. $P_a(f)$ is the power spectrum function of an ideal point target with a 0 dB NRCS, the shape of which is determined by the two-way antenna azimuthal pattern. Further, $f_0$ is the Doppler centroid, $F_r$ refers to the pulse repeat frequency of the SAR system, $N_0$ is the intrinsic noise floor of the SAR system, $\overline{\sigma}(x_0 - nD_x, y_0 - nD_y)$ is the mean NRCS of the pixels located between $\begin{bmatrix} x_0 - nD_x - L/2 & y_0 - nD_y \end{bmatrix}$ and $\begin{bmatrix} x_0 - nD_x + L/2 & y_0 - nD_y \end{bmatrix}$ ($L$ is the data length for calculating the Doppler spectrum), $D_x$ and $D_y$ are the displacements in the flight and look directions, respectively, between the position of the azimuthal ambiguity signal and the real target position. They can be written as:

$$D_x = \frac{R\lambda F_r}{2V}, \; D_y = -\frac{\lambda^2 f_0 F_r R}{4V^2} \tag{4}$$

where $R$ is the slant range of the target, $\lambda$ is the radar wavelength, and $V$ is the velocity of the SAR platform.

In Equation (3), $n = 0$ corresponds to the signal reflected from the mainlobe of the antenna, and $n \neq 0$ corresponds to the contribution from the azimuthal ambiguity effect. In general, among the azimuthal ambiguity signals, only $n = -1$ and 1, which correspond to the azimuthal ambiguity from the first azimuthal antenna sidelobes. Hence, Equation (3) can be simplified as:

$$\begin{aligned} E[p_r(f, x_0, y_0)] &\approx \overline{\sigma}(x_0, y_0) P_a(f - f_0) + \overline{\sigma}(x_0 + D_x, y_0 + D_y) P_a(f - f_0 + F_r) \\ &+ \overline{\sigma}(x_0 - D_x, y_0 - D_y) P_a(f - f_0 - F_r) + \frac{N_0}{F_r} \end{aligned} \tag{5}$$

Equation (5) indicates the shapes of the averaged power spectrum of the backscattering signal are determined by the antenna pattern $P_a(f)$ and $N_0$. To illustrate the shape difference of the power spectrum between system noise, azimuthal ambiguity, and the backscattering signal more clearly, a schematic diagram is given in Figure 1.

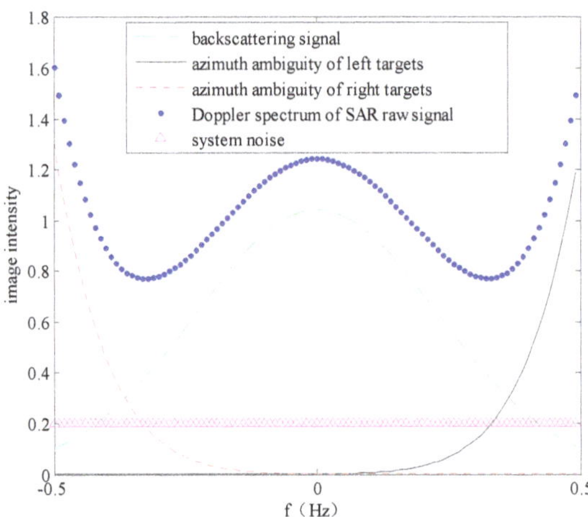


**Figure 1.** Schematic of the Doppler spectrum of synthetic aperture radar (SAR) raw signal and its various components.

Figure 1 is only a schematic diagram. When the image distribution is relatively uniform, the noise is the main disturbance. When there is a strong target around it, the azimuthal ambiguity is mainly from the strong target. From Figure 1, it is clear that the shapes of various components of the power spectrum of the SAR raw signal, composed of the backscattering signal, azimuthal ambiguity, and system noise, are very different. In general, the antenna pattern $P_a(f)$ and system noise $N_0$ can be acquired from the external and internal calibration of the SAR system. Therefore, it is possible to eliminate the effect of the azimuthal ambiguity on the NRCS estimation by taking full advantage of these differences. However, the azimuthal resolution of the SAR raw signal is too coarse for most applications. To increase the azimuthal resolution, the azimuthal matching filter must be applied on the SAR raw signal to convert it to a single-look complex image. However, the unweighted azimuthal matching filter can be used, which only changes the phase of the Doppler spectrum without modifying the amplitude of the Doppler spectrum. Thus, the azimuthal power spectrum of the single-look complex image has the same shape characteristic with that of the SAR raw signal. The relation between the power spectra of single-look complex images and the SAR raw signal is given as:

$$p_s(f, x_0, y_0) = p_r(f, x_0, y_0)|H(f)|^2 = p_r(f, x_0, y_0) \tag{6}$$

where $H(f) = \exp\left(j\pi\frac{\lambda R f^2}{2V^2}\right)$ is the unweighted azimuthal matching filter, $p_s(f, x_0, y_0)$ and $p_r(f, x_0, y_0)$ are the azimuthal power spectra of the single-look complex image and the SAR raw signal, respectively.

The shape patterns shown in Figure 1 are the mathematics expectation of the power spectrum. The real power spectrum of a small patch of the single-look complex images is in fact a stochastic process. As the signal of a single-look complex image is a complex Gaussian process, the probability density function of each sample of the power spectrum can be given by the well-known exponential distribution that follows:

$$g(p_s(f, x_0, y_0)) = \frac{1}{E[p_s(f, x_0, y_0)]}\exp\left\{-\frac{p_s(f, x_0, y_0)}{E[p_s(f, x_0, y_0)]}\right\} \tag{7}$$

Equation (5) indicates that the backscattering signal $\overline{\sigma}(x_0, y_0)$ of a certain area contributes to three spectra: $p_s(f, x_0, y_0)$, $p_s(f, x_0 - D_x, y_0 - D_y)$, and $p_s(f, x_0 + D_x, y_0 + D_y)$. Hence, the joint condition probability density function of all the frequency points is:

$$\begin{aligned} & g\Big(\ p_s(f, x_0, y_0)\quad p_s(f, x_0 - D_x, y_0 - D_y)\quad p_s(f, x_0 + D_x, y_0 + D_y)\ \Big|\overline{\sigma}(x_0, y_0)\Big) \\ & = \prod_{i=1}^{m}\frac{1}{E[p_s(f_i, x_0, y_0)]}\exp\left\{-\frac{p_s(f_i, x_0, y_0)}{E[p_s(f_i, x_0, y_0)]}\right\} \\ & \prod_{i=1}^{m}\frac{1}{E[p_s(f_i, x_0 - D_x, y_0 - D_y)]}\exp\left\{-\frac{p_s(f_i, x_0 - D_x, y_0 - D_y)}{E[p_s(f_i, x_0 - D_x, y_0 - D_y)]}\right\} \\ & \prod_{i=1}^{m}\frac{1}{E[p_s(f_i, x_0 + D_x, y_0 + D_y)]}\exp\left\{-\frac{p_s(f_i, x_0 + D_x, y_0 + D_y)}{E[p_s(f_i, x_0 + D_x, y_0 + D_y)]}\right\} \end{aligned} \tag{8}$$

where $m$ is the point number of the discrete Doppler spectrum.

Moreover, the image intensity is also a stochastic process related to the backscattering signal. In general, if the image pitch is small enough, the probability density function of a multi-look image intensity of an image pitch can be modeled by a gamma distribution as follows:

$$g_p(I(x_0, y_0)) = \begin{cases} \dfrac{M^M I^{M-1}(x_0, y_0)}{\left[E_c\overline{\sigma}(x_0, y_0) + E_l\overline{\sigma}(x_0 - D_x, y_0 - D_y) + E_R\overline{\sigma}(x_0 + D_x, y_0 + D_y)\right]^L\Gamma(M)} \\ \exp\left(-\dfrac{MI(x_0, y_0)}{E_c\overline{\sigma}(x_0, y_0) + E_l\overline{\sigma}(x_0 - D_x, y_0 - D_y) + E_R\overline{\sigma}(x_0 + D_x, y_0 + D_y)}\right) & I(x_0, y_0) > 0 \\ 0 & I(x_0, y_0) \le 0 \end{cases} \tag{9}$$

where $M$ is the look number, $I(x_0, y_0)$ is the mean image intensity of a certain area with the center located at $[x_0, y_0]$. $E_c$, $E_l$, and $E_R$ are the main lobe, left side lobe, and right side lobe factors, respectively:

$$E_c = \int_{-F_r/2}^{F_r/2} P_a(f - f_0)df,\ E_l = \int_{-F_r/2}^{F_r/2} P_a(f - f_0 - F_r)df,\ E_r = \int_{-F_r/2}^{F_r/2} P_a(f - f_0 + F_r)df \tag{10}$$

To estimate higher resolution NRCS from the power spectrum, the single-look complex images are divided into many small pitches and the Fourier transformation is applied on each image patch.

After obtaining $P_a(f)$ and $N_0$, the local NRCS can be further estimated from the Doppler spectrum. In this step, $L$ is selected according to the desired final resolution, but it cannot be significantly larger.

### 2.1.2. Methods and Solutions to Estimate the NRCS from the Doppler Spectrum

Suppose that a SAR single-look complex image has been corrected. For example, the range shift caused by the azimuthal ambiguity has been compensated, and the image has been interpolated $k$ times in azimuth. Meanwhile, the azimuthal shift caused by the azimuthal ambiguity is $X$ times larger than emphL. The corrected image is divided into some small patches. The size of every patch is about $R_m \times R_a$ (range multiplied by azimuthal). Choosing a row of azimuthal patches, and supposing that the scattering coefficient of every patch is $\sigma_1, \sigma_2, \ldots, \sigma_T$, respectively, the Doppler spectrum of the $i$-th block is $f_{i\_m}$.

Estimating NRCS from Doppler spectrum is a typical Bayesian estimation problem [12,13], which is expressed as the following equation.

$$\begin{aligned}\hat{\sigma}_i = \underset{\overline{\sigma}(x_0,y_0)}{\mathrm{argmax}}[g\,(p(f_{i\_1}), p(f_{i\_2}), \ldots, p(f_{i\_m}), p(f_{i-X\_1}), p(f_{i-X\_2}), \ldots,\\ p(f_{i-X\_m}), p(f_{i+X\_1}), p(f_{i+X\_2}), \ldots, p(f_{i+X\_m})|\overline{\sigma}_i) g_p(\overline{\sigma}(x_0, y_0))]\end{aligned} \tag{11}$$

The estimation equation of single patches can be expressed as:

$$\hat{\sigma}(x_0, y_0) = \underset{\overline{\sigma}(x_0,y_0)}{\mathrm{argmax}}\left[g(p(f_1), p(f_2), \ldots, p(f_m)|\overline{\sigma}(x_0, y_0)) g_p(\overline{\sigma}(x_0, y_0))\right] \tag{12}$$

where $\hat{\sigma}(x_0, y_0)$ is the estimation of $\overline{\sigma}(x_0, y_0)$, $f_i$ $(i = 1, 2, \ldots m)$ is the discrete frequency point, $m$ is the point number of the discrete Doppler spectrum, $g(p(f_1), p(f_2), \ldots, p(f_m)|\overline{\sigma}(x_0, y_0))$ refers to the conditional probability density function of the Doppler spectrum, and $g_p(\overline{\sigma}(x_0, y_0))$ refers to the a-priori probability density of $\overline{\sigma}(x_0, y_0)$.

Bayesian estimation is a global optimal estimation method. It increases the estimation precision at the NRCS with high a-priori probability density, but decreases the estimation precision at the NRCS with low a-priori probability density. In general, the a-priori probability density of the NRCS of a SAR image can be expressed by models such as Gamma, inverse Gaussian, or other distribution models [14]. However, in these commonly used models, the probability densities of low-NRCS are relatively low, which will lead to a less accurate estimation result for the low-backscattering areas. Hence, to acquire a higher estimation precision for the low-backscattering areas, the commonly used NRCS distribution models are not adopted, but it is assumed that the a-priori probability density of the NRCS is uniformly distributed. Another point which should be considered is that the NRCS should be greater than zero. Therefore, the a-priori probability density of the NRCS used in this paper is given as:

$$g_p(\sigma) = \begin{cases} 1 & \sigma > 0 \\ 0 & \sigma \leq 0 \end{cases} \tag{13}$$

The Equation (13) is used as the new a-priori probability density in proposed method, then the NRCS is estimated by the maximum likelihood (ML) estimation method. The advantage of the proposed method is that it can avoid meaningless estimation results less than or equal to zero.

Because the a-priori probability density given by Equation (13) is a discontinuous function, which is not convenient for the solving of Equation (11), it is approximated by:

$$g_p(\sigma) = \frac{1}{2} + \frac{1}{\pi}\arctan(\alpha\sigma) \tag{14}$$

where, in order to match Equation (13), $\alpha$ should be more than $10^{16}$ and choose $10^{20}$ in this method.

The estimation of Equation (12) can be obtained by solving the following equation:

$$\begin{aligned}&\frac{\partial \ln g\left( p_n(f_1) \;\; p_n(f_2) \;\; \cdots \;\; p_n(f_m) \;\big|\overline{\sigma}_n\right)}{\partial \overline{\sigma}_n} + \frac{\partial \ln g\left( p_{n-X}(f_1) \;\; p_n(f_2) \;\; \cdots \;\; p_n(f_m) \;\big|\overline{\sigma}_n\right)}{\partial \overline{\sigma}_n} \\ &+ \frac{\partial \ln g\left( p_{n+X}(f_1) \;\; p_{n+X}(f_2) \;\; \cdots \;\; p_{n+X}(f_m) \;\big|\overline{\sigma}_n\right)}{\partial \overline{\sigma}_n} + \frac{\partial \ln g_p(\overline{\sigma}_n|)}{\partial \overline{\sigma}_n} = 0\end{aligned} \tag{15}$$

where $p_n(f_i)$ is the $i$-th frequency of the $n$-th block.

Because the signal of a single-look complex image is a complex Gaussian process, the probability density function of each sample of the Doppler spectrum can be given by the well-known exponential distribution as follows:

$$g(p(f)|\overline{\sigma}(x_0, y_0)) = \frac{1}{E[p(f)]} \exp\left\{ -\frac{p(f)}{E[p(f)]} \right\} \tag{16}$$

Thus, the joint probability density function of all the frequency points is:

$$g(p(f_1)p(f_2)\cdots p(f_m)|\overline{\sigma}(x_0, y_0)) = \prod_{i=1}^{m} \frac{1}{E[p(f_i)]} \exp\left\{ -\frac{p(f_i)}{E[p(f_i)]} \right\} \tag{17}$$

Inserting Equations (5), (16) and (17) into Equation (15), and considering that the Doppler spectrum at different azimuthal locations has different components, thus, the following functions at different azimuthal locations are derived.

When $2X < n \leq T - 2X$

$$\begin{aligned}&-\sum_{i=1}^{m} \frac{P_C(f_i)}{\left[\overline{\sigma}_{n-X}P_L(f_i) + \overline{\sigma}_n P_C(f_i) + \overline{\sigma}_{n+X}P_R(f_i) + \frac{N_0}{F_r}\right]} - \sum_{i=1}^{m} \frac{P_R(f_i)}{\left[\overline{\sigma}_{n-2X}P_L(f_i) + \overline{\sigma}_{n-X}P_C(f_i) + \overline{\sigma}_n P_R(f_i) + \frac{N_0}{F_r}\right]} \\ &-\sum_{i=1}^{m} \frac{P_L(f_i)}{\left[\overline{\sigma}_n P_L(f_i) + \overline{\sigma}_{n+X}P_C(f_i) + \overline{\sigma}_{n+2X}P_R(f_i) + \frac{N_0}{F_r}\right]} + \sum_{i=1}^{m} \frac{p_n(f_i)P_C(f_i)}{\left[\overline{\sigma}_{n-X}P_L(f_i) + \overline{\sigma}_n P_C(f_i) + \overline{\sigma}_{n+X}P_R(f_i) + \frac{N_0}{F_r}\right]^2} \\ &+\sum_{i=1}^{m} \frac{p_{n-X}(f_i)P_R(f_i)}{\left[\overline{\sigma}_{n-2X}P_L(f_i) + \overline{\sigma}_{n-X}P_C(f_i) + \overline{\sigma}_n P_R(f_i) + \frac{N_0}{F_r}\right]^2} + \sum_{i=1}^{m} \frac{p_{n+X}(f_i)P_L(f_i)}{\left[\overline{\sigma}_n P_L(f_i) + \overline{\sigma}_{n+X}P_C(f_i) + \overline{\sigma}_{n+2X}P_R(f_i) + \frac{N_0}{F_r}\right]^2} \\ &+\frac{\alpha}{[1 + (\alpha\overline{\sigma}_n)][\frac{\pi}{2} + \arctan(\alpha\overline{\sigma}_n)]} = 0\end{aligned} \tag{18}$$

where $P_C(f_i) = P_a(f_i - f_0)$, $P_L(f_i) = P_a(f_i - f_0 - F_r)$, and $P_R(f_i) = P_a(f_i - f_0 + F_r)$; $n$ refers to the pixel location in the flight directions. T is the length of the azimuthal data. $X$ is the number of the azimuthal ambiguity.

When $X < n \leq 2X$:

$$\begin{aligned}&-\sum_{i=1}^{m} \frac{P_C(f_i)}{\left[\overline{\sigma}_{n-X}P_L(f_i) + \overline{\sigma}_n P_C(f_i) + \overline{\sigma}_{n+X}P_R(f_i) + \frac{N_0}{F_r}\right]} - \sum_{i=1}^{m} \frac{P_L(f_i)}{\left[\overline{\sigma}_n P_L(f_i) + \overline{\sigma}_{n+X}P_C(f_i) + \overline{\sigma}_{n+2X}P_R(f_i) + \frac{N_0}{F_r}\right]} \\ &+\sum_{i=1}^{m} \frac{p_n(f_i)P_C(f_i)}{\left[\overline{\sigma}_{n-X}P_L(f_i) + \overline{\sigma}_n P_C(f_i) + \overline{\sigma}_{n+X}P_R(f_i) + \frac{N_0}{F_r}\right]^2} + \sum_{i=1}^{m} \frac{p_{n+X}(f_i)P_L(f_i)}{\left[\overline{\sigma}_n P_L(f_i) + \overline{\sigma}_{n+X}P_C(f_i) + \overline{\sigma}_{n+2X}P_R(f_i) + \frac{N_0}{F_r}\right]^2} \\ &+\frac{\alpha}{[1 + (\alpha\overline{\sigma}_n)][\frac{\pi}{2} + \arctan(\alpha\overline{\sigma}_n)]} = 0\end{aligned} \tag{19}$$

When $0 < n \leq X$:

$$-\sum_{i=1}^{m}\frac{P_L(f_i)}{\left[\overline{\sigma}_n P_L(f_i)+\overline{\sigma}_{n+X}P_C(f_i)+\overline{\sigma}_{n+2X}P_R(f_i)+\frac{N_0}{F_r}\right]}+\sum_{i=1}^{m}\frac{p_{n+X}(f_i)P_L(f_i)}{\left[\overline{\sigma}_n P_L(f_i)+\overline{\sigma}_{n+X}P_C(f_i)+\overline{\sigma}_{n+2X}P_R(f_i)+\frac{N_0}{F_r}\right]^2}+\frac{\alpha}{[1+(\alpha\overline{\sigma}_n)][\frac{\pi}{2}+\arctan(\alpha\overline{\sigma}_n)]}=0 \quad (20)$$

When $T-2X < n \leq T-X$:

$$-\sum_{i=1}^{m}\frac{P_C(f_i)}{\left[\overline{\sigma}_{n-X} P_L(f_i)+\overline{\sigma}_{n}P_C(f_i)+\overline{\sigma}_{n+X}P_R(f_i)+\frac{N_0}{F_r}\right]}-\sum_{i=1}^{m}\frac{P_R(f_i)}{\left[\overline{\sigma}_{n-2X} P_L(f_i)+\overline{\sigma}_{n-X}P_C(f_i)+\overline{\sigma}_{n}P_R(f_i)+\frac{N_0}{F_r}\right]}$$
$$+\sum_{i=1}^{m}\frac{p_n(f_i)P_C(f_i)}{\left[\overline{\sigma}_{n-X} P_L(f_i)+\overline{\sigma}_{n}P_C(f_i)+\overline{\sigma}_{n+X}P_R(f_i)+\frac{N_0}{F_r}\right]^2}+\sum_{i=1}^{m}\frac{p_{n-X}(f_i)P_R(f_i)}{\left[\overline{\sigma}_{n-2X} P_L(f_i)+\overline{\sigma}_{n-X}P_C(f_i)+\overline{\sigma}_{n}P_R(f_i)+\frac{N_0}{F_r}\right]^2}$$
$$+\frac{\alpha}{[1+(\alpha\overline{\sigma}_n)][\frac{\pi}{2}+\arctan(\alpha\overline{\sigma}_n)]}=0 \quad (21)$$

When $T-X < n \leq T$:

$$-\sum_{i=1}^{m}\frac{P_R(f_i)}{\left[\overline{\sigma}_{n-2X} P_L(f_i)+\overline{\sigma}_{n-X}P_C(f_i)+\overline{\sigma}_{n}P_R(f_i)+\frac{N_0}{F_r}\right]}+\sum_{i=1}^{m}\frac{p_{n-X}(f_i)P_R(f_i)}{\left[\overline{\sigma}_{n-2X} P_L(f_i)+\sigma_{n-X}P_C(f_i)+\overline{\sigma}_{n}P_R(f_i)+\frac{N_0}{F_r}\right]^2}+\frac{\alpha}{[1+(\alpha\overline{\sigma}_n)][\frac{\pi}{2}+\arctan(\alpha\overline{\sigma}_n)]}=0 \quad (22)$$

Combining all the equations above, there are $n$ equations. To solve all the unknown variables, the Newton iterative method is adopted. The Jacobian matrix of the derived function is in the Appendix A.

To solve $\overline{\sigma}(x_0, y_0)$ from all equations above, $\overline{\sigma}(x_0 - D_x, y_0 - D_y)$ and $\overline{\sigma}(x_0 + D_x, y_0 + D_y)$ should be first known. However, to obtain $\overline{\sigma}(x_0 - D_x, y_0 - D_y)$ or $\overline{\sigma}(x_0 + D_x, y_0 + D_y)$, a known $\overline{\sigma}(x_0, y_0)$ is also needed. This self-contradiction problem can be addressed by using an iterative strategy. In the $n$-th iteration, Equation (17) is written as:

$$\sum_{i=1}^{m}\frac{P_a(f_i-f_0)}{\left[\overline{\sigma}_n(x_0,y_0)P_a(f-f_0)+\overline{\sigma}_{n-1}(x_0+D_x,y_0+D_y)P_a(f-f_0+F_r)+\overline{\sigma}_{n-1}(x_0-D_x,y_0-D_y)P_a(f-f_0-F_r)+\frac{N_0}{F_r}\right]}$$
$$=\sum_{i=1}^{m}\frac{p(f_i)P_a(f_i-f_0)}{\left[\overline{\sigma}_n(x_0,y_0)P_a(f-f_0)+\overline{\sigma}_{n-1}(x_0+D_x,y_0+D_y)P_a(f-f_0+F_r)+\overline{\sigma}_{n-1}(x_0-D_x,y_0-D_y)P_a(f-f_0-F_r)+\frac{N_0}{F_r}\right]^2}$$
$$+\frac{\alpha}{[1+(\alpha\overline{\sigma}_n)][\frac{\pi}{2}+\arctan(\alpha\overline{\sigma}_n)]} \quad (23)$$

where $\overline{\sigma}_n(x_0, y_0)$ is the estimation result in the $n$-th iteration. The initial guess of $\overline{\sigma}(x_0, y_0)$ is given as:

$$\overline{\sigma}_0(x_0,y_0)=\begin{cases}\dfrac{\overline{I}(x_0,y_0)-\overline{I}(x_0+D_x,y_0+D_y)A-\overline{I}(x_0-D_x,y_0-D_y)A}{1-2A}-N_0 \\ ,\overline{I}(x_0,y_0)>\overline{I}(x_0+D_x,y_0+D_y)A+\overline{I}(x_0-D_x,y_0-D_y)A+(1-2A)N_0 \\ \sigma_{\min}, \\ ,\overline{I}(x_0,y_0)\leq\overline{I}(x_0+D_x,y_0+D_y)A+\overline{I}(x_0-D_x,y_0-D_y)A+(1-2A)N_0\end{cases} \quad (24)$$

where $\overline{I}(x_0, y_0)$ is the mean image intensity of the pixels between $\left[\begin{array}{cc} x_0 - L/2 & y_0 \end{array}\right]$ and $\left[\begin{array}{cc} x_0 + L/2 & y_0 \end{array}\right]$, $A$ is the azimuthal ambiguity factor which is given by:

$$A=\int_{-F_r/2}^{-3F_r/2}P_a(\mathrm{f})\mathrm{df} \quad (25)$$

The convergence condition is given by:

$$\frac{\sum\limits_{y_0}\sum\limits_{x_0}|\overline{\sigma}_n(x_0,y_0)-\overline{\sigma}_{n-1}(x_0,y_0)|}{N_xN_y} < \sigma_{\min} \quad (26)$$

where $\sigma_{\min}$ refers to a certain small NRCS value. $N_x$ and $N_y$ are the pixel numbers of the estimated NRCS image in flight and look directions, respectively.

The aforementioned $\overline{\sigma}(x_0,y_0)$ is a relative backscattering intensity rather than an absolute NRCS. If we have the $K$-constant needed in the radiometric calibration, the estimated relative backscattering intensity $\overline{\sigma}(x_0,y_0)$ can be further converted to the absolute NRCS by replacing $I-N_0$ in Equation (1) with $\overline{\sigma}(x_0,y_0)$ estimated by this method, expressed as:

$$\sigma_{cal}(x_0,y_0) = 10\lg[\overline{\sigma}(x_0,y_0)] - K + 10\lg\left[\left(\frac{R}{R_{ref}}\right)^3 \frac{\sin\alpha}{\sin\alpha_{ref}}\frac{G_{ref}}{g(\alpha)G}\right] \quad (27)$$

## 2.2. Algorithm Flow Chart and Summary

The azimuthal matching filters of the standard imaging algorithm of commercial SAR products are generally weighted filters, which does not satisfy the requirements of our method. Thus, our algorithm begins with the SAR raw data product. In the first step, SAR imaging, an unweighted azimuthal matching filter is used. The byproduct of SAR imaging is the Doppler centroids of each range cell, which will be used in the second step. The method shifts the Doppler centroids of the single-look complex images to a zero frequency position. This includes some substeps, such as fast Fourier transform (FFT), inverse FFT, and spectrum shifting. In the last step, the single-look complex image is divided into many subimage patches first. The size of the subimage patches is selected based on the desired resolution. Then, an iteration strategy is used to estimate the signal intensities of each subimage patch. In each iteration, the signal intensities are estimated on the basis of Equation (23), which is solved by the Newton-iteration algorithm. Finally, the estimated relative backscattering intensity $\overline{\sigma}(x_0,y_0)$ is converted to absolute NRCS using Equation (27).

The algorithm used in this method is summarized in Figure 2.

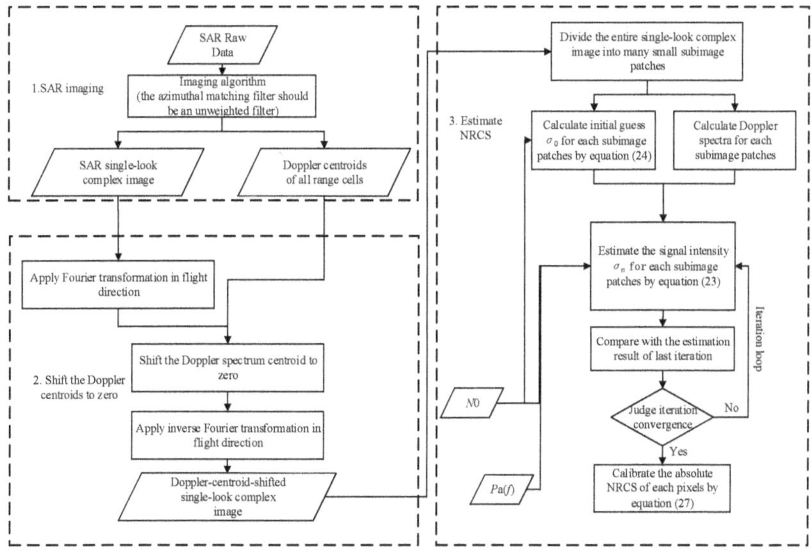


**Figure 2.** Algorithm flow chart.

## 3. Validation of the Proposed Estimation Method

In this section, three examples will be presented to demonstrate the advantages of this method in low NRCS estimation and azimuthal ambiguity filtering.

### 3.1. Example 1: Qualitative Analysis for the Estimation Method in Low NRCS

The first example is an ocean image acquired by ERS-2 (European remote sensing satellite (ERS) was the European Space Agency's first Earth-observing satellite) on 30 April 2005 in the South China Sea as a qualitative analysis, which is shown in Figure 3. There are 4912 pixels in the look direction and 28,695 pixels in the flight direction in the single-look complex image used in this example. Frame 1 in Figure 3 is a subimage for the comparison between a conventional SAR image and the corresponding estimated NRCS image.

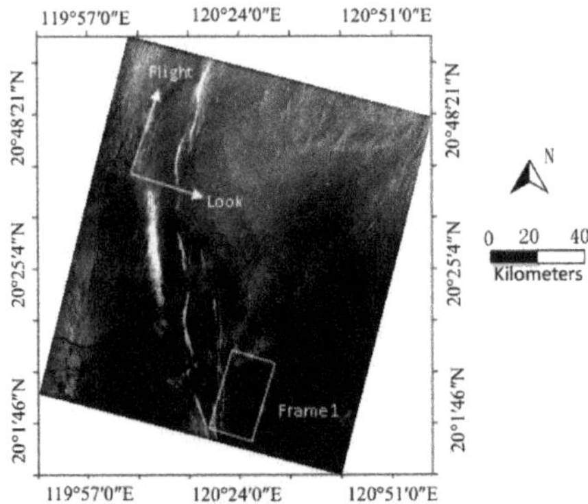


**Figure 3.** ERS-2 (European remote sensing satellite (ERS) was the European Space Agency's first Earth-observing satellite) ocean SAR image of South China Sea collected on 30 April 2005, at 02:28 UTC. Frame 1 is a subimage for the comparison between conventional SAR image and the corresponding estimated NRCS image.

The first step is estimating the Doppler centroid $f_0$ for each range cell [11,15–19], and then shifting the Doppler spectrum centroid of the single-look complex image to zero. Note that ocean currents can lead to an additional local shift of the Doppler centroid [20–22]. However, the Doppler shift resulting from the ocean current is generally less than 5% of the PRF, which can be neglected in the method proposed in this paper. Examples of unshifted and shifted Doppler spectra are shown in Figure 4.

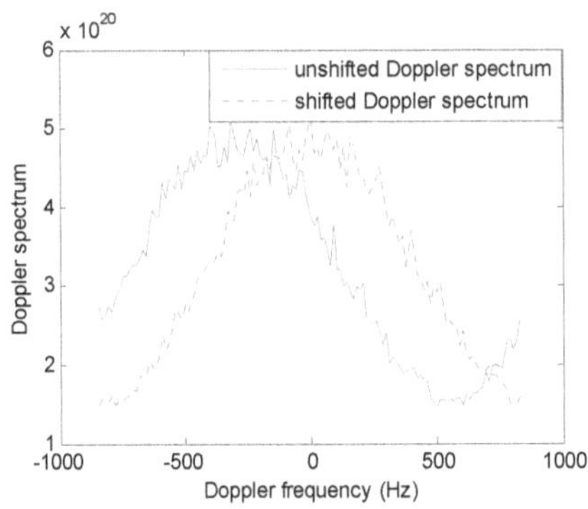


**Figure 4.** Unshifted and shifted Doppler spectra.

The second step is calculating Doppler spectra from the single-look complex image. In this example, each Doppler spectrum is a 128-point discrete spectrum that is averaged by 224 times in the flight direction and 10 times in the look direction. A total of 491 Doppler spectra from the entire SAR image are obtained. The azimuthal length used for calculating one Doppler spectrum is about 121 km (i.e., $L$ = 121 km).

Frame 1 in Figure 3 is chosen to compare the conventional SAR image with the corresponding estimated NRCS image. The result is shown in Figure 5. The pixel size of the single-look complex image is about 21 m (range direction) × 4.2 m (flight direction). Figure 5a is a multi-look SAR image, in which each pixel is averaged by 80 adjacent pixels of the single-look complex image (4 pixels in the look direction × 20 pixels in the flight direction). Figure 5b is the estimated NRCS image, in which each pixel is estimated from 80 pixels of the single-look complex image (in each estimation, the Doppler spectrum is calculated from 20 pixels in the flight direction and averaged by 4 times in the look direction). The pixel sizes of both images in Figure 5 are about 84 m × 84 m. The image intensities of both images are shown by logarithmic grayscaling to display clear texture features of the dark area.

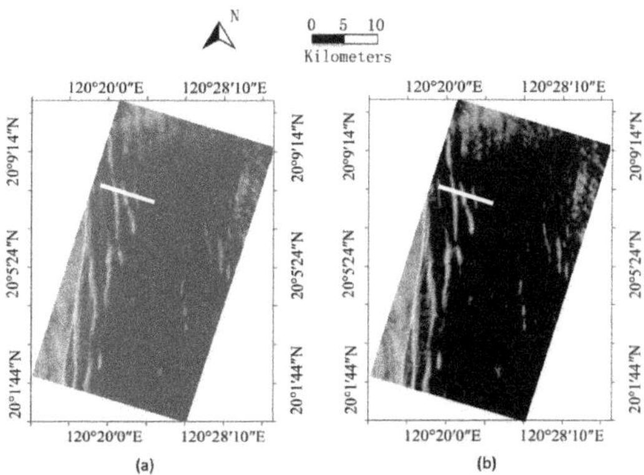


**Figure 5.** Subimages of Frame 1 in Figure 3, (**a**) conventional SAR image; (**b**) estimated normalized radar cross-section (NRCS) image.

The comparison of Figure 5a,b demonstrates that Figure 5b presents the features of the dark area more clearly. To compare these two images qualitatively, the image intensity profiles along the white lines are depicted in Figure 6.

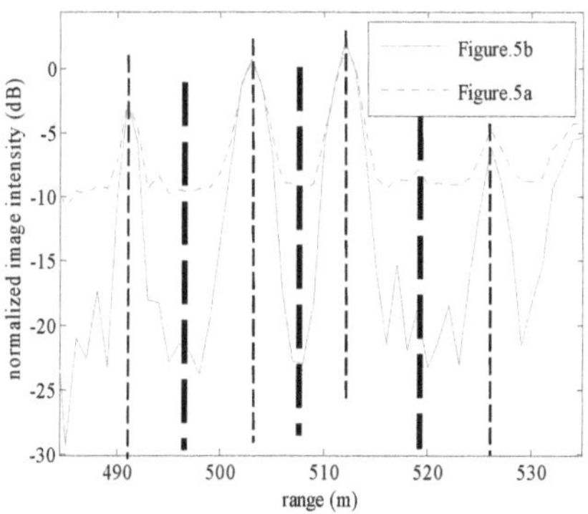


**Figure 6.** Image intensities along the white lines in Figure 5a (red dotted line) and Figure 5b (blue solid line).

In Figure 6, the signal intensity is normalized by the mean intensity of the entire image. The image feature near the white line is an oceanic internal wave. Four peaks of the internal wave are marked by dashed lines, and three troughs are marked by bold dashed lines. At the positions of the peaks, the estimated NRCS intensity is very close to that of conventional SAR image intensity because the SNR of the peaks is sufficiently high. As a comparison, at the positions of the troughs, the SAR signal is buried by the noise floor (about $-10$ dB after normalization), making it hard to judge the exact trough position. In contrast, the estimated NRCS can remove the effect of the noise floor to a large extent, and the trough positions of the estimated NRCS are near the midpoint of the two adjacent peaks, which indirectly validates the correctness of the proposed method.

### 3.2. Example 2: Quantitative Analysis for the Estimation Method in Low NRCS

This example is the atmospheric gravity waves' image acquired by ERS-2 on 11 March 2006 in the East China Sea as a quantitative analysis, which is shown in Figure 7. There are 4912 pixels in the look direction and 28,695 pixels in the flight direction in the single-look complex image used in this example.

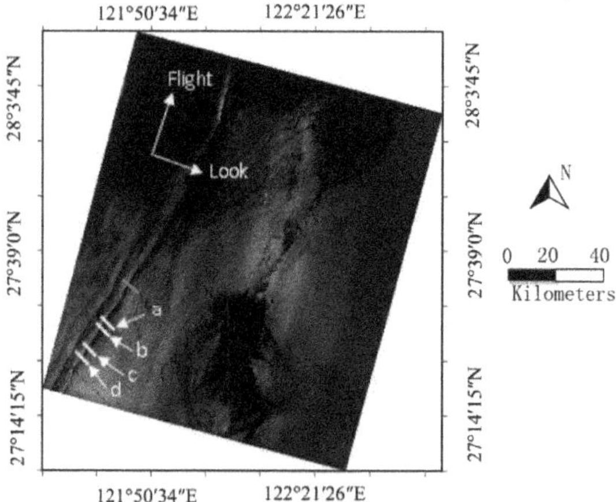


**Figure 7.** ERS-2 Ocean SAR image of the East China Sea collected on 11 March 2006, at 02:24 UTC. The four white data lines a, b, c, and d are the profiles for comparison between the proposed method, the SAR raw image intensity minus $N_0$ and the optimal parameter estimation method of internal waves [23].

In this example, the method of optimal parameter estimation of internal waves in SAR images [23] and the proposed method in this paper will be used to deal with the internal wave in Figure 7.

The optimal parameter estimation is the latest method for estimating the parameter of internal solitary waves. In this article, it is referred to as optimal parameter estimation. In order to verify the feasibility of this method, we found a section at the other location of atmospheric gravity waves (the red solid line region in Figure 7. The estimation result is shown in Figure 8, showing that the optimum estimators are very close to the Cramér–Rao bound (CRB). Therefore, the estimation method in the literature [23] is considered to fit the atmospheric gravity waves' profile in Figure 7.

We selected four profiles from the atmospheric gravity waves in Figure 7, and at the position of the trough, the SAR signal is buried by the noise floor at positions a and b are more obvious than positions c and d. The estimation method in the literature [23], the proposed method and the SAR Raw Image are used in the four profiles for comparison, The results of the comparison are shown in Figure 9.

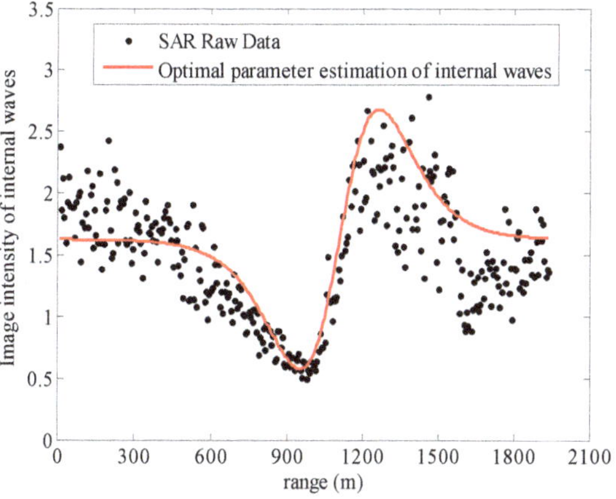


**Figure 8.** The validation of the optimal parameter estimation of internal waves.

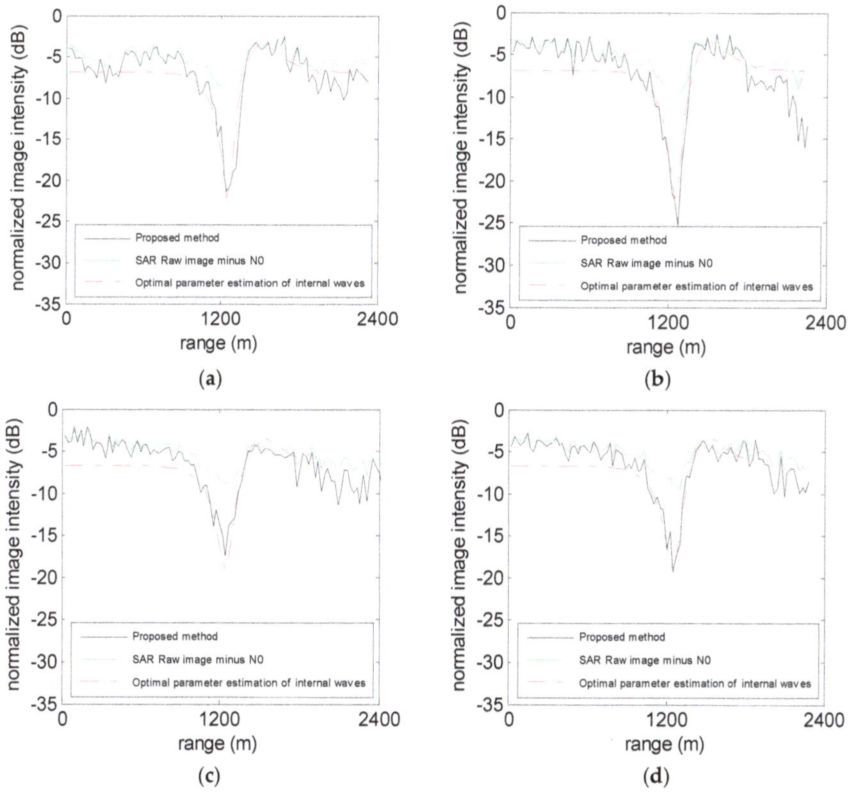


**Figure 9.** Image intensities along the four white lines in Figure 7. (**a**) the profile a in Figure 7; (**b**) the profile b in Figure 7; (**c**) the profile c in Figure 7; (**d**) the profile d in Figure 7; The green dashed line is the SAR raw image intensity minus $N_0$; the black solid line is the proposed method in this paper; the red dot dash line is the optimal parameter estimation method of internal waves.

In Figure 9, the signal intensity is normalized by the mean intensity of the entire image. The image feature near the white lines represent atmospheric gravity waves. At the position of the peak, the estimated NRCS intensity of all the profiles are very close to that of conventional SAR image intensity due to sufficiently high SNR of the peak. As a comparison, at the position of the trough, the SAR signal is buried by the noise floor (about −10 dB after normalization at positions a and b, and about −5 dB to −8 dB at positions c and d), making it hard to judge the exact trough position. In contrast, the estimated NRCS can remove the effect of the noise floor, which can reach −22 dB after normalization at positions a and b, and about −16 dB to −18 dB at positions of c and d. Position a is selected as an example through the method of optimal Parameter Estimation of Internal Waves in SAR images from the literature [23] to estimate the energy intensity value in the wave trough position, which is around −22 dB. From Figure 9, the estimation curve of signal intensity by the proposed method in this paper is very close to the method of optimal Parameter Estimation of Internal Waves in SAR images, which directly validates the accuracy of the proposed method.

### 3.3. Example 3: Validation of the Azimuthal Ambiguity Analysis

The third example is a RADARSAT-1 (RADARSAT is a Canadian remote sensing Earth observation satellite program overseen by the Canadian Space Agency) image of Vancouver, which is shown in Figure 10. The SAR raw data of this example was obtained from the accompanying CD of literature [11].

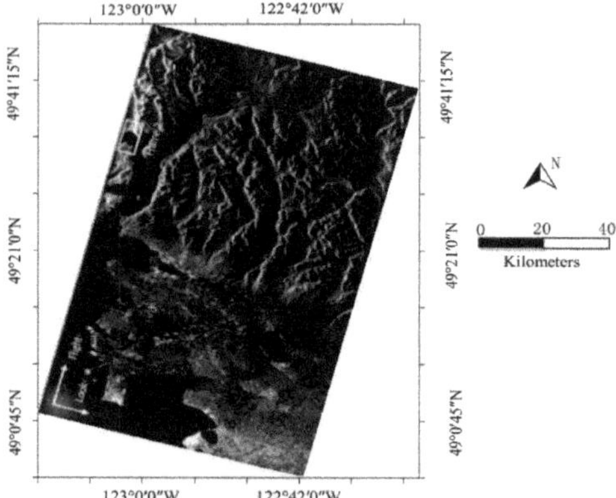


**Figure 10.** RADARSAT-1 (RADARSAT is a Canadian remote sensing Earth observation satellite program overseen by the Canadian Space Agency) SAR image of Vancouver collected on 16 June 2002, at 02:24 UTC.

There are 7940 pixels in the look direction and 19,425 pixels in the flight direction in the single-look complex image of this example. As in the first example, each Doppler spectrum in this example is also a 128-point discrete spectrum, which is averaged by 151 times in the flight direction and 30 times in the look direction. A total of 264 Doppler spectra can be obtained from the entire single-look complex image. The azimuthal length used for calculating one Doppler spectrum is about 109 km (i.e., $L$ = 109 km).

Using Frame 2 in Figure 10 as an example, the conventional SAR image and the corresponding estimated NRCS image are shown in Figure 11. The pixel size of the single-look complex image is about 8 m (look direction) × 5.6 m (flight direction). Figure 11a is a multi-look SAR image, in which each pixel is averaged by 192 adjacent pixels of the single-look complex image (12 pixels in the look direction × 16 pixels in the flight direction). Figure 11b is an estimated NRCS image, in which each pixel is estimated from 192 pixels of the single-look complex image (in each estimation, the Doppler spectrum is calculated from 16 pixels in the flight direction and averaged by 12 times in the look direction). The pixel sizes of both images are about 96 m × 90 m.

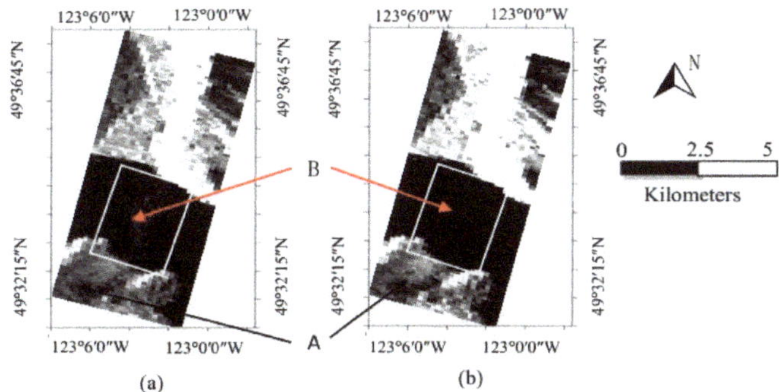


**Figure 11.** Subimage of Frame 2 in Figure 10, (**a**) conventional SAR image; (**b**) estimated NRCS image.

The white frame in Figure 11a is contaminated by the azimuthal ambiguity signal from the strong land targets to the right. As a comparison, the azimuthal ambiguity signal is filtered quite clearly in the same position in Figure 11b.

Points A and B in Figure 11a are selected to illuminate the difference in the Doppler spectrum between the signals contaminated and uncontaminated by the azimuthal ambiguity. The Doppler spectra of points A and B are shown in Figure 12.

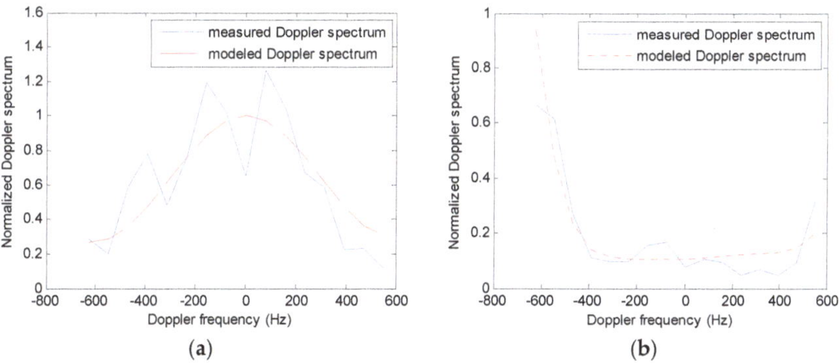


**Figure 12.** (**a**) Doppler spectrum of point A in Figure 11a; (**b**) Doppler spectrum of point B in Figure 11b; Blue solid and red dashed lines are the measured and modeled Doppler spectra, respectively.

The centroids of the Doppler spectra depicted in Figure 12a,b have been shifted to zero. The blue solid lines are the measured Doppler spectra that is calculated directly from the single-look complex image, and the red dashed lines are the Doppler spectra modeled by Equation (5), in which $\overline{\sigma}(x_0, y_0)$, $\overline{\sigma}(x_0 + D_x, y_0 + D_y)$, $\overline{\sigma}(x_0 - D_x, y_0 - D_y)$, $P_a(f)$, and $N_0$ are all known from the raw data. Point A is an uncontaminated target, so its Doppler spectrum satisfies a typical Gaussian function quite well (Figure 12a). The accordance between the measured and modeled spectra validates the accuracy of the proposed method. As a comparison, point B is a target seriously contaminated by the azimuthal ambiguity effect. The low-frequency and high-frequency parts of the Doppler spectrum of B (Figure 12b) are very high; they correspond to the ambiguity signals from the right and left sides of point B, respectively. The spectrum at low frequency is especially high. From Figure 11a, it is known that the left and right sides of point B are all high-backscattering land targets, whereas point B is a water area with a very low NRCS, and the target on the right is much stronger than on the left. The NRCS distribution of Figure 11a agrees with the analysis of the Doppler spectrum of point B and the modeled spectrum satisfies the measured spectrum quite well, which further validates the proposed method.

## 4. Discussion

As SAR imaging is more and more widely used, the radar echo is analyzed in the proposed method, and the relative value of RCS is extracted from the Doppler spectrum, the estimated relative backscattering intensity is converted to absolute NRCS using Equation (27). The comparison between the proposed method and the traditional method is described in detail in Section 3. Three examples show the feasibility and superiority of the proposed method. In the following, the estimation accuracy of the proposed method and the traditional method is analyzed by simulation.

Because the normalized image intensity differs from the NRCS only with a constant offset, the proposed method uses the normalized image intensity to be equivalent to NRCS. In this paper, in order to simplify the calculation without the loss of equivalence, the normalized image intensity is adopted in Figures 6, 8 and 9.

*The Comparative Simulation Analysis of Estimation Accuracy for Different NRCS Estimation Methods*

The simulations were performed under different signal-to-noise ratios (SNR) and azimuthal ambiguity conditions. The parameters of the simulations are given in Table 2, which correspond to low, intermediate, and high azimuthal ambiguity, respectively.

**Table 2.** Simulation parameters.

| | Simulation 1 | Simulation 2 | Simulation 3 |
|---|---|---|---|
| $\overline{\sigma}(x_0, y_0)/N_0$ | 0.01–0.5 | 0.01–0.5 | 0.01–0.5 |
| $\overline{\sigma}(x_0 - D_x, y_0 - D_y)/N_0, \overline{\sigma}(x_0 + D_x, y_0 + D_y)/N_0$ | 2 | 5 | 10 |
| Simulation repeat number | 400 | 400 | 400 |
| Pixel number used for calculating one Doppler spectrum | 80 | 80 | 80 |

Take the estimation precision of ML estimation into consideration; according to mathematical statistics theory, the ML estimation can reach the Cramer–Rao bound [12,13]. That is the root-mean-square (rms) of the estimation which can be expressed as:

$$rms[\hat{\sigma}_n] = \frac{1}{\sqrt{-E\left[\frac{\partial^2 \ln g\left(\begin{array}{cccc} p_{n-X} & p_n & p_{n+X} & I_n \end{array} \middle| \overline{\sigma}_n\right)}{\partial \overline{\sigma}_n^2}\right]}} = \frac{1}{\sqrt{\begin{array}{c}\sum\limits_{i=1}^{m} \frac{P_C^2(f_i)}{\left[\overline{\sigma}_{n-X}P_L(f_i) + \overline{\sigma}_n P_C(f_i) + \overline{\sigma}_{n+X}P_R(f_i) + \frac{N_0}{F_r}\right]^2} + \sum\limits_{i=1}^{m} \frac{P_R^2(f_i)}{\left[\overline{\sigma}_{n-2X}P_L(f_i) + \overline{\sigma}_{n-X}P_C(f_i) + \overline{\sigma}_n P_R(f_i) + \frac{N_0}{F_r}\right]^2} \\ + \sum\limits_{i=1}^{m} \frac{P_L^2(f_i)}{\left[\overline{\sigma}_n P_L(f_i) + \overline{\sigma}_{n+X}P_C(f_i) + \overline{\sigma}_{n+2X}P_R(f_i) + \frac{N_0}{F_r}\right]^2} + \frac{N}{\overline{\sigma}_n^2}\end{array}}} \tag{28}$$

where $\hat{\sigma}(x_0, y_0)$ is the ML estimation of $\overline{\sigma}(x_0, y_0)$, and $rms[\bullet]$ refers to the root–mean–square.

The rms of the modified method is shown in Figure 13. For comparison, the Cramer–Rao bound of ML estimation and the simple estimation of $I - N_0$, which is used in Equation (1), are also depicted in Figure 13.

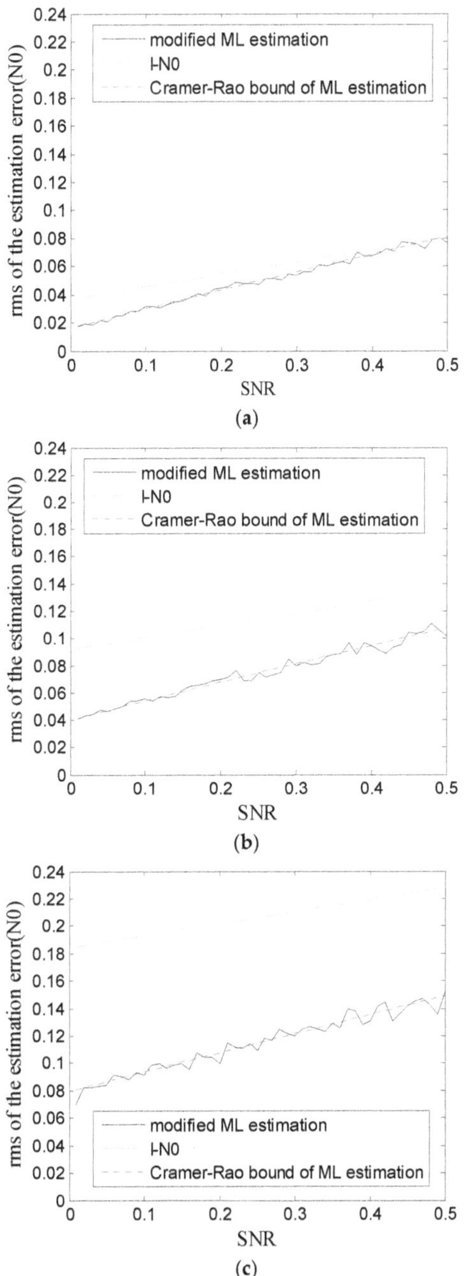


**Figure 13.** The comparison of three kinds of estimation precision, rms of estimation error of modified estimation method (black solid line), $I - N_0$ (green dashed line), and Cramer–Rao bound of maximum likelihood (ML) estimation (red dotted line), (**a**) for simulation 1; (**b**) for simulation 2; and (**c**) for simulation 3.

In Figure 13, the SNR ($x$-axis) refers to $\overline{\sigma}(x_0, y_0)/N_0$, and the estimation error ($y$-axis) is normalized by $N_0$. A comparison of Figure 13a, Figure 13b, and Figure 13c clearly shows that the estimation error of the simple estimation $I - N_0$ increases significantly with increasing azimuthal ambiguity, but the results from the proposed method of estimation can almost maintain the same estimation precision under various azimuthal ambiguity conditions. Even under low azimuthal ambiguity conditions (Figure 13a), the estimation error of the proposed estimation method is also significantly less than that of simple estimation $I - N_0$. In Figure 13a–c, it indicates that the rms of the estimation error of the proposed estimation method is very close to the Cramer–Rao bound with the increase in SNR. This result validates that the proposed estimation method can significantly increase estimation precision under a low SNR or low-scattering area in SAR images.

Take the low SNR condition into account, supposing $\overline{\sigma}(x_0, y_0) << N_0$; neglecting the contribution from the azimuthal ambiguity effect, the rms of the estimation error of the proposed estimation method can be obtained as:

$$rms[\hat{\sigma}(x_0, y_0)] \approx 0.8\frac{N_0}{F_r\sqrt{\sum_{i=1}^{m} P_a^2(f_i - f_0)}} \tag{29}$$

The Doppler power spectrum can often be obtained by incoherently averaging the spectra of several uncorrelated signals. The estimation precision derived so far is also applicable in the case when the number $m$ is replaced by the overall number of pixels contributing to the estimation.

For example, assume that the NESZ of a spaceborne SAR is $-25$ dB, the single-look pixel size is 8 m (look direction) $\times$ 5 m (flight direction), and the number of pixels contributing to one estimation is 240 (in each estimation, the Doppler spectrum is calculated from 20 pixels in the flight direction and incoherently averaged by 12 times in the look direction); an NRCS estimation precision of about $-38$ dB could be acquired in the low-backscattering area at a resolution of 96 m $\times$ 100 m.

## 5. Conclusions

The image intensities of SAR images of low-backscattering areas are often affected by the system noise and azimuthal ambiguity effect. In this paper, a method is proposed for estimating the NRCS of low-backscattering areas. The method can eliminate much of the effect of system noise and azimuthal ambiguity. This method is based on the single-look complex image, and the azimuthal matching filter in the imaging algorithm must be an unweighted filter. The parameters needed for this method can all be estimated from the single-look complex image itself, which makes the method easy to apply. An analysis on the estimation precision demonstrates that, for a typical spaceborne SAR with a NESZ of $-25$ dB and a single-look pixel size of 8 m $\times$ 5 m, the NRCS estimation precision of low-backscattering areas can reach $-38$ dB at a resolution of 96 m $\times$ 100 m.

Three examples are given for validation in Section 3. The first example is a SAR image that is an oceanic internal wave. In the conventional SAR image, the troughs of the internal wave signal intensity are buried by the noise floor, making it hard to judge the exact trough position. In contrast, the NRCS estimated by the proposed method can recover the texture features of the low-scattering area much better, and the recovered troughs of the internal wave are located near the midpoint of the adjacent peaks. The result is a qualitative analysis for the estimation method in low NRCS. In addition, the example in Section 3.2 is atmospheric gravity waves. The estimation values of signal intensity by the proposed method in this paper are very close to the theoretical value of the signal intensity in the low-scattering area of the original image. The result is a quantitative analysis for the estimation method in low NRCS. The third example is a SAR image of a land-water junction, in which the water area is seriously affected by the azimuthal ambiguity signals from high-backscattering land targets. As a comparison, the azimuthal ambiguity signals are filtered out quite clearly in the NRCS image estimated by the proposed method. The Doppler spectra of two points were analyzed, one contaminated and one uncontaminated by the azimuthal ambiguity signal. Analysis proves that the Doppler spectra modeled by the proposed method can satisfy the actual Doppler spectra calculated from the single-look complex image quite well. These three examples both indirectly and directly validate the feasibility of the proposed method in this paper.

This proposed method can be applied to SAR image processing in low-scattering areas in the ocean, such as internal waves, oil spills, low wind speed zones, upwelling, and so on. Conversely, the proposed method can be applied to the data processing of the SAR satellite system with lower NESZ, which can reduce the cost of satellites and improve the bandwidth, resolution, and other indicators of the SAR system.

**Acknowledgments:** This work was supported by the National Natural Science Foundation of China (No. 41276185).

**Author Contributions:** Hui Meng and Xiaoqing Wang conceived and performed the experiments; Xiaoqing Wang and Jinsong Chong supervised and designed the research and contributed to the article's organization; Xiangfei Wei and Weiya Kong provided the data for comparative analysis. Hui Meng and Xiaoqing Wang drafted the manuscript, which was revised by all authors. All authors read and approved the final manuscript.

**Conflicts of Interest:** The authors declare no conflict of interest.

## Appendix A. The Jacobian Matrix of the Derived Function

When $2X < n \leq \mathrm{T} - 2X$

$$
\begin{aligned}
J(n,n-2X) ={}& \sum_{i=1}^{m}\frac{P_L(f_i)P_R(f_i)}{\left[\overline{\sigma}_n P_L(f_i)+\overline{\sigma}_{n+X}P_C(f_i)+\overline{\sigma}_{n+2X}P_R(f_i)+\frac{N_0}{F_r}\right]^2}-2\sum_{i=1}^{m}\frac{p_{n-X}(f_i)P_L(f_i)P_R(f_i)}{\left[\overline{\sigma}_{n-2X}P_L(f_i)+\overline{\sigma}_{n-X}P_C(f_i)+\overline{\sigma}_n P_R(f_i)+\frac{N_0}{F_r}\right]^3}\\
J(n,n-X) = J(n,n) ={}& \sum_{i=1}^{m}\frac{P_C(f_i)P_L(f_i)}{\left[\overline{\sigma}_{n-X}P_L(f_i)+\overline{\sigma}_n P_C(f_i)+\overline{\sigma}_{n+X}P_R(f_i)+\frac{N_0}{F_r}\right]^2}+\sum_{i=1}^{m}\frac{P_C(f_i)P_R(f_i)}{\left[\overline{\sigma}_{n-2X}P_L(f_i)+\overline{\sigma}_{n-X}P_C(f_i)+\overline{\sigma}_n P_R(f_i)+\frac{N_0}{F_r}\right]^2}\\
&-2\sum_{i=1}^{m}\frac{p_n(f_i)P_C(f_i)P_L(f_i)}{\left[\overline{\sigma}_{n-X}P_L(f_i)+\overline{\sigma}_n P_C(f_i)+\overline{\sigma}_{n+X}P_R(f_i)+\frac{N_0}{F_r}\right]^3}-2\sum_{i=1}^{m}\frac{p_{n-X}(f_i)P_C(f_i)P_R(f_i)}{\left[\overline{\sigma}_{n-2X}P_L(f_i)+\overline{\sigma}_{n-X}P_C(f_i)+\overline{\sigma}_n P_R(f_i)+\frac{N_0}{F_r}\right]^3}\\
J(n,n) ={}& \sum_{i=1}^{m}\frac{P_C^2(f_i)}{\left[\overline{\sigma}_{n-X}P_L(f_i)+\overline{\sigma}_n P_C(f_i)+\overline{\sigma}_{n+X}P_R(f_i)+\frac{N_0}{F_r}\right]^2}+\sum_{i=1}^{m}\frac{P_R^2(f_i)}{\left[\overline{\sigma}_{n-2X}P_L(f_i)+\overline{\sigma}_{n-X}P_C(f_i)+\overline{\sigma}_n P_R(f_i)+\frac{N_0}{F_r}\right]^2}\\
&+\sum_{i=1}^{m}\frac{P_L^2(f_i)}{\left[\overline{\sigma}_n P_L(f_i)+\overline{\sigma}_{n+X}P_C(f_i)+\overline{\sigma}_{n+2X}P_R(f_i)+\frac{N_0}{F_r}\right]^2}-2\sum_{i=1}^{m}\frac{p_n(f_i)P_C^2(f_i)}{\left[\overline{\sigma}_{n-X}P_L(f_i)+\overline{\sigma}_n P_C(f_i)+\overline{\sigma}_{n+X}P_R(f_i)+\frac{N_0}{F_r}\right]^3}\\
&-2\sum_{i=1}^{m}\frac{p_{n-X}(f_i)P_R^2(f_i)}{\left[\overline{\sigma}_{n-2X}P_L(f_i)+\overline{\sigma}_{n-X}P_C(f_i)+\overline{\sigma}_n P_R(f_i)+\frac{N_0}{F_r}\right]^3}-2\sum_{i=1}^{m}\frac{p_{n+X}(f_i)P_L^2(f_i)}{\left[\overline{\sigma}_n P_L(f_i)+\overline{\sigma}_{n+X}P_C(f_i)+\overline{\sigma}_{n+2X}P_R(f_i)+\frac{N_0}{F_r}\right]^3}\\
&-\left[\frac{\alpha^2[2\alpha\overline{\sigma}_n(\frac{\pi}{2}+\arctan(\alpha\overline{\sigma}_n))+1]}{[1+(\alpha\overline{\sigma}_n)^2]^2[\frac{\pi}{2}+\arctan(\alpha\overline{\sigma}_n)]^2}\right]\\
J(n,n+X) ={}& \sum_{i=1}^{m}\frac{P_C(f_i)P_R(f_i)}{\left[\overline{\sigma}_{n-X}P_L(f_i)+\overline{\sigma}_n P_C(f_i)+\overline{\sigma}_{n+X}P_R(f_i)+\frac{N_0}{F_r}\right]^2}+\sum_{i=1}^{m}\frac{P_C(f_i)P_L(f_i)}{\left[\overline{\sigma}_n P_L(f_i)+\overline{\sigma}_{n+X}P_C(f_i)+\overline{\sigma}_{n+2X}P_R(f_i)+\frac{N_0}{F_r}\right]^2}\\
&-2\sum_{i=1}^{m}\frac{p_n(f_i)P_C(f_i)P_R(f_i)}{\left[\overline{\sigma}_{n-X}P_L(f_i)+\overline{\sigma}_n P_C(f_i)+\overline{\sigma}_{n+X}P_R(f_i)+\frac{N_0}{F_r}\right]^3}-2\sum_{i=1}^{m}\frac{p_{n+X}(f_i)P_C(f_i)P_L(f_i)}{\left[\overline{\sigma}_n P_L(f_i)+\overline{\sigma}_{n+X}P_C(f_i)+\overline{\sigma}_{n+2X}P_R(f_i)+\frac{N_0}{F_r}\right]^3}\\
J(n,n+2X) ={}& \sum_{i=1}^{m}\frac{P_R(f_i)P_L(f_i)}{\left[\overline{\sigma}_n P_L(f_i)+\overline{\sigma}_{n+X}P_C(f_i)+\overline{\sigma}_{n+2X}P_R(f_i)+\frac{N_0}{F_r}\right]^2}-2\sum_{i=1}^{m}\frac{p_{n+X}(f_i)P_R(f_i)P_L(f_i)}{\left[\overline{\sigma}_n P_L(f_i)+\overline{\sigma}_{n+X}P_C(f_i)+\overline{\sigma}_{n+2X}P_R(f_i)+\frac{N_0}{F_r}\right]^3}
\end{aligned}
\tag{A1}
$$

When $X < n \leq 2X$

$$
\begin{aligned}
J(n,n-X) ={}& \sum_{i=1}^{m}\frac{P_C(f_i)P_L(f_i)}{\left[\overline{\sigma}_{n-X}P_L(f_i)+\overline{\sigma}_nP_C(f_i)+\overline{\sigma}_{n+X}P_R(f_i)+\frac{N_0}{F_r}\right]^2}+\sum_{i=1}^{m}\frac{P_L^2(f_i)}{\left[\overline{\sigma}_nP_L(f_i)+\overline{\sigma}_{n+X}P_C(f_i)+\overline{\sigma}_{n+2X}P_R(f_i)+\frac{N_0}{F_r}\right]^2}\\
&-2\sum_{i=1}^{m}\frac{p_n(f_i)P_C^2(f_i)}{\left[\overline{\sigma}_{n-X}P_L(f_i)+\overline{\sigma}_nP_C(f_i)+\overline{\sigma}_{n+X}P_R(f_i)+\frac{N_0}{F_r}\right]^3}-2\sum_{i=1}^{m}\frac{p_{n+X}(f_i)P_L^2(f_i)}{\left[\overline{\sigma}_nP_L(f_i)+\overline{\sigma}_{n+X}P_C(f_i)+\overline{\sigma}_{n+2X}P_R(f_i)+\frac{N_0}{F_r}\right]^3}\\
J(n,n) ={}& \sum_{i=1}^{m}\frac{P_C^2(f_i)}{\left[\overline{\sigma}_{n-X}P_L(f_i)+\overline{\sigma}_nP_C(f_i)+\overline{\sigma}_{n+X}P_R(f_i)+\frac{N_0}{F_r}\right]^2}+\sum_{i=1}^{m}\frac{P_L^2(f_i)}{\left[\overline{\sigma}_nP_L(f_i)+\overline{\sigma}_{n+X}P_C(f_i)+\overline{\sigma}_{n+2X}P_R(f_i)+\frac{N_0}{F_r}\right]^2}\\
&-2\sum_{i=1}^{m}\frac{p_n(f_i)P_C^2(f_i)}{\left[\overline{\sigma}_{n-X}P_L(f_i)+\overline{\sigma}_nP_C(f_i)+\overline{\sigma}_{n+X}P_R(f_i)+\frac{N_0}{F_r}\right]^3}-2\sum_{i=1}^{m}\frac{p_{n+X}(f_i)P_L^2(f_i)}{\left[\overline{\sigma}_nP_L(f_i)+\overline{\sigma}_{n+X}P_C(f_i)+\overline{\sigma}_{n+2X}P_R(f_i)+\frac{N_0}{F_r}\right]^3}\\
&-\left[\frac{\alpha^2[2\alpha\overline{\sigma}_n(\frac{\pi}{2}+\arctan(\alpha\overline{\sigma}_n))+1]}{[1+(\alpha\overline{\sigma}_n)^2]^2[\frac{\pi}{2}+\arctan(\alpha\overline{\sigma}_n)]^2}\right]\\
J(n,n+X) ={}& \sum_{i=1}^{m}\frac{P_C(f_i)P_R(f_i)}{\left[\overline{\sigma}_{n-X}P_L(f_i)+\overline{\sigma}_nP_C(f_i)+\overline{\sigma}_{n+X}P_R(f_i)+\frac{N_0}{F_r}\right]^2}+\sum_{i=1}^{m}\frac{P_C(f_i)P_L(f_i)}{\left[\overline{\sigma}_nP_L(f_i)+\overline{\sigma}_{n+X}P_C(f_i)+\overline{\sigma}_{n+2X}P_R(f_i)+\frac{N_0}{F_r}\right]^2}\\
&-2\sum_{i=1}^{m}\frac{p_n(f_i)P_C(f_i)P_R(f_i)}{\left[\overline{\sigma}_{n-X}P_L(f_i)+\overline{\sigma}_nP_C(f_i)+\overline{\sigma}_{n+X}P_R(f_i)+\frac{N_0}{F_r}\right]^3}-2\sum_{i=1}^{m}\frac{p_{n+X}(f_i)P_C(f_i)P_L(f_i)}{\left[\overline{\sigma}_nP_L(f_i)+\overline{\sigma}_{n+X}P_C(f_i)+\overline{\sigma}_{n+2X}P_R(f_i)+\frac{N_0}{F_r}\right]^3}\\
J(n,n+2X) ={}& \sum_{i=1}^{m}\frac{P_L(f_i)P_R(f_i)}{\left[\overline{\sigma}_nP_L(f_i)+\overline{\sigma}_{n+X}P_C(f_i)+\overline{\sigma}_{n+2X}P_R(f_i)+\frac{N_0}{F_r}\right]^2}-2\sum_{i=1}^{m}\frac{p_{n+X}(f_i)P_L(f_i)P_R(f_i)}{\left[\overline{\sigma}_nP_L(f_i)+\overline{\sigma}_{n+X}P_C(f_i)+\overline{\sigma}_{n+2X}P_R(f_i)+\frac{N_0}{F_r}\right]^3}
\end{aligned}
\tag{A2}
$$

When $0 < n \leq X$

$$
\begin{aligned}
J(n,n) ={}& \sum_{i=1}^{m}\frac{P_L^2(f_i)}{\left[\overline{\sigma}_nP_L(f_i)+\overline{\sigma}_{n+X}P_C(f_i)+\overline{\sigma}_{n+2X}P_R(f_i)+\frac{N_0}{F_r}\right]^2}-2\sum_{i=1}^{m}\frac{p_{n+X}(f_i)P_L^2(f_i)}{\left[\overline{\sigma}_nP_L(f_i)+\overline{\sigma}_{n+X}P_C(f_i)+\overline{\sigma}_{n+2X}P_R(f_i)+\frac{N_0}{F_r}\right]^3}\\
&-\left[\frac{\alpha^2[2\alpha\overline{\sigma}_n(\frac{\pi}{2}+\arctan(\alpha\overline{\sigma}_n))+1]}{[1+(\alpha\overline{\sigma}_n)^2]^2[\frac{\pi}{2}+\arctan(\alpha\overline{\sigma}_n)]^2}\right]\\
J(n,n+X) ={}& \sum_{i=1}^{m}\frac{P_L(f_i)P_C(f_i)}{\left[\overline{\sigma}_nP_L(f_i)+\overline{\sigma}_{n+X}P_C(f_i)+\overline{\sigma}_{n+2X}P_R(f_i)+\frac{N_0}{F_r}\right]^2}-2\sum_{i=1}^{m}\frac{p_{n+X}(f_i)P_L(f_i)P_C(f_i)}{\left[\overline{\sigma}_nP_L(f_i)+\overline{\sigma}_{n+X}P_C(f_i)+\overline{\sigma}_{n+2X}P_R(f_i)+\frac{N_0}{F_r}\right]^3}\\
J(n,n+2X) ={}& \sum_{i=1}^{m}\frac{P_L(f_i)P_R(f_i)}{\left[\overline{\sigma}_nP_L(f_i)+\overline{\sigma}_{n+X}P_C(f_i)+\overline{\sigma}_{n+2X}P_R(f_i)+\frac{N_0}{F_r}\right]^2}-2\sum_{i=1}^{m}\frac{p_{n+X}(f_i)P_L(f_i)P_R(f_i)}{\left[\overline{\sigma}_nP_L(f_i)+\overline{\sigma}_{n+X}P_C(f_i)+\overline{\sigma}_{n+2X}P_R(f_i)+\frac{N_0}{F_r}\right]^3}
\end{aligned}
\tag{A3}
$$

When $\mathrm{T} - 2X < n \leq \mathrm{T} - X$

$$
\begin{aligned}
J(n,n-2X) ={}& \sum_{i=1}^{m}\frac{P_L(f_i)P_R(f_i)}{\left[\overline{\sigma}_{n-2X}P_L(f_i)+\overline{\sigma}_{n-X}P_C(f_i)+\overline{\sigma}_nP_R(f_i)+\frac{N_0}{F_r}\right]^2}-2\sum_{i=1}^{m}\frac{p_{n-X}(f_i)P_L(f_i)P_R(f_i)}{\left[\overline{\sigma}_{n-2X}P_L(f_i)+\overline{\sigma}_{n-X}P_C(f_i)+\overline{\sigma}_nP_R(f_i)+\frac{N_0}{F_r}\right]^3}\\
J(n,n) ={}& \sum_{i=1}^{m}\frac{P_C^2(f_i)}{\left[\overline{\sigma}_{n-X}P_L(f_i)+\overline{\sigma}_nP_C(f_i)+\overline{\sigma}_{n+X}P_R(f_i)+\frac{N_0}{F_r}\right]^2}+\sum_{i=1}^{m}\frac{P_R^2(f_i)}{\left[\overline{\sigma}_{n-2X}P_L(f_i)+\overline{\sigma}_{n-X}P_C(f_i)+\overline{\sigma}_nP_R(f_i)+\frac{N_0}{F_r}\right]^2}\\
&-2\sum_{i=1}^{m}\frac{p_n(f_i)P_C^2(f_i)}{\left[\overline{\sigma}_{n-X}P_L(f_i)+\overline{\sigma}_nP_C(f_i)+\overline{\sigma}_{n+X}P_R(f_i)+\frac{N_0}{F_r}\right]^3}-2\sum_{i=1}^{m}\frac{p_{n-X}(f_i)P_R^2(f_i)}{\left[\overline{\sigma}_{n-2X}P_L(f_i)+\overline{\sigma}_{n-X}P_C(f_i)+\overline{\sigma}_nP_R(f_i)+\frac{N_0}{F_r}\right]^3}\\
&-\left[\frac{\alpha^2[2\alpha\overline{\sigma}_n(\frac{\pi}{2}+\arctan(\alpha\overline{\sigma}_n))+1]}{[1+(\alpha\overline{\sigma}_n)^2]^2[\frac{\pi}{2}+\arctan(\alpha\overline{\sigma}_n)]^2}\right]\\
J(n,n+X) ={}& \sum_{i=1}^{m}\frac{P_C(f_i)P_R(f_i)}{\left[\overline{\sigma}_{n-X}P_L(f_i)+\overline{\sigma}_nP_C(f_i)+\overline{\sigma}_{n+X}P_R(f_i)+\frac{N_0}{F_r}\right]^2}-2\sum_{i=1}^{m}\frac{p_n(f_i)P_C(f_i)P_R(f_i)}{\left[\overline{\sigma}_{n-X}P_L(f_i)+\overline{\sigma}_nP_C(f_i)+\overline{\sigma}_{n+X}P_R(f_i)+\frac{N_0}{F_r}\right]^3}
\end{aligned}
\tag{A4}
$$

When $T - X < n \leq T$

$$J(n, n-2X) = \sum_{i=1}^{m} \frac{P_L(f_i)P_R(f_i)}{\left[\overline{\sigma}_{n-2X}P_L(f_i) + \overline{\sigma}_{n-X}P_C(f_i) + \overline{\sigma}_n P_R(f_i) + \frac{N_0}{F_r}\right]^2} - 2\sum_{i=1}^{m} \frac{p_{n-X}(f_i)P_L(f_i)P_R(f_i)}{\left[\overline{\sigma}_{n-2X}P_L(f_i) + \overline{\sigma}_{n-X}P_C(f_i) + \overline{\sigma}_n P_R(f_i) + \frac{N_0}{F_r}\right]^3}$$
$$J(n, n-X) = \sum_{i=1}^{m} \frac{P_C(f_i)P_R(f_i)}{\left[\overline{\sigma}_{n-2X}P_L(f_i) + \overline{\sigma}_{n-X}P_C(f_i) + \overline{\sigma}_n P_R(f_i) + \frac{N_0}{F_r}\right]^2} - 2\sum_{i=1}^{m} \frac{p_{n-X}(f_i)P_C(f_i)P_R(f_i)}{\left[\overline{\sigma}_{n-2X}P_L(f_i) + \overline{\sigma}_{n-X}P_C(f_i) + \overline{\sigma}_n P_R(f_i) + \frac{N_0}{F_r}\right]^3} \quad \text{(A5)}$$
$$J(n, n) = \sum_{i=1}^{m} \frac{P_R^2(f_i)}{\left[\overline{\sigma}_{n-2X}P_L(f_i) + \overline{\sigma}_{n-X}P_C(f_i) + \overline{\sigma}_n P_R(f_i) + \frac{N_0}{F_r}\right]^2} - 2\sum_{i=1}^{m} \frac{p_{n-X}(f_i)P_R^2(f_i)}{\left[\overline{\sigma}_{n-2X}P_L(f_i) + \overline{\sigma}_{n-X}P_C(f_i) + \overline{\sigma}_n P_R(f_i) + \frac{N_0}{F_r}\right]^3}$$
$$- \left[\frac{\alpha^2[2\alpha\overline{\sigma}_n(\frac{\pi}{2} + \arctan(\alpha\overline{\sigma}_n)) + 1]}{[1 + (\alpha\overline{\sigma}_n)^2]^2[\frac{\pi}{2} + \arctan(\alpha\overline{\sigma}_n)]^2}\right]$$

## References


1. Kerbaol, V.; Collard, F. SAR-derived coastal and marine applications: From research to operational products. *IEEE J. Ocean. Eng.* **2005**, *30*, 472–486. [CrossRef]
2. Solberg, A.S.; Storvik, G.; Solberg, R.; Volden, E. Automatic detection of oil spills in ERS SAR images. *IEEE Trans. Geosci. Remote Sens.* **1999**, *37*, 1916–1924. [CrossRef]
3. Wei, Z. *Synthetic Aperture Radar Satellite*; Sciences Press: Beijing, China, 2001. (In Chinese)
4. Massonnet, D.; Souyris, J.-C. *Imaging with Synthetic Aperture Radar*; CRC Press: Boca Raton, FL, USA, 2008.
5. Lukowski, T.; St-Jean, R.; Hawkins, R.; Draper, K.; Hamm, P.; Teany, L. Ers-1 SAR Calibration Studies: Preparation for Radarsat. In Proceedings of the International Geoscience and Remote Sensing Symposium, Better Understanding of Earth Environment, Tokyo, Japan, 18–21 August 1993.
6. Lukowski, T.; Hawkins, R.; Moucha, R.; Khandelwal, T.; Neeson, I. Spaceborne SAR calibration studies: Ers-1. In Proceedings of the International Geoscience and Remote Sensing Symposium, Surface and Atmospheric Remote Sensing: Technologies, Data Analysis and Interpretation, Pasadena, CA, USA, 8–12 August 1994.
7. Rosich, B.; Meadows, P. *Absolute Calibration of ASAR Level 1 Products Generated with PF-ASAR*; ESA Document; European Space Agency: Frascati, Italy, 2004.
8. Laur, H.; Bally, P.; Meadows, P.; Sánchez, J.; Schättler, B.; Lopinto, E.; Esteban, D. *ERS SAR Calibration: Derivation of σ0 in ESA ERS SAR PRI Products*; ESA/ESRIN, ES-TN-RS-PM-HL09; European Space Agency/ESRIN: Frascati, Italy, 2003.
9. Lavalle, M.; Wright, T. *Absolute Radiometric and Polarimetric Calibration of Alos Palsar Products*; European Space Agency: Paris, France, 2009.
10. RADARSAT International. *Radarsat-1 Data Product Specification*; RADARSAT International: Richmond, BC, Canada, 2004.
11. Cumming, I.G.; Wong, F.H. *Digital Processing of Synthetic Aperture Radar Data*; Artech House: London, UK, 2005.
12. Zhang, X. *Modern Signal Processing*; Tsinghua University Press: Beijing, China, 2002. (In Chinese)
13. Sage, A.P.; Melsa, J.L. *Estimation Theory with Applications to Communications and Control*; DTIC Document; McGraw-Hill Book Co.: New York, NY, USA, 1971.
14. Gao, G. Statistical modeling of SAR images: A survey. *Sensors* **2010**, *10*, 775–795. [CrossRef] [PubMed]
15. Bamler, R. Doppler frequency estimation and the cramer-rao bound. *IEEE Trans. Geosci. Remote Sens.* **1991**, *29*, 385–390. [CrossRef]
16. Madsen, S.N. Estimating the Doppler centroid of SAR data. *IEEE Trans. Aerosp. Electron. Syst.* **1989**, *25*, 134–140. [CrossRef]
17. Bamler, R.; Runge, H. Prf-ambiguity resolving by wavelength diversity. *IEEE Trans. Geosci. Remote Sens.* **1991**, *29*, 997–1003. [CrossRef]
18. Wong, F.; Cumming, I.G. A combined SAR Doppler centroid estimation scheme based upon signal phase. *IEEE Trans. Geosci. Remote Sens.* **1996**, *34*, 696–707. [CrossRef]
19. Chang, C.-Y.; Curlander, J.C. Application of the multiple PRF technique to resolve Doppler centroid estimation ambiguity for spaceborne SAR. *IEEE Trans. Geosci. Remote Sens.* **1992**, *30*, 941–949. [CrossRef]
20. Chapron, B.; Collard, F.; Ardhuin, F. Direct measurements of ocean surface velocity from space: Interpretation and validation. *J. Geophys. Res. Oceans* **2005**, *110*, 691–706. [CrossRef]

21. Hansen, M.W.; Collard, F.; Dagestad, K.; Johannessen, J.A.; Fabry, P.; Chapron, B. Retrieval of sea surface range velocities from envisat asar doppler centroid measurements. *IEEE Trans. Geosci. Remote Sens.* **2011**, *49*, 3582–3592. [CrossRef]
22. Johannessen, J.A.; Chapron, B.; Collard, F.; Kudryavtsev, V.; Mouche, A.; Akimov, D.; Dagestad, K.F. Direct ocean surface velocity measurements from space: Improved quantitative interpretation of ENVISAT ASAR observations. *Geophys. Res. Lett.* **2008**, *35*, 113–130. [CrossRef]
23. Wang, P.; Wang, X.; Chong, J.; Lu, Y. Optimal parameter estimation method of internal solitary waves in SAR images and the Cramér–Rao bound. *IEEE Trans. Geosci. Remote Sens.* **2016**, *54*, 3143–3150. [CrossRef]

Article

# Performance Analysis of Ocean Surface Topography Altimetry by Ku-Band Near-Nadir Interferometric SAR

**Weiya Kong [1,2,3], Jinsong Chong [1,2,*] and Hong Tan [4,5,*]**

1 National Key Laboratory of Science and Technology on Microwave Imaging, Beijing 100190, China; kongweiya13@mails.ucas.ac.cn
2 Institute of Electronics, Chinese Academy of Sciences, Beijing 100190, China
3 School of Electronics, Electrical and Communication Engineering, University of Chinese Academy of Sciences, Beijing 100190, China
4 Technology and Engineering Center for Space Utilization, Chinese Academy of Sciences, Beijing 100094, China
5 Key Laboratory of Space Utilization, Chinese Academy of Sciences, Beijing 100094, China
* Correspondence: iecas_chong@163.com (J.C.); tanhong@csu.ac.cn (H.T.); Tel.: +86-10-5888-7125 (J.C.)

Received: 14 August 2017; Accepted: 6 September 2017; Published: 9 September 2017

**Abstract:** Interferometric imaging radar altimeter (InIRA) is the first spaceborne Ku-band interferometric synthetic aperture radar (InSAR) which is specially designed for ocean surface topography altimetry. It is on the Tiangong II space laboratory, which was launched on 15 September 2016. Different from any other spaceborne synthetic aperture radar (SAR), InIRA chooses a near-nadir incidence of 1°~8° in order to increase the altimetric precision and swath width. Limited by the size of the Tiangong II capsule, the baseline length of InIRA is only 2.3 m. However, benefitting from the low orbit, the signal-to-noise ratio of InIRA-acquired data is above 10 dB in most of the swath, which, to a certain extent, compensates for the short baseline deficiency. The altimetric precision is simulated based on the system parameters of InIRA. Results show that it is better than 7 cm on a 5-km grid and improves to 3 cm on a 10-km grid when the incidence is below 7.4°. The interferometric data of InIRA are processed to estimate the altimetric precision after a series of procedures (including image coregistration, flat-earth-phase removal, system parameters calibration and phase noise suppression). Results show that the estimated altimetric precision is close to but lower than the simulated precision among most of the swath. The intensity boundary phenomenon is first found between the near range and far range of the SAR images of InIRA. It can be explained by the modulation of ocean internal waves or oil slick, which smooths ocean surface roughness and causes the modulated area to appear either brighter or darker than its surroundings. This intensity boundary phenomenon indicates that the available swath of high altimetric precision will be narrower than expected.

**Keywords:** InSAR; near-nadir incidence; Ku-band; ocean surface topography; altimetry

---

## 1. Introduction

Under the influence of ocean tides, eddies and bathymetry variety, ocean surface topography (OST) changes at a global level [1,2]. Satellite radar altimetry has been utilized for over three decades since the first spaceborne radar altimeter on geodetic satellite (GEOSAT)to acquire high precision OST. Now the latest Poseidon-3B on Jason-3 can achieve a 2-cm precision [3]. The altimeter sends special shaped pulse to the nadir point and times the round trip delay to measure the ocean surface height. The pulse size projected on the ocean surface, also referred to as a footprint, decides the swath width and resolution of the altimeter, which is usually on the order of 2~10 km [4], as shown in Figure 1. An altimeter is none-imaging radar, it only measures one-dimensional OST along the satellite track.

It is difficult for an altimeter to map mesoscale or sub-mesoscale (15~300 km) oceanic processes, since a 200~300 km gap usually exists between two successive tracks. Therefore, an altimeter is normally used to study large-scale (>300 km) oceanic processes [5].

In recent years, there has been a growing demand for scientific research on mesoscale and sub-mesoscale oceanic processes, such as eddies and currents [6,7]. Interferometric synthetic aperture radar (InSAR), which is capable of measuring OST with high precision and a wide swath, is expected to be the next generation of satellite altimeter. InSAR has been successfully used to map terrain elevation, and the Shuttle Radar Topography Mission (SRTM) is one of the most typical examples [8]. InSAR altimetry is performed by acquiring two synthetic aperture radar (SAR) complex images of the same region during the single pass or repeat pass of the satellite [9]. Topography change is then recorded by the interferometric phase, which is the phase difference between two SAR complex images [10]. Different from terrain elevation measurement, ocean surface height changes randomly, hence two SAR antennas should be borne on satellites to acquire the images simultaneously. Moreover, OST altimetry demands centimeter-level precision, which is far beyond the capability of current InSAR systems. Therefore, several key parameters of InSAR need to be adjusted to fulfill the stringent requirements, which include near-nadir incidence, long physical baseline and short radar wavelength, and near-nadir incidence is the most distinctive characteristic from any other existing InSAR systems. Near-nadir InSAR is imaging radar; by using a pulse compressing technique, the image resolution is about a few meters, which is much smaller than its swath, as shown in Figure 1. Therefore, near-nadir InSAR can be used to measure surface water (lakes, rivers, wetlands, etc.) as well [11].

The wide-swath ocean altimeter (WSOA) is the first near-nadir InSAR proposed by National Aeronautics and Space Administration (NASA), and its two antennas work at Ku-band and form a baseline of 6.4 m. The incidence ranges from 0.6° to 4°, and a 5-cm precision is expected on a 15 km grid [12,13]. However, due to the technical limitations at that time, WSOA was finally canceled. The Surface Water and Ocean Topography (SWOT) mission, a collaboration between NASA and Centre National d'Etudes Spatiales (CNES), draws from the heritage of WSOA. KaRIN, as its main instrument, is a Ka-band near-nadir InSAR. A 10-m boom links its antennas to guarantee that the baseline is long enough to achieve a 3-cm precision on a 1-km grid. SWOT is planned to be launched in 2021 [5]. CNES carried out a series of airborne near-nadir InSAR experiments on surface water measurement to imitate SWOT performance [14,15]. Roger Fjørtoft [16] reported that the surface water was severely influenced by layover effect from its surrounding landscape because of the near-nadir incidence. Nevertheless, due to high land/water contrast, which was about 10~20 dB, the layover could be mitigated to some extent. Elizabeth H. Altenau et al. [17] used AirSWOT (an airborne near-nadir InSAR which was developed by Jet Propulsion Laboratory) collected near-nadir InSAR data to retrieve water surface elevation of the Tanana River, Alaska. Verified by in situ measurements, results showed that the root-mean-square error (RMSE) was 9 cm for water surface elevation averaged over 1-km$^2$ areas.

The interferometric imaging radar altimeter (InIRA), as an experimental payload, was developed by the National Space Science Center, Chinese Academy of Sciences. InIRA is borne on Tiangong II, the Chinese space laboratory, which was launched on 15 September 2016 and runs in a non-sun-synchronous orbit at an altitude of about 400 km. InIRA is the second spaceborne InSAR that can perform single-pass interferometry after SRTM. Its two antennas work at Ku-band and form a baseline of 2.3 m. As shown in Figure 1, due to the wide range of the incidence (1°~8°), a swath larger than 40 km can be achieved along the right side of the orbit track. The usual SAR incidence varies from 20° to 60°, which determines that Bragg scattering is prevalent among the received signal [18]. While under the near-nadir incidence, specular-reflected signal from ocean surface dominates, which makes InIRA acquired data quite unique from any other SAR images.

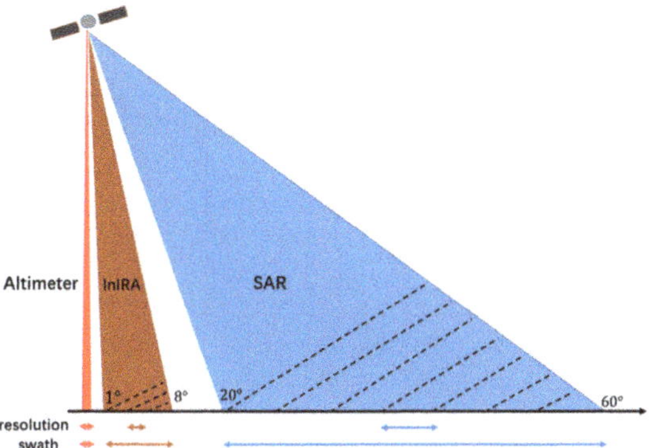


**Figure 1.** Schematic diagram of the resolution and swath of an altimeter, interferometric imaging radar altimeter (InIRA) and synthetic aperture radar (SAR) (the sketch does not represent the actual geometrical size).

In this paper, we analyze the performance of InIRA on ocean surface topography altimetry to evaluate its capabilities on exploiting various oceanic processes; in Section 2, the measurement principle of near-nadir InSAR is briefly introduced and the altimetric precision is simulated based on the system parameters of InIRA; in Section 3, after a series of data processing procedures, the altimetric precision on a 1-km grid is estimated from the multi-look interferometric phase; in Section 4, the intensity boundary phenomenon of near-nadir InSAR image is first found and discussed; some recommendations of parameter design for future near-nadir InSAR are given based on the performance of InIRA in Section 5.

## 2. Altimetry Simulation of OST

### 2.1. System Parameters of InIRA

All system parameters of InIRA are designed in order to increase its altimetric precision and swath width; however, limited by the capsule size of Tiangong II and the compatible requirements with other experiment payloads, some of the system parameters are restricted.

The altimetric precision of InSAR is given by [10]:

$$\Delta h = \frac{\lambda r_1 \sin\theta}{2\pi B \cos(\theta - \alpha)} \Delta\phi \tag{1}$$

where $B$ is the physical baseline length, $\lambda$ is the radar wavelength, $\theta$ is the incidence, $\alpha$ is the baseline roll angle, and $\Delta\phi$ is the phase error. Therefore, a longer baseline, smaller incidence and shorter radar wavelength will improve the altimetric precision significantly.

The antennas are installed outside the capsule of Tiangong II and form a 2.3-m baseline. The baseline length is quite short for the high altimetric precision requirement. In order to compensate for the short baseline deficiency, a near-nadir incidence is chosen. A bonus brought by the near-nadir incidence is that the received signals are mainly from specular reflection, which improves the signal-to-noise ratio (SNR) a lot and further increases the altimetric precision. The relationship between SNR and altimetric precision is explained in Section 2.3.

Theoretically, the shorter the radar wavelength the higher the altimetric precision. However, short radar wavelength has its drawbacks, and rain attenuation will decrease the SNR of reflected signal and eventually worsen the altimetric precision [19]. Moreover, when an electromagnetic wave intersects with the ocean surface, short radar wavelength will lower the possibility of specular reflection because less surface appears smooth [16]. This will definitely worsen the precision as well since most of the reflected signals come from specular reflection under near-nadir incidence. Hence, the Ku-band antennas are chosen for InIRA above all these considerations.

In order to achieve a wide swath, the relative variation of the incidence is quite large. The nominal incidence of InIRA is 1°~8°, as shown in Figure 2a; a 47-km-wide swath can be achieved within the incidence range, while a large relative variation of incidence also makes the ground-range resolution of InIRA change significantly. The ground-range resolution has changed almost eight times from ~213 m in the near range to ~27 m in the far range, as shown in Figure 2b.The variation of ground-range resolution also causes the multi-look number varies significantly among the swath. Averaging N neighboring pixels, also referred to as multi-looking, is normally used to reduce the phase error while processing the interferometric data [20]. The multi-look number is determined by the ratio between grid resolution and ground-range resolution. Supposing that the grid resolution along the ground-range direction is 1 km after multi-looking, then the phase error suppression by multi-looking in the near range of the swath will be much less than in the far range.

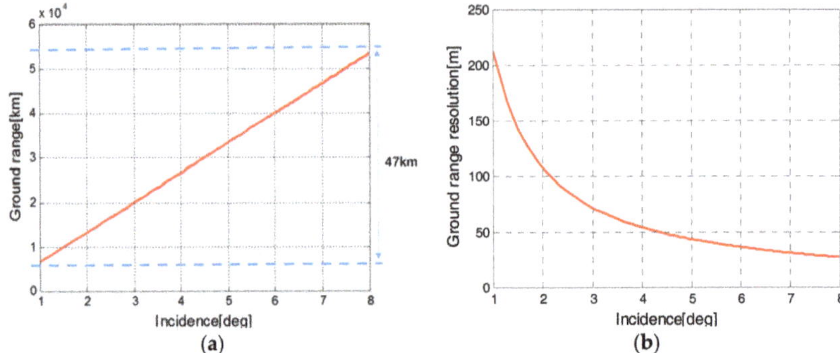


**Figure 2.** (**a**) Swath width and (**b**) ground-range resolution of InIRA within the incidence range.

## 2.2. Altimetry Principle of OST

Before the simulation of altimetric precision, we briefly introduce the OST altimetry principle. As shown in Figure 3, the baseline of InIRA is perpendicular to its orbital track, and $\alpha$ is the baseline tilt angle. Antenna A1 sends pulses to the ocean surface, and both of the antennas receive the reflected pulses; $r_1$ and $r_2$ are the distances between the antennas and point P, respectively. This one-transmit-two-receive working mechanism makes the effective baseline half of its physical length [10]. The received pulses of A1 and A2 are processed into two SAR complex images separately. Because of the return path difference $\Delta r$, which is caused by the existence of baseline, the corresponding phases of point P in the complex images are different. Therefore, the relationship between phase difference $\phi$ and return path difference $\Delta r$ can be expressed by:

$$\phi = \frac{2\pi(r_2 - r_1)}{\lambda} = \frac{2\pi\Delta r}{\lambda} = \frac{2\pi B\sin(\theta - \alpha)}{\lambda} \quad (2)$$

where $\lambda$ is the wavelength, $B$ is the physical baseline length, and $\Delta r$ is the return path difference, which can be approximated by the parallel component of the baseline along the look direction.

After acquiring the phase difference $\phi$, with precise knowledge of $r_1$, the height of point P above the reference plane (could be the local mean sea level) can be deduced by:

$$h = H - r_1\cos\theta \quad (3)$$

where $H$ is the altitude of Tiangong II above the reference plane, $r_1$ is measured by timing the signal round trip delay, and $\theta$ is the incidence which is related to the phase difference $\phi$.

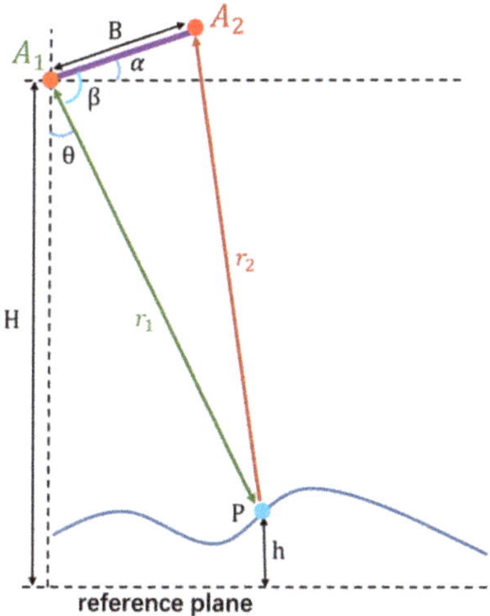


**Figure 3.** Ocean surface topography (OST) measurement principle (slant-range profile).

According to the geometrical relationship in Figure 3, two equations are given:

$$\cos\beta = \frac{r_1^2 + B^2 - {r_2}^2}{2r_1 B} \approx -\frac{\Delta r}{B} \tag{4}$$

$$\beta = \frac{\pi}{2} + \alpha - \theta \tag{5}$$

where $\beta$ is the angle between the baseline and view direction. Combining Equations (2)–(5), the relationship between height $h$ and phase difference $\phi$ is given by:

$$h = H - r_1 \cos(\frac{\pi}{2} + \alpha - \arccos(-\frac{\lambda\phi}{2\pi B})) \tag{6}$$

In fact, Equation (1) is deduced from the differential of Equation (6).

The interferometric phase $\phi_{12}$ which is derived from the phase difference between two SAR complex images, cannot be directly translated into ocean surface height. According to Equation (2), the return path difference $\Delta r$ can be expressed by the number of wavelengths, normally $\Delta r$ is longer than one wavelength. Whereas the phases recorded by complex images are confined within $[-\pi, \pi]$, this means the phases are wrapped. The relationship between unwrapped phase $\phi$ and interferometric phase $\phi_{12}$ is given in Equation (7). Phase wrapping leads to the missing of integer number $n$, so phase calibration must be done before acquiring the absolute ocean surface height.

$$\phi = 2\pi(\mathrm{n} + \frac{\phi_{12}}{\lambda}) \tag{7}$$

### 2.3. Altimetric Precision Simulation

In Equation (1), $\Delta h$ can also be regarded as the relative height change above a reference sea level, $\Delta\phi$ is the corresponding phase change. Since phases are confined within $[-\pi, \pi]$, a too-large relative height change will cause phase ambiguity, also referred to as phase wrapping [10]. The altimetric ambiguity is defined as the relative height change corresponding to a phase ambiguity of $2\pi$. When $\Delta\phi = 2\pi$ in Equation (1), then the altimetric ambiguity is given by:

$$h_{2\pi} = \frac{\lambda r_1 \sin\theta}{B\cos(\theta - \alpha)} \tag{8}$$

Obviously, a longer baseline, smaller incidence and shorter wavelength will decrease the altimetric ambiguity or, in other words, cause the phase to be wrapped more easily by the same relative height change. Phase unwrapping is difficult, hence the altimetric ambiguity of InIRA should be large enough to avoid this problem. Figure 4a shows that the altimetric ambiguity has changed over eight times from ~64 m in the near range to ~518 m in the far range. This indicates that a relative height change larger than ~32 m in the near range will wrap the phase (since the corresponding phase is larger than $\pi$). However, phase wrapping is quite unlikely to happen, since ocean surface height variation is usually only within a few meters.

While a too-large altimetric ambiguity also brings a challenge to InIRA, for a 10-cm relative height change, the corresponding phase change is only 0.01 rad in the near range, then drops down to 0.001 rad in the far range, as shown in Figure 4b. This indicates that the phase error in the near range should be better than 0.01 rad to detect the relative height change effectively, and the phase error in the far range should be even 10 times better than in the near range. The stringent requirement on phase error demands that the system thermal noise be constrained at a certain level, and the data processing must be extremely accurate.

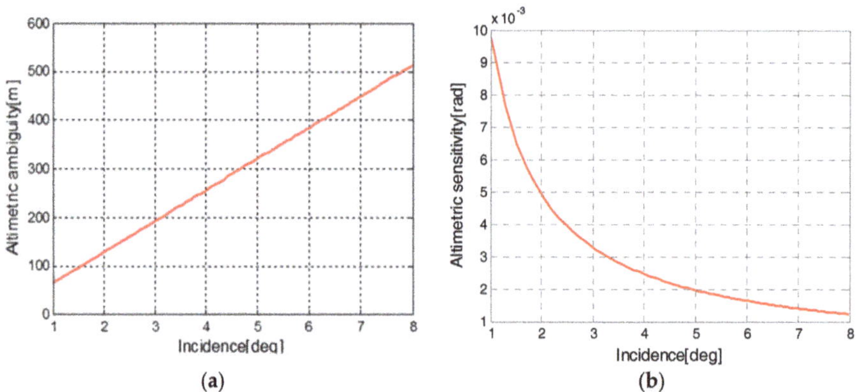


**Figure 4.** (**a**) Altimetric ambiguity of InIRA corresponds to a phase ambiguity of $2\pi$; (**b**) Altimetric sensitivity of InIRA corresponds to a 10-cm ocean surface height change.

According to the analysis mentioned above, phase error will eventually determine the altimetric precision of InIRA. Phase errors are mainly from systematic errors and random errors. Systematic errors are constant or slow drifting, such as differential phase drift error or baseline roll angle error, most of which can be eliminated through calibration. In this paper we focus on the relative altimetric precision of ocean surface height, thus only random errors are considered during the simulation.

Random phase error is introduced in the simulation by multiple decorrelation factors. For any InSAR, thermal noise influences phase error the most [21], the thermal correlation factor $\gamma_N$ is given by:

$$\gamma_N = \frac{SNR}{SNR+1} \tag{9}$$

where $SNR$ is the signal-to-noise ratio. Benefitting from low orbit and near-nadir incidence, the SNR of InIRA received data is very high within most of its swath, as show in Figure 5a (the SNR curve is derived from the acquired data of InIRA), and high SNR guarantees high coherence for the interferometric data. Apart from thermal decorrelation, the presence of ocean waves will introduce a volume-scattering layer on the angle subtended by a range resolution grid, thereby introducing the volumetric decorrelation [22], which is given approximately by:

$$\gamma_v \approx e^{-2(\frac{2\pi\sigma_h B}{r_1 \lambda \tan\theta})^2} \tag{10}$$

where $\sigma_h$ is the ocean surface height standard deviation, related to the significant wave height (SWH) as SWH $= 4\sigma_h$. The simulated volumetric decorrelation coefficient is shown in Figure 5b. It obviously has a much bigger influence in the near range because of the steeper incidence, and it decreases significantly as the SWH increases.

Compared to the above mentioned factors, other decorrelation factors can be neglected during the simulation [22]. The coherence of the interferometric data can be expressed by the multiplication of the decorrelation factors, as given in (11):

$$\gamma = \gamma_N \cdot \gamma_v \tag{11}$$

The coherence of InIRA acquired data is then simulated, as shown in Figure 5c, and the decrease in the far range is mainly caused by the decrease of SNR. By averaging the neighboring pixels, the phase error can be suppressed and follows the Cramer-Rao bound [21], which is given by:

$$\Delta\phi = \sqrt{\frac{1-\gamma^2}{2\gamma^2 N}} \tag{12}$$

where $N$ is the multi-look number, $\gamma$ is the coherence. Though the SNR of InIRA acquired data is very high, a large number of pixels still need to be averaged to achieve centimeter-level precision. We have simulated the altimetric precision on a 1-km grid, as shown in Figure 5d. In the near range, the precision drops due to the volumetric decorrelation. In the far range, when the incidence is above ~7.4°, the precision appears to have a steep drop, this is caused by the increase of the incidence apart from the SNR decline. As mentioned before, the altimetric precision can be increased by multi-looking, thus altimetric precision on a 1-km, 5-km and 10-km grid are simulated and compared, the SWH is set to be 2 m. As shown in Figure 5e, the precision on a 5-km grid is better than 7 cm and improves to 3 cm on a 10-km grid due to larger averaged pixel number.

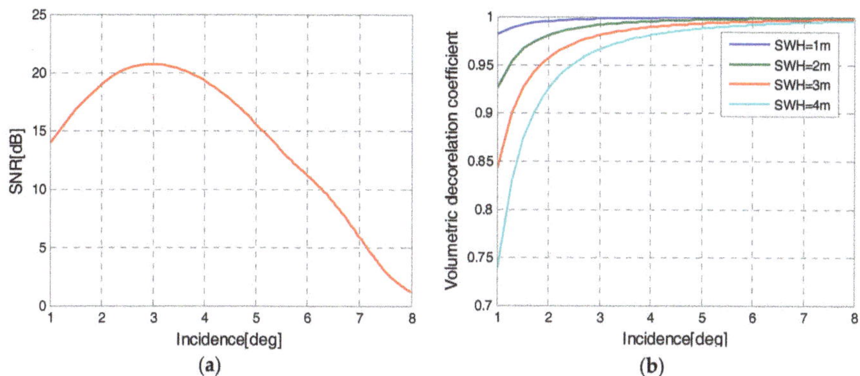


**Figure 5.** *Cont.*

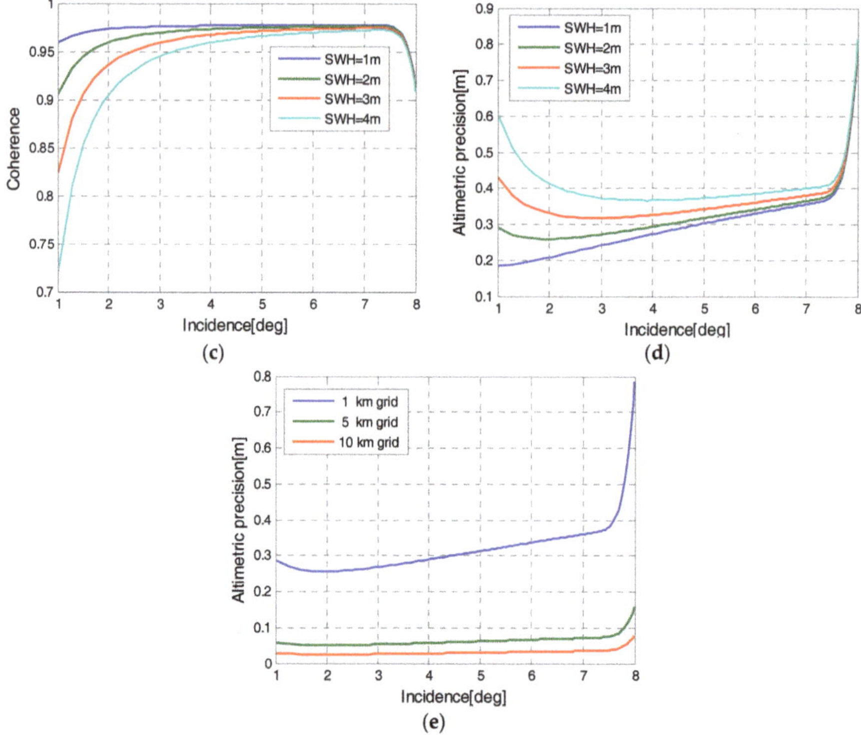


**Figure 5.** (**a**) Signal-to-noise ratio (SNR) curve estimated from InIRA acquired data; (**b**) volumetric decorrelation coefficient; (**c**) coherence, and; (**d**) altimetric precision under a different significant wave height (SWH) setting; (**e**) altimetric precision on a different resolution grid, the SWH is set to be 2 m during the simulation.

## 3. OST Altimetry Results and Analysis of InIRA

InIRA acquired the first batch of data in September 2016; the intensity image shown in Figure 6 is located in the South China Sea, next to Vietnam. The South China Sea is famous for its abundant ocean internal waves [23], and the data has just recorded this phenomenon. Some intensity discontinuity, also referred to as scalloping [24], can be seen in the far range of the ocean along the azimuth direction; this is caused by the burst working mode of InIRA. Contrary to conventional synthetic aperture radar (SAR) images, the intensity of the ocean surface in Figure 6 is much brighter than the land. The reason for this unique feature is that the received signal of InIRA contains both backscattered energy and specular-reflected energy. Compared with rough land surfaces, the signals are much easier to be specular-reflected by the ocean surface, and the specular-reflected signal takes up a much bigger proportion than the backscattered signal. In this section, the processing procedures of the interferometric data are briefly introduced, then the altimetric precision on a 1-km grid is estimated and analyzed by the processed interferometric phase.

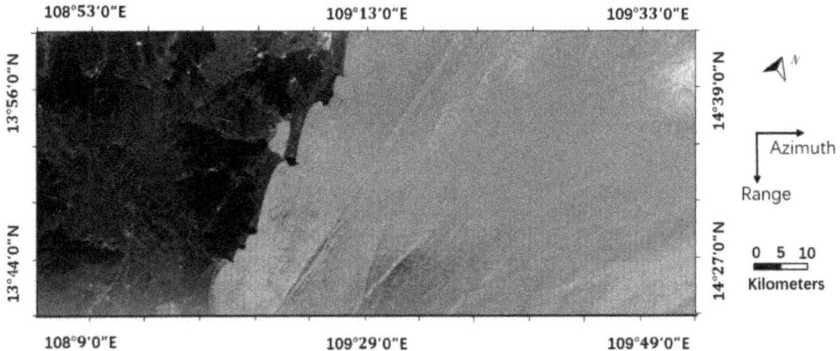


**Figure 6.** SAR intensity image of the South China Sea, acquired by InIRA in September 2016. The image is geocoded, the modulation of antenna pattern along the range direction is corrected. The azimuth and range directions are indicated by arrows.

### 3.1. Interferometric Data Processing

The interferometric data need to be coregistered, flat-earth phase removed and calibrated to finally obtain the absolute ocean surface height [10]. Though the parameter calibration of InIRA has not yet proceeded, we can still utilize the relative phase change to estimate the altimetric precision after the flat-earth phase removal.

Figure 7a is the slant-range image, under the modulation of antenna pattern, ocean surface intensity decreases rapidly in the far range, which indicates the decrease of SNR as well. Figure 7b is the coregistered phase of two SAR complex images, and fringes among the ocean area is the classic flat-earth phase. The flat-earth phase exists and changes regularly in the slant-range direction even if the topography does not change at all [10]. Among the mountainous area, the fringes seem to be distorted, which is caused by the elevation change of the mountains. Due to the decrease of SNR, phases in the far range are much noisier.

Phases in the black box in Figure 7b contains both the flat-earth phase and relative phase, which corresponds to the OST change. To acquire the relative phase, the flat-earth phase should be removed first. The flat-earth phase can be expressed by:

$$\varphi(B,\alpha,n) = \frac{2\pi B\sin(\theta-\alpha)}{\lambda} = \varphi_{fringe} + 2\pi n \tag{13}$$

where the baseline length $B$, baseline tilt angle $\alpha$ and wavelength number $n$ need to be calibrated. $\varphi_{fringe}$ is the wrapped flat-earth phase, which changes regularly along the slant-range direction, as shown in the black box in Figure 7b.

Since the relative phase change is very small according to the altimetric sensitivity simulation in Section 2.1, hence phases in the black box are mostly composed by the wrapped flat-earth phase $\varphi_{fringe}$, which contains precise information of the parameters that need to be calibrated.

Therefore, we utilize the flat-earth phase to calibrate the system parameters. Phases in the black box in Figure 7b are selected to acquire the flat-earth phase curve along the slant-range direction, then the flat-earth phase curve is fitted to the nominal Equation (13). By changing different combination of parameters $(B,\alpha,n)$ and seeking the minimum RMSE between flat-earth phase curve and the nominal equation. The parameters are calibrated in this way, as given in (14), and the nominal flat-earth phase is calculated and eliminated throughout the entire image. Besides, most of the systematic errors, such as differential phase drift [22], are also eliminated during this process.

$$\overset{\wedge}{(B,\alpha,n)} = \text{argmin}[\varphi_{fringe} - \varphi(B,\alpha,n)] \tag{14}$$

The relative phases which correspond to the topography are shown in Figure 7c. The phases vary notably among the mountainous area due to the large terrain variation. However, among the ocean area, the phases hardly change and tend to be near 0 rad. In the far range, where the internal waves exists, the phase deviates from 0 rad and looks much noisier; this phenomenon will be analyzed in Section 3.2.

Figure 7d is the coherence between two SAR complex images, and most of the ocean surface area is above 0.95. However, the coherence drops in the far range because of the decrease of SNR, especially in the internal wave area. Coherence of the mountainous area is between 0.3 and 0.7, and this relatively low coherence is mainly caused by volumetric decorrelation of the Ku-band signal, since mountains in Vietnam are covered by lush vegetation.

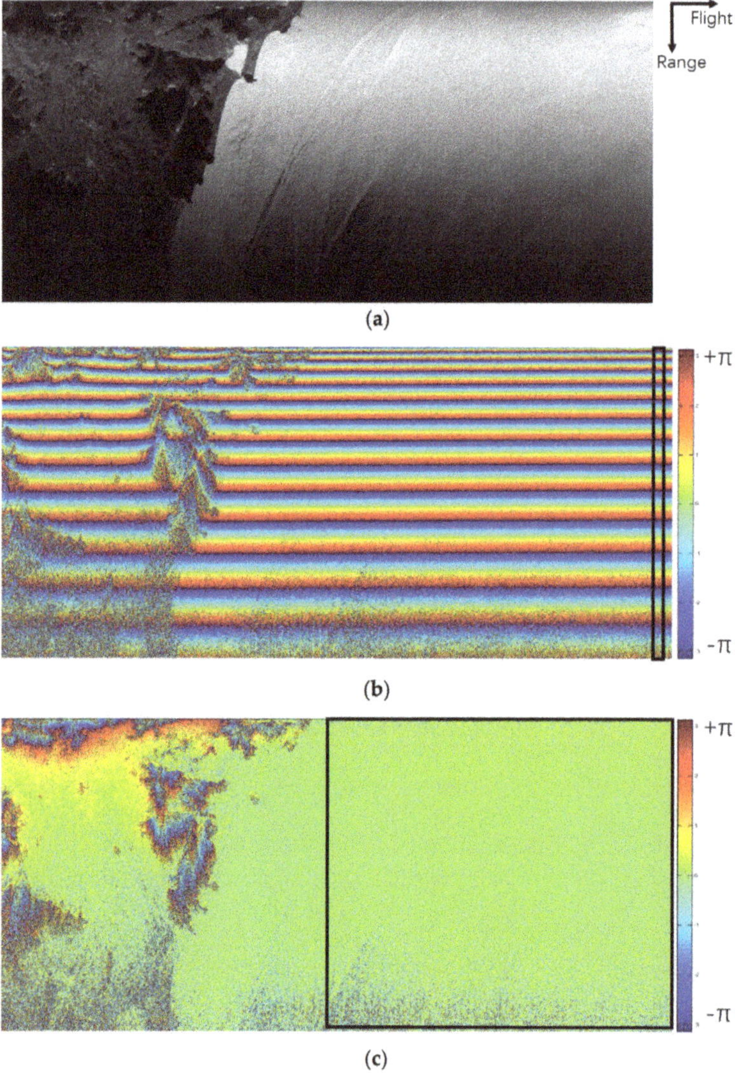


**Figure 7.** *Cont.*

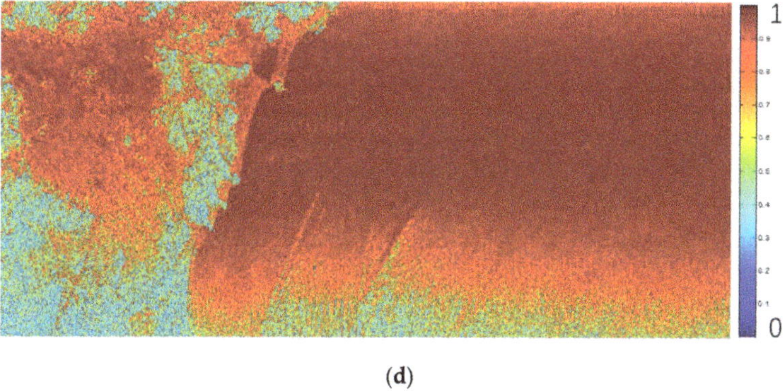


**Figure 7.** (**a**) Intensity image in slant-range direction without antenna pattern correction; (**b**) coregistered phase of two SAR complex images; (**c**) relative phase after flat-earth phase removal; (**d**) coherence between two SAR complex images. The azimuth and slant-range directions are indicated by arrows.

### 3.2. Altimetric Precision Estimation

According to the altimetric sensitivity analysis in Section 2.3, a 10-cm height change causes a mere 0.001~0.01 rad phase change for InIRA. The phase change is too small compared to the dynamic phase range, which is between [-π, π]. Moreover, the phase noise level is too high for OST-associated phase change to emerge on Figure 7c.

Phases in the black box in Figure 7c are multi-looked to lower the phase noise level; the phase dynamic range decreases significantly and the OST associated phases finally emerges on the 1-km grid, as shown in Figure 8.

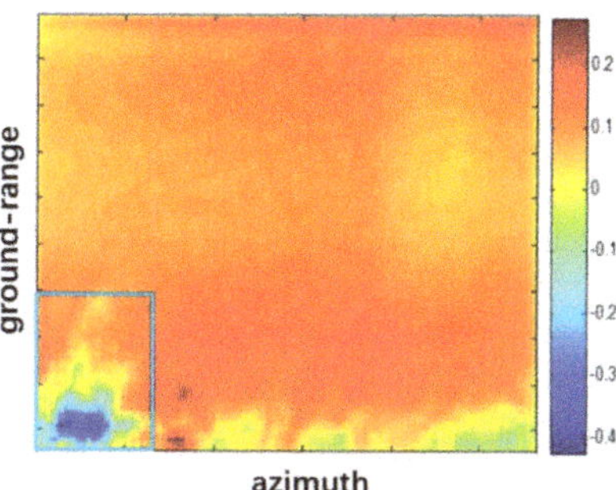


**Figure 8.** Multi-look interferometric phase on a 1-km grid. Phases in the blue box corresponds to the internal wave area.

The altimetric precision on a 1-km and 5-km grid are estimated by calculating the RMSE of the phase in the same ground-range bin. The simulated altimetric precision is also presented to compare with the estimated results (the SWH is set to be 1.8 m according to the meteorological data during the image acquisition), as shown in Figure 9a. Since the altimetric precision within 7°~8° incidence range is too large, the precision axis is on logarithmic scale to present the curves more clearly. Throughout the entire swath, the estimated precision on a 5-km grid (the magenta solid line) is higher than that of a 1-km grid (the red solid line) due to larger multi-look number. The estimated precision is lower than the simulated results (the dashed lines), and to analyze the reason for this difference, the precision difference curves within 1°~7° incidence range are shown in Figure 9b. Due to the random error suppression by multi-looking, the precision difference on a 5-km grid is lower than that of a 1-km grid. However, a considerable residual difference still exits and, apart from the fact that the OST changes slightly in the area, the residual difference is mainly caused by the residual calibration error and data processing induced error. Most of these are systematic errors and cannot be further removed by multi-looking (within the 4°~5° incidence range, the difference is almost the same as the 1-km curve, which indicates that the precision difference is mainly from systematic errors).

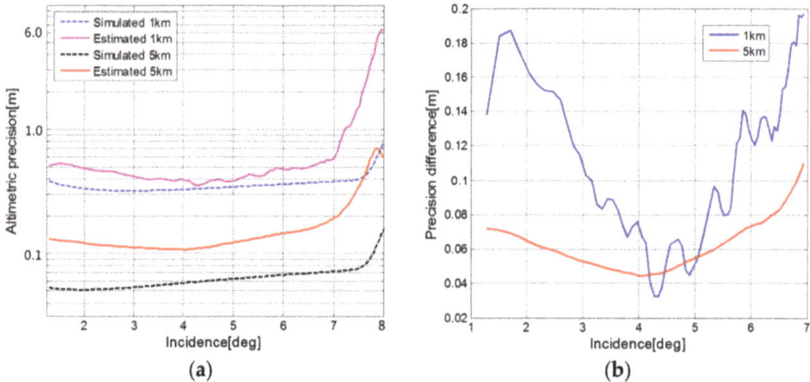


**Figure 9.** (**a**) Altimetric precision of the estimated and simulated results and (**b**) precision difference between the estimated and simulated results on a 1-km and 5-km grid, respectively.

In Figure 8, phases in the blue box are much lower than the surrounding phases. This anomaly causes the estimated precision drops sharply and deviates notably from the simulated precision when the incidence is above ~7°, as shown in Figure 9a. According to Figure 7a, ocean internal waves happen to exist in this area. In Figure 10, by comparing the intensity, coherence and phase images of the interval waves, we deduce that the altimetric precision decline is caused by the modulation of internal waves on ocean surface roughness. In the far range, the modulated area appears darker than the surrounding area, which indicates the decline of the SNR. As the SNR decreases, so does the coherence of the modulated area. The coherence decline increases the phase noise and worsens the altimetric precision.

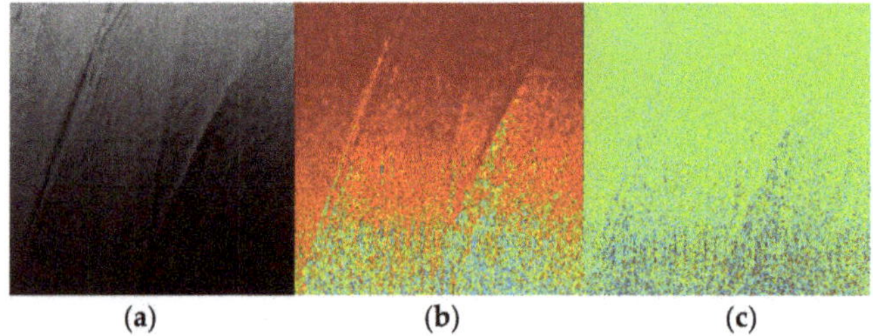


**Figure 10.** (**a**) Intensity (**b**) coherence and (**c**) phase image of the internal wave area.

## 4. Discussion

The modulation of internal waves on ocean surface roughness leads to a sharp decline in altimetric precision. However, the modulation effect seems to be quite different between the near and far range of the image. Classical SAR ocean imaging theory presumes that under 20°~60° incidence, the SAR received signal is mainly from Bragg scattering. When the modulated area gets smoother, the backscattered signal will decrease and the internal waves will appear as dark stripes, which coincides with the situation that occurred in the far range. While different from conventional SAR, the near-nadir incidence of InIRA determines its received signals are mainly from specular reflection, so internal waves are supposed to be bright stripes, which coincides with the near range situation, and a similar phenomenon observed by satellite altimeter has also been reported in [25]. Obviously, there is an intensity boundary around ~5° incidence, as shown in Figure 11a.

The intensity boundary phenomenon also happens around the area where the ocean surface is covered by oil slick, as shown in Figure 11b. In the near range, the slicks look brighter than the surrounding area, while darker in the far range. Since the modulation of oil slick also makes the ocean surface smoother, so the intensity boundary can be explained by the same reason as of the internal waves'.

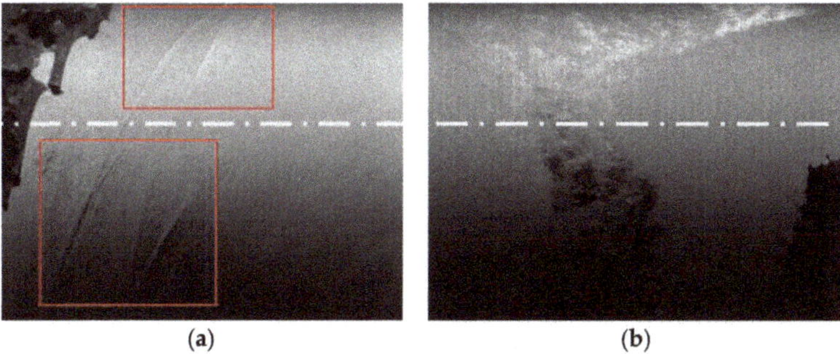

**Figure 11.** Intensity boundary phenomenon of (**a**) internal waves (as shown in the red boxes); (**b**) oil slick.

Considering the near-nadir incidence of InIRA, the intensity boundary can be explained as follows.

In the near range, the incidence is almost vertical, meanwhile, a small surface slope induced by ocean waves also exists, as shown in Figure 12. When radar signals illuminate the ocean surface, most of the signals will reflect straight back into the receiver. Since under the modulation of internal waves or oil slick, the ocean surface get smoother, so reflected signals from modulated ocean area become stronger. This makes the internal waves or oil slick appear brighter than the surrounding area.

In the far range, the incidence increases, however, the surface slope does not increase since it is largely dominated by the local sea state. Though most of the radar signals are also specular-reflected, only few of them can head back into the receiver. It gets even worse around the modulated area since the backscattered signal also decreases because of the smoother surface roughness. This makes the internal waves or oil slick appear darker than the surrounding area.

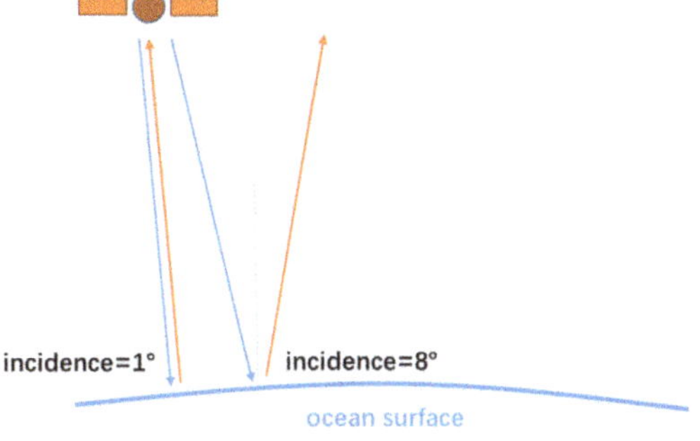


**Figure 12.** Schematic diagram of the intensity boundary phenomenon.

The position where the intensity boundary locates is determined by ocean wave slope, and the ocean wave slope is dominated by the local sea state. So future near-nadir InSAR should consider this phenomenon while designing the incidence range. Our future work will derive the relationship between the intensity boundary and sea state from the statistics of the acquired data.

## 5. Conclusions

As an experimental payload, even though some of the system parameters are limited by the current technique, InIRA has shown its great potential in measuring the OST with high precision that have never been resolved before. Based on the performance analysis of InIRA in this paper, here are some suggestions on parameter design for the future near-nadir InSAR.

(1) According to the intensity boundary phenomenon, the available swath of high altimetric precision will be narrower than expected. As for InIRA, when the incidence range is $1^{\circ}$~$5^{\circ}$, the swath will decrease to only 20 km. This narrow swath can hardly fulfill the altimetry on sub-mesoscale oceanic processes. Therefore, future near-nadir InSAR could fly on a higher orbit to increase the swath width.

(2) Flying on a higher orbit means the SNR of the received signal will decrease significantly, thereby definitely increasing the phase noise level. As mentioned before, multi-looking is one of the most effective ways to lower the phase noise, and the 3-cm altimetric precision on a 10-km grid is much higher than that of a 1-km grid, according to the simulation. However, it is quite challenging for the 10-km grid to resolve mesoscale or sub-mesoscale oceanic processes whose physical dimension is only about 15~300 km. Therefore, future near-nadir InSAR should guarantee an even higher intrinsic image resolution so that a larger multi-look number could be acquired on the same grid.

(3) Baseline length is the key parameter which guarantees high altimetric precision for near-nadir InSAR. Meanwhile, a longer baseline will mitigate the stringent requirement for a systematic phase noise level and data processing precision, allowing the near-nadir InSAR to achieve better performance much more easily.

**Acknowledgments:** The authors would like to thank the Key Laboratory of Space Utilization, Technology and Engineering Center for Space Utilization, Chinese Academy of Sciences for providing the InIRA data. The authors would also like to thank the anonymous reviewers and the editors for their comments and suggestions that improved the manuscript.

**Author Contributions:** Weiya Kong conceived and performed the experiments; Jinsong Chong and Weiya Kong designed the research and contributed to the article's organization; Hong Tan provided the data for comparative analysis. Weiya Kong drafted the manuscript, which was revised by all authors. All authors read and approved the final manuscript.

**Conflicts of Interest:** The authors declare no conflict of interest.

## References

1. Pitz, I.; Otto, A.; Schmidt, M. Ocean surface topography constellation: The next 15 years in satellite altimetry. *Kybern. Praha* **2010**, *2*, 490–493.
2. Coleman, R. Satellite altimetry and earth sciences: A handbook of techniques and applications. *Eos Trans. Am. Geophys. Union* **2013**, *82*, 376. [CrossRef]
3. Zawadzki, L.; Ablain, M. Accuracy of the mean sea level continuous record with future altimetric missions: Jason-3 vs. Sentinel-3a. *Ocean Sci. Discuss.* **2016**, *12*, 1511–1536. [CrossRef]
4. Fu, L.L.; Stammer, D.; Leben, R.R.; Chelton, D.B. Improved spatial resolution of ocean surface topography from the t/p-jason-1 altimeter mission. *Eos Trans. Am. Geophys. Union* **2003**, *84*, 241–248. [CrossRef]
5. Fu, L.L.; Alsdorf, D.; Morrow, R.; Rodriguez, E.; Mognard, N. *Swot: The Surface Water and Ocean Topography Mission: Wide-Swath Altimetric Elevation on Earth*; Jet Propulsion Lab: Pasadena, CA, USA, 2012.
6. Fu, L.L.; Ferrari, R. Observing oceanic submesoscale processes from space. *Eos Trans. Am. Geophys. Union* **2008**, *89*, 488. [CrossRef]

7. Kim, S.Y. Observations of submesoscale eddies using high-frequency radar-derived kinematic and dynamic quantities. *Cont. Shelf Res.* **2010**, *30*, 1639–1655. [CrossRef]
8. Rodriguez, E.; Morris, C.S.; Belz, E.J. A global assessment of srtm performance. *Photogramm. Eng. Remote Sens.* **2006**, *72*, 249–260. [CrossRef]
9. Ulander, L.M.H.; Askne, J. Repeat-pass sar interferometry over forested terrain. *IEEE Trans. Geosci. Remote Sens.* **1995**, *33*, 331–340.
10. Rodriguez, E.; Martin, J.M. Theory and design of interferometric synthetic aperture radars. *Radar Signal Process. IEE Proc.* **1992**, *39*, 147–159. [CrossRef]
11. Musa, Z.N.; Popescu, I.; Mynett, A. A review of applications of satellite sar, optical, altimetry and dem data for surface water modelling, mapping and parameter estimation. *Hydrol. Earth Syst. Sci. Discuss.* **2015**, *12*, 4857–4878. [CrossRef]
12. Fu, L.L.; Rodriguez, E. Wide-swath altimetric measurement of ocean surface topography. In Proceedings of the 2003 International Union of Geodesy and Geophysics (IUGG) Conference, Sapporo, Japan, 30 June–11 July 2003.
13. Pollard, B.D.; Rodriguez, E.; Veilleux, L.; Akins, T.; Brown, P.; Kitiyakara, A.; Zawadski, M.; Datthanasombat, S.; Prata, A. The Wide Swath Ocean Altimeter: Radar Interferometry for Global Ocean Mapping with Centimetric Accuracy. In Proceedings of the IEEE Aerospace Conference, Big Sky, MT, USA, 9–16 March 2002; Volume 1002, pp. 1007–1020.
14. Fjørtoft, R.; Gaudin, J.M.; Pourthie, N.; Lion, C. Karin—The ka-band radar interferometer on swot: Measurement principle, processing and data specificities. In Proceedings of the IEEE International Geoscience and Remote Sensing Symposium, Honolulu, HI, USA, 25–30 July 2010; pp. 4823–4826.
15. Wu, X.; Hensley, S.; Rodriguez, E.; Moller, D.; Muellerschoen, R.; Michel, T. Near nadir ka-band sar interferometry: Swot airborne experiment. In Proceedings of the Geoscience and Remote Sensing Symposium, Vancouver, BC, Canada, 24–29 July 2011; pp. 2681–2684.
16. Fjortoft, R.; Gaudin, J.M.; Pourthie, N.; Lalaurie, J.C.; Mallet, A.; Nouvel, J.F.; Martinot-Lagarde, J.; Oriot, H.; Borderies, P.; Ruiz, C. Karin on swot: Characteristics of near-nadir ka-band interferometric sar imagery. *IEEE Trans. Geosci. Remote Sens.* **2014**, *52*, 2172–2185. [CrossRef]
17. Altenau, E.H.; Pavelsky, T.M.; Moller, D.; Lion, C.; Pitcher, L.H.; Allen, G.H.; Bates, P.D.; Calmant, S.; Durand, M.; Smith, L.C. Airswot measurements of river water surface elevation and slope: Tanana River, AK. *Geophys. Res. Lett.* **2017**, *44*, 181–189. [CrossRef]
18. Jackson, C. *Synthetic Aperture Radar Marine User's Manual*; U.S. Department of Commerce: Washington, DC, USA, 2004; pp. 279–360.
19. Doerry, W.A. Atmospheric attenuation and sar operating frequency selection. In Proceedings of the Workshop on Synthetic Aperture Radar Technology, Huntsville, AL, USA, 22–23 October 2003.
20. Lee, J.S.; Hoppel, K.W.; Mango, S.A.; Miller, A.R. Intensity and phase statistics of multilook polarimetric and interferometric sar imagery. *IEEE Trans. Geosci. Remote Sens.* **1994**, *32*, 1017–1028.
21. Just, D.; Bamler, R. Phase statistics of interferograms with applications to synthetic aperture radar. *Appl. Opt.* **1994**, *33*, 4361–4368. [CrossRef] [PubMed]
22. Esteban Fernandez, D.; Pollard, B.; Vaze, P.; Abelson, R. Swot Project Mission Performance and Error Budget. Available online: https://swot.oceansciences.org/documents.htm (accessed on 8 September 2017).
23. Duda, T.F.; Lynch, J.F.; Irish, J.D.; Beardsley, R.C.; Ramp, S.R.; Chiu, C.S.; Tang, T.Y.; Yang, Y.J. Internal tide and nonlinear internal wave behavior at the continental slope in the northern south china sea. *IEEE J. Ocean. Eng.* **2004**, *29*, 1105–1130. [CrossRef]
24. Holzner, J.; Bamler, R. Burst-mode and scansar interferometry. *IEEE Trans. Geosci. Remote Sens.* **2002**, *40*, 1917–1934. [CrossRef]
25. Magalhaes, J.M.; Silva, J.C.B.D. Satellite altimetry observations of large-scale internal solitary waves. *IEEE Geosci. Remote Sens. Lett.* **2017**, *14*, 534–538. [CrossRef]

*Article*

# Azimuth Ambiguities Removal in Littoral Zones Based on Multi-Temporal SAR Images

**Xiangguang Leng , Kefeng Ji *, Shilin Zhou and Huanxin Zou**

School of Electronic Science and Engineering, National University of Defense Technology, Sanyi Avenue, Changsha 410073, China; luckight@163.com (X.L.); slzhou@nudt.edu.cn (S.Z.); hxz@nudt.edu.cn (H.Z.)
* Correspondence: jikefeng@nudt.edu.cn; Tel.: +86-138-0731-2017

Academic Editors: Xiaofeng Yang, Xiaofeng Li, Ferdinando Nunziata and Alexis Mouche
Received: 31 May 2017; Accepted: 17 August 2017; Published: 22 August 2017

**Abstract:** Synthetic aperture radar (SAR) is one of the most important techniques for ocean monitoring. Azimuth ambiguities are a real problem in SAR images today, which can cause performance degradation in SAR ocean applications. In particular, littoral zones can be strongly affected by land-based sources, whereas they are usually regions of interest (ROI). Given the presence of complexity and diversity in littoral zones, azimuth ambiguities removal is a tough problem. As SAR sensors can have a repeat cycle, multi-temporal SAR images provide new insight into this problem. A method for azimuth ambiguities removal in littoral zones based on multi-temporal SAR images is proposed in this paper. The proposed processing chain includes co-registration, local correlation, binarization, masking, and restoration steps. It is designed to remove azimuth ambiguities caused by fixed land-based sources. The idea underlying the proposed method is that sea surface is dynamic, whereas azimuth ambiguities caused by land-based sources are constant. Thus, the temporal consistence of azimuth ambiguities is higher than sea clutter. It opens up the possibilities to use multi-temporal SAR data to remove azimuth ambiguities. The design of the method and the experimental procedure are based on images from the Sentinel data hub of Europe Space Agency (ESA). Both Interferometric Wide Swath (IW) and Stripmap (SM) mode images are taken into account to validate the proposed method. This paper also presents two RGB composition methods for better azimuth ambiguities visualization. Experimental results show that the proposed method can remove azimuth ambiguities in littoral zones effectively.

**Keywords:** azimuth ambiguities; littoral zones; multi-temporal; RGB composition; synthetic aperture radar (SAR); Sentinel-1

---

## 1. Introduction

Synthetic Aperture Radar (SAR) is a microwave remote sensing technology providing 2D images with 24-h all-weather sensing capability [1]. Since the launch of the Seasat in 1978, the number of spaceborne SAR sensors has drastically increased [2,3]. It is a mature and successful discipline for global ocean monitoring [3–6]. Advanced spaceborne SAR sensors today provide fine resolutions, such as TerraSAR-X, COSMOS-SkyMed, RADARSAT-2, GF-3, and Sentinel-1 [4–9]. As more than 90% of the world's trade goods, and more than 70% of global crude oil are transported by sea, maritime surveillance is of utmost importance [10].

Azimuth ambiguities are inherent artifacts in current SAR systems. They are visible in most spaceborne SAR images. SAR images often suffer from azimuth ambiguities. Experience shows that azimuth ambiguities are today a real problem for ocean monitoring [11,12]. SAR images can be strongly affected by land-based sources. If azimuth ambiguities are unrecognized, they can give rise to false alarms and errors in SAR image interpretation.

Littoral zones are 'hard-hit areas' of azimuth ambiguities. There are three reasons for this. The first one is that littoral zones are usually areas of low wind speed condition. The neighboring land can provide a barrier to generate a weak-wind region. Azimuth ambiguities become prominent when the calm sea surface appears darker. The second one is that there are always strong sources at the coast. Coastal regions are always the most developed areas, e.g., New York, Shanghai, Singapore, etc. There are many tall buildings, bridges, ports, and other artifacts. They are all strong sources which can generate azimuth ambiguities. The final one is that the typical spaceborne azimuth ambiguity displacements are 4~8 km. Azimuth ambiguities of land-based sources will occur exactly in sea areas near the coast. Unfortunately, littoral zones are also regions of interest (ROI) in most cases. There are many ships in the harbors, straits, and other important littoral zones usually. The pressure on the security of the ocean is also increasing as shipping traffic grows. Littoral zones bear the brunt of the pressure. They are exposed to various hazards like pollution, terrorism, or privacy. Thus, efficient methods for monitoring, prediction, and visualization of littoral zones are of great importance.

Many investigations have been carried out for azimuth ambiguities removal in SAR imagery. The postprocessing techniques which this paper focuses on can be divided into three categories based on SAR data type used, i.e., methods based on detected (intensity/amplitude) products, single-look complex (SLC) products, and quad-pol (QP) products. Calculating the azimuth displacement to identify azimuth ambiguities is one of the most important methods used in detected products [4,12,13]. This method checks whether there is a stronger source target present at the azimuth displacement. If either side has a detection of stronger power than the source target, it is considered to be an ambiguity. As SLC or QP products can be used to construct detected products, detected products are more common than other type of products. Generally this method is one of the most tractable methods, especially for azimuth ambiguities caused by ship targets. It has been widely used in existed SAR ship detection systems [4,13]. However this method fails to discard azimuth ambiguities when the source is flooded by strong targets in littoral zones or outside of the image. Besides, azimuth ambiguities have unpredictable variations leading to a difficult correlation between sources and ambiguities. Finally, azimuth displacements can be different in ScanSAR or TOPSAR (Terrain Observation with Progressive Scans SAR) images. Methods based on SLC products usually adopt filtering techniques in the frequency domain [14,15], and remove azimuth ambiguities by filtering azimuth signals. They can take advantage of both real and imagery parts in SLC products. Filtering is effective in suppressing azimuth ambiguities, but can result in additional speckle noise and small ship loss. Methods based on QP products employ the fact that the two cross-polarized channels are each other's complex conjugate for azimuth ambiguities [16]. They consider that a proper combination of two cross-polarized channels [17] or polarimetric decomposition can cancel out azimuth ambiguities [18]. Methods based on QP products are quite effective, whereas QP products are not available in practice usually.

The increasing of multi-temporal SAR images today provides new insight into this problem. For example, Sentinel-1 has a repeat cycle of 12 days by one satellite and 6 days by a pair of satellites [8,19]. A scene on the littoral zones can be re-imaged every 12 days or 6 days. Thus, image pixels of the same scene can be made to exactly coincide. This can identify fixed artifacts in littoral zones. This paper opens up the possibilities to use multi-temporal SAR data to remove azimuth ambiguities. A method for azimuth ambiguities removal in littoral zones based on multi-temporal SAR images is proposed in this paper. The proposed processing chain includes co-registration, local correlation, binarization, masking, and restoration steps applied to images. It is designed to remove azimuth ambiguities caused by fixed land-based sources in littoral zones. The basic idea underlying the proposed method is that the sea surface is dynamic whereas azimuth ambiguities caused by land-based sources are constant. The temporal consistence of azimuth ambiguities is higher than sea clutter. Thus, azimuth ambiguities can be identified by the correlation of multi-temporal SAR images. The design of the method and the experimental procedure are based on detected products from the Sentinel data hub of Europe Space Agency (ESA) [8]. Both Interferometric Wide Swath (IW) and Stripmap (SM) mode images are taken into account to validate the algorithms. Experimental results show that the proposed method can remove azimuth ambiguities in littoral zones effectively when no QP data is available. To obtain better visual results of azimuth ambiguities than original images, this paper also presents two RGB composition methods for azimuth ambiguities based on the number of multi-temporal SAR images available.

The rest of this paper is organized as follows. Section 2 is a review on azimuth ambiguities. Section 3 proposes the method for azimuth ambiguities removal in littoral zones based on multi-temporal SAR images. Section 4 presents two RGB composition methods for azimuth ambiguities. Section 5 validates and discusses the method proposed in this paper by testing on Sentinel-1 IW and SM images. Finally, Section 6 concludes this paper.

## 2. Review on Azimuth Ambiguities

Azimuth ambiguities are caused by finite sampling of azimuth Doppler signals. A too low pulse repetition frequency (PRF) may cause those Doppler frequencies higher than PRF are folded into the central part of the azimuth spectrum. Thus, aliased signals are produced in this case [17]. As shown in Figure 1, targets A and B have equal Doppler histories due to aliasing. If target B is much stronger than target A, then it is possible that a ghost image of target B will be visible in the position of A. This ghost is called azimuth ambiguity. The Doppler shift as a function of a pointing error, increases linearly with frequency. Azimuth ambiguities in SAR images are spatially displaced in azimuth directions at approximate locations [11,14], as

$$\Delta D_{AZ} \approx n \frac{f_{PRF}}{f_{DR}} V \tag{1}$$

where $\Delta D_{AZ}$ is the azimuth displacement, $V$ is the satellite velocity, $f_{PRF}$ is the PRF, $f_{DR}$ is the Doppler rate, $n$ is the order.

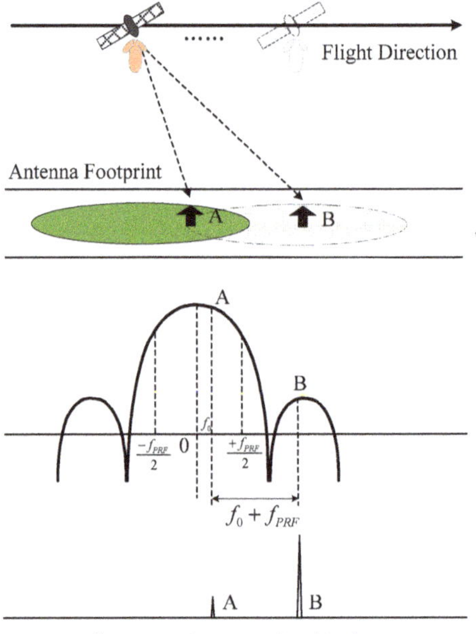


**Figure 1.** Illustration of azimuth ambiguity formation in SAR images, targets A and B have equal Doppler histories due to aliasing [20].

As

$$f_{DR} \approx \frac{2V^2}{\lambda R_s} \tag{2}$$

where $\lambda$ is the wavelength and $R_s$ is the slant range, an equivalent representation of Formula (1) is

$$\Delta D_{AZ} \approx n \frac{\lambda R_s}{2V} f_{PRF} \tag{3}$$

Typical azimuth ambiguity displacements are 4~8 km. In cases where the sea clutter level is low, e.g., weak-wind areas, azimuth ambiguities can occur more easily. A quality measure for azimuth ambiguities is the Azimuth Ambiguity to Signal Ratio (AASR). The requirement of AASR for ship detection and oil spill detection is −25 dB and −23 dB, respectively [12]. Azimuth ambiguities are inherent artifacts in current SAR systems, which appear in most spaceborne SAR images. As shown in Figure 2, azimuth ambiguities and their azimuth displacements in TerraSAR-X, COSMOS-SkyMed, RADARSAT-2, and Sentinel-1 images are presented. Azimuth ambiguities in the littoral zones affect SAR image interpretation severely.

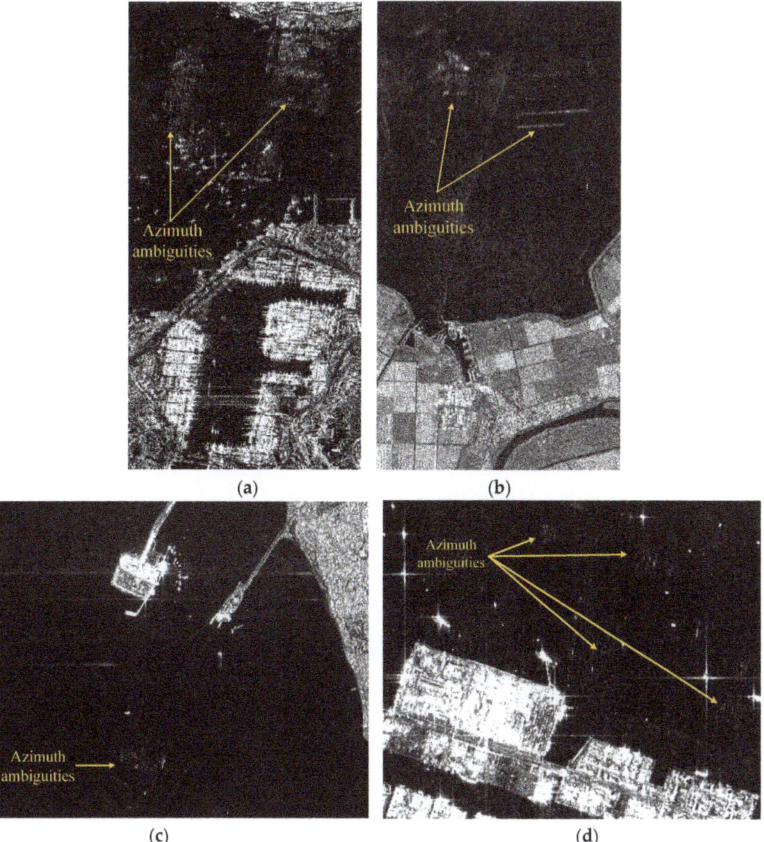


**Figure 2.** Examples of azimuth ambiguities in current synthetic aperture radar (SAR) imagery, (**a**) TerraSAR-X StripMap image, azimuth displacement is ~4.4 km (**b**) COSMOS-SkyMed StripMap image, azimuth displacement is ~4.4 km (**c**) RADARSAT-2 Standard mode image, azimuth displacement is ~5.2 km (**d**) Sentinel-1 IW image, azimuth displacement is ~4.7 km (beam IW2).

The important features of azimuth ambiguities can be concluded into 5 aspects:

(1) They occur in azimuth direction at determinate multiple distances at two sides of the source target. They may also have a small shift in range direction. Thus, the theoretical distance can be less than the measured distance.
(2) The displacement rises as the wavelength increases, other parameters remaining the same. The ambiguities are less severe in SAR images of a long wavelength than a short wavelength.
(3) As the order $n$ increases, the displacements of the related ambiguities increase and their intensities decrease usually. The power of the 1st order azimuth ambiguity is weaker than the source target.
(4) They appear to be not well focused or not very strong in intensities. However they are similar to the source target in some features, e.g., shape, texture, size. Thus, it is not easy to identify them by methods based on features.
(5) HV and VH channels have approximately the same magnitude, whereas their azimuth ambiguities are shifted in phase by about 180°.

## 3. Proposed Method

In the proposed method, multi-temporal SAR images of the same scene in the same configuration should be available first. The same configuration indicates that the multi-temporal SAR images should be acquired by a same sensor with same the imaging mode and pass way (ascending or descending) at least. Figure 3 shows the detailed flowchart of the proposed method which can be divided into 6 steps.

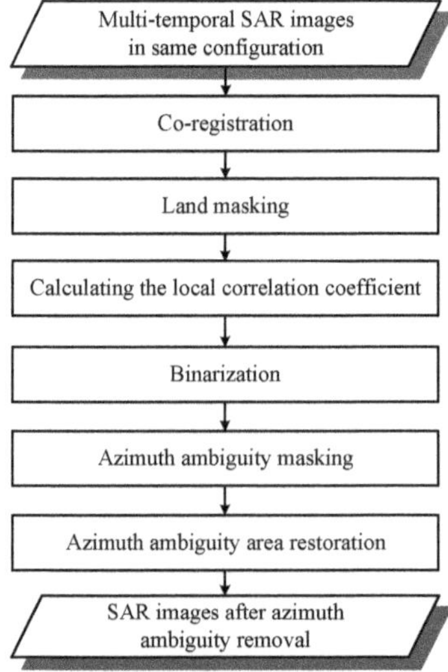


**Figure 3.** The flowchart of the proposed method. It includes co-registration, local correlation, binarization, masking, and restoration steps. The idea underlying the proposed method is that sea surface is dynamic whereas azimuth ambiguities caused by land-based sources are constant. It is designed to remove azimuth ambiguities caused by fixed land-based sources.

### 3.1. Co-Registration

To support the mapping of multi-temporal SAR images, some relevant preprocessing operations must be performed. These mainly include the co-registration of SAR data. Co-registration is accomplished through cross correlation and warp usually. Then we abstract the same geographical areas from multi-temporal SAR data. Thus, corresponding pixels in multi-temporal SAR images exactly coincide and represent the same point on the Earth surface. It should be noted that radiometric correction and speckle filtering are not needed in the preprocessing. This simplifies the processing chain. The reason will be presented in the subsequent steps.

### 3.2. Land Masking

As only the ocean is of interest, land masking is important for multi-temporal SAR images. There are two popular approaches for land masking. One is to register SAR images with existing geospatial databases, e.g., the GSHHS (Global Self-consistent, Hierarchical, High-resolution Shoreline) [21]. Another one is to use automatic algorithms to detect coastlines or land areas [22,23]. In this paper, land will be removed for better results in subsequent steps.

### 3.3. Calculating the Local Correlation Coefficient

As sea surface is dynamic, whereas azimuth ambiguities caused by land-based sources are fixed. The temporal consistence of azimuth ambiguities is higher than sea clutter. Thus, we can use correlation operation to identify azimuth ambiguities from sea clutter. Two-temporal images in the same configuration are needed at least. Generally, more images seem to be better when applying the correlation operation. However there is an increasing risk that land-based sources changed in a long period. Thus, two-temporal images acquired in continuous periods are preferred in this paper.

The local correlation coefficient between each image is calculated in a local sliding window. The size of the sliding window can be decided by a reference size, e.g., the smallest ship one wants to detect. If there are two-temporal images available, named $I_1$, $I_2$, then

$$R = I_1 \otimes I_2 \tag{4}$$

where $R$ is the local correlation coefficient map, $\otimes$ indicates the local correlation operation here, i.e.,

$$r = \frac{\sum_m \sum_n \left(I_1(m,n) - \overline{I}_1\right)\left(I_2(m,n) - \overline{I}_2\right)}{\sqrt{\left(\sum_m \sum_n \left(I_1(m,n) - \overline{I}_1\right)^2\right)\left(\sum_m \sum_n \left(I_2(m,n) - \overline{I}_2\right)^2\right)}} \tag{5}$$

where $r$ is the local correlation coefficient value at sliding window center, $I_1(m,n)$ and $I_2(m,n)$ are the pixel values at location $(m,n)$ in the sliding window respectively, $\overline{I}_1$ and $\overline{I}_2$ are the mean of all pixels in the sliding window respectively.

As azimuth ambiguities are temporally more consistent than sea clutter and ships, the correlation coefficient of azimuth ambiguities are much higher than sea clutter and ships. Thus, azimuth ambiguities and other constant artifacts are prominent in the local correlation coefficient map. Note that the local correlation operation does not need radiometric correction and speckle filtering. The reason is that the correlation operation is applied in a local window. The radiometric scaling can be considered as a constant in a local window, while the correlation operation is invariant to this constant scaling. Meanwhile, speckle filtering reduces granularity in SAR images. This may increase the consistence of sea clutter. Thus, both radiometric correction and speckle filtering are not needed.

### 3.4. Binarization

Azimuth ambiguities are prominent, whereas sea clutter is suppressed in the local correlation coefficient map. Thus, azimuth ambiguities can be identified by a binarization operation. Binarization can be achieved by a thresholding method. There are many thresholding methods. In this paper, the thresholding method based on maximum entropy [24] is employed. In this method, two probability distributions ${h_i}^{(1)}$ and ${h_i}^{(2)}$ (e.g., object and background) are derived from the original gray distribution of the image. If $t$ is the threshold, then the entropy of the object and background are defined as

$$
\begin{aligned}
H_b &= -\sum_{i=1}^{t} h_i^{(1)} \lg h_i^{(1)} \\
H_w &= -\sum_{i=t+1}^{L} h_i^{(2)} \lg h_i^{(2)}
\end{aligned} \tag{6}
$$

where $L$ is the gray level. Then the optimal threshold $t^*$ is defined as the gray level, which maximizes $H_b + H_w$, i.e.,

$$t^* = \arg\max \{H_b(t) + H_w(t)\} \tag{7}$$

The idea underlying this entropic method is that the optimal threshold should make the object and background areas more homogenous with each other. This can make sense for the local correlation coefficient map which is like a blurred 'salt and pepper' map. It should be noted that Otsu method [25] or Mode method [26] that require two peaks in the histogram are not appropriate in this case as two peaks may not occur in the histogram.

### 3.5. Azimuth Ambiguity Masking

The binarization result can be considered as an azimuth ambiguity mask. This mask can be used to remove azimuth ambiguities in multi-temporal SAR images. Thus, multi-temporal SAR images are freed from azimuth ambiguities in littoral zones. On the other hand, an azimuth ambiguity mask can also be used to locate the strong land-based sources. It is an inversion of the method by calculating the displacement to identify azimuth ambiguities. The inversion results can also be used to distinguish 1st order azimuth ambiguities and higher order azimuth ambiguities. If the shifted mask is on the land, the mask can be recognized as 1st order azimuth ambiguities. However if the shifted mask is on the sea, the mask can be recognized as higher order azimuth ambiguities or fixed artifacts on the sea. This is because only the mask of 1st order azimuth ambiguities can be shifted on the land by an azimuth displacement.

### 3.6. Azimuth Ambiguity Area Restoration

Azimuth ambiguities have little help for the ocean environment and maritime traffic monitoring, though Liu, C. et al. [16] mentioned that azimuth ambiguities may indicate moving objects by the subtraction of two cross-polarized channels. Thus, this paper restores the azimuth ambiguity patches masked in the multi-temporal SAR images. If sea clutter is homogeneous, azimuth ambiguity patches are filled with pixels randomly from a Gaussian distribution. The parameters of the Gaussian distribution are estimated from sea clutter based on methods of moments. If sea clutter is heterogeneous, given the complex textures and structures, the Exemplar-based inpainting method [27] is employed for filling the lost patches of images. The homogeneity is estimated by the equivalent number of looks (ENL). It is defined as follows

$$ENL = \left(\frac{\mu}{\sigma}\right)^2 \tag{8}$$

where $\mu$ and $\sigma$ are the mean and standard deviation of sea clutter. ENL can be estimated from a homogenous portion of the image. Larger ENL means lower speckle noise and more homogeneous sea clutter. The empirical threshold of ENL is set as 15 to decide whether sea clutter is homogeneous or not. This is based on the fact that ENL is 4.4 for the Sentinel-1 detected IW products, whereas it is 29.7 for SM products [19]. This means that SM products are more homogeneous than IW products. We set the approximate median value as the threshold to distinguish homogeneous areas and heterogeneous areas. The threshold has been validated by several experiments.

In the Exemplar-based inpainting method, a patch $\psi_q$ is searched for patch $\psi_{\widehat{p}}$ in the source region $\Phi$ which is the most similar to $\psi_{\widehat{p}}$. The similarity between the patches is measured as [27]

$$\psi_{\widehat{q}} = \arg\min d\left(\psi_{\widehat{p}}, \psi_q\right), \ \psi_q \in \Phi \tag{9}$$

where the distance $d(\psi_a, \psi_b)$ between two generic patches $\psi_a$ and $\psi_b$ is defined as the sum of squared differences of the already filled pixels in the two patches. In this method, the fill order is based on a defined priority parameter and the patch on front with the highest priority is filled first. Its priority is defined as [27]

$$R(\alpha) = C(\alpha)D(\alpha) \tag{10}$$

where $C(\alpha)$ and $D(\alpha)$ represent the confidence term and data term, respectively. For more details about this technique, the readers are referred to [27].

## 4. RGB Composition Method for Azimuth Ambiguities

The use of RGB composition in multi-temporal SAR image analysis has been already presented for flood mapping [28,29]. However, there has rarely been a study on RGB composition for azimuth ambiguities [11]. This section presents and compares two RGB composition methods for better azimuth ambiguities visualization based on the number of SAR images available.

### *4.1. Two or More Than Two-Temporal Images Available*

This method is based on temporal consistency of azimuth ambiguities. It also requires same configuration and co-registration. Otherwise the temporal consistence can be lost. In this case, three images are composed as RGB channels directly (one image can be used twice if only two-temporal images available). Thus, azimuth ambiguities and land pixels that are in a good relationship appear in white. Color pixels are associated with low correlation, e.g., moving ship targets in different temporal images.

### *4.2. Only One Image Available*

This method is based on the strong spatial correlation existing between the ambiguity and the related source target. In this case, one image should be used three times in RGB composition. The R and G channels are associated to the original image. The B channel is associated to a copy generated by the original image. The copy is the original image translated in an azimuth ambiguity displacement given by Formula (1) [11]. Similarly, azimuth ambiguities and strong source target pixels that are in good relationship appear in white. Yellow and blue pixels are associated with a low correlation.

In sum, as the consistence of azimuth ambiguities, they always appear in white, whereas other targets on the ocean can appear in color. Thus, they can be used for azimuth ambiguities enhancement and identification.

## 5. Experimental Results and Discussion

The proposed method is tested on Sentinel-1A IW data first. There are three-temporal images available. The data subsets are presented in Figure 4. They are acquired at 6, 18 February, and 2 March 2017, on Napoli, Italian coast separately. Their polarization modes are VV mode. Azimuth ambiguities are visible in three images, whereas their source targets cannot easily be identified. Thus, it is intractable by using the method by calculating the displacement. QP data is also not available as Sentinel-1 is a dual-polarized system.

In order to have an overview of the location of the subimage, the optical image (taken from Google Earth) of Napoli is presented in Figure 5a, where the red trectangle indicates the subimage location. Figure 5b is the corresponding optical image of the subimage, where the yellow rectangles 1 and 2 represent the typical ports in Figure 5c,d respectively. The dihedrals of the typical ports in Figure 5c,d can cause strong reflection leading to azimuth ambiguities.

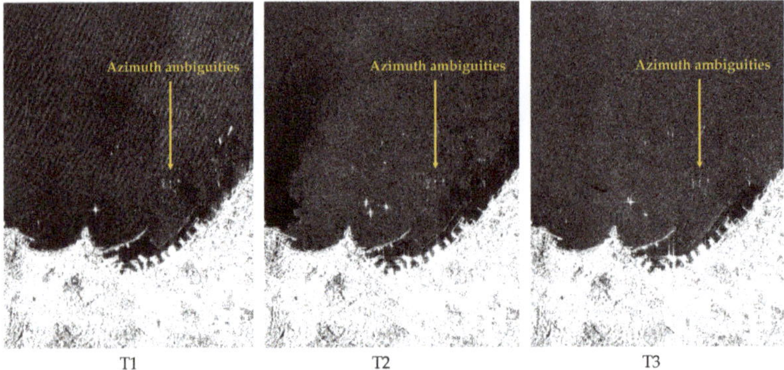


**Figure 4.** The Sentinel-1A multi-temporal dataset. The similar clusters of dots in T1, T2 and T3 at the same location are prominent azimuth ambiguities from the strong source targets on land. T1: S1A_IW_GRDH_1SDV_20170206T165649_20170206T165714_015166_018D05_8064; T2: S1A_ IW_GRDH_1SDV_20170218T165649_20170218T165714_015341_01927E_D691; T3: S1A_IW _GRDH_1SDV_20170302T165648_20170302T165713_015516_0197CF_75D4.

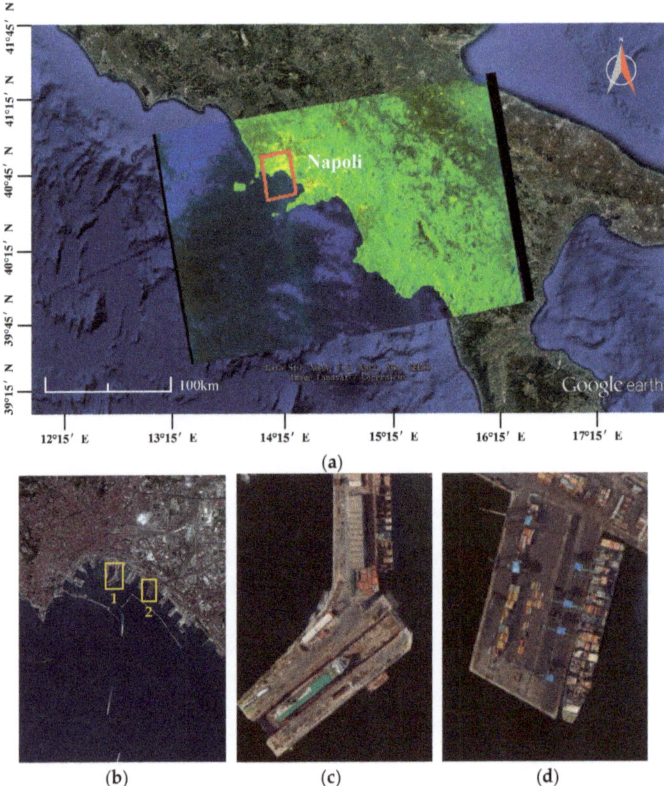


**Figure 5.** Google Earth image indicating the Sentinel-1A images. The red rectangle in (**a**) indicates the subimage location in Figure 4, (**b**) is the corresponding optical image of the subimage, yellow rectangles 1 and 2 represent the typical ports (**c**) and (**d**) which can cause azimuth ambiguities respectively.

RGB composition results are shown in Figure 6. The results are obtained by different methods presented in Section 4. From the results, it can be seen that white azimuth ambiguities can be recognized more easily than in the original images. The colorful moving ships can also catch one's eye. Different color represents ships in different period. Compared to the RGB composition methods based on more than two images, the result based on only one image would have a smaller valid size because of the shift operation, as shown in Figure 6c. In this case, strong ship targets also present in white because of its strong spatial correlation between ambiguities. This is distinctive to the method based on more than two images. In sum, azimuth ambiguities always appear in white in all methods. This can help azimuth ambiguity and ship identification in multi-temporal SAR images.

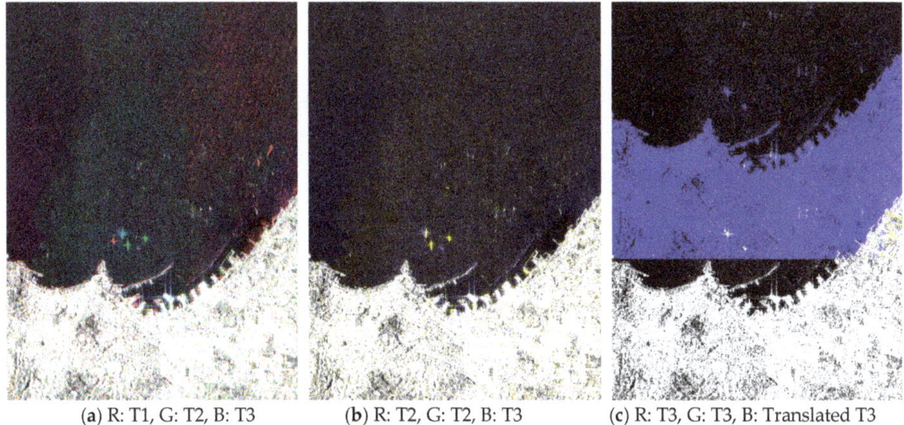

(**a**) R: T1, G: T2, B: T3 (**b**) R: T2, G: T2, B: T3 (**c**) R: T3, G: T3, B: Translated T3

**Figure 6.** RGB composition results of the Sentinel-1A multi-temporal data. Azimuth ambiguities appearing white in (**a**–**c**) can be recognized more easily than in the original images. A ship target in (**b**) seems to be white as there is a blue target near it. (**c**) has a smaller valid size because of the shift operation.

T2 and T3 images are chosen to test the proposed method in this paper. The local window size here is 7. ENL estimated is 7.7. The ambiguity map is shown in Figure 7. Figure 7a is the local correlation coefficient map, Figure 7b is the binarization result, Figure 7c is the T3 ambiguity masks. It can be seen that the proposed method can identify very small azimuth ambiguities. Statistical results show that the azimuth ambiguities with more than 20 pixels can be identified by the proposed method. Some masks may be larger than the actual ambiguities. This is because the local correlation value is not only high for azimuth ambiguities, but also for the surrounding pixels. Figure 8 shows the restored images by the proposed method, which shows a significant enhancement as compared to the original images. The zoomed patches before and after restoration in yellow rectangles are also shown in Figure 8, on the side. Azimuth ambiguities are mitigated effectively. Figure 9 is the inversion to mask strong land sources. The strong land sources are masked as red. Surprisingly, this approach can also distinguish 1st and 2nd order azimuth ambiguities. The red mask on the land can be recognized as 1st order azimuth ambiguities, whereas the green mask on the sea can be recognized as 2nd order azimuth ambiguities or fixed artifacts on the sea. It should be noted that the azimuth displacement in different beams is different. It is ~5.3 km in beam IW1 and 4.8 km in beam IW2 for these images.

The proposed method is also validated by the Sentinel-1b SM data acquired on Boston, as shown in Figure 10. They are acquired at the 20 March and 1 April 2017, separately. Their polarization modes are HH mode. The zoomed patches before and after restoration in yellow rectangles are also shown in Figure 10. The experimental results show that the proposed method can mitigate most azimuth ambiguities effectively.

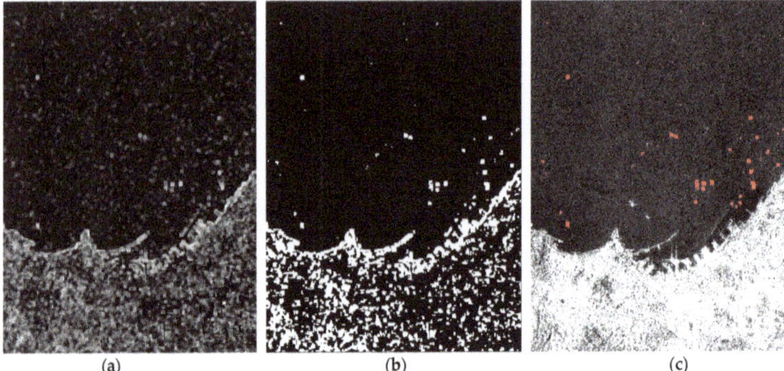
(a) (b) (c)

**Figure 7.** Azimuth ambiguity mask, (**a**) local correlation coefficient map, (**b**) binarization result, (**c**) T3 ambiguity mask.

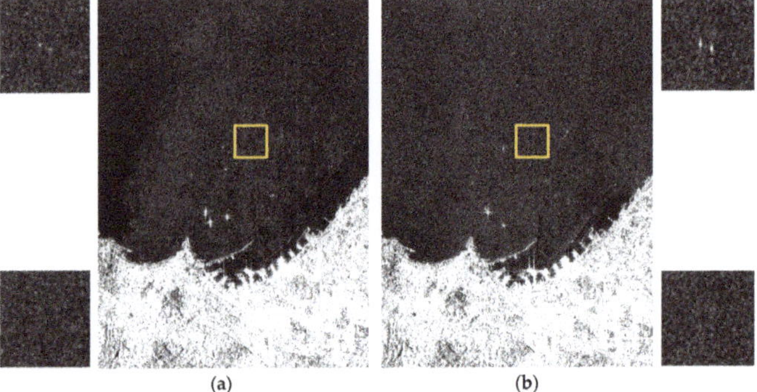
(a) (b)

**Figure 8.** Restored images, (**a**) T2 restored image, (**b**) T3 restored image. The zoomed patches before and after restoration in yellow rectangles are shown on the side.

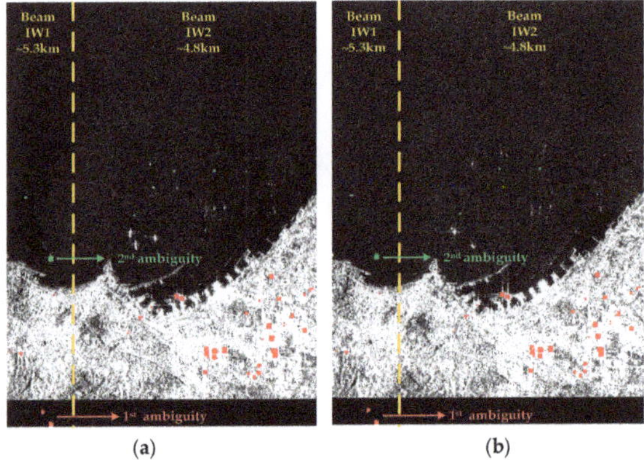

(a) (b)

**Figure 9.** Inversion to mask strong land sources, (**a**) T2 image, (**b**) T3 image. The strong land sources are masked as red. Note that the azimuth displacement in different beams is different, ~5.3 km in beam IW1 and 4.8 km in beam IW2. Red masks on the land can be recognized as 1st order azimuth ambiguities whereas green masks can be recognized as 2nd order azimuth ambiguities or fixed artifacts on the sea.

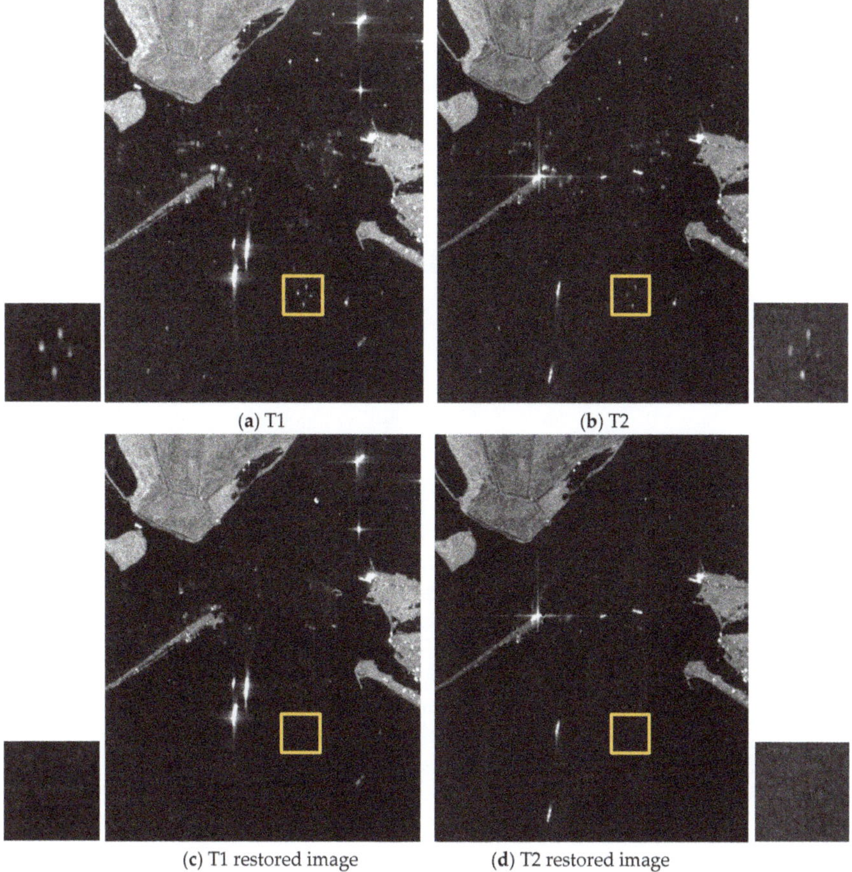


**Figure 10.** Azimuth ambiguities removal results. (**a**,**b**) Original Sentinel 1-b images, (**c**,**d**) Restored Sentinel-1b images. The zoomed patches before and after restoration in yellow rectangles are shown on the side. T1: S1B_S6_GRDH_1SDH_20170320T002541_20170320T002604_004785_0085BB_9204; T2: S1B_S6_GRDH_1SDH_20170401T002541_20170401T002605_004960_008AC4_D8E2.

The important advantages of the proposed method can be concluded from the experimental results:

(1) The method by calculating the displacement fails to identify azimuth ambiguities in littoral zones because of the complicated scenes, whereas the proposed method is able to.
(2) When no QP data is available, the proposed method is a successful candidate for azimuth ambiguities removal in littoral zones.
(3) The proposed method is a quite tractable method, in which both radiometric correction and speckle filtering are not needed.
(4) The proposed method only restores azimuth ambiguities areas. There is no change for other areas. Thus, it will not degrade image quality as methods based on SLC products.
(5) The proposed method can identify very small azimuth ambiguities in detected products, whereas the method by calculating the displacement cannot.

Obviously, though the proposed method is designed to identify azimuth ambiguities, it can also identify other fixed image features, e.g., range ambiguities or fixed artificial structures in multi-temporal images as their temporal consistence. However, small islands in the sea will not be identified by the proposed method because they are removed by the land masking step. We believe that the proposed method would be a practical candidate for ambiguities removal with more multi-temporal images available. The proposed method would be very helpful for ocean monitoring.

The proposed method assumes that the strong land sources are fixed in the period. If the strong land sources changed in the period, the temporal consistence of azimuth ambiguities would decrease. Thus, the proposed method may fail to identify azimuth ambiguities. The method by calculating the azimuth displacement can identify azimuth ambiguities caused by ships on the sea effectively, but fail to identify azimuth ambiguities caused by land sources. Thus, the proposed method can be complementary for it.

## 6. Conclusions

This paper opens up the possibilities to use multi-temporal SAR data to remove azimuth ambiguities by proposing an azimuth ambiguities removal method in littoral zones based on multi-temporal SAR images. Azimuth ambiguities are identified through an effective technique by the correlation operation of two temporal images. Azimuth ambiguities areas are restored by either a Gaussian distribution or Exemplar-based inpainting techniques based on the homogeneity of sea clutter. Experimental results based on multi-temporal Sentinel-1 IW and SM data validate the effectiveness of the proposed method. It is a successful candidate for azimuth ambiguities removal in littoral zones when no QP product is available. Azimuth ambiguities with more than 20 pixels in IW mode images can be identified by the proposed method. This paper also presents two RGB composition methods for azimuth ambiguities displayed in multi-temporal SAR images. Experimental results show that the RGB composition results can help to enhance azimuth ambiguities for visualization.

Despite the effectiveness of the proposed method, further investigations about how more accurate local correlation needs to be done, which may significantly improve the accuracy of azimuth ambiguities identification.

**Acknowledgments:** This work is supported partly by the National Natural Science Foundation of China under Grant 61372163 and 61331015. The authors would like to thank European Space Agency (ESA) for providing free Sentinel-1 data online. The authors would also like to thank the anonymous reviewers for their very competent comments and helpful suggestions.

**Author Contributions:** X.L. conceived and designed the experiments; L.X. performed the experiments and analyzed the data; K.J., S.Z. and Z. H. contributed materials; X. L. and K.J. wrote the paper.

**Conflicts of Interest:** The authors declare no conflict of interest.

## References

1. Curlander, J.C.; McDonoug, R.N. *Synthetic Aperture Radar Systems and Signal Processing*; John Wiley& Sons: New York, NY, USA, 1991.
2. Crisp, D.J. *The State-of-the-Art in Ship Detection in Synthetic Aperture Radar Imagery*; DSTO Information Sciences Laboratory: Edinburgh, Australia, 2004.
3. Greidanus, H.; Alvarez, M.; Santamaria, C.; Thoorens, F.; Kourti, N.; Argentieri, P. The SUMO ship detector algorithm for satellite radar images. *Remote Sens.* **2017**, *9*, 246. [CrossRef]
4. Brusch, S.; Lehner, S.; Fritz, T.; Soccorsi, M. Ship suveillance with TerraSAR-X. *IEEE Trans. Geosci. Remote Sens.* **2011**, *49*, 1092–1103. [CrossRef]
5. Crisp, D.J. A Ship Detection System for Radarsat-2 Dual-Pol Multi-Look Imagery Implemented in the ADSS. In Proceedings of the 2013 International Conference on Radar (Radar), Adelaide, Australia, 9–12 September 2013; pp. 318–323.
6. Martorella, M.; Pastina, D.; Berizzi, F.; Lombardo, P. Spaceborne radar imaging of maritime moving targets with the Cosmo-SkyMed SAR system. *IEEE J. Sel. Top. Appl. Earth Obs. Remote Sens.* **2014**, *7*, 2797–2810. [CrossRef]

7. GF-3. Available online: https://www.chinaspaceflight.com/satellite/Gaofen/GF-3/gaofen-3.html (accessed on 1 March 2017).
8. Sentinels Scientific Data Hub. Available online: https://scihub.copernicus.eu/ (accessed on 1 September 2014).
9. Leng, X.; Ji, K.; Yang, K.; Zou, H. A Bilateral CFAR algorithm for Ship Detection in SAR Images. *IEEE Geosci. Remote Sens. Lett.* **2015**, *12*, 1536–1540. [CrossRef]
10. *TerraSAR-X Applications Guide. Extract: Maritime Monitoring: Ship Detection. Airbus Defence and Space Geo-Intelligence Programme Line*; Airbus Defence and Space: Toulouse, France, 2015.
11. Dellepiane, S.; Martino, M.D.; Toma, M. A data fusion approach for the analysis of azimuth ambiguities. In Proceedings of the 2013 IEEE International Geoscience and Remote Sensing Symposium (IGARSS), Melbourne, Australia, 21–26 July 2013; pp. 4138–4141.
12. Vespe, M.; Greidanus, H. SAR image quality assessment and indicators for vessel and oil spill detection. *IEEE Trans. Geosci. Remote Sens.* **2012**, *50*, 4726–4734. [CrossRef]
13. Stastny, J.; Hughes, M.; Garcia, D.; Bagnall, B. A novel adaptive synthetic aperture radar ship detection system. In Proceedings of the OCEANS 2011, Waikoloa, HI, USA, 19–22 September 2011; pp. 1–7.
14. Guarnieri, A.M. Adaptive removal of azimuth ambiguities in SAR images. *IEEE Trans. Geosci. Remote Sens.* **2005**, *43*, 625–633. [CrossRef]
15. Martino, G.D.; Iodice, A.; Riccio, D.; Ruello, G. Filtering of Azimuth Ambiguity in Stripmap Synthetic Aperture Radar Images. *IEEE J. Sel. Top. Appl. Earth Obs. Remote Sens.* **2014**, *7*, 3967–3978. [CrossRef]
16. Liu, C.; Gierull, C.H. A new application for PolSAR imageryin the field of moving target indication/ship detection. *IEEE Trans. Geosci. Remote Sens.* **2007**, *45*, 3426–3436. [CrossRef]
17. Velotto, D.; Soccorsi, M.; Lehner, S. Azimuth ambiguities removal for ship detection using full polarimetric X-Band SAR data. *IEEE Trans. Geosci. Remote Sens.* **2014**, *52*, 76–88. [CrossRef]
18. Wang, C.; Wang, Y.; Liao, M. Removal of azimuth ambiguities and detection of a ship: using polarimetric airborne C-band SAR images. *Int. J. Remote Sens.* **2012**, *33*, 3197–3210. [CrossRef]
19. Greidanus, H.; Santamaria, C. First Analyses of Sentinel-1 Images for Maritime Surveillance. In *JRC Science and Policy Reports*; Publications Office of the European Union: Luxembourg, 2014.
20. Leng, X.; Ji, K.; Zhou, S.; Xing, X.; Zou, H. An adaptive ship detection scheme for spaceborne SAR imagery. *Sensors* **2016**, *16*, 1345. [CrossRef] [PubMed]
21. Wessel, P. GSHHG—A Global Self-Consistent, Hierarchical, High-Resolution Geography Database. Available online: http://www.soest.hawaii.edu/pwessel/gshhg (accessed on 1 September 2014).
22. Baselice, F.; Ferraioli, G. Unsupervised coastal line extraction from SAR images. *IEEE Geosci. Remote Sens. Lett.* **2013**, *10*, 1350–1354. [CrossRef]
23. Niedermeier, A.; Romaneessen, E.; Lehner, S. Detection of coastlines in SAR images using wavelet methods. *IEEE Trans. Geosci. Remote Sens.* **2000**, *38*, 2270–2281. [CrossRef]
24. Kapur, J.N.; Sahoo, P.K.; Wong, A.K.C. A new method for gray-level picture thresholding using the entropy of the histogram. *Comput. Vis. Graph. Image Process.* **1985**, *29*, 273–285. [CrossRef]
25. Otsu, N. A Threshold Selection Method from Gray-Level Histogram. *IEEE Trans. Syst. Man Cybern.* **1979**, *9*, 62–66. [CrossRef]
26. Sahoo, P.K.; Soltani, S.; Wong, A.K.C. A survey of thresholding techniques. *Comput. Vis. Graph. Image Process.* **1988**, *41*, 233–260. [CrossRef]
27. Criminisi, A.; Perez, P.; Toyama, K. Region filling and object removal by exemplar-based image inpainting. *IEEE Trans. Image Process.* **2004**, *13*, 1200–1212. [CrossRef] [PubMed]
28. Cossu, R.; Schoepfer, E.; Bally, P.; Fusco, L. Near real-time SAR-based processing to support flood monitoring. *J. Real-Time Image Process.* **2009**, *4*, 205–218. [CrossRef]
29. Dellepiane, S.G.; Angiati, E. A New Method for Cross-Normalization and Multitemporal Visualization of SAR Images for the Detection of Flooded Areas. *IEEE Trans. Geosci. Remote Sens.* **2012**, *50*, 2765–2779. [CrossRef]

*Letter*

# GF-3 SAR Ocean Wind Retrieval: The First View and Preliminary Assessment

**He Wang [1,*], Jingsong Yang [2], Alexis Mouche [3], Weizeng Shao [4], Jianhua Zhu [1], Lin Ren [2] and Chunhua Xie [5]**

1. National Ocean Technology Center, State Oceanic Administration, Tianjin 300112, China; besmile@263.net
2. State Key Laboratory of Satellite Ocean Environment Dynamics, Second Institute of Oceanography, State Oceanic Administration, Hangzhou 310012, China; jsyang@sio.org.cn (J.Y.); renlin210@sio.org.cn (L.R.)
3. Laboratoire d'Océanographie Physique et Spatiale, Institut Français de Recherche pour l'Exploitation de la Mer, Brest 29280, France; alexis.mouche@ifremer.fr
4. Marine Acoustics and Remote Sensing Laboratory, Zhejiang Ocean University, Zhoushan 316000, China; shaoweizeng@zjou.edu.cn
5. National Satellite Ocean Application Service, State Oceanic Administration, Beijing 100081, China; chxie@mail.nsoas.org.cn

* Correspondence: wanghe_sio@126.com; Tel.: +86-22-2753-6515

Academic Editors: Xiaofeng Yang, Xiaofeng Li and Ferdinando Nunziata
Received: 4 June 2017; Accepted: 4 July 2017; Published: 5 July 2017

**Abstract:** Gaofen-3 (GF-3) is the first Chinese civil C-band synthetic aperture radar (SAR) launched on 10 August 2016 by the China Academy of Space Technology (CAST), which operates in 12 imaging modes with a fine spatial resolution up to 1 m. As one of the primary users, the State Oceanic Administration (SOA) operationally processes GF-3 SAR Level-1 products into ocean surface wind vector and plans to officially release the near real-time SAR wind products in the near future. In this paper, the methodology of wind retrieval at C-band SAR is introduced and the first results of GF-3 SAR-derived winds are presented. In particular, the case of the coastal katabatic wind off the west coast of the U.S. captured by GF-3 is discussed. The preliminary accuracy assessment of wind speed and direction retrievals from GF-3 SAR is carried out against in situ measurements from National Data Buoy Center (NDBC) buoy measurements of National Oceanic and Atmospheric Administration (NOAA). Only the buoys located inside the GF-3 SAR wind cell (1 km) were considered as co-located in space, while the time interval between observations of SAR and buoy was limited to less the 30 min. These criteria yielded 56 co-locations during the period from January to April 2017, showing the Root Mean Square Error (RMSE) of 2.46 m/s and 22.22° for wind speed and direction, respectively. Different performances due to geophysical model function (GMF) and Polarization Ratio (PR) are discussed. The preliminary results indicate that GF-3 wind retrievals are encouraging for operational implementation.

**Keywords:** GF-3; synthetic aperture radar (SAR); ocean surface wind; validation

---

## 1. Introduction

Space-borne synthetic aperture radar (SAR) sensors operating in C-band (~5.3 GHz) have the capability to detect the sea surface at high spatial resolution under all-weather conditions, even in hurricanes. Normalized radar cross-section (NRCS) observations from SAR are directly related to the sea surface roughness of short surface waves. Therefore, assuming the surface wind is the first order contributor to these waves, coastal ocean surface winds at high resolution can be retrieved from SAR images. In recent decades, C-band SARs aboard Canadian satellite RadarSAT-1/2, and European satellites ERS-1/2, Envisat, and Sentinel-1A/B have demonstrated the capability of SAR to provide remote sensed ocean surface winds at a kilometer scale [1–3]. Moreover, these SAR-derived ocean surface winds are used in various applications, such as coastal wind energy resource assessment (e.g., [4]).

To date, Geophysical Model Functions (GMFs), which were originally developed to relate NRCS measured by scatterometers to ocean surface wind vector, are widely used for the estimations of ocean winds from C-band SARs. For VV-polarization, various empirical GMFs, describing the relationship between radar NRCS and the 10-m height ocean surface wind vector relative to radar viewing geometry, have been developed (e.g., [5–7]). When such GMFs are applied to HH-polarized SAR images, polarization ratio (PR) models have to be used to convert HH-NRCS into VV-NRCS before wind inversion. As for VV-GMF, different PR models have been developed (e.g., [8–10]). Studies indicate different performances depending on the GMF models (e.g., [11–13]). To date, ocean surface wind products derived from different satellite missions (scatterometers or SARs operating in C-band) rely on different GMFs. For instance, CMOD5.N [7] is currently used in MetOp-ASCAT scatterometer ocean wind processor; meanwhile, CMOD-IFR2 [6] is applied for the production of level 2 SAR ocean winds from Sentinel-1A/B [14]. Note also that different teams can use different GMFs. For instance, National Oceanic and Atmospheric Administration (NOAA) relies on a different GMF called CMOD5.H [15]. Thus, the choice of the GMF and PR models to be used for wind inversion is critical for GF-3 C-band SAR missions.

On 10 August 2016, carrying the first Chinese civil multi-polarization SAR operating at C-band, Gaofen-3 (GF-3) satellite was successfully launched into a polar sun-synchronous orbit of 755 km altitude with a 26-day repeat cycle. Following several months of in-orbit commissioning phase, GF-3 SAR has now been in operation since January 2017. Recently, a case of internal waves in the Yellow Sea captured by GF-3 was reported in Reference [16], implying promising marine application of the GF-3 SAR mission. As one of the primary users, the State Oceanic Administration (SOA) is operationally processing GF-3 SAR Level-1 products into Level-2 ocean surface wind vector and plan to operationally release the near real-time GF-3 SAR wind products. The aim of this paper is to present the first results of GF-3 SAR-derived winds and the preliminary assessment using the buoy measurements.

The remainder of the paper is organized as follows. Section 2 introduces the match-ups for GF-3 SAR images and in situ winds. Methodology for ocean wind retrieval with GF-3 SAR is presented in Section 3. The results of GF-3-derived winds and the assessment are given and discussed in Section 4. Finally, conclusions are given in Section 5.

## 2. Description of Collocated Data Sets

Validations of SAR-derived winds mostly lie on collocations with observations from buoys (e.g., [11]), offshore meteorological masts (e.g., [17–19]), and scatterometers (e.g., [3]). To date, besides HY-2A SCAT (exceeded its three-year design lifetime since 2015) providing degraded wind products, the available scatterometers are ASCATs onboard MetOp-A/B. Unfortunately, GF-3 and MetOp are in sun-synchronous orbits with the local equator crossing time at the ascending node of 6:00 am and 9:30 am, respectively. The large local time difference could limit the comparison of ocean winds derived from GF-3 and ASCATs. Hence, in this study, the validation of GF-3 wind retrieval was performed using in situ buoys.

### *2.1. GF-3 SAR Data*

SAR data of high radiometric quality are essential for accurate ocean wind inversion. For GF-3 SAR, the external calibration campaign was carried out using active radar transponders deployed in Inner Mongolia, during the commissioning phase (from September to November 2016). The results from the in-orbit external calibration experiment reveal the NRCS radiometric accuracy of 1.3~1.4 dB ($3\sigma$), and Noise Equal Sigma Zero (NESZ) of $-20 \sim -22$ dB [20] for different GF-3 SAR operating modes.

In this study, a total of 37 GF-3 SAR scenes were co-located with buoys over the period of January through April 2017. These GF-3 data have been collected in five GF-3 imaging modes (see Table 1 for details) amongst the 12 modes of the sensor, including Standard Strip, Quad-Polarization Strip I, Quad-Polarization Strip II, Fine Strip I, and Narrow ScanSAR imaging mode with different polarization (dual-polarization or quad-polarization), pixel spacing, and SAR scene swath.

**Table 1.** Main parameters for GF-3 imaging mode used here.

| Imaging Mode [1] | Polarization | Resolution (m) | Swath (Km) | Num of Synthetic Aperture Radar Scenes Used |
|---|---|---|---|---|
| SS | VV + VH or HH + HV | 25 | 130 | 26 |
| QPSI | VV + HH + VH + HV | 8 | 30 | 3 |
| QPSII | VV + HH + VH + HV | 25 | 40 | 5 |
| FSI | HH + HV | 5 | 50 | 2 |
| NSC | VV + VH | 50 | 300 | 1 |

[1] SS, QPSI, QPSII, FSI and NSC stand for Standard Strip, Quad-Polarization Strip I, Quad-Polarization Strip II, Fine Strip I, and Narrow ScanSAR imaging mode of GF-3, respectively.

### 2.2. Buoy Data

Buoy measurements are generally assumed to be of high quality up to 30 m/s [21,22] and hence were used as ground truth for the validation of wind retrieval from GF-3 SAR here. Wind observations of buoys have been collected from the National Data Buoy Center (NDBC) of NOAA. Most of them are off the west coast of the U.S., while some of them are in the region of Hawaiian Islands, southeast Pacific (Stratus buoy station), and Korean coastal sea.

In this study, only the buoys located inside the GF-3 SAR scene were considered as co-located in space. The buoy winds from NDBC are measured hourly by averaging the wind speed and direction over 10 min. Therefore, the time interval between observations of SAR and buoy was limited to less the 30 min. During the period from January to April 2017, these criteria yielded 53 match-ups, whose locations are depicted in Figure 1.

Since the anemometers on the buoys measure the wind at different heights above the sea surface level, buoy wind speeds had to be converted to the equivalent neutral winds at 10 m for comparison between GF-3 retrieval and buoy measurements. Here, the correction was performed using the simple semi-rational formula of assuming a logarithmically varying wind profile [23]; the wind speed given at any elevation z can be calculated from the friction velocity $u^*$ by:

$$U(z) = \frac{u_*}{\kappa} \ln\left(\frac{z}{z_0}\right) \tag{1}$$

where $\kappa = 0.4$ is the constant of von Kármán, and U(z) is the wind speed at a height of z. The friction velocity $u^*$ is related to the sea surface roughness length $z_0$ by [24]:

$$z_0 = \frac{0.11\upsilon}{u^*} + \frac{\alpha u^{*2}}{g} \tag{2}$$

Here, $\alpha = 0.011$ is the Charnock parameter [23], $\upsilon\left(1.5 \times 10^{-5}\ \mathrm{m}^2/\mathrm{s}\right)$ is the kinematic viscosity of air, and g is the gravitational acceleration.

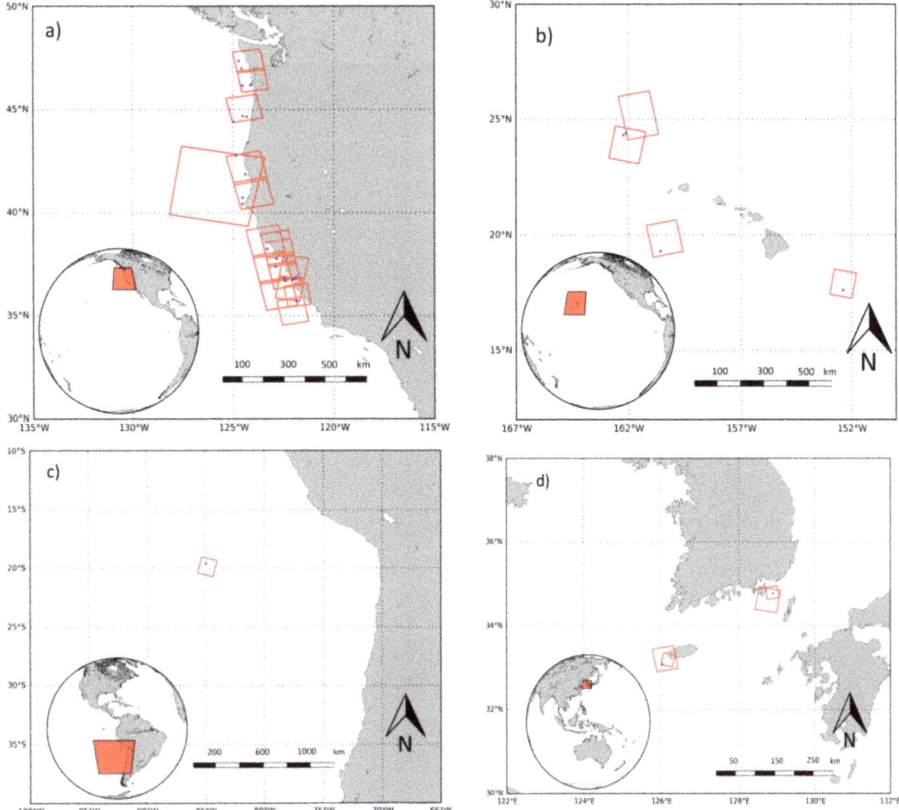


**Figure 1.** Map of footprints of GF-3 SAR acquisitions (red boxes) in the area of (**a**) U.S. west coast; (**b**) Hawaiian region; (**c**) South Pacific Ocean; (**d**) Korean coast. The blue dots denote the location of the National Data Buoy Center (NDBC) buoys.

## 3. GF-3 SAR Wind Retrieval Scheme

Both wind speed and direction could be inverted from NRCS measurements if multiple acquisitions over the same area are obtained simultaneously with different radar geometries, as for the scatterometers. In contrast, because SARs operate with only one radar viewing angle, the inversion of the NRCS leads to multiple solutions of wind vector. In this context, SAR wind retrieval schemes usually assume the wind direction as known from atmospheric models (i.e., [3,4]), or directly extracted from the SAR image if wind-induced streaks exist (i.e., [25]).

Before wind inversion, the pre-processing of GF-3 images requires radiometric calibration, land masking, and resampling of NRCS, incidence angle, and azimuth angle at a 1000 m resolution cell. The inversion methodology proposed here combines SAR NRCS with a priori wind from the European Centre for Medium-range Weather Forecasts (ECMWF) model, taking into account that both SAR observations and atmospheric models may contain errors [26,27].

In principle, in order to invert to the wind vector $(\mathrm{u}, \mathrm{v}$ for two components$)$, the cost function $\mathrm{J}(\mathrm{u}, \mathrm{v})$ defined as follows is minimized.

$$\mathrm{J}(\mathrm{u},\mathrm{v}) = \left(\frac{\sigma^0{}_{\mathrm{obs}} - \sigma^0{}_{\mathrm{GMF}}(\mathrm{u},\mathrm{v})}{\Delta\sigma^0}\right)^2 + \left(\frac{\mathrm{u}_{\mathrm{model}} - \mathrm{u}}{\Delta \mathrm{u}}\right)^2 + \left(\frac{\mathrm{v}_{\mathrm{model}} - \mathrm{v}}{\Delta \mathrm{v}}\right)^2 \quad (3)$$

where $\sigma^0{}_{obs}$ is the calibrated GF-3 SAR NRCS at co-polarized channel (VV or HH). And $u_{model}$ and $v_{model}$ are the U,V wind components from ECMWF model. These a priori winds provided from ECMWF at a 0.125° spatial and a three-hourly temporal resolution were interpolated into the GF-3 wind retrieval cell at 1000 m resolution as $\sigma^0{}_{obs}$. $\Delta\sigma^0$, $\Delta u$ and $\Delta v$ are the Gaussian standard deviation errors for the NRCS and the ECMWF model wind vector, respectively. $\sigma^0{}_{GMF}$ is the NRCS prediction from the GMF model. A general form of GMF could be expressed in functions of second-order cosine harmonic, as below:

$$\sigma^0 = A_0[1 + A_1 \cos\varnothing + A_2 \cos 2\varnothing]^B \tag{4}$$

where $A_0$, $A_1$, $A_2$ and B are the function of the 10-m-height wind speed and radar incidence angle, and $\varnothing$ is the wind direction relative to the radar look direction. For the HH channel, an additional PR model is used.

In order to make the inversion scheme more efficient, the minimization of cost function of Equation (3) is implemented with the help of the pre-computed look-up tables (LUT) from GMF models. Here, different LUT from models are used to test the GF-3 wind retrieval performance, including CMOD-like GMFs of CMOD-IFR2 adopted by Sentinel-1A/B and CMOD5.N used by MetOp-ASCAT, and PR models of Mouche et al. used by Sentinel-1A/B [9] and Zhang et al. [10]. Moreover, we also investigate the performances of C-band GMF called C-SARMOD recently proposed by Reference [28] for both VV- and HH-polarizations.

## 4. Results and Discussion

### *4.1. GF-3 SAR Wind: The Case of Katabatic Wind*

Figure 2a shows a VV-polarized GF-3 SAR scene in Narrow Scan mode over the Californian coast. The image was acquired at 14:31 Universal Time Coordinated (UTC) (07:31 in local time) on 6 January 2017, with an incidence angle ranging from 22° (near range) to 37° (far range). The wind field at 1000 m resolution retrieved from Figure 2a using the scheme proposed in Section 3 with the GMF of CMOD5.N is presented in Figure 2b.

In this case, brighter NRCS and corresponding higher sea surface wind speed are manifested along the coast, as shown in Figure 2a,b. This signature captured by GF-3 could be interpreted as the coastal katabatic wind. This kind of wind, blowing toward the coastal waters from the mountains, is the gravitational cold-air flow due to the temperature difference between the sea water and the shore. In addition to a meso-scale background wind, the katabatic wind increases the short surface waves, and thus the NRCS. This produces brighter areas near the coastline visible in SAR imagery as reported by References [29,30]. Compared with the ECMWF winds in 0.125° grid, as shown in Figure 2c, finer scale for the katabatic wind could be found the GF-3 wind retrievals in Figure 2b.

Three NDBC buoys are located inside this GF-3 SAR image, as the dots depicted in Figure 2a. For these three match-ups, ocean winds (speed at 10 m and direction) from GF-3 SAR retrieval and buoy observation are listed in in Table 2.

**Table 2.** GF-3 SAR winds against NDBC measurements for the case shown in Figure 2.

| Buoy ID | Latitude (°N)/Longitude (°W) | Buoy U10 (m/s) | GF-3 U10 (m/s) | Buoy Wind Direction (°) | GF-3 Wind Direction (°) |
|---|---|---|---|---|---|
| 46015 | 42.764/124.832 | 6.16 | 6.1 | 143 | 150.4 |
| 46027 | 41.852/124.382 | 6.16 | 4.2 | 104 | 112.2 |
| 46022 | 40.720/124.531 | 7.2 | 5.2 | 114 | 145.5 |

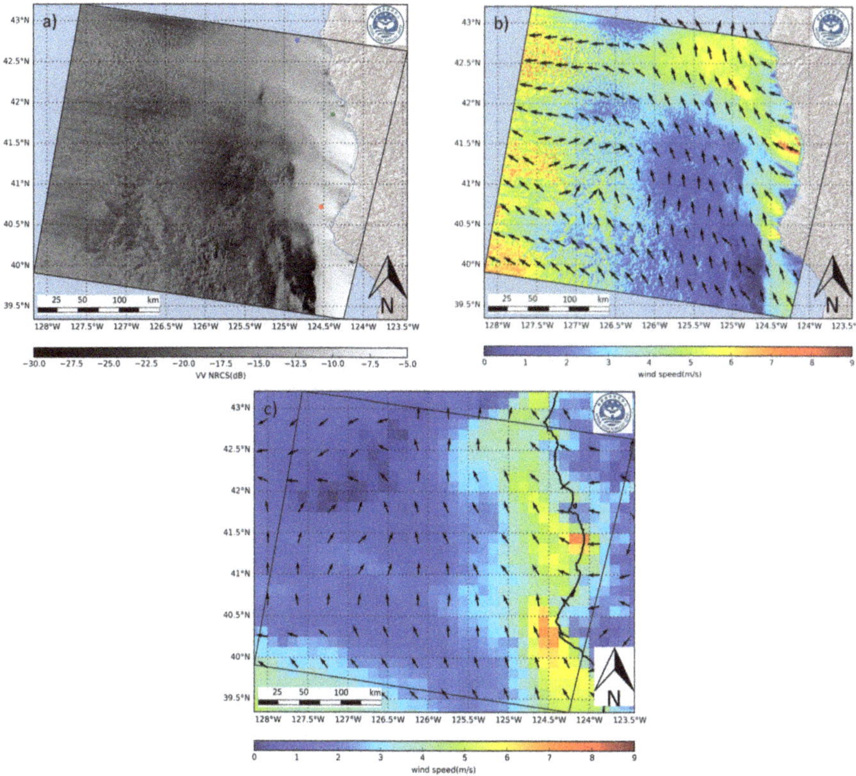


**Figure 2.** A katabatic wind case of GF-3 SAR image taken at 14:31 Universal Time Coordinated (UTC) (07:31 in local time) on 6 January 2017 in the Western U.S. coastal region: (**a**) Normalized radar cross-section (NRCS) at VV-polarization; (**b**) ocean surface wind inverted from VV NRCS using Geophysical Model Function (GMF) of CMOD5.N; (**c**) ocean surface wind from European Centre for Medium-range Weather Forecasts (ECMWF). The NDBC buoys of 46015, 46027, and 46022 are depicted as dots in blue, green, and red respectively.

### 4.2. GF-3 SAR Winds from VV-Polarization

For VV-polarization, 14 match-ups between GF-3 in SS, QPSI, QPSII, and NSC imaging mode and the buoys are found. In situ wind speed ranges between 5 m/s and 15 m/s. The comparison of winds derived from VV-polarized GF-3 SAR using different GMFs against those measured from NDBC buoys are presented in Figure 3. In each scatterplot from Figure 3, colors represent the incidence angle of GF-3, ranging from 23° to 45°. For wind speed, the application of GMF models of CMOD5.N, CMOD-IFR2, and C-SARMOD results in negative bias, with the smallest of -0.15 m/s using CMOD5.N for our retrieval. It is also shown that the Root Mean Square Error (RMSE) of the GF-3 retrieval of 10-m-height wind speed with respect to the buoy data is 2.34 m/s, 2.60 m/s, and 2.53 m/s using GMF models of CMOD5.N, CMOD-IFR2, and C-SARMOD, respectively. In terms of wind direction, a slight difference could be found for VV-polarization using the three GMFs, with bias and RMSE around 5° and 21.9°, respectively.

Thus, among the three GMF models used, CMOD5.N shows the overall best performance for VV-polarized GF-3 SAR.

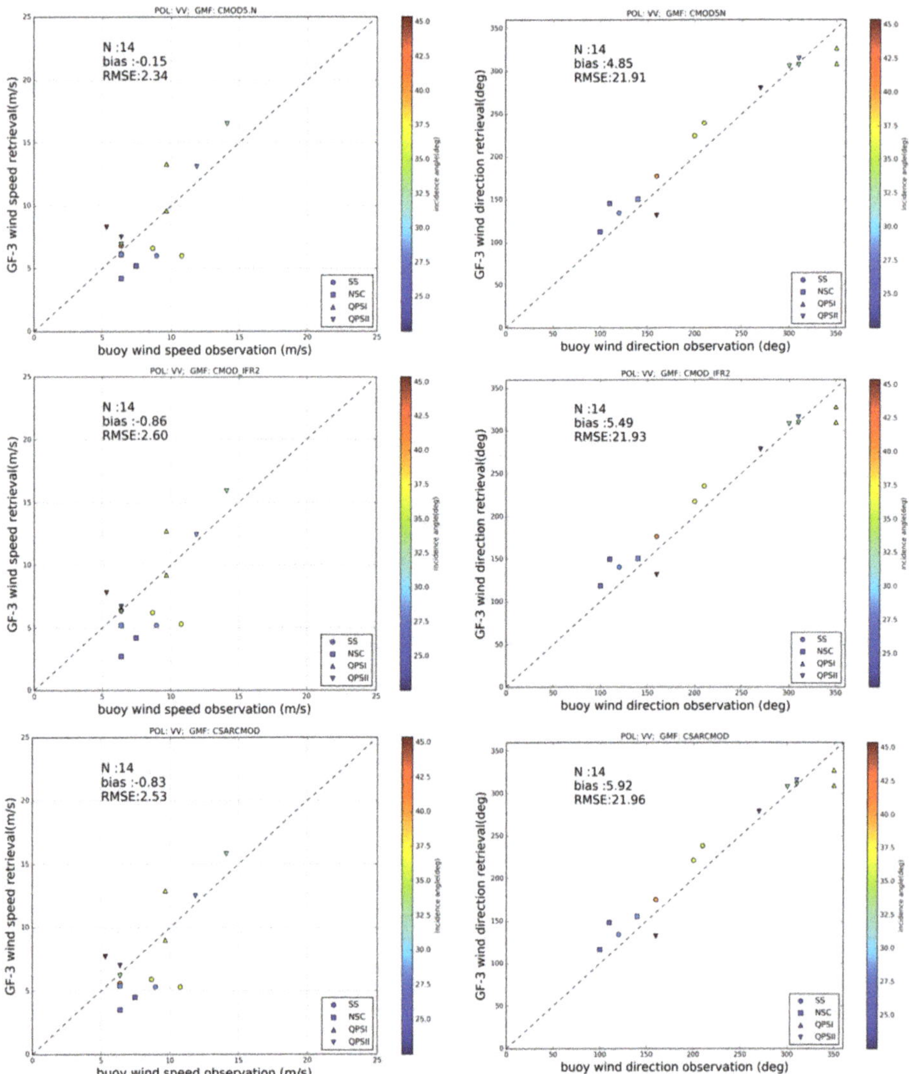


**Figure 3.** Scatter plots of wind retrievals from GF-3 SAR (VV-polarization) against NDBC buoy observations for wind speed (**left panels**) and wind direction (**right panels**). GF-3 winds estimated using GMF of CMOD5.N, CMOD-IFR2, CSARMOD_VV are presented from upper to bottom panels. Points with color indicate the incidence angle, and different markers present different imaging modes of GF-3, according to the legend.

### *4.3. GF-3 SAR Winds from HH-Polarization*

Figure 4 illustrates the scatter plots of GF-3 SAR winds from HH-polarization against NDBC observations. Three different GMFs are used: CMOD5.N + PR model of Zhang et al. [10], CMOD5.N + Mouche et al. [9], and C-SARMOD for HH-polarization [28]. Respectively, 42 and 31 match-ups of HH-polarized wind retrievals against in situ buoys for wind speed and wind direction are compared here; some of the buoy wind direction data are missing.

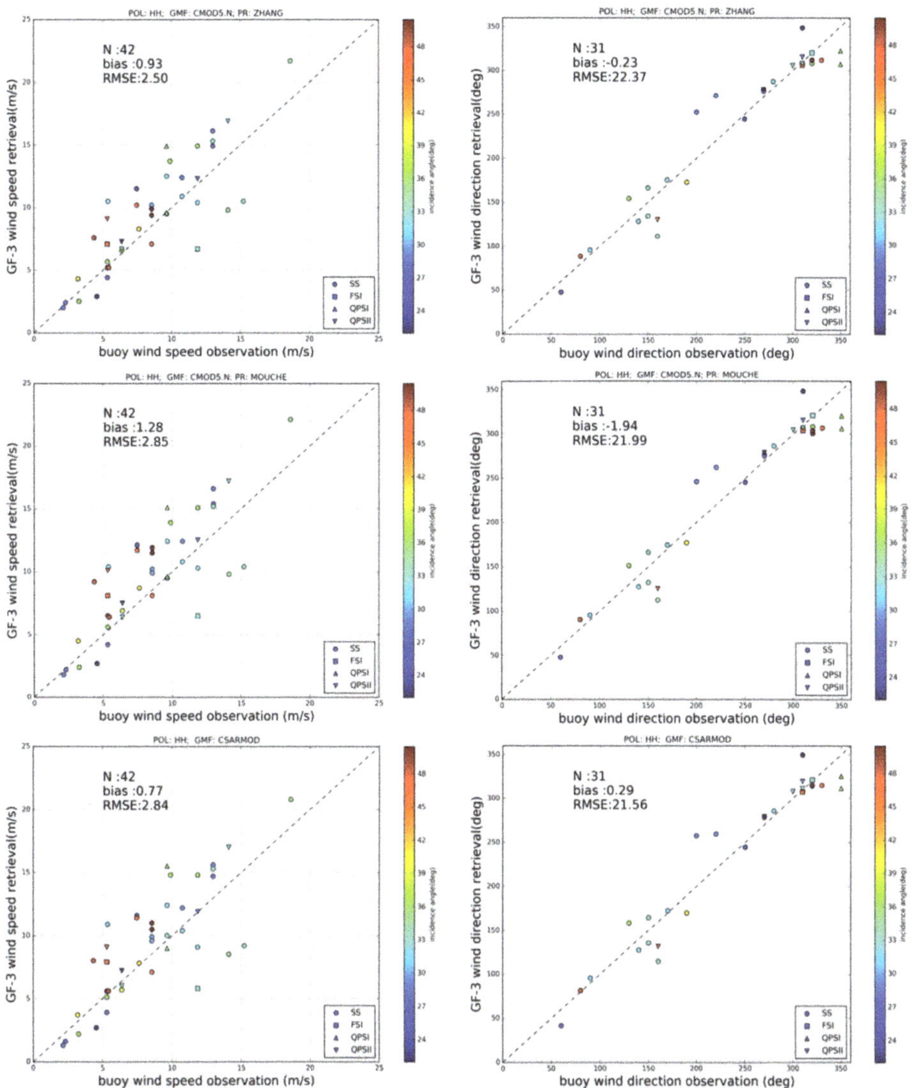


**Figure 4.** As in Figure 3, but for HH-polarized GF-3 SAR wind retrievals. GF-3 wind estimated using GMF of CMOD-5.N with of Zhang et al. [10], GMF of CMOD5.N with of Mouche et al. [9], and GMF of C-SARMOD_HH are presented from upper to bottom panels.

The wind direction retrievals show a slight difference for HH-polarization using the three GMFs. The RMSEs are very close between HH- and VV-polarized GF-3 retrievals, with a smaller bias for HH-polarized retrievals.

In terms of wind speed from HH-polarized GF-3 SAR, all the three GMFs used here present a positive bias. Also, larger RMSEs could be found compared to the results using CMOD5.N (see Figure 3) for VV-polarized SAR images, indicating that the GMF models used for HH-polarization could be improved in the future. Among the three models of GMF and PR applied to HH-polarized GF-3 SAR in this study, the combination of CMOD5.n and the PR model from Zhang et al. [10] leads to the smallest RMSE of 2.50 m/s, though the bias of 0.93 m/s is a litter bit larger than the 0.77 m/s of C-SARMOD. For the combination model of CMOD5.n and PR from Mouche et al. [9], and the empirical GMF C-SARMOD [28], very close RMSEs (around 2.85 m/s) could be found, while the latter GMF model presents smaller bias (0.77 m/s) than those of the former(1.28 m/s).

The new model of C-SARMOD is expressed by an explicit GMF so that the conversion of HH- to VV-polarization NRCS are no longer required, making it more applicable for HH-polarized SAR images. Although the smallest bias is found here, a larger RMSE is found for this new model, indicating that careful tuning may be needed in order to make it applicable for HH-polarized GF-3. Generally, the PR model proposed by Zhang et al. [10] illustrates good performances, due to the fact that it includes both the dependence of wind speed and incidence angle. Taking this into account for the preliminary comparison of different models, we selected the combination of CMOD5.n and PR model from Zhang et al. [10] for the HH-polarized GF-3 SAR wind retrieval in our scheme.

### 4.4. GF-3 SAR Winds General Validation Statistics

From the analysis in Sections 4.2 and 4.3, the GMF of CMOD5.N [7] and the PR of Zhang et al. [10] resulted in the best performance for the GF-3 wind retrieval scheme used here. As a consequence, they were chosen for the production of GF-3 ocean wind retrievals. Based on these GMF models, the GF-3 wind retrievals were assessed against NDBC winds, with the scatter plots depicted in Figure 5 and general statistics listed in Table 3. In general, an RMSE of 2.46 m/s and 22.22°, and a bias of 0.66 m/s and 1.35° could be found for the comparison of GF-3-derived wind against buoy winds in this preliminary assessment.

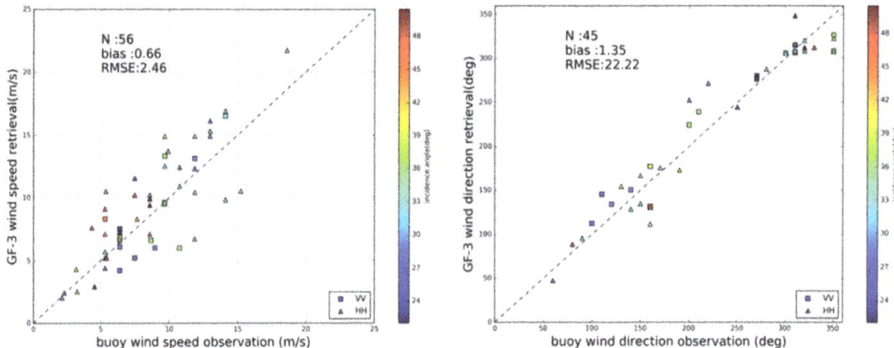


**Figure 5.** Scatter plots of GF-3 SAR wind retrievals against NDBC buoy observations for wind speed (**left**) and wind direction (**right**). GF-3 winds are estimated using the GMF of CMOD5.N and the PR of Zhang et al. [10], for HH-polarization. Points with color indicate the incidence angle, and different markers present different polarizations of GF-3, according to the legend.

**Table 3.** Statistics for GF-3 SAR winds against NDBC measurements.

| Polarization of GF-3 Data | Wind Speed | | | Wind Direction | | | GMF Used |
|---|---|---|---|---|---|---|---|
| | N | Bias (m/s) | RMSE (m/s) | N | Bias (°) | RMSE (°) | |
| All co-polarization | 56 | 0.66 | 2.46 | 45 | 1.35 | 22.22 | VV: CMOD5.n<br>HH:CMOD5.n + PR model from Zhang et al. [10] |
| VV | 14 | −0.15 | 2.34 | 14 | 4.85 | 21.91 | |
| HH | 42 | 0.93 | 2.50 | 31 | −0.23 | 22.37 | |

## 5. Conclusions

In this paper, the proposed GF-3 SAR wind inversion methodology combines SAR-observed NRCS at a co-polarized channel with a priori wind information from ECMWF winds, taking into account that both NRCS observations and models may contain errors. In order to extract the wind speed and direction, the cost function is minimized with the help of look-up tables computed from geophysical model functions (GMFs), making the inversion scheme more efficient. Using this inversion scheme, coastal winds at 1 km resolution were estimated from the GF-3 SAR. One case of the coastal katabatic wind off the west coast of the U.S. captured by GF-3 is presented.

The first accuracy assessment of the ocean surface wind vector from GF-3 was carried out through a comparison with in situ observations from moored NDBC buoys, over the period from January to April 2017.

To select the GMF and PR models for the GF-3 wind inversion scheme, retrieval performances were compared for models including CMOD-like GMFs of CMOD-IFR2 and CMOD5.N, PR models of Mouche et al. [9], Zhang et al. [10], and the recently proposed C-SARMOD. The results indicate that the GMF of CMOD5.N and the PR of Zhang et al. [10] present the best performance for GF-3 wind inversion. Thus, these two models were chosen for the production of GF-3 ocean wind retrievals. A validation of GF-3-derived winds against NDBC measurements shows an RMSE of 2.46 m/s and 22.22° for wind speed and direction, respectively. These preliminary results indicate that GF-3 wind retrievals are encouraging for operational products, which will be released from the State Oceanic Administration in the near future.

Future work will be dedicated to collect more data over buoys to increase the representativeness of our dataset and refine the performance assessment such as the error with respect to incidence angle, wind speed, or NESZ. In addition, the possibility of estimating the wind direction directly from GF-3 SAR images (e.g., [31]) will be investigated.

**Acknowledgments:** The GF-3 SAR data has been provided by the National Satellite Ocean Application Service in the framework of National High-Resolution Project of China via website of http://dds.nsoas.org.cn/ (registration required), and we would like to thank Limin Cui and Wentao An for the GF-3 SAR data ordering and delivering. The authors are also grateful for the free access to buoy wind measurements provided by NDBC via http://www.ndbc.noaa.gov/. This work is supported by the National Natural Science Foundation of China (41506205), National Key R&D Program of China (2016YFC1401003, 2016YFC1401007), National High-Resolution Project of China (Y20A14-9001-15/16) and Dragon-4 project (32249).

**Author Contributions:** He Wang wrote the manuscript; Alexis Mouche and He Wang devised and implemented the SAR ocean wind retrieval scheme; He Wang, Weizeng Shao and Lin Ren processed the GF-3 SAR data. All authors contributed to the discussion and revision of the manuscript.

**Conflicts of Interest:** The authors declare no conflict of interest.

## References

1. Dagestad, K.-F.; Horstmann, J.; Mouche, A.; Perrie, W.; Shen, H.; Zhang, B.; Li, X.; Monaldo, F.; Pichel, W.; Lehner, S.; et al. Wind Retrieval from Synthetic Aperture Radar—An Overview. In Proceedings of the 4th SAR Oceanography Workshop (SEASAR 2012), Tromsø, Norway, 18–22 June 2012; pp. 1–24.
2. Horstmann, J.; Koch, W. Measurement of ocean surface winds using synthetic aperture radars. *IEEE J. Ocean. Eng.* **2005**, *30*, 508–515. [CrossRef]
3. Monaldo, F.; Jackson, C.; Li, X.; Pichel, W.G. Preliminary evaluation of Sentinel-1A wind speed retrievals. *IEEE J. Sel. Top. Appl. Earth Obs. Remote Sens.* **2016**, *9*, 2638–2642. [CrossRef]
4. Chang, R.; Zhu, R.; Badger, M.; Hasager, C.B.; Xing, X.; Jiang, Y. Offshore wind resources assessment from multiple satellite data and WRF modeling over South China Sea. *Remote Sens.* **2015**, *7*, 467–487. [CrossRef]
5. Stoffelen, A.; Anderson, D. Scatterometer data interpretation: Estimation and validation of the transfer function CMOD4. *J. Geophys. Res.* **1997**, *102*, 5767–5780. [CrossRef]
6. Quilfen, Y.; Chapron, B.; Elfouhaily, T.; Katsaros, K.; Tournadre, J. Observation of tropical cyclones by high-resolution scatterometry. *J. Geophys. Res.* **1998**, *103*, 7767–7786. [CrossRef]

7. Hersbach, H. Comparison of C-band scatterometer CMOD5.N equivalent neutral winds with ECMWF. *J. Atmos. Ocean. Tech.* **2010**, *27*, 721–736. [CrossRef]
8. Thompson, D.R.; Elfouhaily, T.M.; Chapron, B. Polarization Ratio for Microwave Backscattering from the Ocean Surface at Low to Moderate Incidence Angles. In Proceedings of the 1998 Geoscience and Remote Sensing Symposium Proceedings (IGARSS'98), Seattle, WA, USA, 6–10 July 1998; pp. 1671–1673.
9. Mouche, A.; Hauser, D.; Daloze, J.F.; Guerin, C. Dual-polarization measurements at C-band over the ocean: Results from airborne radar observations and comparison with ENVISAT ASAR data. *IEEE Trans. Geosci. Remote Sens.* **2005**, *43*, 753–769. [CrossRef]
10. Zhang, B.; Perrie, W.; He, Y. Wind speed retrieval from RADARSAT-2 quad-polarization images using a new polarization ratio model. *J. Geophys. Res.* **2011**, *116*, C08008. [CrossRef]
11. Takeyama, Y.; Ohsawa, T.; Kozai, K.; Hasager, C.B.; Badger, M. Comparison of geophysical model functions for SAR wind speed retrieval in Japanese coastal waters. *Remote Sens.* **2013**, *5*, 1956–1973. [CrossRef]
12. Bergeron, T.; Bernier, M.; Chokmani, K.; Lessard-Fontaine, A.; Lafrance, G.; Beaucage, P. Wind speed estimation using polarimetric RADARSAT-2 images: Finding the best polarization and polarization ratio. *IEEE J. Sel. Top. Appl. Earth Obs. Remote Sens.* **2011**, *4*, 896–904. [CrossRef]
13. Liu, G.; Yang, X.; Li, X.; Zhang, B.; Pichel, W.; Li, Z.; Zhou, X. A systematic comparison of the effect of polarization ratio models on sea surface wind retrieval from C-band synthetic aperture radar. *IEEE J. Sel. Top. Appl. Earth Obs. Remote Sens.* **2013**, *6*, 1100–1108. [CrossRef]
14. Mouche, A. Sentinel-1 ocean wind fields (OWI) algorithm definition. In *Sentinel-1 IPF Reference: (S1-TN-CLS-52-9049) Report*; CLS: Brest, France, 2010; pp. 1–75.
15. Soisuvarn, S.; Jelenak, Z.; Chang, P.S.; Alsweiss, S.O.; Zhu, Q. CMOD5.H—A High Wind Geophysical Model Function for C-Band Vertically Polarized Satellite Scatterometer Measurements. *IEEE Trans. Geosci. Remote Sens.* **2013**, *51*, 3744–3760. [CrossRef]
16. Yang, J.S.; Wang, J.; Ren, L. The first quantitative remote sensing of ocean internal waves by Chinese GF-3 SAR satellite. *Acta Oceanol. Sin.* **2017**, *36*. [CrossRef]
17. Hasager, C.B.; Christiansen, M.B.; Peña, A.; Larsén, X.G.; Bingöl, F. SAR-Based wind resource statistics in the Baltic Sea. *Remote Sens.* **2011**, *3*, 117–144. [CrossRef]
18. Badger, M.; Badger, J.; Nielsen, M.; Hasager, C.B.; Peña, A. Wind class sampling of satellite SAR imagery for offshore wind resource mapping. *J. Appl. Meteorol. Climatol.* **2010**, *49*, 2474–2491. [CrossRef]
19. Beaucage, P.; Bernier, M.; Lafrance, G.; Choisnard, J. Regional mapping of the offshore wind resource: Towards a significant contribution from space-borne synthetic aperture radars. *IEEE J. Sel. Top. Appl. Earth Obs. Remote Sens.* **2008**, *1*, 48–56. [CrossRef]
20. Zhang, Q. System Design and Key Technologies of the GF-3 Satellite. *Acta Geodaetica et Cartographica Sinica* **2017**, *46*, 269–277. (In Chinese) [CrossRef]
21. Zeng, L.; Brown, R.A. Scatterometer observations at high wind speeds. *J. Appl. Meteorol.* **1998**, *37*, 1412–1420. [CrossRef]
22. Weller, R.A.; Rudnick, D.L.; Payne, R.E.; Dean, J.P.; Pennington, N.J.; Trask, R.P. Measuring near-surface meteorology over the ocean from an array of surface moorings in the subtropical convergence zone. *J. Atmos. Ocean. Technol.* **1990**, *7*, 85–103. [CrossRef]
23. Charnock, H. Wind stress on a water surface. *Quart. J. R. Meteorol. Soc.* **1955**, *81*, 639–640. [CrossRef]
24. Smith, S.D. Coefficients for sea surface wind stress, heat flux, and wind profiles as a function of wind speed and temperature. *J. Geophys. Res.* **1988**, *93*, 15467–15472. [CrossRef]
25. Leite, G.C.; Ushizima, D.M.; Medeiros, F.N.S.; De Lima, G.G. Wavelet Analysis for Wind Fields Estimation. *Sensors* **2010**, *10*, 5994–6016. [CrossRef] [PubMed]
26. Portabella, M.; Stoffelen, A.; Johannessen, J.A. Toward an optimal inversion method for synthetic aperture radar wind retrieval. *J. Geophys. Res.* **2002**, *107*, 3086. [CrossRef]
27. Mouche, A.; Collard, F.; Chapron, B.; Dagestad, K.; Guitton, G.; Johannessen, J.A.; Kerbaol, V.; Hansen, M.W. On the use of doppler shift for sea surface wind retrieval from SAR. *IEEE Trans. Geosci. Remote Sens.* **2012**, *50*, 2901–2909. [CrossRef]
28. Mouche, A.; Chapron, B. Global C-Band Envisat, RADARSAT-2 and Sentinel-1 SAR measurements in copolarization and cross-polarization. *J. Geophys. Res.* **2015**, *120*, 7195–7207. [CrossRef]
29. Alpers, W.; Pahl, U.; Gross, G. Katabatic wind fields in coastal areas studied by ERS-1 synthetic aperture radar imagery and numerical modeling. *J. Geophys. Res.* **1998**, *103*, 7875–7886. [CrossRef]

30. Li, X.; Zheng, W.; Pichel, W.G.; Zou, C.-Z.; Clemente-Colon, P. Coastal katabatic winds imaged by SAR. *Geophys. Res. Lett.* **2007**, *34*, L03804. [CrossRef]
31. Koch, W. Directional analysis of SAR images aiming at wind direction. *IEEE Trans. Geosci. Remote Sens.* **2004**, *42*, 702–710. [CrossRef]

*Article*

# Ku-Band Sea Surface Radar Backscatter at Low Incidence Angles under Extreme Wind Conditions

**Xiuzhong Li [1,2], Biao Zhang [1,2,*], Alexis Mouche [1,3,*], Yijun He [1,2] and William Perrie [1,4]**

1 School of Marine Sciences, Nanjing University of Information Science and Technology, Nanjing 210044, China; lixiuzhong@nuist.edu.cn (X.L.); yjhe@nuist.edu.cn (Y.H.); William.Perrie@dfo-mpo.gc.ca (W.P.)
2 Jiangsu Research Center for Ocean Survey and Technology, Nanjing 210044, China
3 Laboratoire d'Oceanographie Spatiale, Centre de Brest, IFREMER, Plouzané 29280, France
4 Fisheries and Oceans Canada, Bedford Institute of Oceanography, Dartmouth, Nova Scotia, NS B2Y4A2, Canada
* Correspondence: zhangbiao@nuist.edu.cn (B.Z.); Alexis.Mouche@ifremer.fr (A.M.); Tel.: +86-25-5869-5692 (B.Z.)

Academic Editors: Xiaofeng Yang, Xiaofeng Li, Ferdinando Nunziata, Alexis Mouche and Prasad S. Thenkabail
Received: 14 March 2017; Accepted: 9 May 2017; Published: 12 May 2017

**Abstract:** This paper reports Ku-band normalized radar cross section (NRCS) at low incidence angles ranging from 0° to 18° and in the wind speed range from 6 to 70 m/s. The precipitation radar onboard the tropical rainfall measuring mission and Jason-1 and 2 have provided 152 hurricanes observations between 2008 and 2013 that were collocated with stepped-frequency microwave radiometer measurements. It is found that the NRCS decreases with increasing incidence angle. The decrease is more dramatic in the 40–70 m/s range of wind speeds than in the 6–20 m/s range, indicating that the NRCS is very sensitive to low incidence angles under extreme wind conditions and insensitive to the extreme wind speed. Consequently, the sea surface appears relatively "smooth" to Ku-band electromagnetic microwaves. This phenomenon validates the observed drag coefficient reduction under extreme wind conditions, from a remote sensing viewpoint. Using the NRCS dependence on incidence angle under extreme wind conditions, we also present an empirical linear relationship between NRCS and incidence angles, which may assist future-satellites missions operating at small incidence angles to measure sea surface wind and wave field.

**Keywords:** NRCS; extreme wind conditions; low incidence angles

---

## 1. Introduction

Tropical cyclones (also called hurricanes or typhoons) are always associated with high wind speeds. The wind speed plays a crucial role in air–sea interactions, numerical forecasting models, and tropical cyclone intensity. Active microwave remote sensors have advantages for estimating wind speeds because the normalized radar cross section (NRCS, $\sigma^0$) of the sea surface varies with the roughness and whitecaps or the foam effects that are driven by high wind speeds. The NRCS is a critical parameter for ocean remote sensors such as synthetic aperture radar (SAR) devices, scatterometers, and altimeters. Besides wind speeds at the sea surface, the NRCS can be used to detect ship targets at the sea surface [1], ocean surface currents [2], ocean wave spectra [3], and other ocean surface information.

However, most NRCS applications are limited to low or moderate wind conditions. Several researchers have analyzed the C-band NRCS saturation at wind speeds up to 35 m/s [4–6]. Donnelly et al. [7] and Carswell et al. [8] studied the Ku- and C-band NRCS up to wind speeds of 45 m/s. Thus far, only Fernandez et al. [9] have presented the Ku and C-band NRCS up to wind speeds of 65 m/s. In all of these reports, the co-polarization (VV) NRCS saturated at high wind speeds up to 25 m/s. Non-saturated cross-polarization (HV or VH) NRCS variation at high wind speeds has also been reported [10–14].

NRCS analyses at low incidence angles and low wind conditions have been frequently reported [15–18]. Some researchers consider that the NRCS under high wind conditions can be extrapolated from the NRCS characteristics in moderate wind conditions [19–21]. At high winds (up to 35 m/s), Quilfen et al. [22] empirically derived an inverse relationship between the sea surface roughness and altimeter backscatter. However, at wind speeds exceeding 35 m/s, the collocated data are too sparse to analyze the $\sigma^0$ characteristics with statistical confidence.

The paper collocates the NRCS from the precipitation radar (PR) of the tropical rainfall measuring mission (TRMM) and the Jason-1 and 2 altimeter observations, and the stepped-frequency microwave radiometer (SFMR) wind speed measurements up to 70 m/s. Subsequently, we derive a new NRCS dependence on wind speed at small incidence angles. Section 2 introduces the datasets, and Section 3 presents the data acquisition methodology. Then Section 4 gives results of relatively common moderately intense wind conditions and the newly proposed NRCS characteristic. The relationship between NRCS and incidence angle is also presented in this section. Discussion and conclusions are given in Sections 5 and 6, respectively.

## 2. Dataset

To acquire the NRCS at extremely high wind conditions, NRCS data at low incidence angles were collected from the standard products 2A21 (version-7) of PR and the geophysical data record (GDR) of Jason-1 and 2. The wind speeds were obtained from the National Oceanic Atmospheric Administration (NOAA) Hurricane Research Division (HRD) SFMR along-track measurements in hurricanes. The datasets are summarized and analyzed as follows.

### 2.1. TRMM PR

PR is a spaceborne Ku-band horizontal polarization radar onboard TRMM that has measured precipitation since 1997. The PR antenna is an electronically scanned phased array that scans a planar cross-track area through the nadir (in fact, as the satellite travels very quickly, the scanning track is not strictly vertical to the along-track direction). The nadir spatial resolution of the PR is 5.0 km (cross-track) $\times$ 4.1 km (along-track). Since the boost of the satellite orbit in 2001 to higher altitude, the swath width has been extended to 250 km. The 49 incidence angles vary from approximately $-18^\circ$ to $18^\circ$.

PR is the first instrument designed to measure the reflectivity of rain from space. The technique is that short pulses are transmitted at Ku-band, and the time delay and strength of the echo gives the distance and intensity of the rain. One of the key points with this technique is to determine the rain attenuation effect at a location. In order to solve this problem, one needs to know the strength of the sea surface reflectance, which varies according to the prevailing wind conditions [23]. Liao and Meneghini [24] compared the performances of the version-6 PR product before and after the satellite boost with a ground-based weather surveillance radar (WSR-88D) in Melbourne, FL. The PR product maintained its calibration accuracy of about 1 dB. Although the orbit boost reduced the NRCS by 0.67–0.88 dB, the accuracy of the PR $\sigma^0$ remained within 1 dB [25].

### 2.2. Jason-1 and 2

Jason-1 and Jason-2 were launched in 2001 and 2008, respectively. Jason-1 has expired but Jason-2 remains in orbit. This study uses the 0° NRCS of Jason-1 and 2 from the Ku bands, acquired from 2008 to 2012. The temporal and spatial intervals during the matchup were required to be within 60 min and 100 km, respectively. The product GDR–c of Jason-1 contains 1-Hz measurements, so points with less than 10 valid NRCSs were eliminated. For Jason-2, the "qual_alt_1hz_sig0_ku" (a quality flag variable in the Ku-band NRCS product) of GDRs are flagged as "good" and these values are retained.

Both GDRs flagged "no rain" are selected. In total, Jason-1 and Jason-2 were able to contribute 1209 and 3046 collocated pairs, respectively.

### 2.3. SFMR

The SFMR onboard NOAA WP-3D aircraft has been the prototype for a new generation of operational airborne remote sensing instruments designed for measuring the surface winds and rainfall in hurricanes since 1984 [26]. The SFMR operates at 4.6–7.2 GHz and measures meteorological parameters such as wind speed and rain rate [27]. The temporal and spatial resolutions of the measured wind speed are 1 s and 1.5 km, respectively. The wind speeds are estimated at 10 m above the sea surface by SFMR; they have been validated by the dropwindsonde measurements and the root-mean-square error (RMSE) is approximately 4 m/s, or 5%–25% [28]. Since 2007, the SFMR onboard the Air Force Reserve Command has measured the wind speed using the model developed by Uhlhorn and Black [28]. The SFMR-derived wind speeds in hurricanes have also been validated against those retrieved by cross-polarized SAR image, with a bias and RMSE of −0.89 m/s and 3.24 m/s, respectively [10]. Recently, the SFMR measurements of the surface winds in tropical cyclones with heavy precipitation have been further improved using a new relationship between microwave absorption and rain rate [29]. This relationship significantly decreased the surface-wind retrieval bias in the presence of rain at weak hurricane wind speeds.

## 3. Methodology

152 hurricanes observations were collected from the TRMM PR between 2008 and 2013, which were collocated with SFMR measurements. The temporal and spatial intervals for collocations were 30 min and 50 km, respectively. For each collocated data pair (denoting 1 s of SFMR data and one incidence angle of PR), the wind speed and rain rate from SFMR and the NRCS from the PR were recorded. The collocated dataset contains 29,798,600 points, of which 87.22% and 58.11% are below 5 mm/h and 2 mm/h, respectively. Only 140,795 points correspond to wind speeds above 40 m/s. The distribution of the wind speeds between 40 m/s and 70 m/s is shown in Figure 1. The numbers in each of the wind speed bin do not have large discrepancies among one another.

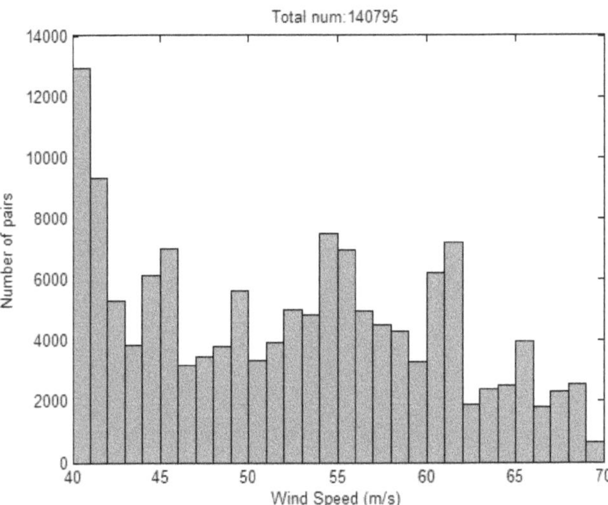


**Figure 1.** Wind speed histogram of collocated data pairs (140,795) between precipitation radar (PR)-observed normalized radar cross section (NRCS) and stepped-frequency microwave radiometer (SFMR)-measured wind speed ranging from 40 to 70 m/s.

Typically, strong winds in intense storms bring intense rainfall. To reduce the rain impacts on the NRCS and obtain sufficient data pairs for our analysis at extremely high winds, we selected pairs with rain rates below 5 mm/h. These pairs are gridded by discretizing the wind speeds and incidence angles into 1 m/s bins and 1° bins, respectively. In the two-dimensional bins, data pairs whose NRCS deviates by more than one standard deviation (STD) from the NRCS average are discarded. This simple quality control criterion yields a new NRCS. Figures 2 and 3 show the STD distributions of the old and new NRCSs, respectively. Panels a and b of each figure display the binned STDs in the wind speed ranges 6–20 m/s and 40–70 m/s, respectively.

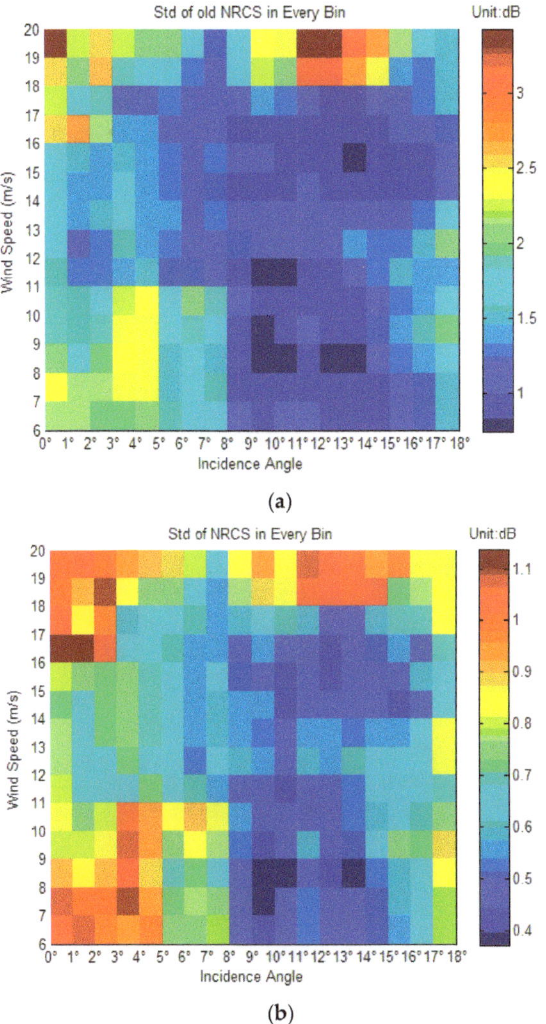


**Figure 2.** (**a**) Distribution of NRCS standard deviations (STDs) in each bin before quality control at wind speed 6–20 m/s. The wind speed is from SFMR and NRCS is from PR. Each collocated data pair contains 1 s of SFMR data and one NRCS and incidence angle of PR. (**b**) Distribution of NRCS STDs in each bin after quality control at wind speed 6–20 m/s. The wind speed is from SFMR and NRCS is from PR. Each collocated data pair contains 1 s of SFMR data and one NRCS and incidence angle of PR.

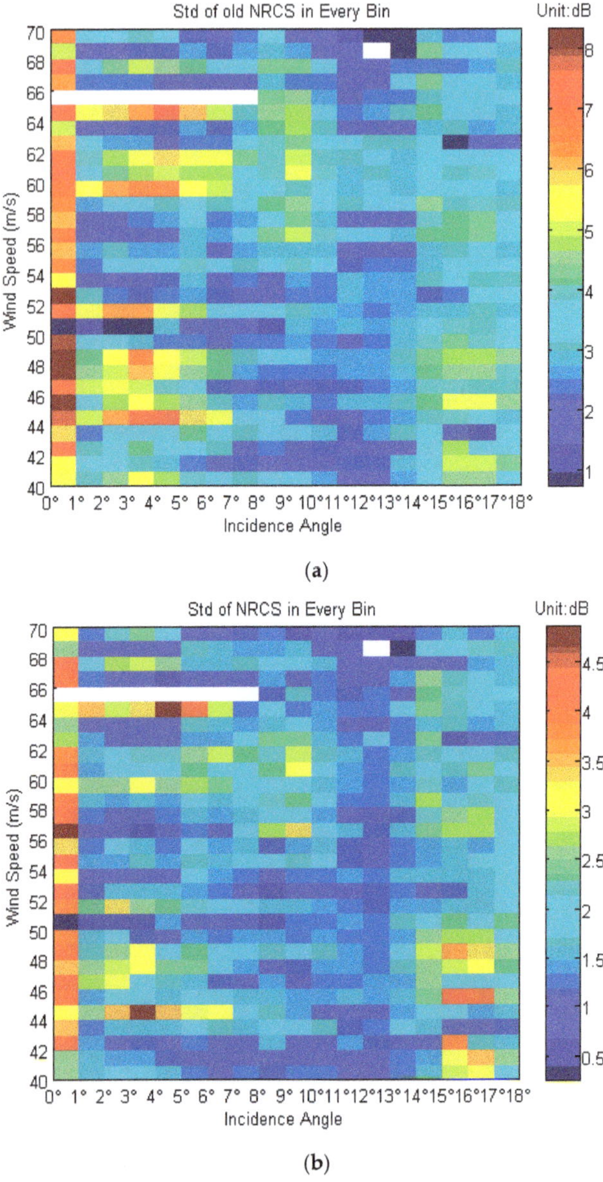


**Figure 3.** (**a**) Distribution of NRCS STDs in each bin before quality control at wind speed 40–70 m/s. The wind speed is from SFMR and NRCS is from PR. Each collocated data pair contains 1 s of SFMR data and one NRCS and incidence angle of PR. (**b**) Distribution of NRCS STD in each bin after quality control at wind speed 40–70 m/s. The wind speed is from SFMR and NRCS is from PR. Each collocated data pair contains 1 s of SFMR data and one NRCS and incidence angle of PR.

As confirmed in Figures 2 and 3, the simple quality control criterion drastically decreases the STDs. As wind speeds are between 6 and 20 m/s (Figure 2), the new STDs are constrained to be within 1.1 dB, whereas the non-controlled STDs are approximately within 3.5 dB. For high wind speeds ranging from

40 to 70 m/s (Figure 3), the NRCS is rendered noisier by the high wind and reduced number of points. In this case, the simple control criterion reduces the largest STDs from approximately 8 dB to 4.7 dB. Although 4.7 dB is still below the required accuracy, most of the STDs are within 1.5 dB (the largest errors occur at 0°). The data is acceptable for our analysis, as the PR is accurate to 1 dB.

## 4. Results

To show the results of NRCS from PR clearly, four figures are presented. Figures 4 and 6 study the dependence of NRCS on incidence angle at different wind speeds. Figures 5 and 7 study the dependence of NRCS on wind speed at different incidence angles. In this condition, Figures 4 and 5 illustrate dependence of NRCS on moderate wind condition while Figures 6 and 7 correspond to high wind condition.

### 4.1. PR NRCS at Small and Moderate Wind Speeds

At low incidence angles and wind speeds below 20 m/s, the NRCS is dominated by a quasi-specular reflection mechanism, and can be simulated by simple formulas [30,31]. Here we validate the NRCSs measured by the TRMM PR by using a geometrical optics model based on non-Gaussian probability density distribution [25]. This model was derived by the NRCS from PR and the wind speed from NDBC buoys, and is formulated as follows:

$$\sigma_{sp}^{0}(\theta) = \frac{|R(0)|^2}{MSS \cdot \cos^4\theta} \exp\left[-\frac{\tan^2\theta}{svp}\right]\left[a_1\frac{\tan^4\theta}{MSS^2} - a_2\frac{\tan^2\theta}{MSS} + \frac{a_2}{4} + 1\right] \quad (1)$$

In Equation (1), coefficients $a_1$ and $a_2$ are related to the kurtosis of the sea surface slopes. $|R(0)|^2$ is effective nadir reflection coefficient. $MSS$ is the mean square slope of sea surface. The four parameters are the function of wind speed, which are given in Li et al. [25]. The NRCS dependences on low incidence angle and low to moderate wind speed are illustrated in Figures 4 and 5, respectively.

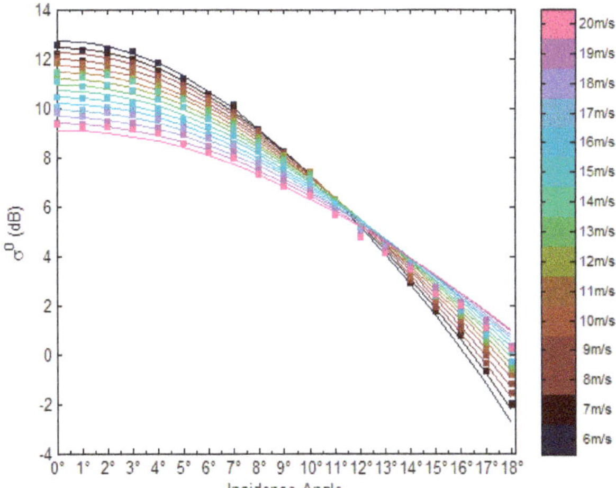


**Figure 4.** Comparison between collocated NRCS and simulated NRCS, showing NRCS dependence on incidence angle. The squared colored dots are the NRCS collocated from PR and SFMR. The lines are plotted according to the geometrical optics model. The different colors denote different wind speeds.

As shown in Figure 4, the NRCS decreases with increasing incidence angle and is generally consistent with the geometrical optics model predication. Small discrepancies appear at 20 m/s wind speed and incidence angles of 0° and 15°–18°. The geometrical optics model was derived from a limited number of wind speeds above 15 m/s, and may not be sufficiently accurate under high wind conditions. Alternatively, the discrepancy could arise from the different wind speed measurement accuracies of SFMR and buoys. Here, the small discrepancies can be neglected. The NRCS becomes independent of wind speed at approximately 12° incidence angle, slightly larger than reported 10° in previous studies [32].

Figure 5 shows the dependence of NRCS on wind speed, analyzed with the same dataset as Figure 4. The NRCSs destabilize at near-nadir angles (as evidenced from the STDs in Figure 2b), and the discrepancy increases at 18°. However, the differences between the measured NRCSs and those simulated with the geometrical optics model [25] are less than 1 dB. Therefore, the results are considered to be consistent (given the 1 dB uncertainty in PR) [24].

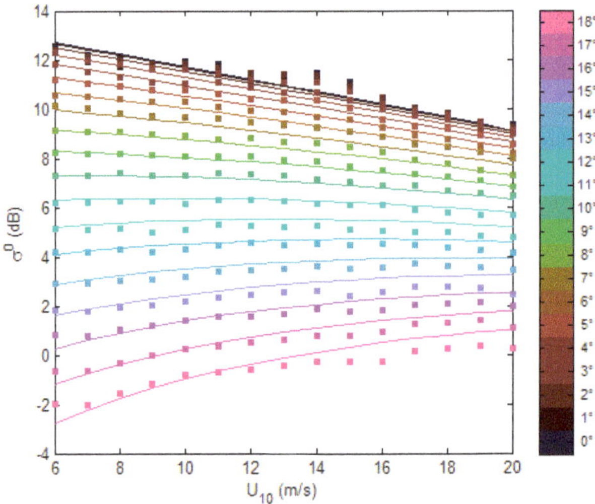


**Figure 5.** Comparison between collocated NRCS and simulated NRCS, showing NRCS dependence on wind speed. The squared colored dots are the NRCS collocated from PR and SFMR. The lines are plotted according to the geometrical optics model. The different colors denote different wind speeds.

In summary, the NRCS values collocated from PR in the present study display the expected dependences on incidence angle and wind speed in moderate wind conditions.

*4.2. Analysis at Extremely High Wind Speeds*

However, at extremely high wind speeds, the NRCS characteristics deviate from those in moderate wind conditions. To distinguish the NRCS characteristics between the two different wind conditions, we plot the NRCS dependencies on wind speeds and incidence angles, which are shown in Figures 6 and 7, respectively. The results are derived over the full wind condition that is, sea-surface wind speeds ranging from 6 to 70 m/s.

Since the rain rates in the data in Figures 6 and 7 are low (0–5 mm/h), the rainfall impacts on the NRCS can be ignored. The same figures were also plotted at very low rain rates of 0–2 mm/h (results were not shown). The NRCS trends were identical at the low rain rate, except the number of collocated pairs is reduced. Although the plots in Figure 6 appear chaotic, the colors clearly reveal that the wind speeds decrease with increasing incidence angles more dramatically for winds at 40–70 m/s than at 6–20 m/s. Figure 7 clarifies that at wind speeds exceeding 40 m/s, the NRCS characteristics distinctly differ from those of moderate wind conditions. In the 40–70 m/s range, the NRCS is disturbed and exhibits no clear dependence on wind speed. In extreme high wind conditions, the NRCS is generally very sensitive to low incidence angles and not sensitive to extreme wind speeds.

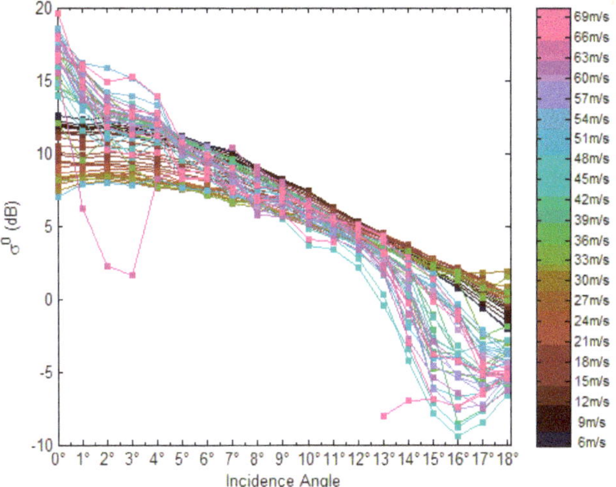


**Figure 6.** NRCS dependence on incidence angle at different wind speeds. The squared colored dots are the NRCS values collocated from PR and SFMR. The different colors denote different wind speeds.

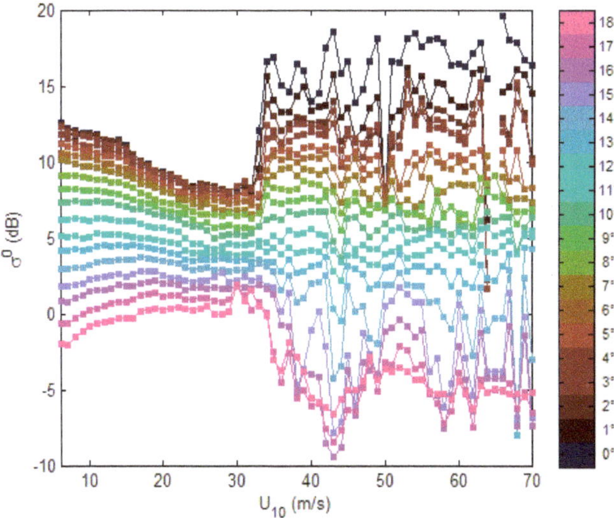


**Figure 7.** NRCS dependence on wind speed at different incidence angles. The squared colored dots are the NRCS values collocated from PR and SFMR. The different colors denote different incidence angles.

### *4.3. Results of Altimeter NRCS*

To determine whether the altimeter NRCS data yield similar results or not, we plot the dependence of NRCS acquired by Jason-1 and 2 on wind speed. The results are shown in Figure 8. Here, the wind speeds are also collocated from SFMR. The model of Quilfen [22] is also illustrated for comparison. The model coefficients were derived by orthogonal regression between the QuikSCAT and Jason-2 data at wind speeds between 18 and 30 m/s. The formula of the model mentioned above is given as follows:

$$U_{10} = 96.68 - 7.32 \cdot \left(\sigma^0 + offset\right) \quad for \ \sigma^0 < 10.7896 \tag{2}$$

In Figures 8 and 9, the temporal and spatial intervals decrease from 60 min to 30 min and from 100 km to 50 km respectively. As the wind speed increases, the number of collocated pairs also decreases. Moreover, both plots present the same tendency of NRCS dependence on wind speed. In general, our results are consistent with Quilfen (green line in the two figures) [22] in the wind speed range from 18 to 30 m/s. It is important to note that the NRCS becomes large at extremely high wind speeds, similarly to the NRCS values from PR, although the points are few and scattered at wind speeds above 40 m/s.

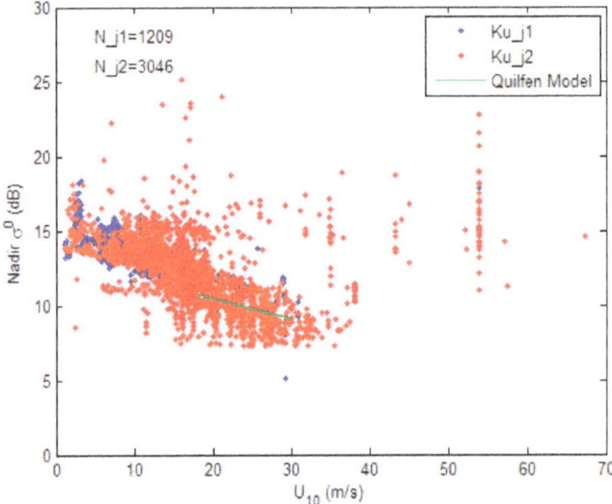


**Figure 8.** NRCS dependence on wind speed. Blue dots are from Jason-1 Ku-band NRCS while the red ones are Jason-2. The green line is plotted according to Quilfen Model (2011). Here the temporal and spatial distances are 60 min and 100 km, respectively. N_j1 and N_j2 show that the number of collocation points are 1209 and 3046, respectively.

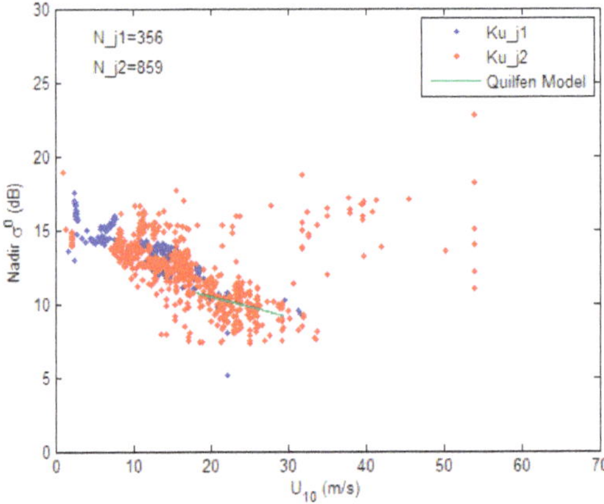


**Figure 9.** NRCS dependence on wind speed. Blue dots are from Jason-1 Ku-band NRCS while the red ones are Jason-2. The green line is plotted according to Quilfen Model (2011). Here the temporal and spatial distances are 30 min and 50 km, respectively. N_j1 and N_j2 show that the number of collocation points are 356 and 859, respectively.

## 5. Discussion

This study newly presents the NRCS characteristics at extremely high wind speeds. To further describe the phenomenon in detail, we assume that NRCS are constant in the wind speed range from 46 to 70 m/s for a given incidence angle. Because NRCS is not sensitive to extreme high winds, we average the NRCSs at different incidence angles and plot them versus incidence angle in Figure 10. Then the result is fitted to the following linear equation:

$$\sigma^0_{dB}(0) = 10 \cdot \log_{10}\left(\sigma^0\right) = a \cdot \theta + b \tag{3}$$

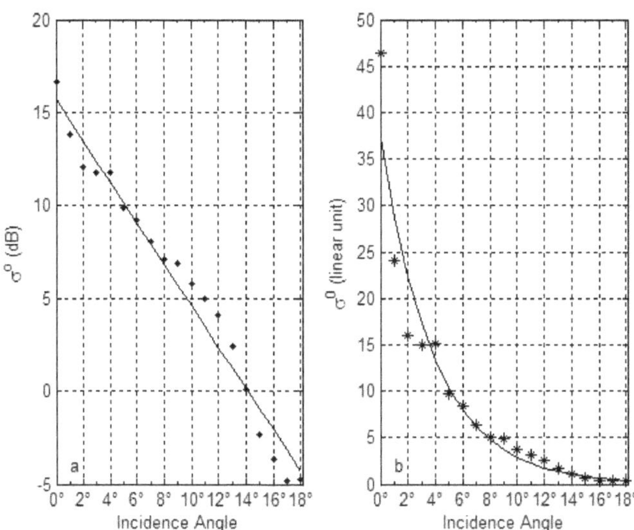


**Figure 10.** NRCS dependence on incidence angle under extreme wind conditions. (**Left**) in units of dB; (**Right**) in linear units.

The linear fitting yields $a = -1.1097$ and $b = 15.6995$. The NRCS and $\theta$ are expressed in dB and degrees, respectively. Near the nadir angle, the NRCS is very large and decreases drastically with small increments in incidence angle. Generally, the decrease in NRCS with increasing incidence angle obeys the simple Equation (3). This newly found phenomenon may be useful in the following research directions.

### 5.1. Validation of $C_d$

From the late 1960s through the 1990s, the drag coefficient $C_d$ was considered to be positively correlated with the sea surface wind speed. However, this universally accepted relationship is only confirmed for wind speeds under 26 m/s, because no high wind speed observations can be acquired especially in regions over the deep ocean [33].

In 2003, Powell et al. showed that $C_d$ decreases at $U_{10} > 33$ m/s. The team hypothetically attributed this reduction to sea foam, spray and bubbles resulting from breakage of steep wave faces, which forms a slip surface [34]. Donelan and his team [35] observed that the drag coefficient is saturated as sea surface wind speeds are above 33 m/s. They concluded that as the wind speed over the open ocean increases from gale to hurricane force values, continuous intense wave-breaking occurs and the crests are essentially blown away by the strong winds. This mechanism fills the air with sea spray and the surface with spume, altering its frictional and roughness characteristics [35].

Many subsequent papers have studied the nonlinearity and parametrization of $C_d$ in these conditions (e.g., [36–38]).

For moderate wind speeds as shown in Figure 4, the dependence of NRCS on incidence angle complies with quasi-specular backscattering theory. As shown in Figure 7, the decrease of NRCS with increasing incidence angles in the wind speed range from 40 to 70 m/s is more dramatic than in that from 6 to 20 m/s. The distinction indicates that the sea surface at extreme wind speed is relatively "smooth" to Ku-band electromagnetic microwaves, in contrast to at low and moderate wind speeds. That is, the NRCS is very sensitive to low incidence angles under extreme wind conditions and insensitive to the extreme wind speed. Due to this relative "smooth" sea surface, the $C_d$ under high wind conditions should be smaller than that of in moderate wind circumstances. Therefore, the phenomenon revealed in this paper validates the $C_d$ decreases at extremely high wind conditions, in contrast to moderate wind conditions. Hwang [39,40] derived an experimental formula for $C_d$ based on previous studies, but this formula is limited to wind speeds below 50 m/s. According to Figure 7, the sea surface remains "smooth" at wind speeds up to 70 m/s.

In previous studies, the $C_d$ levels off at 32–33 m/s; in others, saturation occurs at 22–23 m/s [33]. Figure 7 is consistent with $C_d$ saturation at 32–33 m/s; otherwise, the NRCS should begin dramatically varying at 22–23 m/s. Besides validating the remote sensing data, the new characteristics may potentially assist anticipated satellites missions in the future.

### 5.2. CFOSAT

The CFOSAT satellite is scheduled for launch in 2018. The payload instruments are Surface Waves Investigation and Monitoring (SWIM) and a rotating scatterometer. SWIM operates in the Ku-band, with VV polarization, at small incidence angles. At low wind speeds, the interface between the atmosphere and sea surface is well-clarified. Moreover, the VV-polarized NRCS is similar to the HH-polarized NRCS at small incidence angles for the quasi-specular scattering mechanism. Therefore, at high wind conditions, the NRCS acquired from VV polarization probably shares similar dependence on incidence angle to those acquired from HH.

In extreme wind conditions, the atmosphere ocean interface essentially disappears and the sea surface becomes relatively "smooth" for the Ku-band (or the NRCS will conform to the quasi-specular backscattering results, such as Equation (1)). Beneath the interface are the water balls surrounded by the atmosphere and beneath the water balls may be the ocean water fluctuation. CFOSAT SWIM detects the ocean wave spectra based on the tilted ocean surface modulation, that is, the variations of backscattering cross section of long waves are linearly proportional to the sea surface slopes:

$$\partial\sigma/\sigma = \alpha \cdot \partial\xi/\partial x \tag{4}$$

Here, $\sigma$ is backscattering cross section of long waves, $\alpha$ the modulation transfer function (related to the incidence angle), and $\partial\xi/\partial x$ the sea surface slope. After spectral estimation, the sea wave spectra can be acquired [41]. However, in extreme wind conditions the information of ocean water fluctuation at the bottom cannot be acquired by the Ku-band EM waves because the relatively "smooth" surface backscatters EM energy and thus no fluctuation signals can be used to retrieve the ocean water fluctuation. The dramatic decrement of NRCS with increasing incidence angle can flag the presence of a high sea state; that is, can qualitatively indicate the severity of the sea-wave spectra.

### 5.3. Altimeter

In the near future, wide swath ocean altimeters (WSOA), HY-2B etc. will be launched. They all operate at small incidence angles. Here, a preliminary NRCS dependence on high wind speed is shown, and more validation is needed at nadir incidence angles.

The potential impacts of the NRCS characteristics at extremely high wind speeds on altimetry may be the distorted waveforms reflected from the "smooth" sea surface. The waveforms will peak with very steep leading and sharp trailing edges, just like a delta function, which appears when the nadir direction surface is land occupied by ice. This phenomenon will lead to errors for estimates of the sea surface height or the wrong land flag. Thus, this influence cannot be neglected.

## 6. Conclusions

In this paper, in order to analyze the NRCS at low incidence angles at extreme wind condition, the wind speed from SFMR and NRCS from PR on TRMM or altimeters are collocated. After the data statistics are computed and a simple quality control criterion is specified, the NRCS dependencies on incidence angle from 0° to 18° and wind speed from 6 m/s to 70 m/s are shown.

The NRCS values at moderate and low wind speeds are compared with those from a geometrical optical model, which confirms the consistency among them. Figures 4 and 5 show that there are only small differences at wind speed 20 m/s and incidence angles 0°, 15°–18°. The optical model is developed by the NRCS from PR and wind speeds from NDBC. Taking into account the SFMR wind speed error in contrast to that from NDBC, the NRCS dependence is generally consistent with that of the optical model.

At extremely high wind speeds, the characteristics of the NRCS are shown in Figures 6 and 7. These results have not been reported previously in the literature. We have found that the NRCS increases drastically with increasing wind speeds from about 32 m/s to 40 m/s at near-nadir incidence angles and then becomes generally invariable, for higher winds. At about 11°–13°, the NRCS is nearly invariant for all kinds of wind conditions (wind speeds from 6 m/s to 70 m/s). When the incidence angles increase to about 18°, the saturation is shown to occur at around 30 m/s–32 m/s, as expected, which is the same as that of scatterometers. However, the NRCS at near 18° decreases drastically from 32 m/s to 40 m/s and then becomes nearly constant for higher winds. Generally, the NRCS values at extremely high wind speeds have large values at near-nadir incidence angels and then decrease drastically with increasing incidence angles. In Figures 8 and 9, the matchups from Jason-1 and 2 and SFMR are collected; the same NRCS characteristics at nadir incidence angles as found for PR can also confirm the results, although the number of points at extremely high wind speeds is too small to be statistically significant.

The NRCS dependence on incidence angles at extremely high wind speed is generally presented in Figure 10 and a simple formula is given by Equation (3). In physical terms, at extremely high wind speeds for Ku-band EM wave and HH polarization, the sea surface becomes very "smooth", just like a "mirror", and only this assumption can explain the characteristics of the NRCS values in Figure 10. This is consistent with the $C_d$ decrement at high wind speed, which has been discovered by many researchers with in-situ measurements.

The new discovery may have potential use for the future missions, such as CFOSAT, WSOA, etc. Because of the scarcity of the data at high wind speeds, the data points are not enough for validation at 0° incidence angle and the phenomenon can be verified in more detail by NRCS at nadir angles and medium angles.

**Acknowledgments:** This work was supported in part by the National Key Research and Development Program of China under Grant 2016YFC1401005 and 2016YFC1401001, in part by the National Science Foundation of China for Outstanding Young Scientist under Grant 41622604, in part by the National Science Foundation of China under Grant 41476158, in part by the Excellent Youth Science Foundation of Jiangsu Province under Grant BK20160090, in part by the Startup for Introducing Talent of NUIST, and the CFOSAT project. We would also like to acknowledge the Goddard Earth Sciences (GES) Data and Information Services Center (DISC) for the straightforward and rapid electronic access to the TRMM PR data, Jason-1 and 2 data from AVISO website and HRD, NOAA for the SFMR data. Canadian Space Agency program DUAP contributed to the support of W. Perrie.

**Author Contributions:** Xiuzhong Li carried out the research and prepared the manuscript under the guidance of Biao Zhang and Alexis Mouche. Yijun He and Will Perrie contributed to the preparation of the paper through editing.

**Conflicts of Interest:** The authors declare no conflict of interest.

## References


1. Dai, H.; Du, L.; Wang, Y.; Wang, Z. A Modified CFAR Algorithm Based on Object Proposals for Ship Target Detection in SAR Images. *IEEE Geosci. Remote Sens. Lett.* **2016**, *13*, 1925–1929. [CrossRef]
2. Martin, A.C.; Gommenginger, C.; Marquez, J.; Doody, S.; Navarro, V.; Buck, C. Wind-Wave induced velocity in ATI SAR Ocean Surface Currents: First experimental evidence from an airborne campaign. *J. Geophys. Res. Oceans* **2016**, *121*, 1640–1653. [CrossRef]
3. Caudal, G.; Hauser, D.; Valentin, R.; Gac, C.L. KuROS: A New Airborne Ku-Band Doppler Radar for Observation of Surfaces. *J. Atmos. Ocean. Technol.* **2014**, *31*, 2223–2245. [CrossRef]
4. Stoffelen, A.; Anderson, D. Scatterometer data interpretation: Estimation and validation of the transfer function CMOD4. *J. Geophys. Res. Oceans* **1997**, *102*, 7767–7786. [CrossRef]
5. Quilfen, Y.; Chapron, B.; Elfouhaily, T.; Katsaros, K.; Tournadre, J. Observation of tropical cyclones by high-resolution scatterometry. *J. Geophys. Res. Oceans* **1998**, *103*, 7767–7786. [CrossRef]
6. Jones, W.L.; Cardone, V.; Pierson, W.; Zec, J.; Rice, L.; Cox, A.; Sylvester, W. NSCAT high-resolution surface wind measurements in Typhoon Violet. *J. Geophys. Res. Oceans* **1999**, *104*, 11247–11260. [CrossRef]
7. Donnelly, W.J.; Carswell, J.R.; McIntosh, R.E.; Chang, P.S.; Wilkerson, J.; Marks, F.; Black, P.G. Revised ocean backscatter models at C and Ku band under high-wind conditions. *J. Geophys. Res. Oceans* **1999**, *104*, 11485–11497. [CrossRef]
8. Carswell, J.R.; Knapp, E.J.; Chang, P.S.; Black, P.D.; Marks, F.D. Limitations of scatterometry high wind speed retrieval. In Proceedings of the IEEE 2000 International Geoscience and Remote Sensing Symposium (IGARSS), Honolulu, HI, USA, 24–28 July 2000.
9. Fernandez, D.E.; Carswell, J.R.; Frasier, S.; Chang, P.S.; Black, P.G.; Marks, F.D. Dual-polarized C- and Ku-band ocean backscatter response to hurricane-force winds. *J. Geophys. Res. Oceans* **2006**. [CrossRef]
10. Zhang, B.; Perrie, W. Cross-polarized synthetic aperture radar: A new potential measurement technique for hurricanes. *Bull. Am. Meteorol. Soc.* **2012**, *93*, 531–541. [CrossRef]
11. Zhang, B.; Perrie, W.; Zhang, J.A.; Uhlhorn, E.W.; He, Y. High-resolution hurricane vector winds from C-band dual-polarization SAR observations. *J. Atmos. Ocean. Technol.* **2014**, *31*, 272–286. [CrossRef]
12. Horstmann, J.; Wackerman, C.; Falchetti, S.; Maresca, S. Tropical cyclone winds retrieved from synthetic aperture radar. *Oceanography* **2013**, *26*, 46–57. [CrossRef]
13. Fois, F.; Hoogeboom, P.; Le Chevalier, F.; Stoffelen, A. Future Ocean Scatterometry: On the Use of Cross-Polar Scattering to Observe Very High Winds. *IEEE Trans. Geosci. Remote Sens.* **2015**, *53*, 5009–5020. [CrossRef]
14. Sapp, J.; Chang, P.; Jelenak, Z.; Frasier, S.; Hartley, T. Cross-polarized C-band sea-surface NRCS observations in extreme winds. In Proceedings of the 2016 IEEE International Geoscience and Remote Sensing Symposium (IGARSS), Beijing, China, 10–15 July 2016.
15. Freilich, M.H.; Vanhoff, B.A. The Relationship between Winds, Surface Roughness, Radar Backscatter at Low Incidence Angles from TRMM Precipitation Radar Measurements. *J. Atmos. Ocean. Technol.* **2003**, *20*, 549–562. [CrossRef]
16. Chu, X.; He, Y.; Karaev, V.Y. Relationships Between Ku-Band Radar Backscatter and Integrated Wind and Wave Parameters at Low Incidence Angles. *IEEE Trans. Geosci. Remote Sens.* **2012**, *50*, 4599–4609. [CrossRef]
17. Chu, X.; He, Y.; Chen, G. Asymmetry and anisotropy of microwave backscatter at low incidence angles. *IEEE Trans. Geosci. Remote Sens.* **2012**, *50*, 4014–4024. [CrossRef]
18. Boisot, O.; Nouguier, F.; Chapron, B.; Guérin, C.A. The GO4 model in near-nadir microwave scattering from the sea surface. *IEEE Trans. Geosci. Remote Sens.* **2015**, *53*, 5889–5900. [CrossRef]
19. Young, I.R. An estimate of the Geosat altimeter wind speed algorithm at high wind speeds. *J. Geophys. Res. Oceans* **1993**, *98*, 20275–20285. [CrossRef]
20. Quilfen, Y.; Chapron, B.; Tournadre, J. Altimeter dualfrequency observations of surface winds, waves, and rain rate in tropical cyclone Isabel. *J. Geophys. Res. Oceans* **2006**, *111*, C01004. [CrossRef]
21. Quilfen, Y.; Chapron, B.; Tournadre, J. Satellite Microwave Surface Observations in Tropical Cyclones. *Mon. Weather Rev.* **2010**, *138*, 421–437. [CrossRef]
22. Quilfen, Y.; Vandemark, D.; Chapron, B.; Feng, H.; Sienkiewicz, J. Estimating Gale to Hurricane Force Winds Using the Satellite Altimeter. *J. Atmos. Ocean. Technol.* **2011**, *28*, 453–458. [CrossRef]
23. Quartly, G.; Guymer, T.; Srokosz, M. Back to basics: Measuring rainfall at sea: Part 2-Space-borne sensors. *Weather* **2002**, *57*, 363–366. [CrossRef]

24. Liao, L.; Meneghini, R. Changes in the TRMM Version-5 and Version-6 Precipitation Radar Products Due to Orbit Boost. *J. Meteorol. Soc. Jpn.* **2009**, *87*, 93–107. [CrossRef]
25. Li, X.; He, Y.; Zhang, B.; Ge, J.; Chu, X. A geometrical optics model based on the non-Gaussian probability density distribution of sea surface slopes for wind speed retrieval at low incidence angles. *Int. J. Remote Sens.* **2016**, *37*, 537–550. [CrossRef]
26. Uhlhorn, E.W.; Black, P.G. Verification of remotely sensed sea surface winds in hurricanes. *J. Atmos. Ocean. Technol.* **2003**, *20*, 99–116. [CrossRef]
27. Knapp, E.; Carswell, J.; Swift, C. A dual polarization multi-frequency microwave radiometer. Paper Presented at the 2000 IEEE International Geoscience and Remote Sensing Symposium, New York, NY, USA, 24–28 July 2000.
28. Uhlhorn, E.W.; Black, P.G.; Franklin, J.L.; Goodberlet, M.; Carswell, J.; Goldstein, A.S. Hurricane surface wind measurements from an operational stepped frequency microwave radiometer. *Mon. Weather Rev.* **2007**, *135*, 3070–3085. [CrossRef]
29. Bradley, W.K.; Uhlhorn, E.W. Improved stepped frequency microwave radiometer tropical cyclone surface winds in heavy precipitation. *J. Atmos. Ocean. Technol.* **2014**, *31*, 2392–2408.
30. Barrick, D.E. Rough surface scattering based on specular point theory. *IEEE Trans. Antennas Propag.* **1968**, *16*, 449–454. [CrossRef]
31. Bringer, A.; Guerin, C.; Chapron, B.; Mouche, A.A. Peakedness effects in near-nadir radar observations of the sea surface. *IEEE Trans. Geosci. Remote Sens.* **2012**, *50*, 3293–3301. [CrossRef]
32. Hesany, V.; Plant, W.J.; Keller, W.C. The normalized radar cross section of the sea at 10 incidence. *IEEE Trans. Geosci. Remote Sens.* **2000**, *38*, 64–72. [CrossRef]
33. Bryant, K.M.; Akbar, M. An Exploration of Wind Stress Calculation Techniques in Hurricane Storm Surge Modeling. *J. Mar. Sci. Eng.* **2016**, *4*, 58. [CrossRef]
34. Powell, M.D.; Vickery, P.J.; Reinhold, T.A. Reduced drag coefficient for high wind speeds in tropical cyclones. *Nature* **2003**, *422*, 279–283. [CrossRef] [PubMed]
35. Donelan, M.A.; Haus, B.K.; Reul, N.; Plant, W.J.; Stiassnie, M.; Graber, H.C.; Brown, O.B.; Saltzman, E.S. On the limiting aerodynamic roughness of the ocean in very strong winds. *Geophys. Res. Lett.* **2004**. [CrossRef]
36. Moon, I.J.; Ginis, I.; Hara, T.; Thomas, B. A physics-based parameterization of air–sea momentum flux at high wind speeds and its impact on hurricane intensity predictions. *Mon. Weather Rev.* **2007**, *135*, 2869–2878. [CrossRef]
37. Zhao, W.; Liu, Z.; Dai, C.; Song, G.; Lv, Q. Typhoon air-sea drag coefficient in coastal regions. *J. Geophys. Res. Oceans* **2015**, *120*, 716–727. [CrossRef]
38. Bi, X.; Gao, Z.; Liu, Y.; Liu, F.; Song, Q.; Huang, J.; Huang, H.; Mao, W.; Liu, C. Observed drag coefficients in high winds in the near offshore of the South China Sea. *J. Geophys. Res. Atmos.* **2015**, *120*, 6444–6459. [CrossRef]
39. Hwang, P.A. A note on the ocean surface roughness spectrum. *J. Atmos. Ocean. Technol.* **2011**, *28*, 436–443. [CrossRef]
40. Hwang, P.A. Foam and roughness effects on passive microwave remote sensing of the ocean. *IEEE Trans. Geosci. Remote Sens.* **2012**, *50*, 2978–2985. [CrossRef]
41. Hauser, D.; Soussi, E.; Thouvenot, E.; Rey, L. SWIMSAT: A real-aperture radar to measure directional spectra of ocean waves from space—Main characteristics and performance simulation. *J. Atmos. Ocean. Technol.* **2001**, *18*, 421–437. [CrossRef]

*Article*

# Technical Evaluation of Sentinel-1 IW Mode Cross-Pol Radar Backscattering from the Ocean Surface in Moderate Wind Condition

**Lanqing Huang [1], Bin Liu [1,*], Xiaofeng Li [2], Zenghui Zhang [1] and Wenxian Yu [1]**

1 Shanghai Key Laboratory of Intelligent Sensing and Recognition, Shanghai Jiao Tong University, Shanghai 200240, China; huanglanqing@sjtu.edu.cn (L.H.); zenghui.zhang@sjtu.edu.cn (Z.Z.); wxyu@sjtu.edu.cn (W.Y.)

2 Global Science and Technology, National Oceanic and Atmospheric Administration (NOAA)/NOAA Satellite and Information Service, College Park, MD 20740, USA; xiaofeng.li@noaa.gov

* Correspondence: bliu.rsti@sjtu.edu.cn; Tel.: +86-186-0178-9768

Academic Editors: Xiaofeng Yang, Ferdinando Nunziata and Alexis Mouche
Received: 30 June 2017; Accepted: 16 August 2017; Published: 17 August 2017

**Abstract:** The Sentinel-1 synthetic aperture radar (SAR) allows sufficient resources for cross-pol wind speed retrievals over the ocean. In this paper, we present technical evaluation on wind retrieval from both Sentinel-1A and Sentinel-1B IW cross-pol images. Algorithms are based on the existing theoretical and empirical ones derived from the RADARSAT-2 cross-pol data. First, to better understand the Sentinel-1 observed normalized radar cross section (NRCS) values under various environmental conditions, we constructed a dataset that integrates SAR images with wind field information from scatterometer measurements. There are 11,883 matchup data in the experimental dataset. We then calculated the systemic noise floor of Sentinel-1 IW mode, and presented its unique noise characteristics among different sub-bands. Based on the calculated NESZ measurements, the noise is removed for all matchup data. Empirical relationships among the noise free NRCS $\sigma^0_{VH}$, wind speed, wind direction, and radar incidence angle are analyzed for each sub-band, and a piecewise model is proposed. We showed that a larger correlation coefficient, $r$, is achieved by including both wind direction and incidence terms in the model. Validation against scatterometer measurements showed the suitability of the proposed model.

**Keywords:** cross-pol; Sentinel-1; radar backscattering; wind retrieval

---

## 1. Introduction

Synthetic aperture radar (SAR) images provide deep knowledge of the characteristics of wind fields over both the ocean and littoral zones with high spatial resolution. Due to the complex electromagnetic scattering mechanism of SAR, empirical approaches of wind field retrievals based on geophysical model functions (GMF), which describe the dependence of the normalized radar cross section (NRCS) with respect to wind vectors and incidence angles, are widely investigated. For C-band co-polarization (co-pol) SAR images, along with a polarization ratio model for HH polarizations, many GMF models are developed [1–6]. However, the NRCS of co-pol exhibits data saturated when wind speed exceeds about 16 m/s for incidence angle under 35° [7]. Otherwise, the cross-polarization (cross-pol) shows that the signal increases with wind speed and does not saturate even at very high speeds. Therefore, cross-pol data has been widely used for high wind speed retrievals [7–11].

Some related assessments and evaluations of cross-pol images have been done to RADARSAT-2 quad polarization (quad-pol) which have lower noise (about −36 dB). By collocating quad-pol RADARSAT-2 SAR images and buoy measurements, Vachon and Wolfe in 2011 have shown that the cross-pol backscattering is independent of incidence angle and wind direction, and developed a monotonic linear relationship between cross-pol backscattering and wind speed [12]. In [7], Hwang et al. have shown that the linearly increasing relationship between the NRCS $\sigma^0_{\mathrm{VH}}$ in moderate wind conditions (about <7 m/s) gradually transits to cubic in high wind conditions (about >10m/s). Bergeron et al. [13] compared the wind retrieval performances of two models proposed by Vachon [12] and Hwang [7] for wind speed retrieval in cross-pol. These researches describe the relationship between the NRCS $\sigma^0_{\mathrm{VH}}$ and wind speeds assuming that cross-pol is independent on the variations of incidence angles. In [14], the incidence angle and wind direction dependences of cross-pol radar backscattering are analyzed and theoretically explained.

Some researches concentrated on wind retrieval using C-band high noise data in dual polarization (dual-pol) such as RADARSAT-2 ScanSAR mode, Sentinel-1 and planned RADARSAT Constellation. Differ from RADARSAT-2 quad-pol data which have lower noise (about −36 dB), the wide swath RADARSAT-2 ScanSAR have higher noise (about −30 dB) [14]. In [15], Shen et al. indicate that dual-pol and quad-pol exhibit different relationships with wind speed, therefore a distinct model particularly for high noise dual-pol data wind retrieval is required. For RADARSAT-2 ScanSAR mode images, existing wind retrieval models only consider wind speed dependence [15,16]. In [16], the NRCS $\sigma^0_{\mathrm{VH}}$ is found to be a linear function of wind speed, thus the C-2POD model was derived for hurricane wind retrieval. A two-piecewise linear GMF model was developed to quantify the relationship between the NRCS $\sigma^0_{\mathrm{VH}}$ and wind speeds [15]. In [17], the wind direction dependence of NRCS $\sigma^0_{\mathrm{VH}}$ is analyzed. Zhang et al. [11] proposed a hybrid backscattering model to carry out theoretically analyze the relationship among both VH and VV polarized NRCS, wind speed and incidence angle. Using this hybrid model, a hurricane wind retrieval model named C-3PO was developed, by including dependence on incidence angle.

With the free and open data policy of Sentinel-1 satellites, VH dual polarization SAR images have become publicly available, allowing sufficient resources for cross-pol wind speed retrievals. In [18], based on Sentinel-1 VV polarization images, the spatial characteristics of the SAR-derived winds are presented, and the detailed descriptions of the wind fields provided by SAR are analyzed. Monaldo et al. [19] evaluate the performance of Sentinel-1A wind speed retrievals using GMFs (CMOD4, CMOD_IF2, CMOD5,CMOD5.N), and conclude good agreement of Advanced SCATterometer (ASCAT) wind speeds and Sentinel-1 VV and HH measurements. However, for Sentinel-1 cross-pol SAR images, their radar backscattering from the ocean surface and technical performance for wind speed retrievals are yet to be evaluated.

The planned RADARSAT Constellation Mission (RCM), which will be launched in 2018, is designed primarily to be dedicated to regular monitoring with various modes and resolutions [20]. Among the main operational modes, the Medium Resolution 50 m mode was designed for general-purpose wide area surveillance [21,22]. It is 4 looks and a 350 km imaging swath, and is comparable to the RADARSAT-2 ScanSAR mode, thus it is promising to be employed in marine wind application [21,22]. We expect that there will be more available C-band high noise data such as Sentinel-1 and RCM in the coming years. Therefore, wind retrieval model designed for C-band high noise data is required.

In this paper, with the rapidly growing availability of Sentinel-1 data, we revisit the empirical relationships between cross-pol radar backscattering $\sigma^0_{\mathrm{VH}}$ and wind speeds with attention to evaluate contribution from incidence angle and wind direction terms. A model for moderate wind speed retrieval is proposed, by including the dependences on wind speed, incidence angle, and wind direction. The scopes of this paper are to give a technical evaluation of Sentinel-1 cross-pol images based on the theoretical and empirical analyses derived from RADARSAT-2 data, and provide a wind retrieval model based on Sentinel-1 cross-pol images.

First, in order to better understand the Sentinel-1 cross-pol NRCS under various environmental conditions, we construct a dataset dedicated to Sentinel-1A and 1B SAR ocean environment by integrating SAR images with wind vectors from ASCAT scatterometer. The matchup dataset contains data from 90 VH dual-pol SAR images acquired in 2016. Next, all the matchup data are integrated with precipitation estimates from the Tropical Rainfall Measuring Mission (TRMM). 11,883 cross-pol data with 0 mm/h are selected from the matchup dataset. We calculate the systemic noise floor of Sentinel-1 interferometric mode (IW) mode, and point out its unique noise characteristics among different sub-bands. The noise is removed for the matchup data. Empirical relationships among the noise removal NRCS, wind speed, wind direction, and radar incidence angle are analyzed for each sub-band. The lowest boundary wind speed for retrievals is specified. Finally, a piecewise model for wind retrieval is proposed. We showed that a larger correlation coefficient, $r$, is achieved by including wind speed, wind direction and incidence terms in the model. Validation against scatterometer measurements showed the suitability of the proposed model.

The remainder of this paper is organized as follows. Section 2 introduces the construction of the matchup dataset. The relationship between NRCS and wind speeds with attention to evaluate contribution from incidence angle and wind direction terms are analyzed in Section 3. In Section 4 an empirical model is developed and validated. Discussion of model is given in Section 5. The conclusions are drawn and the future work is introduced in Section 6.

## 2. The Matchup Dataset

In order to better understand the Sentinel-1 NRCS under various environmental conditions, a dataset which integrates SAR images with wind information from ASCAT scatterometer [23], is constructed. The construction and organization of the dataset are summarized as follows.

### 2.1. Data Collection

#### 2.1.1. Sentinel-1A and 1B

Sentinel-1 satellites, which are designed by the European Space Agency (ESA), operate C-band imaging in four exclusive imaging modes with different resolutions (down to 5 m) and coverage (up to 400 km) [24]. Sentinel-1 satellites achieve short revisit times (6 days) and rapid data delivery. Besides, Sentinel-1 are the first satellites exploiting the Terrain Observation with Progressive ScanSAR (TOPSAR) technique for the IW mode. The IW mode, with VV and VH dual polarization, is the default Sentinel-1 acquisition mode [25]. Although several other acquisition modes for specific aims are available: the Stripmap Mode (SM) which only be used on request for emergency management, the Extra-wide Swath Mode (EW) which is primarily applied to sea-ice monitoring over high latitude areas, and the Wave Mode (WV) which is designed for ocean wind field and swell spectra, these modes are usually only acquired for some specific areas with limited quantity. Therefore, in order to build up a consistent long-term dataset with various coverage, we focus on the IW mode that almost covers global littoral zones. Both Sentinel-1A and Sentinel-1B products are freely and opening available from the ESA's Sentinels Scientific Data Hub [26]. Our experiments concentrate on the Ground Range Detected (GRD) products. The details of the GRD products are listed in Table 1.

**Table 1.** Parameters for Sentinel-1 IW GRD products [25].

| Product Type | Resolution (rg × az) (m) | Pixel Spacing (rg × az) (m) | Swath Width (km) | Looks (rg × az) | Equivalent Number of Independent Looks |
|---|---|---|---|---|---|
| GRD | 20 × 22 | 10 × 10 | 250 | 5 × 1 | 4.9 |

#### 2.1.2. ASCAT

The ASCAT is one of the instruments carried on the Meteorological Operational (Metop) satellites launched by ESA. The ASCAT wind product is produced by the European Organization for the Exploitation of Meteorological Satellites (EUMETSAT) Ocean and Sea Ice Satellite Application Facility (OSI SAF) and provided through the Royal Netherlands Meteorological Institute (KNMI) [23].

Here, we focus on the ASCAT on Metop-A and B Level 2 coastal (ASCAT-L2-coastal) dataset, which contains operational coastal ocean surface wind vector retrieval products of global coverage, with 1800 km swath width. The ASCAT-L2-coastal wind products are at 25 km resolution and 12.5 km cell spacing, with 1.7 h (minimum) or 5 days (maximum) temporal repeat [23]. The ASCAT-L2-coastal wind products used in the matchup dataset are provided by Physical Oceanography Distributed Active Archive Center (PO.DAAC), Jet Propulsion Laboratory [27].

### 2.2. Construction of the Dataset

To construct the matchup dataset, Sentinel-1A and 1B SAR products, the ASCAT-L2-coastal products are matched together within a 2 h window. Specifically, the data match-up processing of the matchup dataset consists of the following steps:

1. *Sentinel-1 data preprocessing.*
   The original Sentinel-1A and 1B SAR products [26] are downloaded. In the preprocessing, we employ SNAP 4.0 [28] to perform the Radiometric Calibration.
2. *ASCAT data preprocessing.*
   The ASCAT-L2-coastal products [27] are downloaded. The products are in the NetCDF format. The information of wind vector, geographical position, and acquisition time is extracted by MATLAB programming.
3. *Integration of the Sentinel-1 and the ASCAT-L2-coastal products.*
   For each SAR image, determine whether the corresponding ASCAT-L2-coastal product is acquired at the same zone within 2 h window whose center is the SAR acquisition time. If yes, the latitude and longitude coordinates of all the Sentinel-ASCAT match-up points are recorded.
4. *Selection of the matchup points with* 0mm/h *precipitation.*
   According to the SAR acquisition time, the corresponding TRMM-3B42 product within 3 h temporal difference is downloaded [29]. Because of the global 0.25° × 0.25° (lat./lon) -averaged of TRMM-3B42 product [30], all the Sentinel-ASCAT matchup points are integrated with precipitation information. In order to avoid the effect of rainfall on the NRCS $\sigma^0_{VH}$, only the 0mm/h matchup points are selected in the dataset and utilized in further analyses.
5. *Post-processing.*
   The SAR data integrated with wind information are generated. Besides, a XML file is generated for the convenience of retrieving the integrated information of each matchup data.

Following the steps above, the experimental dataset was constructed.

### 2.3. Noise Removal

Since dual-pol measurements may suffer crosstalk among channels and thus perform lower signal-to-noise ratio (SNR), the noise removal for Sentinel-1 IW dual-pol images is necessary [12,15].

For the Sentinel-1 IW mode products, the nominal instrumental noise floors (i.e., the noise equivalent sigma naught (NESZ)) are calculated according to the formula provided in the Sentinel-1 product specification [31] and the metadata XML file contained in each product. The NESZ measures of IW mode VH polarization images with respect to incidence angles are plotted in Figure 1.

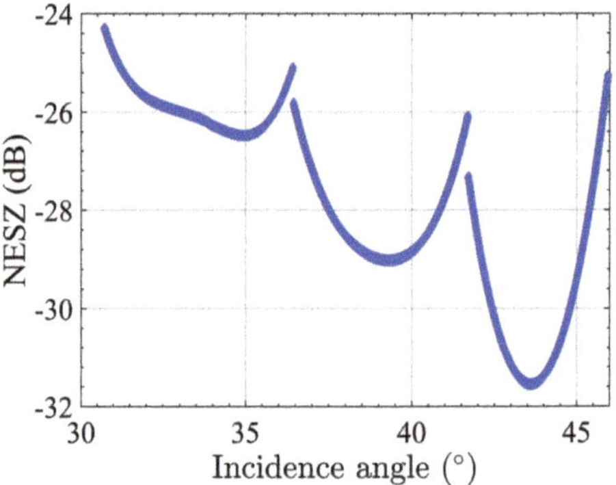


**Figure 1.** NESZ measures for IW mode VH images.

The noise removal is processed in two steps.

1. Based on the calculated NESZ measures, data do not satisfy Equation (1) are removed.

$$\frac{N_{\mathrm{obs}_{\mathrm{dB}}}}{N_{\mathrm{noise}_{\mathrm{dB}}}} > 0.6\,\mathrm{dB} \tag{1}$$

   where $N_{\mathrm{obs}_{\mathrm{dB}}}$ denotes the NRCS in dB, and $N_{\mathrm{noise}_{\mathrm{dB}}}$ is the measured NESZ in dB. Empirical threshold is set to 0.6 dB to guarantee the noise free signal is above −40 dB, which is applicable for cross-pol wind analyses illustrated in [17]. Please note that, we do not apply higher cut-off level since we aimed at finding the minimum cross-pol NRCS for wind retrieval. Thus, some relatively low SNR data are reserved for further analyses.
2. NRCS is composed of signal and noise which is particularly high for low SNR cross-pol observations. We attempt to subtract the noise component based on the calculated NESZ measures according to Equations (2)–(5) [15].

$$N_{\mathrm{obs}_{\mathrm{linear}}} = 10^{N_{\mathrm{obs}_{\mathrm{dB}}}/10} \tag{2}$$

$$N_{\mathrm{noise}_{\mathrm{linear}}} = 10^{N_{\mathrm{noise}_{\mathrm{dB}}}/10} \tag{3}$$

$$N_{\mathrm{sig}_{\mathrm{linear}}} = N_{\mathrm{obs}_{\mathrm{linear}}} - N_{\mathrm{noise}_{\mathrm{linear}}} \tag{4}$$

$$N_{\mathrm{sig}_{\mathrm{dB}}} = 10 \times \log_{10} N_{\mathrm{sig}_{\mathrm{linear}}} \tag{5}$$

   where $N_{\mathrm{obs}_{\mathrm{linear}}}$, $N_{\mathrm{noise}_{\mathrm{linear}}}$, and $N_{\mathrm{sig}_{\mathrm{linear}}}$ denote the NRCS, the NESZ measures, and the noise free signal NRCS in linear scale. $N_{\mathrm{sig}_{\mathrm{dB}}}$ denotes the noise free signal NRCS in dB.

Based on the introduced steps of noise removal, an example of noise removal for Sentinel-IW GRD product is illustrated in Figure 2.

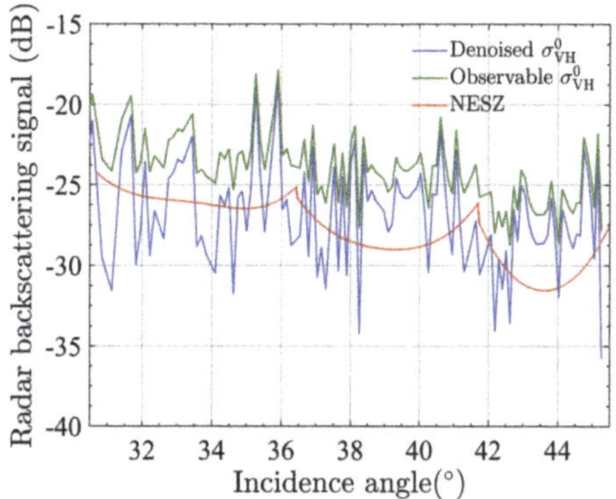


**Figure 2.** NRCS in radar range direction (blue) with and (green) without noise removal.

### 2.4. Experimental Data

The 11,883 matchup data (noise removal) covering 90 Sentinel-1A and 1B VH dual-pol GRD images with 0 (mm/h) precipitation. We divided the whole data into training and testing sets in two steps. First, the mathchup data are randomly divided into 85% training set and 15% testing set. Then in order to guarantee that both training and testing sets cover full ranges of incidence angles, wind directions and wind speeds, we manually adjust a small number of samples in the training and testing sets. After the two steps, the number of samples in training and testing sets are 10,258 and 1625, respectively. The distributions of wind speeds, wind directions, and incidence angles are shown in Figures 3–5, respectively. The spatial distance error of the matchup data are shown in Figure 6. Here the spatial distance error is defined as the distance between the geographic position of the matchup ASCAT and the center of the corresponding SAR image.

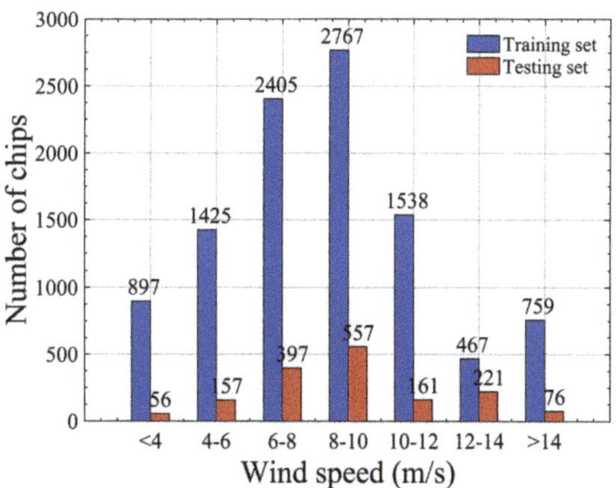


**Figure 3.** Wind speed histogram of the matchup dataset.

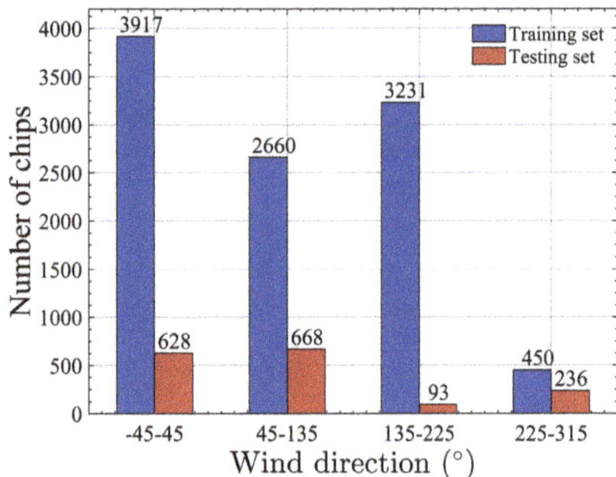


**Figure 4.** Wind direction histogram of the matchup dataset.

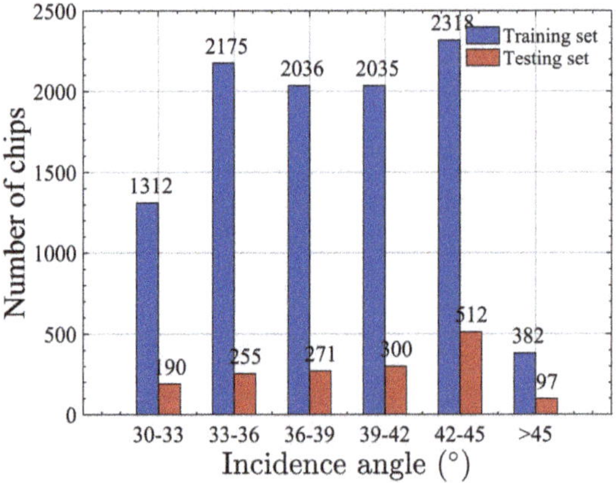


**Figure 5.** Incidence angle histogram of the matchup dataset.

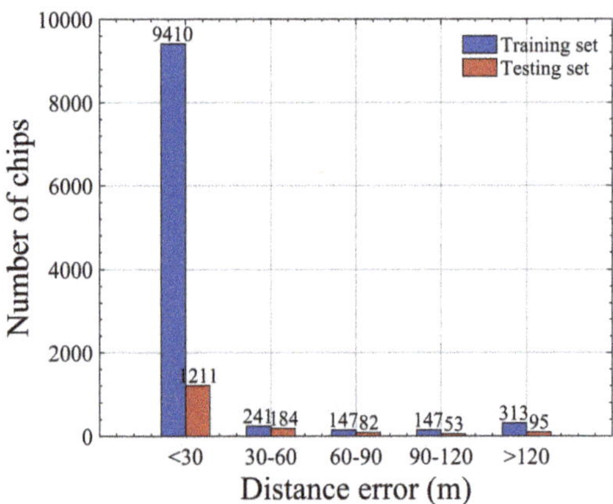


**Figure 6.** Spatial distance error histogram of the matchup dataset.

## 3. Experiments and Analyses

### *3.1. Relationship between NRCS and Incidence Angles*

For the 12,058 training data and 1625 testing data, each SAR sub-image is integrated with environmental information, with the size of 50 × 50 pixels (500 m resolution). The averaged value of the NRCS of SAR sub-image is calculated, denoted as $\sigma^0_{\mathrm{VH}}$. 500 m resolution is applicable and widely employed for SAR wind analyses [4,5,19]. Both Figures 7 and 8 show the empirical relationship between the NRCS $\sigma^0_{\mathrm{VH}}$ and incidence angle based on the 12,058 training data. In Figure 7, different colors represent the different values of wind speeds. Here, the wind direction is defined as the angle that lies between the true wind direction and the radar look direction. In addition, in Figure 8, different colors represent the different values of wind directions.

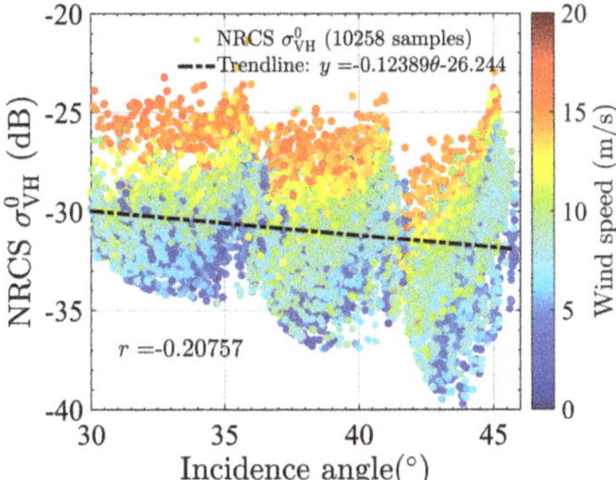


**Figure 7.** Relationship between NRCS and incidence angle (different colors represent different wind speeds).

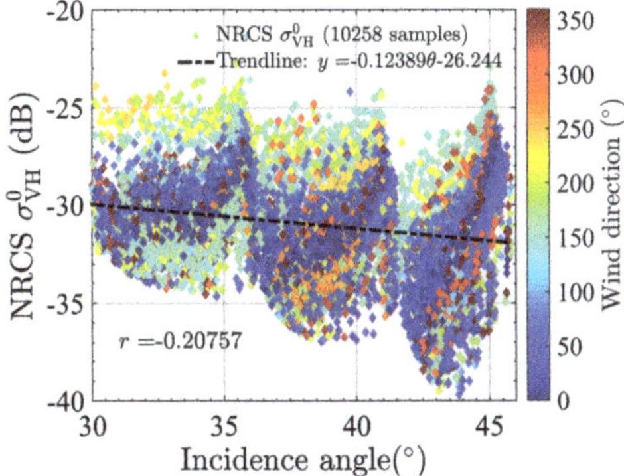


**Figure 8.** Relationship between NRCS and incidence angle (different colors represent different wind directions).

As illustrated in Figures 7 and 8, because of the low SNR in cross-pol images, the radar backscattering is interfered with the instrumental noise. As such, the NRCS $\sigma^0_{\mathrm{VH}}$ is fluctuated and negatively correlated to the incidence angles. The maximum wind speed in this dataset is 17.9 m/s.

From Figure 7, under every bin of incidence angles, the colors of data which represent different values of wind speeds are distinguishable. Known that the radar backscattering can reflect the ocean clutter more accurately when exceeding NESZ, from Figures 1 and 7, it can be seen that the NRCS $\sigma^0_{\mathrm{VH}}$

exceeds NESZ when the wind speed is roughly in the interval of about 8–10 m/s. Moreover, under every bin of incidence angles, the colors of data are clearly distinguishable when wind speeds exceed the interval of 8–10 m/s. While, the colors of the data are mixed-up when speed is below the interval of 8–10 m/s, because in this condition, the radar backscattering of cross-pol are too low to reflect the ocean clutter. The accurate value of this wind speed threshold will be specified and discussed in Section 3.2.

From Figure 8, under every bin of incidence angles, the colors of data which represent different values of wind directions are scattered and mixed-up. However, we could not observe the wind direction dependence directly from this plot, since the wind direction dependence is relatively small for cross-pol [14,17]. The wind direction dependence will be quantitatively analyzed in Section 3.3.

### *3.2. Relationship between NRCS and Wind Speeds*

In this section, we further analyzed the relationship between the NRCS $\sigma^0_{VH}$ and wind speeds. Due to the unique TOPSAR technique for the IW mode, three sub-swaths are captured, denoted as IW1-band (with incidence angles roughly ranging from 30° to 36°), IW2-band (with incidence angles roughly ranging from 36° to 41°), and IW3-band (with incidence angles roughly ranging from 41° to 46°), respectively. Therefore, we categorized the 12,058 data according to their bands: IW1-band, IW2-band, and IW3-band.

#### 3.2.1. IW1-Band

The relationship between the NRCS $\sigma^0_{VH}$ and wind speeds for IW1-band is illustrated in Figure 9. For the IW1-band, the range of incidence angles is 30–36°. Clearly, NRCS $\sigma^0_{VH}$ monotonically increase with wind speed. In the three regions (IW1-G1, IW1-G2 and IW1-G3) separated by dotted line representing 8 m/s and 12.3 m/s, the slope is constant and increases as wind speed increases. To quantify this relationship, a three piecewise-linear fitting is carried out, and the results of fitting are illustrated in Table 2.

For the IW1-G1 (wind speed is lower than 8 m/s), the NRCS $\sigma^0_{VH}$ are scattered with large variation. This variation arises because the radar return signals are low, the NRCS $\sigma^0_{VH}$ can not reflect the backscattering of ocean clutter. In addition, from Table 2, the slope of fitting is 0.13.

For the IW1-G2 (wind speed is between 8 m/s and 12.3 m/s), the variation of the $\sigma^0_{VH}$ decreases obviously. This indicates that when wind speed is higher than 8 m/s, for the Sentinel-1 cross-pol IW1 products, the radar backscattering is sensitive enough to reflect ocean clutter signatures when wind speed exceeds 8 m/s, and thus the wind speed retrieval from cross-pol observations is valid. This threshold is accordance with the analyses in the last paragraph of Section 3.1, in which we estimate the threshold to be 8–10 m/s based on results Figures 1 and 7. From Table 2, the slope of fitting is 0.46.

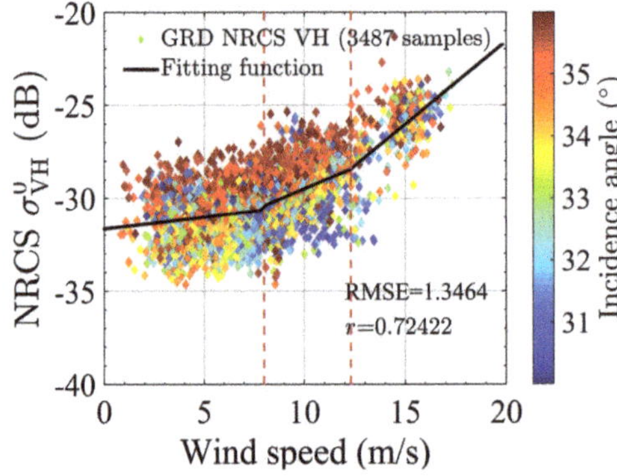


**Figure 9.** Relationship between NRCS and wind speeds for IW1-band.

**Table 2.** Relationship of NRCS and wind speed for different sub-bands.

| | Number of Samples | RMSE | *r* | Fitting Function |
|---|---|---|---|---|
| IW1-band | 3487 | 1.35 | 0.72 | $\sigma^0_{VH} = \begin{cases} 0.13\,v - 31.66\ (v \leq 8\,\mathrm{m/s}) \\ 0.46\,v - 34.06\ (8 < v \leq 12.3\,\mathrm{m/s}) \\ 0.89\,v - 39.36\ (v > 12.3\,\mathrm{m/s}) \end{cases}$ |
| IW2-band | 3671 | 1.64 | 0.70 | $\sigma^0_{VH} = \begin{cases} 0.23\,v - 33.65\ (v \leq 9.2\,\mathrm{m/s}) \\ 0.73\,v - 38.08\ (v > 9.2\,\mathrm{m/s}) \end{cases}$ |
| IW3-band | 3100 | 2.69 | 0.43 | $\sigma^0_{VH} = 0.44\,v - 35.67$ |

For IW1-G3 (wind speed is above 12.3 m/s), the slope of the fitting function obviously increases. From Table 2, the slope of fitting is 0.89, indicating the higher rate of increase of the NRCS $\sigma^0_{VH}$ with wind speeds. In the research of RADARSAT-2 VH dual-polarization images [15], Shen et al. also referred to this turning point estimated at about 10 m/s. This phenomena is theoretically explained as the wave-breaking contributions by non-bragg surface scattering mechanisms or volume scattering from breaking generated foamy layers, by Hwang et al. in [7,14]. Therefore, the wind speed sensitivity, as reflected in slope, increases with higher wind speeds, which suggests the potential of Sentinel-1 cross-pol for higher wind retrievals. Here, noted that due to the limitation of maximum wind speed of ASCAT products, the signal saturation of Sentinel-1 cross-pol will be further investigated in the future.

### 3.2.2. IW2-Band

The relationship between the NRCS $\sigma^0_{VH}$ and wind speeds for IW2-band is illustrated in Figure 10. For the IW2-band, the range of incidence angles is 36–41°. Similarly, the relationship between the NRCS $\sigma^0_{VH}$ and wind speeds has the same trends as these in IW-1 band. Therefore, according to the slopes, the data can be divided into two groups, denoted as IW2-G1, IW2-G2, which are separated by wind speed at 9.2 m/s, marked in the dotted line. To quantify this relationship, a two piecewise-linear fitting is carried out, and the results of fitting are illustrated in Table 2.

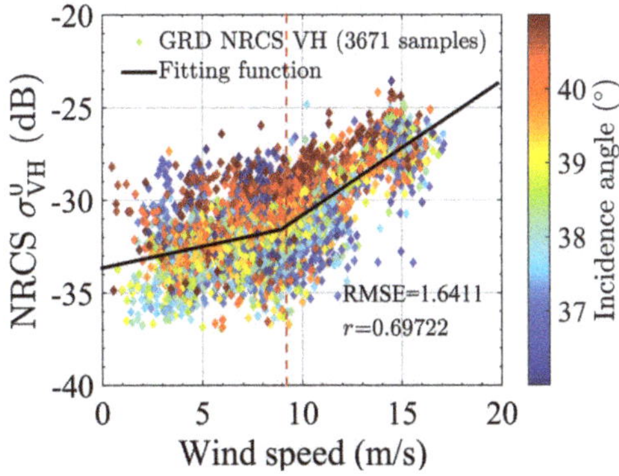


**Figure 10.** Relationship between NRCS and wind speeds for IW2-band.

For IW2-G1, when the wind speed is lower than 9.2 m/s, the $\sigma^0_{VH}$ are scattered with large variation, and from Table 2, the slope of fitting is 0.23. As discussed in Section 3.2.1, this variation arises because the radar return signals are too low to reflect the backscattering of ocean clutter signatures accurately. For IW2-G2, when the wind speed is higher than 9.2 m/s, the variation of the $\sigma^0_{VH}$ decreases obviously. This indicates that when wind speed is higher than 9.2 m/s, the radar return signals are sensitive enough to reflect ocean clutter signatures, and thus the wind speed retrieval from cross-pol observations is valid. Due to the limitation of the maximum wind speed is under 20 m/s, there suggests no turning point of significant change of slope. Because of the higher incidence angle of IW2-band, this point may occur at larger wind speed than that of IW1-band (12.3 m/s).

#### 3.2.3. IW3-Band

The relationship between the NRCS $\sigma^0_{VH}$ and wind speeds for IW3-band is illustrated in Figure 11. For the IW3-band, the range of incidence angles is 41–46°. To quantify the relationship, a linear fitting is carried out, and the results of fitting are illustrated in Table 2. For IW3-band, due to the higher incidence angles, the NRCS $\sigma^0_{VH}$ are relatively lower. The data are scattered, and the correlation coefficient $r$ is very low. This observation of Sentinel-1 is consistent with [32]. Based on the data of ASAR AP mode, compared with higher incidence angles, Vachon et al. conclude that the cross-pol radar returns can better reflect the signatures of clutter in lower incidence angles [32]. Therefore, we infer that for the IW3-band, the NRCS $\sigma^0_{VH}$ can not reflect the radar backscattering of ocean clutter signatures, and the accuracy of wind retrieval can not be guaranteed in moderate wind conditions (<20 m/s).

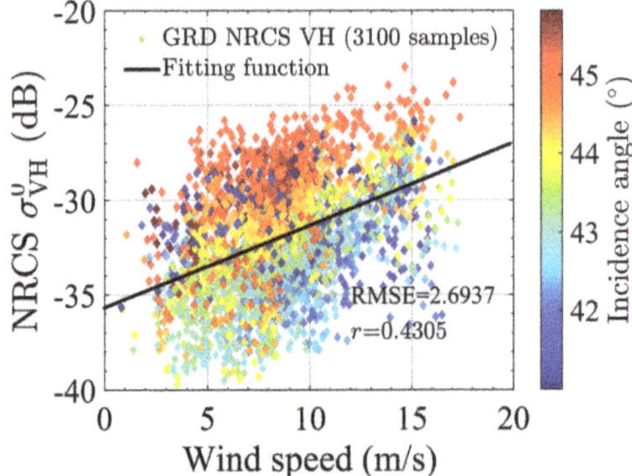


**Figure 11.** Relationship between NRCS and wind speeds for IW3-band.

### 3.3. *Relationship between NRCS and Wind Directions*

As illustrated in Section 3.1, wind speed performs more significant effects on NRCS $\sigma^0_{VH}$ than wind direction. Therefore, to determine the dependence of NRCS $\sigma^0_{VH}$ with respect to wind direction, the effects of wind speed should be isolated [17]. In this section, the NRCS $\sigma^0_{VH}$ are investigated for wind speeds at 3, 5, 7, 9, 11 and 13 m/s bounded by $\pm 1$ m/s interval. The relationships between NRCS $\sigma^0_{VH}$ and wind direction for different wind speed intervals are illustrated in Figure 12. For each subfigure, the mean values of NRCS $\sigma^0_{VH}$ are calculated for wind directions at 0°, 90°, 180°, 270° and 360° bounded by $\pm 10°$ interval. The mean values are illustrated in Table 3. In addition, in Figure 12, the mean values are presented in red points and connected into the red lines to show the trend of NRCS $\sigma^0_{VH}$ variation with respect to different wind directions. Note that the range of wind direction is $[0°, 360°]$, thus the mean values of NRCS $\sigma^0_{VH}$ at 0° and 360° are calculated in $[0°, 10°]$ and $[350°, 360°]$ intervals, respectively. Besides, because of the deficiency of data at $13 \pm 1$ m/s for $[0°, 10°]$ and $[350°, 360°]$ intervals, their mean values of NRCS $\sigma^0_{VH}$ are not calculated.

From Figure 12, the dependence on wind directions is visible. The values of NRCS $\sigma^0_{VH}$ reach local maxima at the up- and downwind directions (0°, 180° and 360°) and local minima at the crosswind directions (90° and 270°). This observation is consistent with the recent studies [14,17], which indicated wind direction dependence for cross-pol NRCS. It should be noted that for the interval of wind speed at $3 \pm 1$ m/s and $5 \pm 1$ m/s, the wind dependences of NRCS $\sigma^0_{VH}$ are not obvious. This can be explained as the relatively low SNR of radar signal returns in low wind conditions, which has been detailed in Sections 3.1 and 3.2. Based on Table 3, the average variation between the NRCS $\sigma^0_{VH}$ at the up-/downwind directions and the crosswind is about 1 dB for wind speed exceeds 5 m/s. Horstmann et al. [17] indicate that the wind direction dependence decreases with the increasing wind, and the direction dependence loses when speed exceeds 22.5 m/s. In this section, because of the limited data in high wind condition, this observation is not presented. We will handle with this problem in the future work.

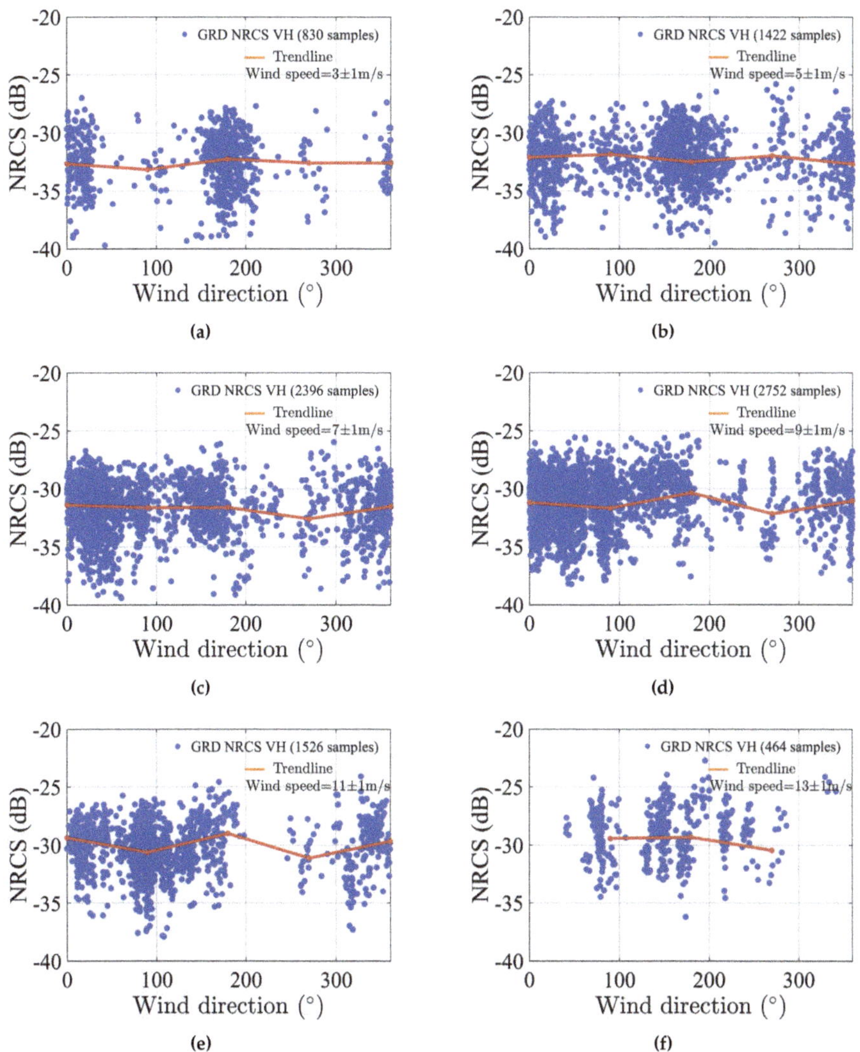


**Figure 12.** Dependencies of the NRCS on wind direction. (**a–f**) show wind speed intervals of ±1 m/s at 3, 5, 7, 9, 11, and 13 m/s, respectively. The red line represents the trendline.

**Table 3.** Relationship of NRCS and wind direction.

| NRCS $\sigma^0_{VH}$ (dB) Direction / Speed | $[0°, 10°]$ | $90° \pm 10°$ | $180° \pm 10°$ | $270° \pm 10°$ | $[350°, 360°]$ |
|---|---|---|---|---|---|
| $3 \pm 1$ m/s | −32.65 | −33.15 | −32.24 | −32.59 | −32.58 |
| $5 \pm 1$ m/s | −32.10 | −31.84 | −32.49 | −31.96 | −32.67 |
| $7 \pm 1$ m/s | −31.40 | −31.63 | −31.62 | −32.59 | −31.53 |
| $9 \pm 1$ m/s | −31.20 | −31.67 | −30.36 | −32.15 | −31.05 |
| $11 \pm 1$ m/s | −29.37 | −30.62 | −28.99 | −31.12 | −29.67 |
| $13 \pm 1$ m/s | NaN | −29.43 | −29.32 | −30.44 | NaN |

### 3.4. Summary

Based on the theoretical and empirical researches developed on RADARSAT-2, and considering the unique characteristics of Sentinel-1 IW mode, the relationship between the NRCS $\sigma^0_{VH}$ and wind speeds with attention to evaluate contribution from incidence angle and wind direction terms are analyzed in Sections 3.1–3.3. After analyses, six observations are made.

1. For Sentinel-1 cross-pol images, the radar backscattering NRCS $\sigma^0_{VH}$ are fluctuated and negatively correlated with the incidence angles.
2. For Sentinel-1 cross-pol images, the values of NRCS $\sigma^0_{VH}$ reach local maxima at the up- and downwind directions ($0°$, $180°$ and $360°$) and local minima at the crosswind directions ($90°$ and $270°$). In addition, the average variation between the NRCS $\sigma^0_{VH}$ at the up-/downwind directions and the crosswind is about 1 dB in our experiment.
3. Due to the unique TOPSAR technique for the IW mode (three sub-swaths IW1, IW2, IW3), the relationship between the NRCS $\sigma^0_{VH}$ and wind speeds for three sub-swaths are different and should be analyzed respectively.
4. For IW1-band, the relationship between the NRCS $\sigma^0_{VH}$ and wind speeds is monotonically linear increase, and the slope increases with higher wind speeds. The data can be divided into three groups. When the wind speed is lower than 8 m/s, the NRCS $\sigma^0_{VH}$ are scattered with large variation because the radar returns are low. When the wind speed is between 8 m/s and 12.3 m/s, the variation of the NRCS $\sigma^0_{VH}$ decreases obviously. This indicates the radar backscattering is sensitive enough to reflect ocean clutter signatures, and thus the wind speed retrieval from cross-pol observations is valid. When the wind speed is above 12.3 m/s, the wind speed sensitivity, as reflected in slope, increases with higher wind speeds, suggesting the potential of Sentinel-1 cross-pol to high wind retrievals.
5. For the IW2-band, the relationship between the NRCS $\sigma^0_{VH}$ and wind speeds is monotonically linear increase. When the wind speed is lower than 9.2 m/s, the NRCS $\sigma^0_{VH}$ are scattered with large variation because the radar return signals are low. When the wind speed is higher than 9.2 m/s, the variation of the NRCS $\sigma^0_{VH}$ decreases obviously. The radar backscattering is sensitive enough to reflect ocean clutter signatures, and thus the wind speed retrieval from cross-pol observations is valid.
6. For the IW3-band, due to the higher incidence angles, we infer that for IW3-band, the NRCS $\sigma^0_{VH}$ can not reflect the radar backscattering of ocean clutter signatures, and the accuracy of wind retrieval can not be guaranteed in moderate wind condition (<20 m/s).

The meaning of these analyses is that by revisiting the theoretical and empirical relationship derived from RADARSAT-2 data, we assess the Sentinel-1 IW cross-pol images, and provide a technical evaluation on ocean wind retrieval from Sentinel-1 cross-pol images. In summary, for Sentinel-1 IW products, the NRCS $\sigma^0_{VH}$ can better reflect the ocean clutter signatures in relatively lower incidence angles, which is IW1-band and IW2-band, therefore are suitable for wind field retrieval when wind speed exceeds the low boundary (8 m/s and 9.2 m/s for IW1 and IW2 band, respectively). An empirical model based on this is proposed in the next section.

## 4. Proposed Model for Sentinel-1 Cross-Pol Wind Retrieval

Based on the analyses in Section 3, we infer the NRCS $\sigma^0_{\mathrm{VH}}$ is dependent on incidence angle, wind speed and wind direction. Therefore, we propose a model to describe this relationship for Sentinel-1 cross-pol IW images.

$$\sigma^0_{\mathrm{VH}} = f_1(v)(1 + \omega \, \|f_2(\theta)\|) + C + A \quad (6)$$

where $\sigma^0_{\mathrm{VH}}$ denotes the NRCS values of radar return signals in dB, $\omega$ is defined as weight parameter, $C$ is the constant, $f_1(v)$ is the wind speed function which describes the relationship between the NRCS $\sigma^0_{\mathrm{VH}}$ and wind speed, $\|f_2(\theta)\|$ is the incidence angle function normalized into $[-1, 1]$, $A$ is to compensate the effects of wind direction. Both $f_1(v)$ and $f_2(\theta)$ are obtained from the observational data by fitting functions. The applicable condition of the proposed model will be given in Equations (7) and (9) in next sections.

### 4.1. Wind Speed Function

Based on the analyses in Section 3.2, under specific condition of incidence angles and wind speeds, the radar return signals can be utilized to retrieve wind filed. Specifically, for IW1-band (incidence angle between 30° and 36°) and for IW2-band (incidence angle between 36° and 41°), wind speed retrievals can be performed when wind speeds exceed 8 m/s and 9.2 m/s, respectively. Base on Table 2, the wind speed function is proposed as

$$\begin{cases} 0.46\,v - 34.06 \ (30^\circ < \theta \leq 36^\circ,\ 8 < v \leq 12.3\ \mathrm{m/s}) \\ 0.89\,v - 39.36 \ (30^\circ < \theta \leq 36^\circ,\ v > 12.3\ \mathrm{m/s}) \\ 0.73\,v - 38.08 \ (36^\circ < \theta \leq 41^\circ,\ v > 9.2\ \mathrm{m/s}) \end{cases} \quad (7)$$

### 4.2. Wind Direction Compensation

For Sentinel-1 cross-pol images, the values of NRCS $\sigma^0_{\mathrm{VH}}$ reach local maxima at the up- and downwind directions (0°, 180° and 360°) and local minima at the crosswind directions (90° and 270°). In addition, the average variation between the NRCS $\sigma^0_{\mathrm{VH}}$ at the up-/downwind directions and the crosswind is about 1 dB in our experiment. Therefore, for simplicity, $A$ is empirically set according to Equation (8) to compensate the effects of wind direction.

$$A = \begin{cases} 0.5 \ \ (\text{up-/downwind direction}) \\ -0.5 \ \ (\text{crosswind direction}) \end{cases} \quad (8)$$

### 4.3. Incidence Angle Function

Based on the analyses of Section 3.1, for IW1-band (incidence angle between 30° and 36°) wind speed retrieval can be performed when wind speed exceeds 8 m/s, and the slope increase obviously when wind speed exceed 12.3 m/s. For the IW2-band (incidence angle between 36° and 41°), wind speed retrieval can be performed when wind speed exceeds 9 m/s. Therefore, for the three subgroups of data, 2 degree polynomial functions are employed to fitting the data, illustrated in Figures 13 and 14.

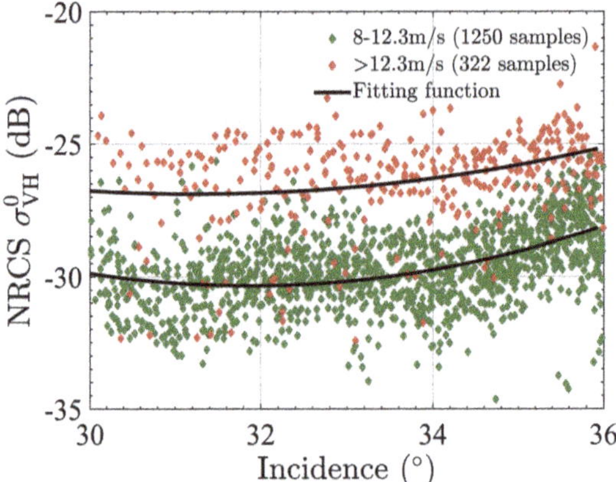


**Figure 13.** Incidence angle function for IW1-band.

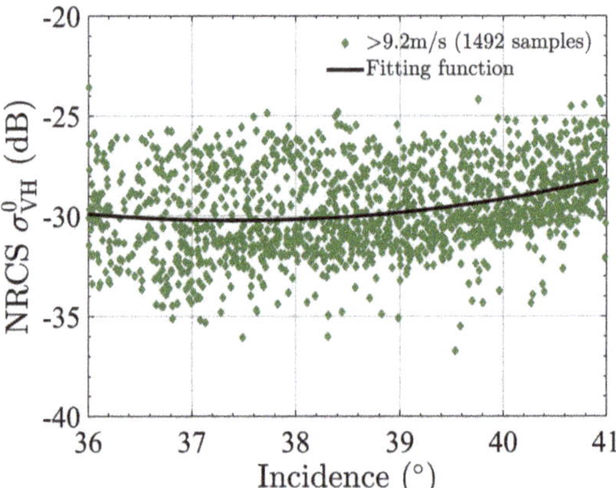


**Figure 14.** Incidence angle function for IW2-band.

The incidence angle function is proposed as

$$\begin{cases} 0.13\,\theta^2 - 8.42\,\theta + 103.88 \; (30^\circ < \theta \leq 36^\circ,\ 8 < v \leq 12.3\,\mathrm{m/s}) \\ 0.08\,\theta^2 - 4.86\,\theta + 48.97 \; (30^\circ < \theta \leq 36^\circ,\ v > 12.3\,\mathrm{m/s}) \\ 0.16\,\theta^2 - 12.10\,\theta + 195.98 \; (36^\circ < \theta \leq 41^\circ,\ v > 9.2\,\mathrm{m/s}) \end{cases} \tag{9}$$

By substituting Equations (7) and (9) into Equation (6), we get the model to describe this relationship for Sentinel-1 IW cross-pol images. The parameters of the proposed model corresponding to different situations are listed in Table 4. Figures 15 and 16 illustrate the performance of the proposed model. The correlation coefficient $r_{\mathrm{IW1}}$ and $r_{\mathrm{IW2}}$ are 0.80 and 0.82 for IW1 and IW2 band, respectively, which are improved by taking the dependence on incidence angle, wind speed and wind direction into consideration.

**Table 4.** Parameters for the proposed model.

| | Incidence Angle $\theta$ | Wind Speed $v$ (m/s) | $\omega$ | $C$ |
|---|---|---|---|---|
| IW1-band | 30–36° | 8–12.3 | −0.039 | 0.32 |
| IW1-band | 30–36° | >12.3 | −0.039 | 0.32 |
| IW2-band | 36–41° | >9.2 | −0.045 | 0.68 |

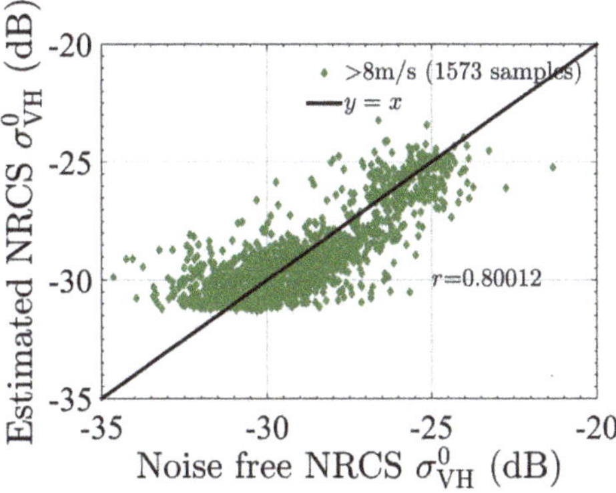


**Figure 15.** Estimated NRCS of the proposed model for IW1-band.

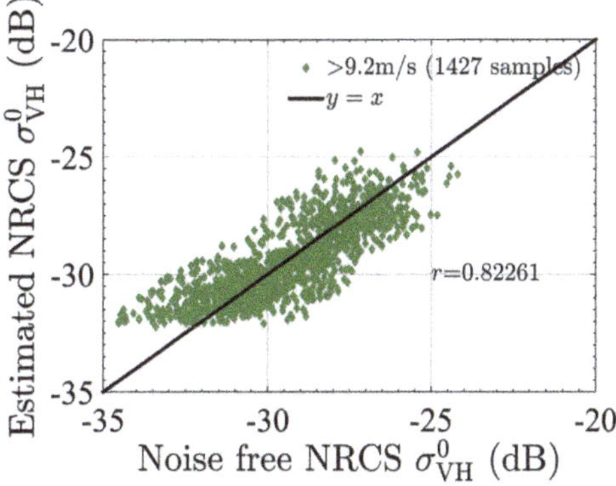


**Figure 16.** Estimated NRCS of the proposed model for IW2-band.

### *4.4. Model Validation*

As introduced in Section 2.4, the testing dataset which were not used to derive Equation (6) contains 1625 data covering full ranges of incidence angles, wind speeds and wind directions. Since the applicable condition of the proposed model are given in Equations (7) and (9), 230 data which satisfy this condition are selected from the testing dataset. Based on Equation (6), the retrieved wind speeds from Sentinel-1 IW cross-pol images compared with the ASCAT measurements are illustrated in Figure 17. The bias of the retrieved wind speed is 0.42 m/s, and the root mean square error (RMSE) is 1.26 m/s. Besides, from Figure 17, the accuracy of retrievals increases in high speed condition. For wind speed higher than 13 m/s, the bias and RMSE are 0.38 m/s and 0.97 m/s, respectively. This may be due to less accurate radar return signals in low-to-moderate wind condition. As Vachon and Wolf suggested in [12], the higher noise floor is, the larger wind speed is required to guarantee the useful radar backscattering which accurately reflect ocean clutter signatures.

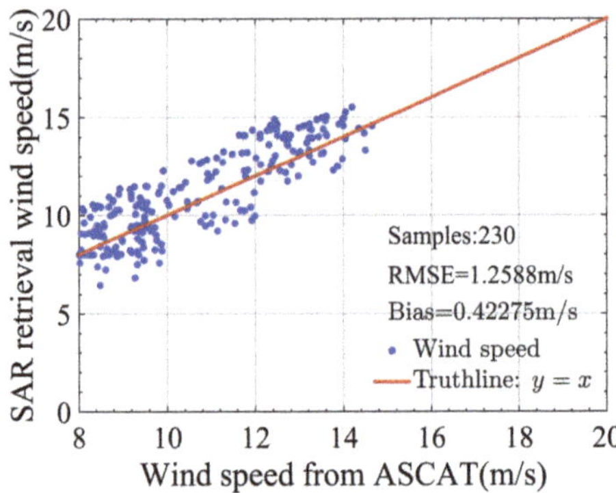


**Figure 17.** Comparison of the estimated wind speed with ASCAT wind speed.

In order to further demonstrate the suitability of the proposed model, a model proposed by Shen et al. based on RADARSAT-2 ScanSAR cross-pol data is carried out for comparison [15]. Shen's model is a two piecewise-linear function proposed as

$$\sigma^0_{\mathrm{VH}} = \begin{cases} 0.16 \times v - 28.49 \ (v \leq 10.1 \ \mathrm{m/s}) \\ 0.42 \times v - 30.98 \ (v > 10.1 \ \mathrm{m/s}) \end{cases} \tag{10}$$

where $v$ is the wind speed. and the two line segments meet when the wind speed is 10.1 m/s. Similarly, we first estimate NRCS by Shen's model and calculate the correlation coefficient $r_{\mathrm{IW1}}$ and $r_{\mathrm{IW2}}$ for IW1 and IW2 band, respectively. Then, the wind filed retrieval is carried out on same testing dataset and the bias and RMSE are calculated. The detailed performances comparison are shown in Table 5.

**Table 5.** Performances of the proposed and compared model.

| | $r_{\mathrm{IW1}}$ | $r_{\mathrm{IW2}}$ | **Bias (m/s)** | **RMSE(m/s)** |
|---|---|---|---|---|
| The proposed model | 0.80 | 0.82 | 0.42 | 1.26 |
| The compared model | 0.73 | 0.68 | −3.63 | 4.52 |

As we can see from Table 5, the proposed model shows better performance, for larger values of $r_{\mathrm{IW1}}$ and $r_{\mathrm{IW2}}$ and smaller values of Bias and RMSE. In [15], Shen et al. indicate that $\sigma^0_{\mathrm{VH}}$ show no evident dependence on incidence angles or wind direction, therefore the model only took wind speed into consideration, and performed very good wind retrieval ability for hurricane. In this experiment, for moderate wind condition, the $\sigma^0_{\mathrm{VH}}$ show dependences on incidence angles and wind directions. Therefore, the proposed model which considers the $\sigma^0_{\mathrm{VH}}$ dependences on incidence angle, wind speed and wind direction performs better.

## 5. Discussion

### 5.1. Influence of Different Samples of Training and Testing Sets

Based on the 10,258 matchup data as training dataset, we analyzed the dependences of $\sigma^0_{\mathrm{VH}}$ with respect to incidence angles, wind speeds, and wind directions, and proposed a model for wind retrieval. However, different sets of samples for training and testing might cause undesirable effects on fitting results, thus the stability and performance of the model might be affected. In this section, in order to evaluate this effect, we carried out all the experiments in Sections 3 and 4 with different samples in training dataset and testing dataset. The results are summarized in Tables 6 and 7.

As shown in Table 6, dataset 1 is composed of 12,058 for training and 1625 for testing, which is employed in the former sections of this paper. Both the training and testing data cover full ranges of incidence angles, wind speeds and wind directions. Dataset 2 is composed of 9285 for training and 2598 for testing, which is employed for comparison. Both the training and testing data cover full ranges of incidence angles, wind speeds and wind directions. Dataset 3 is composed of 8134 for training and 3797 for testing, which is employed for comparison. Both the training and testing data cover full ranges of incidence angles, wind speeds and wind directions. Table 6 summarized the parameters of the proposed model for different sets of samples. As we can observe, most of the coefficients in the model maintain relatively stable for three different datasets. Thus, we infer that the effects of training samples on model deriving are not significant.

Based on Equation (6), the performances of the proposed model for the three different datasets are summarized in Table 7. For dataset 1–3, 230, 335, 917 data which satisfy the applicable condition (given in Equations (7) and (9)) are selected from the testing dataset and performed wind field retrievals, respectively. Table 7 summarized the performances of the proposed model for three different datasets. As we can observe, most of the measurements maintain relative stability. Thus, we infer that the effects of testing samples on wind retrieval are not significant.

**Table 6.** Parameters of the model for datasets with different training and testing samples.

| Dataset | Incidence Angle | Wind Speed | Wind Speed Function | Incidence Angle Function | $\omega$ | $C$ |
|---|---|---|---|---|---|---|
| | 30–36° | 8–12.3 m/s | $0.46\,v - 34.06$ | $0.13\,\theta^2 - 8.42\,\theta + 103.88$ | −0.039 | 0.32 |
| 1 | 30–36° | >12.3 m/s | $0.89\,v - 39.36$ | $0.08\,\theta^2 - 4.86\,\theta + 48.97$ | −0.039 | 0.32 |
| | 36–41° | >9.2 m/s | $0.73\,v - 38.08$ | $0.16\,\theta^2 - 12.10\,\theta + 195.98$ | −0.045 | 0.68 |
| | 30–36° | 8–12.3 m/s | $0.44\,v - 34.04$ | $0.13\,\theta^2 - 8.46\,\theta + 104.44$ | −0.039 | 0.32 |
| 2 | 30–36° | >12.3 m/s | $0.89\,v - 39.36$ | $0.08\,\theta^2 - 4.86\,\theta + 48.97$ | −0.039 | 0.32 |
| | 36–41° | >9.2 m/s | $0.72\,v - 38.08$ | $0.18\,\theta^2 - 13.60\,\theta + 225.65$ | −0.045 | 0.68 |
| | 30–36° | 8–12.3 m/s | $0.41\,v - 33.54$ | $0.13\,\theta^2 - 8.12\,\theta + 98.84$ | −0.039 | 0.32 |
| 3 | 30–36° | >12.3 m/s | $0.89\,v - 39.29$ | $0.07\,\theta^2 - 4.64\,\theta + 45.45$ | −0.039 | 0.32 |
| | 36–41° | >9.2 m/s | $0.69\,v - 37.52$ | $0.14\,\theta^2 - 10.80\,\theta + 171.85$ | −0.045 | 0.68 |

**Table 7.** Performances of the model for datasets with different training and testing samples.

| Dataset | $r_{IW1}$ | $r_{IW2}$ | Bias | RMSE |
|---|---|---|---|---|
| 1 | 0.80 | 0.82 | 0.42 | 1.26 |
| 2 | 0.82 | 0.82 | 0.49 | 1.14 |
| 3 | 0.81 | 0.81 | 0.31 | 1.96 |

### 5.2. *Influence of Different Number of Pixels for SAR Chip*

The experiments in Sections 3 and 4 are carried out on SAR chips with the size of 50 × 50 pixels (500 m resolution). Therefore, for each matchup data, $\sigma^0_{VH}$ is calculated as the averaged value of the NRCS of 50 × 50 pixels. However, the different number of samples for average might cause undesirable effects on fitting results, thus the stability and performance of the model might be affected. In this section, in order to evaluate this effect, we carried out all the experiments in Sections 3 and 4 with different number of samples for average. The results are summarized in Tables 8 and 9.

Table 8 summarized the parameters of the proposed model for different number of averaged pixels: 25 × 25, 50 × 50 and 100 × 100 pixels. As we can observe, most of the coefficients in the model maintain relatively stable with different pixel numbers. Thus, we infer that the effects of pixel numbers on model deriving are not significant. Based on Equation (6), the performances of the proposed model for different number of averaged pixels are summarized in Table 9. As we can observe, most of the measurements maintain relatively stable. Thus, we infer that the effects of pixel numbers on wind retrieval are not significant.

These inferences are consistent with the observation indicated by Bergeron et al. [13], which demonstrates the exactly same results for SAR cross-pol wind speed retrieval at 1 km and 400 m resolution. In addition, 500 m resolution for SAR wind speed retrieval are widely used [4,5,19].

**Table 8.** Parameters of the model for datasets with different number of pixels.

| Pixel Number | Incidence Angle | Wind Speed | Wind Speed Function | Incidence Angle Function | $\omega$ | $C$ |
|---|---|---|---|---|---|---|
| | 30–36° | 8–12.3 m/s | $0.46\,v - 34.03$ | $0.15\,\theta^2 - 9.91\,\theta + 128.95$ | −0.039 | 0.32 |
| 25 × 25 | 30–36° | >12.3 m/s | $0.92\,v - 39.85$ | $0.08\,\theta^2 - 5.10\,\theta + 52.69$ | −0.039 | 0.32 |
| | 36–41° | >9.2 m/s | $0.71\,v - 37.91$ | $0.18\,\theta^2 - 13.58\,\theta + 225.04$ | −0.045 | 0.68 |
| | 30–36° | 8–12.3 m/s | $0.46\,v - 34.06$ | $0.13\,\theta^2 - 8.42\,\theta + 103.88$ | −0.039 | 0.32 |
| 50 × 50 | 30–36° | >12.3 m/s | $0.89\,v - 39.36$ | $0.08\,\theta^2 - 4.86\,\theta + 48.97$ | −0.039 | 0.32 |
| | 36–41° | >9.2 m/s | $0.73\,v - 38.08$ | $0.16\,\theta^2 - 12.10\,\theta + 195.98$ | −0.045 | 0.68 |
| | 30–36° | 8–12.3 m/s | $0.48\,v - 34.27$ | $0.13\,\theta^2 - 8.05\,\theta + 97.59$ | −0.039 | 0.32 |
| 100 × 100 | 30–36° | >12.3 m/s | $0.90\,v - 39.56$ | $0.06\,\theta^2 - 3.50\,\theta + 46.26$ | −0.039 | 0.32 |
| | 36–41° | >9.2 m/s | $0.73\,v - 38.07$ | $0.17\,\theta^2 - 12.40\,\theta + 201.72$ | −0.045 | 0.68 |

**Table 9.** Performances of the model for datasets with different number of pixels.

| Pixel Number | $r_{IW1}$ | $r_{IW2}$ | Bias | RMSE |
|---|---|---|---|---|
| 25 × 25 | 0.80 | 0.81 | 0.50 | 1.27 |
| 50 × 50 | 0.80 | 0.82 | 0.42 | 1.26 |
| 100 × 100 | 0.81 | 0.82 | 0.41 | 1.33 |

## 6. Conclusions

In this paper, we revisit the empirical relationships between cross-pol radar backscattering $\sigma^0_{VH}$ and wind speeds with attention to evaluate contribution from incidence angle and wind direction terms. The scopes of this paper are to give a technical evaluation of Sentinel-1 cross-pol images based on the theoretical and empirical analyses derived from RADARSAT-2 data, and provide a wind retrieval model based on Sentinel-1 cross-pol images.

In order to better understand the Sentinel-1 cross-pol NRCS values under various environmental conditions, we construct a dataset by matching SAR winds with near coincide wind vectors from ASCAT scatterometer. 11,883 data with 0 mm/h precipitation are utilized as training and testing data in the following analyses.

Next, we calculate the NESZ of Sentinel-1 IW mode, and perform noise removal for all the matchup data. Empirical relationships among the noise free NRCS $\sigma^0_{VH}$, wind speed, wind direction, and radar incidence angle are analyzed. The lowest boundary wind speed for retrievals is specified.

A piecewise model particularly fitting the Sentinel-1 IW mode is proposed to describe the relationships between the NRCS $\sigma^0_{VH}$ and factors. The model is composed of the incidence angle function, wind speed function and wind direction compensation, derived from data fitting. Compared with the analyses of incidence angle and wind speed separately in Section 3, the larger $r$ suggests the necessity of taking incidence angle, wind speed and wind direction into consideration for describing this relationship. In addition, the model is validated using 230 data. The low bias and RMSE show the suitability of the proposed model for Sentinel-1 IW cross-pol images. Furthermore, the retrieval accuracy of the model increases in high speed condition; for smaller bias and RMSE with higher wind speeds. This indicates the potential of apply Sentinel-1 cross-pol images to wind retrievals in extreme weather.

Future work involves further analyzing the Sentinel-1 cross-pol radar backscattering with incidence angles and wind speeds in very high wind conditions. The signal saturation of Sentinel-1 cross-pol images will be further investigated, because there are not enough high-wind matchup data available as of now. However, we expect that there will be more Sentinel-1 images covering hurricanes/typhoons as shown in [33] in the coming years, and we will be able to validate the model against airborne and spaceborne sensors special designed for high-wind measurements. Moreover, we will match Sentinel data with buoys and do more validation and modification of the model against multiple wind products.

**Acknowledgments:** This work was supported by the State Key Program of the National Natural Science Foundation of China under Grant 61331015. The authors would also like to thank the European Space Agency for providing the Sentinel-1 data and the SNAP 4.0, JPL for the ASCAT data, and Precipitation Measurement Missions, NASA for the TRMM data. The views, opinions, and findings contained in this paper are those of the authors and should not be construed as an official NOAA or U.S. government position, policy, or decision.

**Author Contributions:** Lanqing Huang and Bin Liu conceived and performed the experiments. Bin Liu and Xiaofeng Li supervised and designed the research and contributed to the organization of article. Lanqing Huang drafted the manuscript, and all authors revised and approved the final version of the manuscript.

**Conflicts of Interest:** The authors declare no conflict of interest.

## References

1. Hersbach, H. *CMOD5: An Improved Geophysical Model Function for ERS C-Band Scatterometry*; European Centre for Medium-Range Weather Forecasts: Reading, UK, 2003.
2. Xu, Q.; Lin, H.; Li, X.; Zuo, J.; Zheng, Q.; Pichel, W.G.; Liu, Y. Assessment of an analytical model for sea surface wind speed retrieval from spaceborne SAR. *Int. J. Remote Sens.* **2010**, *31*, 993–1008.
3. Wackerman, C.C.; Clemente-Colón, P.; Pichel, W.G.; Li, X. A two-scale model to predict C-band VV and HH normalized radar cross section values over the ocean. *Can. J. Remote Sens.* **2002**, *28*, 367–384.
4. Yang, X.; Li, X.; Pichel, W.G.; Li, Z. Comparison of ocean surface winds from ENVISAT ASAR, MetOp ASCAT scatterometer, buoy measurements, and NOGAPS model. *IEEE Trans. Geosci. Remote Sens.* **2011**, *49*, 4743–4750.
5. Yang, X.; Li, X.; Zheng, Q.; Gu, X.; Pichel, W.G.; Li, Z. Comparison of ocean-surface winds retrieved from QuikSCAT scatterometer and Radarsat-1 SAR in offshore waters of the US west coast. *IEEE Geosci. Remote Sens. Lett.* **2011**, *8*, 163–167.
6. Liu, G.; Yang, X.; Li, X.; Zhang, B.; Pichel, W.; Li, Z.; Zhou, X. A systematic comparison of the effect of polarization ratio models on sea surface wind retrieval from C-band synthetic aperture radar. *IEEE J. Sel. Top. Appl. Earth Obs. Remote Sens.* **2013**, *6*, 1100–1108.
7. Hwang, P.A.; Zhang, B.; Toporkov, J.V.; Perrie, W. Comparison of composite Bragg theory and quad-polarization radar backscatter from RADARSAT-2: With applications to wave breaking and high wind retrieval. *J. Geophys. Res. Oceans* **2010**, *115*, doi:10.1029/2009JC005995.
8. Zhang, B.; Perrie, W.; Vachon, P.W.; Li, X.; Pichel, W.G.; Guo, J.; He, Y. Ocean vector winds retrieval from C-band fully polarimetric SAR measurements. *IEEE Trans. Geosci. Remote Sens.* **2012**, *50*, 4252–4261.
9. Zhang, B.; Li, X.; Perrie, W.; He, Y. Synergistic measurements of ocean winds and waves from SAR. *J. Geophys. Res. Oceans* **2015**, *120*, 6164–6184.
10. Kim, T.S.; Park, K.A.; Li, X.; Mouche, A.A.; Chapron, B.; Lee, M. Observation of Wind Direction Change on the Sea Surface Temperature Front Using High-Resolution Full Polarimetric SAR Data. *IEEE J. Sel. Top. Appl. Earth Obs. Remote Sens.* **2017**, *10*, 2599–2607.
11. Zhang, G.; Li, X.; Perrie, W.; Hwang, P.A.; Zhang, B.; Yang, X. A Hurricane Wind Speed Retrieval Model for C-Band RADARSAT-2 Cross-Polarization ScanSAR Images. *IEEE Trans. Geosci. Remote Sens.* **2017**, *55*, 4766–4774.
12. Vachon, P.W.; Wolfe, J. C-band cross-polarization wind speed retrieval. *IEEE Geosci. Remote Sens. Lett.* **2011**, *8*, 456–459.
13. Bergeron, T.; Bernier, M.; Chokmani, K.; Lessard-Fontaine, A.; Lafrance, G.; Beaucage, P. Wind speed estimation using polarimetric RADARSAT-2 images: Finding the best polarization and polarization ratio. *IEEE J. Sel. Top. Appl. Earth Obs. Remote Sens.* **2011**, *4*, 896–904.
14. Hwang, P.A.; Perrie, W.; Zhang, B. Cross-polarization radar backscattering from the ocean surface and its dependence on wind velocity. *IEEE Geosci. Remote Sens. Lett.* **2014**, *11*, 2188–2192.
15. Shen, H.; Perrie, W.; He, Y.; Liu, G. Wind speed retrieval from VH dual-polarization RADARSAT-2 SAR images. *IEEE Trans. Geosci. Remote Sens.* **2014**, *52*, 5820–5826.
16. Zhang, B.; Perrie, W.; Zhang, J.A.; Uhlhorn, E.W.; He, Y. High-resolution hurricane vector winds from C-band dual-polarization SAR observations. *J. Atmos. Ocean. Technol.* **2014**, *31*, 272–286.

17. Horstmann, J.; Falchetti, S.; Wackerman, C.; Maresca, S.; Caruso, M.J.; Graber, H.C. Tropical cyclone winds retrieved from C-band cross-polarized synthetic aperture radar. *IEEE Trans. Geosci. Remote Sens.* **2015**, *53*, 2887–2898.
18. Zecchetto, S.; De Biasio, F.; della Valle, A.; Quattrocchi, G.; Cadau, E.; Cucco, A. Wind Fields From C-and X-Band SAR Images at VV Polarization in Coastal Area (Gulf of Oristano, Italy). *IEEE J. Sel. Top. Appl. Earth Obs. Remote Sens.* **2016**, *9*, 2643–2650.
19. Monaldo, F.; Jackson, C.; Li, X.; Pichel, W.G. Preliminary evaluation of Sentinel-1A wind speed retrievals. *IEEE J. Sel. Top. Appl. Earth Obs. Remote Sens.* **2016**, *9*, 2638–2642.
20. Agency, C.S. RADARSAT Constellation. Available online: http://www.asc-csa.gc.ca/eng/satellites/radarsat/default.asp (accessed on 1 September 2015).
21. Flett, D.; Crevier, Y.; Girard, R. The RADARSAT Constellation Mission: Meeting the government of Canada'S needs and requirements. In Proceedings of the 2009 IEEE International Geoscience and Remote Sensing Symposium, Cape Town, South Africa, 12–17 July 2009; Volume 2, pp. 907–910.
22. Thompson, A.A. Innovative capabilities of the RADARSAT constellation mission. In Proceedings of the 8th European Conference on Synthetic Aperture Radar, Aachen, Germany, 7–10 June 2010; pp. 1–3.
23. Global, OSI. ASCAT Wind Product User Manual. Available online: http://projects.knmi.nl/scatterometer/publications/pdf/ASCAT_Product_Manual.pdf (accessed on 1 September 2013).
24. Torres, R.; Snoeij, P.; Geudtner, D.; Bibby, D.; Davidson, M.; Attema, E.; Potin, P.; Rommen, B.; Floury, N.; Brown, M.; et al. GMES Sentinel-1 mission. *Remote Sens. Environ.* **2012**, *120*, 9–24.
25. European Space Agency, Sentinel-1 Team. Sentinel-1 User Handbook. Available online: http://sentinel.esa.int/ (accessed on 1 September 2013).
26. ESA. Copernicus Open Access Hub. Available online: https://scihub.copernicus.eu/ (accessed on 1 September 2014).
27. NASA EOSDIS PO.DAAC. Physical Oceanography Distributed Active Archive Center (PO.DAAC). Available online: https://podaac.jpl.nasa.gov/ (accessed on 1 September 2015).
28. ESA. Step Science Toolbox Explotiation Platform. Available online: http://step.esa.int/main/toolboxes/snap/ (accessed on 1 September 2014).
29. Missions, P.M. TRMM Data Download. Available online: https://pmm.nasa.gov/index.php?q=data-access/downloads/trmm (accessed on 1 September 2015).
30. Huffman, G.J.; Bolvin, D.T. TRMM and Other Data Precipitation Data Set Documentation. Available online: https://pmm.nasa.gov/sites/default/files/document_files/3B42_3B43_doc_V7_4_19_17.pdf (accessed on 1 September 2017).
31. European Space Agency, Sentinel-1 Team. Sentinel-1 Product Specification. Available online: https://sentinels.copernicus.eu/web/sentinel/user-guides/sentinel-1-sar/document-library/-/asset_publisher/1dO7RF5fJMbd/content/sentinel-1-product-specification (accessed on 1 November 2016).
32. Vachon, P.W.; Wolfe, J. *GMES Sentinel-1 Analysis of Marine Applications Potential (AMAP)*; DRDC Ottawa ECR: Ottawa, ON, Canada, 2008; Volume 218.
33. Li, X. The first Sentinel-1 SAR image of a typhoon. *Acta Oceanol. Sin.* **2015**, *34*, 1–2.

Article

# An Improved Local Gradient Method for Sea Surface Wind Direction Retrieval from SAR Imagery

Lizhang Zhou [1], Gang Zheng [1,*], Xiaofeng Li [2], Jingsong Yang [1], Lin Ren [1], Peng Chen [1], Huaguo Zhang [1] and Xiulin Lou [1]

1 State Key Laboratory of Satellite Ocean Environment Dynamics, Second Institute of Oceanography, State Oceanic Administration, Hangzhou 310012, China; zlz19921122@163.com (L.Z.); jsyang@sio.org.cn (J.Y.); renlin210@sio.org.cn (L.R.); chenpeng@sio.org.cn (P.C.); zhanghg@sio.org.cn (H.Z.); lxl@sio.org.cn (X.L.)

2 GST at National Oceanic and Atmospheric Administration (NOAA)-National Environmental Satellite, Data, and Information Service (NESDIS), College Park, MD 20740, USA; xiaofeng.li@noaa.gov

* Correspondence: zhenggang@sio.org.cn

Received: 21 April 2017; Accepted: 26 June 2017; Published: 30 June 2017

**Abstract:** Sea surface wind affects the fluxes of energy, mass and momentum between the atmosphere and ocean, and therefore regional and global weather and climate. With various satellite microwave sensors, sea surface wind can be measured with large spatial coverage in almost all-weather conditions, day or night. Like any other remote sensing measurements, sea surface wind measurement is also indirect. Therefore, it is important to develop appropriate wind speed and direction retrieval models for different types of microwave instruments. In this paper, a new sea surface wind direction retrieval method from synthetic aperture radar (SAR) imagery is developed. In the method, local gradients are computed in frequency domain by combining the operation of smoothing and computing local gradients in one step to simplify the process and avoid the difference approximation. This improved local gradients (ILG) method is compared with the traditional two-dimensional fast Fourier transform (2D FFT) method and local gradients (LG) method, using interpolating wind directions from the European Centre for Medium-Range Weather Forecast (ECMWF) reanalysis data and the Cross-Calibrated Multi-Platform (CCMP) wind vector product. The sensitivities to the salt-and-pepper noise, the additive noise and the multiplicative noise are analyzed. The ILG method shows a better performance of retrieval wind directions than the other two methods.

**Keywords:** local gradients method; retrieval; SAR; wind direction

## 1. Introduction

Surface wind field over the oceans is needed for weather forecasts, wind resources assessment, numerical modeling of waves, oil spill monitoring, and so on [1–10]. Traditional measurements of sea surface wind from ships, buoys and land stations are far from meeting the growing demand for human beings, as these approaches can only provide data with limited spatial and temporal coverage. Spaceborne microwave radiometers and scatterometers have provided global-scale observations for sea surface wind, but the wind data in coastal regions are missing due to contamination from signal reflection of land [11–13]. Also, the spatial resolution of radiometers or scatterometers is relatively low (about 25 km). This is suitable for open ocean studies. However, wind fields on a much finer scale can be provided by spaceborne synthetic aperture radar (SAR) due to its relatively high spatial resolution. As a result, SARs can be useful tools when high-resolution ocean surface wind fields are needed, especially in coastal regions, and the retrieval of wind field from SAR images is widely researched [14–21].

Many satellites have been launched with SAR onboard, e.g., Seasat, ERS-1/2, Envisat, Radarsat-1/2, and so on [14–18]. These sensors have acquired abundant SAR images containing many interesting features including coastal upwelling, typhoon/hurricane, atmospheric waves, atmospheric vortex street, and so on [19–31]. Some of SAR imaged features, e.g., Langmuir cells, boundary layer rolls, surfactant streaks, foam and water blown from breaking waves, or wind shadowing, align with wind direction.

Wind direction (with an ambiguity of 180°) can be retrieved from these features by different methods including Fourier transforms, wavelet analysis, local gradients and so on. The 180° ambiguity can be removed by referencing to weather model output, Doppler shift or land shadows [3,16,32]. The accuracy of various methods ranges from 15–40° [33]. After wind direction is obtained, wind speed can be retrieved by physical or empirical models. Accordingly, estimating wind vectors directly from SAR images becomes feasible [34]. Both cross-polarization and co-polarization SAR images can be used for the retrieval of wind vectors [35–37]. There are two conventional methods of retrieving wind directions (with an ambiguity of 180° which is removed later) from SAR images, namely, two-dimensional fast Fourier transform (2D FFT) method and local gradients (LG) method [34,38,39]. In the FFT method, the Fourier spectrum of SAR images is computed and the main spectral energy is located perpendicular to the orientation of the wind streaks. The reported standard deviations of FFT method are between 10–37° and the method works fine on large image areas, e.g., 20 km by 20 km [34]. In the LG method, local gradients are computed with standard image processing algorithms, and the orthogonal of the most frequent gradient direction is chosen to be the likely wind direction. The reported directional error of LG method was about 20° for ERS-1/2 images and the most frequent spatial sampling used was 20 km by 20 km and 10 km by 10 km [34].

The tests in [34] indicated that the LG method could provide a higher resolution of retrieved wind field than the FFT method. However, the current LG method has a problem; that is, the local gradients in the conventional LG method are computed with difference approximation like Sobel operators, and this process is easily affected by noise, e.g., speckle noise. Thus, SAR images are usually first smoothed before computing local gradients. In this study, we develop an improved local gradients (ILG) method for sea surface wind retrieval by combining smoothing and computation of the local gradients together in the frequency domain. In the method, the computation of local gradients is analytical as the Gaussian function can be expressed analytically in both spatial domain and frequency domain and it can avoid the errors of difference approximation which can be easily affected by noise. The new method is tested on the images acquired by the advanced synthetic aperture radar (ASAR) onboard Envisat, and its retrieved results show better agreement with both following kinds of interpolating wind directions than the other two methods. The interpolating wind directions from the European Centre for Medium-Range Weather Forecast (ECMWF) reanalysis data are obtained by interpolating the ECMWF reanalysis data to the SAR imaging times. The same procedure is applied to the Cross-Calibrated Multi-Platform (CCMP) data to obtain the interpolating wind directions from the CCMP data.

The remaining sections are organized as follows. In Section 2, the ILG method is described in detail. In Section 3, the data sets used are introduced. In Section 4, we compare the three wind direction retrieval methods using the wind directions from the ECMWF reanalysis data and the CCMP wind products. The ILG method is also tested by using small images (thus high resolution) in this section. The performance of each retrieval method is analyzed while the images are corrupted by different types of noise in Section 5. Conclusions are given in Section 6.

## 2. Improved Local Gradients Method

The direction of the gradient should be the same as the direction of the strongest change in an image, and an ideal image of streaks should have nearly no change along the direction of streaks, and show the strongest variation in the orthogonal direction of streaks. Thus, the wind direction, which is assumed to be parallel to the wind streaks [40], is also perpendicular to the direction of the gradient.

Usually, the components of the gradient are computed with the optimized Sobel operators pixel by pixel after smoothing, and the most frequent gradient direction is chosen to be the local wind direction.

As mentioned in the introduction, the computing method described above, however, can lead to some problems. So, we try to combine the operation of smoothing and computation of local gradients on the basis of a conventional LG method which is given by Koch in 2004. The smoothing of an image can be realized by convolution, and the Gaussian function is commonly used for smoothing, as shown in

$$s'(x',y') = \iint s(x,y)f(x-x',y-y')dxdy \tag{1}$$

where $s'(x',y')$ is the image after smoothing, and $s(x,y)$ is the original image. $f(\cdot,\cdot)$ is the Gaussian function for smoothing, with the form

$$f(x,y) = \frac{1}{2\pi\sigma^2}e^{-\frac{x^2+y^2}{2\sigma^2}} \tag{2}$$

where $\sigma$ determines the width of smoothing window in the ILG method. The choice of the value of the parameter is empirical, and in this paper the parameter is set to 15 for all SAR images. The benefit of the choice of the Gaussian is that it can be expressed analytically in both spatial and frequency domains. The gradients of the image can be expressed as

$$\nabla s'(x',y') = \frac{\partial}{\partial x'}s'(x',y')\hat{e}_x + \frac{\partial}{\partial y'}s'(x',y')\hat{e}_y \tag{3}$$

where $\hat{e}_x$ and $\hat{e}_y$ are the unit vectors in the $x$ and $y$ directions, respectively. Using Equation (1)–(3) and exchanging the order of integral and partial derivative, $\frac{\partial}{\partial x'}s'(x',y')$ can be expressed as

$$\frac{\partial}{\partial x'}s'(x',y') = -s(x,y) * h_x(x,y) \tag{4}$$

where * denotes convolution, and $h_x(x,y) = \frac{\partial}{\partial x}f(x,y)$. We perform Fourier transform on both sides of Equation (4), and get

$$F[\frac{\partial}{\partial x'}s'(x',y')] = -F[s(x,y)] \cdot F[h_x(x,y)] \tag{5}$$

where $F[\cdot]$ denotes the 2D Fourier transform.

The first term on the right hand side of Equation (5) can be obtained from the Fourier transform of the image directly. The second term, represented by $H_x$ for convenience of expression, can be expressed as

$$H_x = F[h_x(x,y)] = -\frac{ik_x}{\sigma^2}e^{-2\sigma^2\pi^2(k_x^2+k_y^2)} \tag{6}$$

where $i$ indicates the imaginary part. So, $\frac{\partial}{\partial x'}s'(x',y')$ can be expressed as

$$\frac{\partial}{\partial x'}s'(x',y') = -F^{-1}\{F[s(x,y)] \cdot H_x\} \tag{7}$$

where $F^{-1}[\cdot]$ denotes the 2D inverse Fast Fourier transform (2D IFFT). Similarly, $\frac{\partial}{\partial y'}s'(x',y')$ can be expressed as

$$\frac{\partial}{\partial y'}s'(x',y') = -F^{-1}\{F[s(x,y)] \cdot H_y\} \tag{8}$$

where $H_y$ can be expressed by Equation (9).

$$H_y = F[h_y(x,y)] = -\frac{ik_y}{\sigma^2}e^{-2\sigma^2\pi^2(k_x^2+k_y^2)} \tag{9}$$

$\frac{\partial}{\partial x'}s'(x',y')$ and $\frac{\partial}{\partial y'}s'(x',y')$ are denoted as $g_x$ and $g_y$, respectively, for convenience, so the gradients can be stored as complex numbers in the form of

$$g = g_x + ig_y \tag{10}$$

Now, the gradients are in the same form as those in Koch's LG method [34]. Figure 1 shows the flowchart of the ILG method for wind direction retrieval. The remaining steps are consistent with those in Koch's method, including discarding unusable points and extracting the main gradient direction. The 180° wind direction ambiguity is removed by referencing coincident wind direction from the ECMWF model result [41].

The image acquired by the advanced synthetic aperture radar (ASAR) onboard Envisat at 02:06 on 14 October 2007 is shown in Figure 2a. Figure 2b shows the wind direction computed on a 30 km by 30 km subimage which is indicated by red box in Figure 2a. Wind streaks are visible all over the subimage. The white arrow (211°) in Figure 2b indicates the wind direction computed from 75 m pixels by using the ILG method. The wind direction from the ECMWF was 222°, indicated by the red arrow in Figure 2b. The wind direction from the CCMP was 225° (the green arrow in Figure 2b). Note that the wind directions here are placed in the same coordinate. In this coordinate, the direction of the northward wind vector is 0° (or 360°), the direction of the eastward wind vector is 90°, the direction of the southward wind vector is 180°, and the direction of the westward wind vector is 270°.

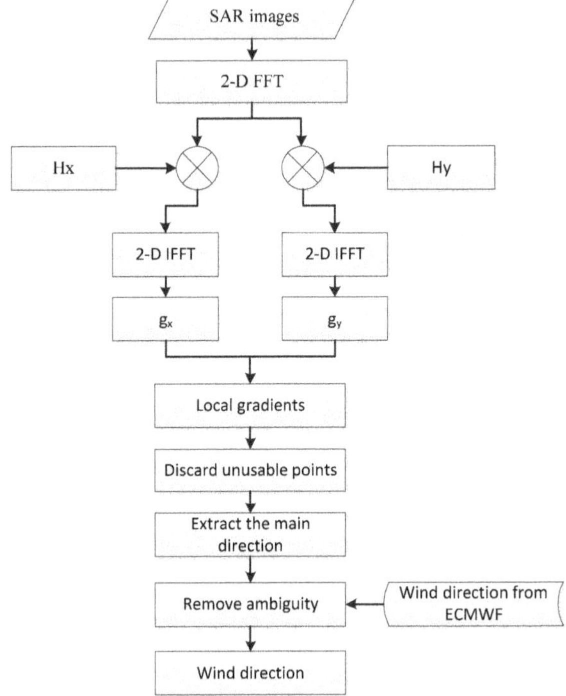


**Figure 1.** Flowchart of the improved local gradients method for wind direction retrieval.

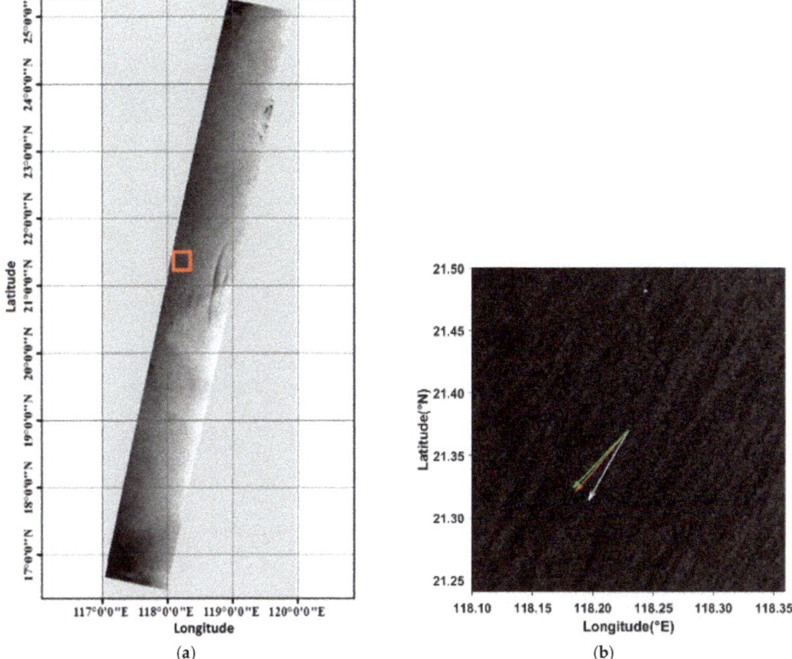


**Figure 2.** (**a**) An Envisat advanced synthetic aperture radar (ASAR) image acquired at 02:06 on 14 October 2007 in VV polarization with a resolution of 75 m. (**b**) The subimage which is indicated by the red box in (**a**). White, red and green arrows indicate wind directions obtained from improved local gradients (ILG) method, European Centre for Medium-Range Weather Forecast (ECMWF) and Cross-Calibrated Multi-Platform (CCMP), respectively.

$H_x$ and $H_y$ are shown in Figure 3 in frequency domain, and their amplitudes are indicated by different colors. The computed $g_x$ and $g_y$ of the image in Figure 2b are shown in Figure 4 in spatial domain and different colors indicate different values of the gradient component.

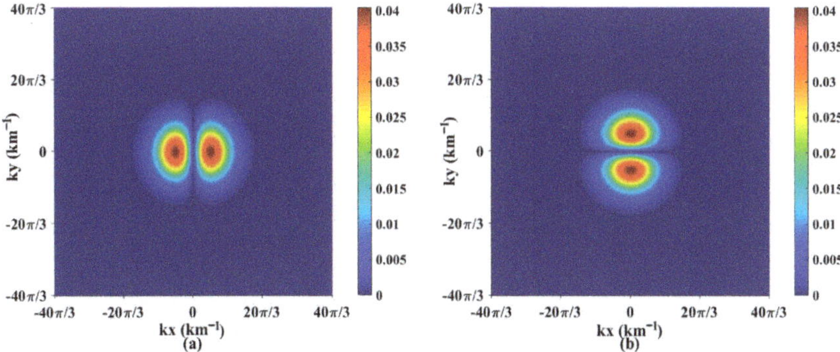


**Figure 3.** The amplitude distributions of $H_x$ (**a**) and $H_y$ (**b**), $\sigma$ is 15.

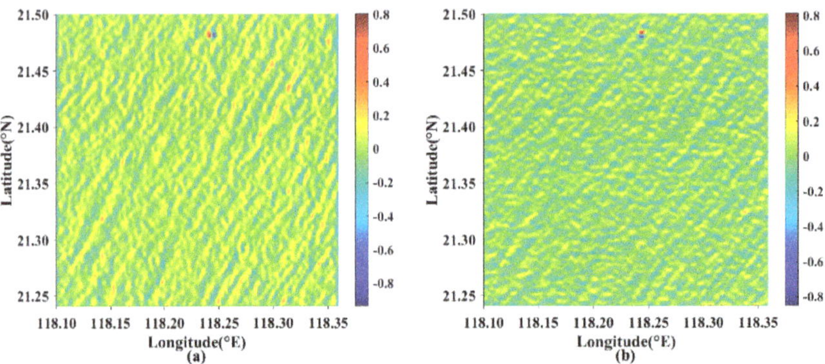


**Figure 4.** Computed $g_x$ (**a**) and $g_y$ (**b**) of the SAR image in Figure 2b.

## 3. Data Sets

### 3.1. SAR Data

The wind streaks do not appear in all regions of the full-frame SAR image, so we find the wind streak patterns in a SAR image and apply the retrieval methods to the sub-images at these locations. After that, we can validate the retrieval results against the interpolating wind directions at these locations. To compare the retrieved results of different methods with the CCMP products on 0.25° grid, we test all methods over the 30 km by 30 km sub-images, uniformly. Finally, we acquired 62 Envisat ASAR sub-images between 2004 and 2011. Most of the images are located between 15°N to 35°N and 113°E to 129°E. ASAR operates in the C band in a wide variety of acquisition modes. The incidence angles of ASAR range from 15° to 45°. The ASAR images used here are in VV polarization with a spatial resolution of 75 m by 75 m in range and azimuth. Geometric calibrations are carried out by the Next ESA SAR Toolbox provided by the European Space Agency (ESA). The distribution of wind directions of all ASAR images used in the study is shown in Figure 5. The radius of each fan indicates the amount of wind directions in each interval.

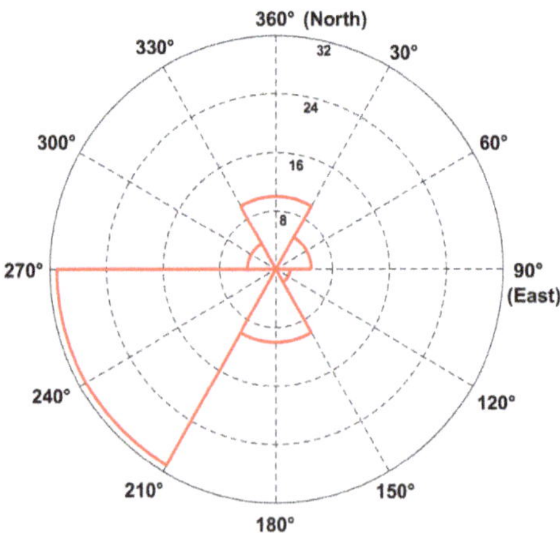


**Figure 5.** The distribution of wind directions of the images.

### 3.2. ECMWF Data

The ERA-Interim is the latest global atmospheric reanalysis produced by ECMWF. The ERA-Interim reanalysis is produced with data assimilation, advancing forward using 12-hourly analysis cycles in time. In each cycle, available observations and prior information from a forecast model are combined to obtain the evolving state of the global atmosphere and its underlying surface. Thus, the data assimilation produces 6-hourly products at different spatial resolutions, which include various kinds of global surface parameters including $u$ and $v$ wind components at 10 m height [42]. It covers a long period starting in 1979 and continues updating in near real time. The detailed method for extracting the wind direction of a specific SAR image from the ECMWF data (0.25°) is as follows. First, the nearest location is found by the latitude and longitude of the center of the SAR image. Then, the $u$ and $v$ components at imaging time are obtained by interpolating the reanalysis data of the nearest two times. At last, the wind direction is computed by using the $u$ and $v$ components from interpolation.

### 3.3. CCMP Data

The wind directions from the CCMP wind product are used for validation. The CCMP data set combines cross-calibrated satellite microwave winds and instrument observations to produce high-resolution (0.25°) gridded analysis using a 4-dimensional variational analysis method with the ECWMF ERA-Interim Reanalysis model wind field as a background wind. Each daily data file contains three arrays of size 1440 (longitude) by 628 (from 78.375°S to 78.375°N) by 4 (times of 0, 6, 12 and 18 UTC) [9]. Two of the arrays are $u$ and $v$ wind components in meters per second at 10 m height. Another array in the file is the number of observations (satellites or buoys) used to derive the wind components for each grid and a number of observations value of zero means that the wind vector for this grid was obtained from the background field only as no satellite or moored buoy wind data were available. We prefer the data from satellite or instruments, so only the locations where the number of observations is more than zero are taken into consideration. The steps for extracting the wind direction of a specific SAR image from the CCMP data are similar to that from the ECMWF data.

## 4. Results and Discussion

To compare the results of different wind direction retrieval methods, the wind directions interpolated from ECMWF and CCMP data are used, respectively. For each 30 km by 30 km image, the interpolating wind direction was obtained by the nearest grid from the center of the image. Furthermore, the performance of the ILG and the other two methods for small regions (e.g., 7.5 km, 6 km and 3 km) is discussed visually. We implemented the FFT and LG method based on algorithms described in the papers [34,38]. Therefore, we specify these methods as FFT-like method and LG-like method to suggest that they are our implementation of the original wind direction retrieval techniques, respectively.

### 4.1. Comparison with ECMWF

In this part, the wind directions interpolated from ECMWF data are used to compare the retrieved results of different methods. Figure 6 shows the comparison between the retrieved result of each method and ECMWF wind directions for all 62 images, and each dot represents the retrieved wind direction and the wind direction interpolated from ECMWF data.

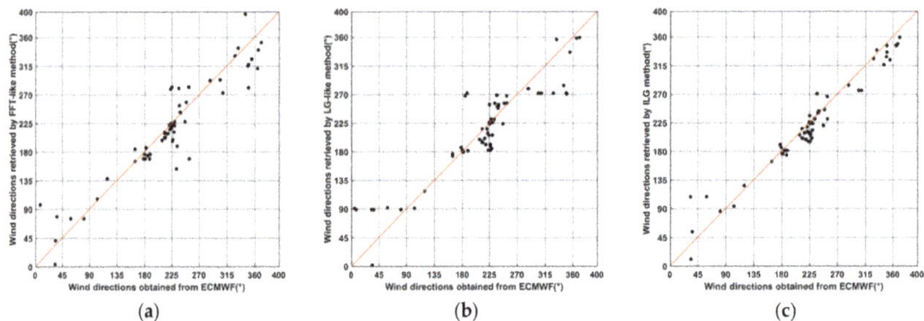


**Figure 6.** Comparison of wind directions between each retrieval method and ECMWF: (**a**) fast Fourier transform (FFT)-like method, (**b**) local gradients (LG)-like method, (**c**) ILG method.

From Figure 6, we can see that the wind directions retrieved by each method are obviously correlated with those from the ECMWF reanalysis data for these images. Table 1 shows the root-mean-square (RMS) error, correlation coefficient (R) and *p*-value (*p*-value demonstrates whether the two sets of data are correlated and the significance of correlation is larger when its value is smaller) of wind direction comparisons obtained by different retrieval methods and the ECMWF reanalysis data. In addition, if the absolute value of the angle difference between the retrieved direction and the ECMWF reanalysis data is bigger than 90°, the retrieved direction should add or subtract 360° before computing these statistics, e.g., the retrieved direction is 2° and the direction from ECMWF is 358°, so the retrieved direction should be converted to 362°. The process is similar for the computation with the CCMP data. We can see that the wind directions retrieved by the ILG method are closer to those of ECMWF reanalysis with the least RMS error of 21.57° and the largest correlation coefficient of 0.9765. In addition, the *p*-value of ILG method is the smallest among the three methods. These statistics indicate that the retrieved results of the ILG method are the closest to the ECMWF reanalysis data among the three methods. By the way, the directional error of LG method in Koch's paper is 17.6° for ERS-1/2 images, but there are no details about how to obtain the value. In fact, the validation results could be affected by several factors (e.g., the ground truth data, difference of different SAR sensors). In our study, we tested the different methods using ASAR images and the ECMWF (or CCMP) data, in order to compare their performances fairly. As a result, the statistics are different.

**Table 1.** The results of general statistics between wind directions obtained by different retrieval methods and ECMWF reanalysis data for 62 images.

| | FFT-Like Method | LG-Like Method | ILG Method |
|---|---|---|---|
| RMS(°) | 30.22 | 36.10 | 21.57 |
| R | 0.9344 | 0.9314 | 0.9765 |
| *p*-value | $1.38 \times 10^{-28}$ | $4.30 \times 10^{-25}$ | $1.03 \times 10^{-41}$ |

### 4.2. Comparison with CCMP

We prefer the CCMP data with observations (satellites or buoys), and we matched up SAR images with the CCMP data which have observations. Finally, 22 images are matched. Figure 7 shows the comparison between the results of each method and CCMP wind directions for all matched images.

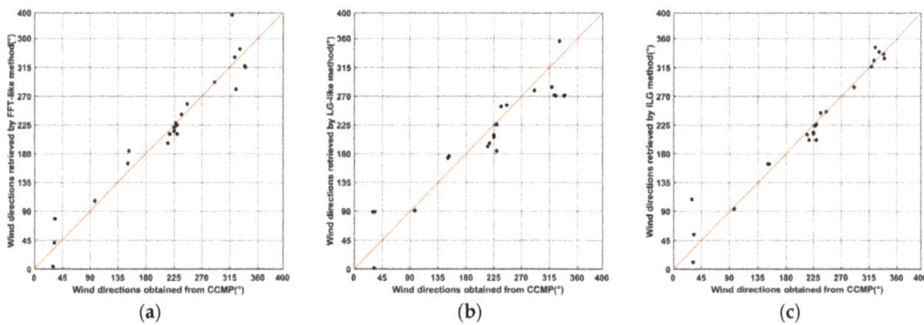


**Figure 7.** Comparison of wind directions between each retrieval method and CCMP: (**a**) FFT-like method, (**b**) LG-like method and (**c**) ILG method.

From Figure 7, we can see that the wind directions retrieved by each method show an obvious correlation with those from CCMP data for the matched images. Table 2 shows the results of statistics between wind directions obtained by each retrieval method and CCMP for these matched images. We can see that wind directions retrieved by the ILG method is the closest to the CCMP data with the least RMS error of 21.61° and the largest correlation coefficient of 0.9771. The $p$-value of ILG method is also the smallest. Consequently, the wind directions retrieved by the ILG method are the closest to CCMP data among the three methods.

**Table 2.** The results of general statistics between wind directions obtained by different retrieval methods and CCMP data for 22 images.

| | FFT-Like Method | LG-Like Method | ILG Method |
|---|---|---|---|
| RMS(°) | 25.41 | 36.65 | 21.61 |
| R | 0.9659 | 0.9413 | 0.9771 |
| $p$-value | $3.29 \times 10^{-13}$ | $6.92 \times 10^{-11}$ | $6.55 \times 10^{-15}$ |

### 4.3. Discussion

The wind directions discussed above are retrieved from 30 km by 30 km images with a spatial resolution of 75 m in both range and azimuth. To test the performance of the ILG method in small regions, wind directions of the example image in Figure 2b are computed on 7.5 km, 6 km and 3 km grids, respectively. The other two retrieval methods are also carried out on the same grids. Figure 8 shows the results of the three retrieval methods of the image on different grids. Yellow, magenta and cyan arrows indicate wind directions retrieved by the FFT-like, LG-like and ILG methods, respectively. The general wind direction of the image is from upper right to lower left. On the three kinds of grids in Figure 8, the retrieved results of the FFT-like method does not work well, and the retrieved results of the LG-like and ILG methods agree well with the streaks in the most areas. However, the retrieved results of the ILG method show better agreement with the streaks than that of the ILG method on the grids where the retrieved results of the two methods differ, visually. Another ASAR image acquired at 00:45, 25 March 2005 in VV polarization with a spatial resolution of 75 m was also tested. The image was along the coastal area of Iwate Prefecture, Japan. Figure 9 shows the image and the retrieved results of the three methods on 7.5 km, 6 km and 3 km grids, respectively. Yellow, magenta and cyan arrows indicate the wind directions retrieved by the FFT-like method, LG-like and ILG methods, respectively. From the SAR image, it can be seen that the general wind blows from upper left to lower right. From Figure 9a–c, we can see that the retrieved wind directions of the LG-like and ILG methods show better agreement with the streaks in most regions than the results of the FFT-like method. The retrieved wind directions of the ILG method still agree better with the general wind direction than the other two methods at the bottom of the image with the influence of the land.

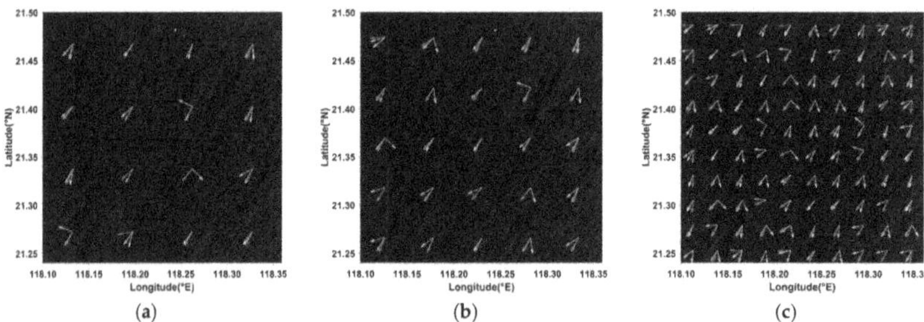

**Figure 8.** Wind directions retrieved by the three methods from 75 m pixels on different grids of the ASAR image in Figure 2b: (**a**) 7.5 km grid, (**b**) 6 km grid, (**c**) 3 km grid. Yellow, magenta and cyan arrows indicate wind directions retrieved by the FFT-like, LG-like and ILG methods, respectively.

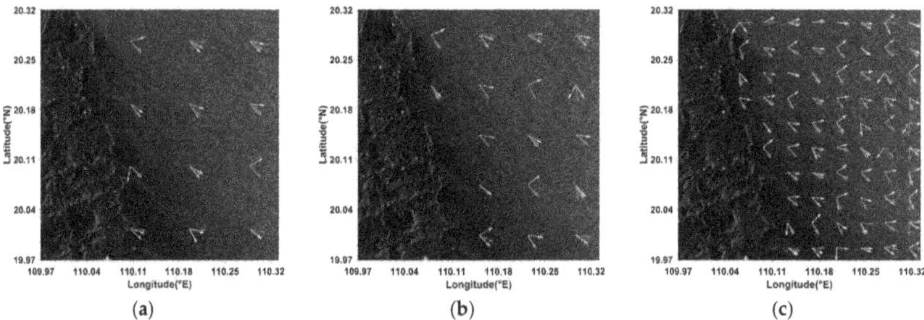

**Figure 9.** Wind directions retrieved by the three methods on different grids of the ASAR image acquired at 00:45, 25 March 2005 in VV polarization: (**a**) 7.5 km grid, (**b**) 6 km grid, (**c**) 3 km grid. Yellow, magenta and cyan arrows indicate wind directions retrieved by the FFT-like, LG-like and ILG methods, respectively.

We have tried to match all the ASAR sub-images we have with buoys, but none are matched. However, a Radarsat-2 image is matched with the buoy (46073) of National Data Buoy Center (NDBC). The Radarsat-2 image was acquired at 17:58, 11 February 2011 in VV polarization, and it was reduced to 80 m by 80 m. A region (32 km by 32 km) where the wind streaks are visible was cut out to perform the test. Figure 10 shows the results of the three retrieval methods of the sub-image on different grids (8 km, 6.4 km and 3.2 km), and the buoy is marked as a green dot in the SAR image. Again, the yellow, magenta and cyan arrows indicate wind directions retrieved by the FFT-like, LG-like and ILG methods, respectively. The wind direction (263°) measured by the buoy is indicated by the green arrow. From Figure 10, it can be seen that the retrieved wind directions of the LG-like and ILG methods agree better with the wind direction measured by the buoy than the FFT-like method. From Figure 10c, it can be seen that the retrieved results of the LG-like and ILG methods agree with each other in most regions, but the retrieved results of the ILG method show better agreement with the streaks than that of the LG method, especially at the edges of the SAR image.

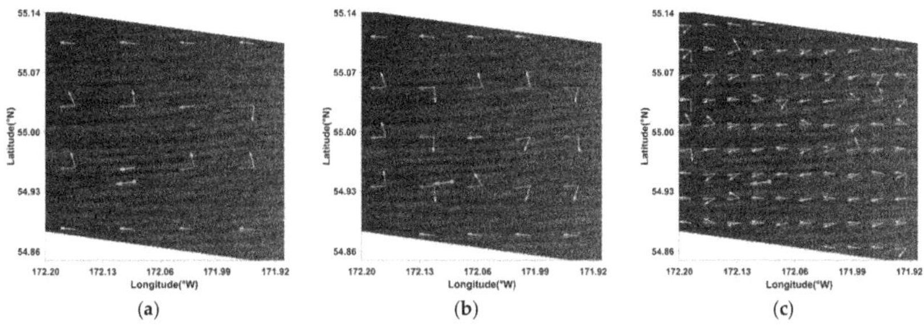


(a) (b) (c)

**Figure 10.** Wind directions retrieved by the three methods on different grids of the Radarsat-2 image acquired at 17:58, 11 February 2011 in VV polarization: (**a**) 8 km grid, (**b**) 6.4 km grid, (**c**) 3.2 km grid. Yellow, magenta and cyan arrows indicate wind directions retrieved by the FFT-like, LG-like and ILG methods, respectively. The buoy (46073) is marked as the green dot and the in situ measurement is indicated by the green arrow.

## 5. Sensitivities to Different Noise

In the conventional LG method, the local gradients are computed using difference operators (e.g., Sobel operators), which is easily affected by noise. However, the ILG method does not need any difference approximation. Therefore, the sensitivities of the three retrieval methods to different types of noise are analyzed, including the salt-and-pepper noise, the additive noise and the multiplicative noise. It is assumed that there is no noise in the original SAR images before they are corrupted by the noise in the tests. Figure 11 shows the image corrupted by different noise, taking the image in Figure 2b as an example. For a specific kind of noise with a specific intensity, the RMS of a specific method was calculated from the differences between the retrieved results of the 62 corrupted images by the method and those interpolated from the ECMWF data. After we calculate the RMS differences of the method for different intensity, we get the RMS curve of the method with respect to intensity for the kind of noise. The RMS curves are used to compare the sensitivity of each method. The same procedure was done for the 22 images matched with CCMP data.

### 5.1. Salt-and-Pepper Noise

Figure 12 shows the RMS differences between each of the three retrieval methods and interpolating wind directions of ECMWF or CCMP data when the salt-and-pepper noise exists in each image with different intensities. The intensity of the salt-and-pepper noise is decided by the percentage of contaminated pixels in the image. It is obvious that the quality of wind direction retrieval roughly decreases with the increase of noise level. The curve of the ILG method is below the curves of the other two methods in Figure 12a,b. It indicates that the ILG method can achieve better retrieved results than the other two methods comparing with ECMWF (or CCMP) data when the salt-and-pepper noise exists.

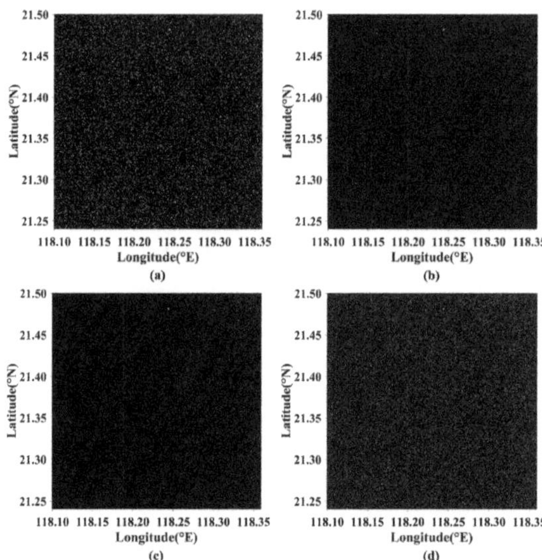


**Figure 11.** The image in Figure 2b is corrupted by each kind of noise: (**a**) salt-and-pepper noise (10% of contaminated pixels), (**b**) additive noise (SNR is −10 dB), (**c**) multiplicative noise (variance of n′ is 0.4), (**d**) speckle noise.

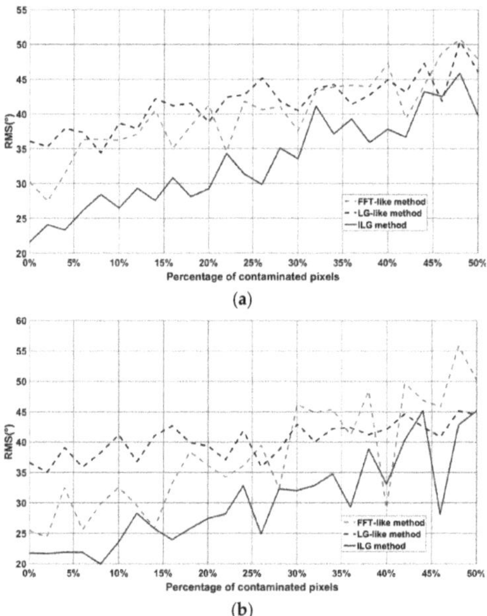


**Figure 12.** Root-mean-square (RMS) differences between the retrieved results of each method and the interpolating wind directions when the salt-and-pepper noise exists with different intensity: (**a**) ECMWF and (**b**) CCMP.

### 5.2. Additive Noise

The additive noise is added to the image $I$ using the equation

$$J = I + n \tag{11}$$

where $n$ is the Gaussian white noise with zero mean and a specific variance. As mentioned above, it is assumed that there is no noise in the original SAR images before they are corrupted by the noise in the tests. Therefore, the signal-to-noise ratio (SNR) of an image corrupted by noise can be considered as the ratio of the variance of the original image and the variance of $n$. The intensity of the noise is lower when the SNR is larger. Figure 13 shows the RMS differences between the results of the three retrieval methods and interpolating wind directions when the images are corrupted by the additive noise with different SNR. We can find that all curves decline with increasing SNR, roughly. Furthermore, the ILG method can achieve better retrieved results than the other two methods comparing with ECMWF (or CCMP) data when the additive noise is present.

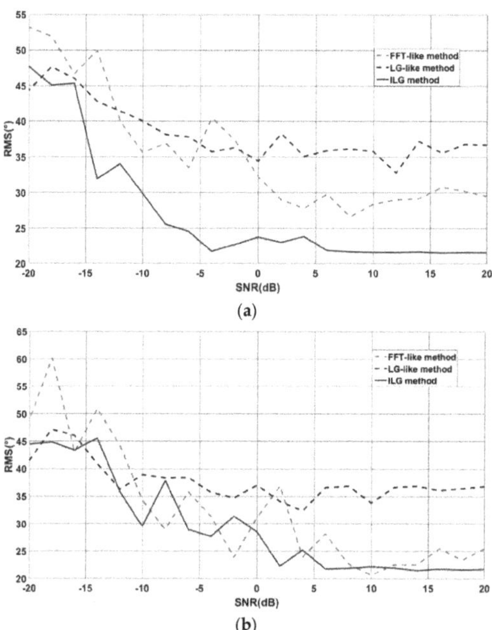


**Figure 13.** RMS differences between the retrieved results of each method and the interpolating wind directions when the additive noise exists with a different signal-to-noise ratio (SNR): (**a**) ECMWF and (**b**) CCMP.

### 5.3. Multiplicative Noise

The multiplicative noise is added to the image $I$ using the equation

$$J = (1 + n)I \tag{12}$$

where $n$ is uniformly distributed random noise with zero mean and a given variance. The multiplicative noise intensity is decided by the given variance. The noise intensity increases with the increase of the variance. Figure 14 shows the RMS differences between the results of the three retrieval methods and interpolating wind directions while the multiplicative noise changes with different variance.

The retrieved results of the three methods become worse roughly while the noise becomes greater. However, the ILG method always gives better retrieved results than the other two methods when the multiplicative noise is present comparing with ECMWF (or CCMP) data.

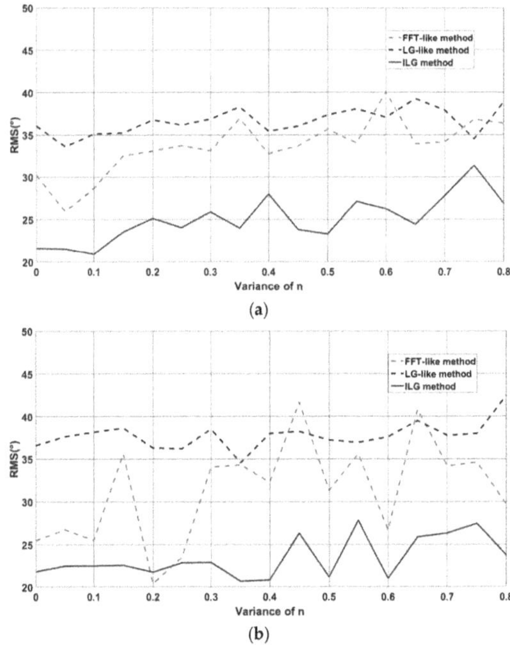


**Figure 14.** RMS differences between the retrieved results of each method and the interpolating wind directions when the multiplicative noise exists with different variance: (**a**) ECMWF and (**b**) CCMP.

The speckle noise is a kind of multiplicative noise as well, but the equation used to add the speckle noise to the image $I$ is a little different from Equation (12), which can be described as:

$$J = nI \tag{13}$$

where $n$ is the noise with an exponential distribution whose expectation and variance are 1, as the SAR intensity images are corrupted. The RMS difference of each method between retrieved results of the images corrupted by this noise and wind directions obtained from the ECMWF (or CCMP) data is shown in Table 3. From Table 3, it can be seen that the retrieved results of the corrupted images of each method become worse than the results in Tables 1 and 2, as the speckle noise affects the retrieved results. However, the retrieved results of the corrupted images by the ILG method are better than the other two methods comparing with ECMWF (or CCMP) data when the speckle noise exists.

**Table 3.** The RMS differences between retrieved results of each method of the images corrupted by the speckle noise and wind directions obtained from the ECMWF (or CCMP) data.

| | FFT-Like Method | LG-Like Method | ILG Method |
| --- | --- | --- | --- |
| ECMWF | 46.73° | 45.29° | 37.06° |
| CCMP | 46.39° | 42.14° | 37.19° |

Form the analysis above, it can be found that the sensitivity of the ILG method to each kind of noise is lower than the other two methods.

## 6. Conclusions

In this study, the conventional LG method was improved by using the new approach to compute the local gradients. With the new approach, we can avoid the difference approximation which can be easily affected by noise. Comparing with ECMWF and CCMP data, we found that the RMS difference of the ILG method is smaller (about 15°) than the FFT-like and LG-like methods. The correlation coefficient of ILG method is the largest (about 0.98) among the three methods, suggesting that its retrieved results fit the ECMWF and CCMP data better than the other two methods. The ILG method could retrieve wind directions from small images (thus high resolution) with the tests on different grids, and the retrieved results of the ILG method on small grids (e.g., 3 km) agree better with the wind streaks than the other two methods. When the SAR images containing both sea and land portions are processed, the ILG method still can work well. Furthermore, the sensitivity of the ILG method to different noises is lower than the other two methods, as the new method can avoid the use of difference approximation operators. All of these indicate that the ILG method is feasible.

**Acknowledgments:** This work is supported by the National Natural Science Foundation of China (Nos. 41676167, 41306192, 41476088, 41576174 and 41621064), the National Key R&D Program of China (Nos. 2016YFC1401007, 2016YFC1401001 and 2016YFC1401005), and the Project of State Key Laboratory of Satellite Ocean Environment Dynamics, Second Institute of Oceanography (No. SOEDZZ1503), and the Key Research and Development Program of Hainan Province (No. ZDYF2017167). ASAR data is provided by ESA. The ERA-Interim reanalysis is produced by the European Centre for Medium-Range Weather Forecasts, and the data from the ECMWF is available at www.ecmwf.int. CCMP Version-2.0 vector wind analyses, which are available at www.remss.com, are produced by the Remote Sensing Systems. The NDBC data is available at http://www.ndbc.noaa.gov. I also want to thank Zuojun Yu for her suggestions on my English expressions. The views, opinions, and findings contained in this report are those of the authors and should not be construed as an official NOAA or U.S. Government position, policy or decision.

**Author Contributions:** Lizhang Zhou and Gang Zheng contributed the main idea and wrote the manuscript; Xiaofeng Li contributed to the design of the research and the article's organization; Jingsong Yang, Lin Ren and Peng Chen contributed to the collection the SAR images and data analysis; Huaguo Zhang and Xiulin Lou provided the data for comparative analysis and revised the manuscript. All authors have read and approved the submitted manuscript.

**Conflicts of Interest:** The authors declare no conflict of interest.

## References

1. Friedman, K.S.; Sikora, T.D.; Pichel, W.G.; Clementecolón, P.; Hufford, G. Using spaceborne synthetic aperture radar to improve marine surface analyses. *Weather Forecast.* **2010**, *16*, 270–276. [CrossRef]
2. Von Ahn, J.M.; Sienkiewicz, J.M.; Chang, P.S. Operational impact of QuikSCAT winds at the NOAA ocean prediction center. *Weather Forecast.* **2006**, *21*, 523–539. [CrossRef]
3. Christiansen, M.B.; Koch, W.; Horstmann, J.; Hasager, C.B.; Nielsen, M. Wind resource assessment from C-band SAR. *Remote Sens. Environ.* **2006**, *105*, 68–81. [CrossRef]
4. Chang, R.; Zhu, R.; Badger, M.; Hasager, C.B.; Xing, X.; Jiang, Y. Offshore wind resources assessment from multiple satellite data and WRF modeling over south China sea. *Remote Sens.* **2015**, *7*, 467–487. [CrossRef]
5. Cavaleri, L.; Alves, J.H.; Ardhuin, F.; Babanin, A.; Banner, M.; Belibassakis, K.; Benoit, M.; Donelan, M.; Groeneweg, J.; Herbers, T.H.C. Wave modeling—The state of the art. *Prog. Oceanogr.* **2007**, *74*, 603–674. [CrossRef]
6. Sullivan, P.P.; Mcwilliams, J.C. Dynamics of winds and currents coupled to surface waves. *Annu. Rev. Fluid Mech.* **2009**, *42*, 19–42. [CrossRef]
7. Xu, Q.; Li, X.; Wei, Y.; Tang, Z.; Cheng, Y.; Pichel, W.G. Satellite observations and modeling of oil spill trajectories in the Bohai sea. *Mar. Pollut. Bull.* **2013**, *71*, 107–116. [CrossRef] [PubMed]
8. Cheng, Y.; Liu, B.; Li, X.; Nunziata, F. Monitoring of oil spill trajectories with Cosmo-SkyMed X-band SAR images and model simulation. *IEEE J. Sel. Top. Appl. Earth Obs. Remote Sens.* **2014**, *7*, 2895–2901. [CrossRef]

9. Atlas, R.; Hoffman, R.N.; Ardizzone, J.; Leidner, S.M.; Jusem, J.C.; Smith, D.K.; Gombos, D. A cross-calibrated, multiplatform ocean surface wind velocity product for meteorological and oceanographic applications. *Bull. Am. Meteorol. Soc.* **2011**, *92*, 157–174. [CrossRef]
10. Salonen, K.; Niemelä, S.; Fortelius, C. Application of radar wind observations for low-level NWP wind forecast validation. *J. Appl. Meteorol.* **2011**, *50*, 1362–1371. [CrossRef]
11. Yang, X.; Li, X.; Zheng, Q.; Gu, X.F. Comparison of ocean-surface winds retrieved from QuikSCAT scatterometer and Radarsat-1 SAR in offshore waters of the US west coast. *IEEE Geosci. Remote Sens. Lett.* **2011**, *8*, 163–167. [CrossRef]
12. Wentz, F.J. Measurement of oceanic wind vector using satellite microwave radiometers. *IEEE Trans. Geosci. Remote Sens.* **1992**, *30*, 960–972. [CrossRef]
13. Naderi, F.M.; Freilich, M.H.; Long, D. Spaceborne radar measurement of wind velocity over the ocean—An overview of the NSCAT scatterometer system. *Proc. IEEE* **1991**, *79*, 850–866. [CrossRef]
14. Gerling, T.W. Structure of the surface wind field from the Seasat SAR. *J. Geophys. Res.* **1986**, *91*, 2308–2320. [CrossRef]
15. Mastenbroek, K. High-resolution wind fields from ERS SAR. *Earth Obs. Quart.* **1998**, *59*, 20–22.
16. Horstmann, J.; Koch, W. Ocean wind field retrieval using Envisat ASAR data. In Proceedings of the Geoscience and Remote Sensing Symposium, Toulouse, France, 21–25 July 2003; pp. 3102–3104.
17. Zhang, B.; Perrie, W.; He, Y. Wind speed retrieval from Radarsat-2 quad-polarization images using a new polarization ratio model. *J. Geophys. Res. Atmos.* **2011**, *116*, 1318–1323. [CrossRef]
18. Li, X. The first Sentinel-1 SAR image of a typhoon. *Acta Ocean. Sin.* **2015**, *34*, 1–2. [CrossRef]
19. Kim, T.S.; Park, K.A.; Li, X.; Hong, S. SAR-derived wind fields at the coastal region in the East/Japan sea and relation to coastal upwelling. *Int. J. Remote Sens.* **2014**, *35*, 3947–3965. [CrossRef]
20. Xuan, Z.; Yang, X.F.; Li, Z.W.; Yang, Y.; Bi, H.B.; Sheng, M.; Li, X.F. Estimation of tropical cyclone parameters and wind fields from SAR images. *Sci. China Earth Sci.* **2013**, *56*, 1977–1987.
21. Li, X.; Zhang, J.A.; Yang, X.; Pichel, W.G.; Demaria, M.; Long, D.; Li, Z. Tropical cyclone morphology from spaceborne synthetic aperture radar. *Bull. Am. Meteorol. Soc.* **2013**, *94*, 215. [CrossRef]
22. Li, X.; Pichel, W.G.; He, M.; Wu, S.Y. Observation of hurricane-generated ocean swell refraction at the Gulf Stream north wall with the Radarsat-1 synthetic aperture radar. *IEEE Trans. Geosci. Remote Sens.* **2002**, *40*, 2131–2142.
23. Friedman, K.S.; Li, X. Monitoring hurricanes over the ocean with wide swath SAR. *Johns Hopkins Apl Tech. Dig.* **2000**, *21*, 80–85.
24. Li, X.; Zheng, W.; Pichel, W.G.; Zou, C.Z.; Clemente-Colón, P. Coastal katabatic winds imaged by SAR. *Geophys. Res. Lett.* **2007**, *34*, 300–315. [CrossRef]
25. Li, X.; Zheng, W.; Yang, X.; Zhang, J.A.; Pichel, W.G.; Li, Z. Coexistence of atmospheric gravity waves and boundary layer rolls observed by SAR. *J. Atmos. Sci.* **2013**, *70*, 3448–3459. [CrossRef]
26. Liu, S.; Li, Z.; Yang, X.; Pichel William, G.; Yu, Y.; Zheng, Q.; Li, X. Atmospheric frontal gravity waves observed in satellite SAR images of the Bohai sea and Huanghai sea. *Acta Ocean. Sin.* **2010**, *29*, 35–43. [CrossRef]
27. Li, X.; Dong, C.; Clemente-Colón, P.; Pichel, W.G.; Friedman, K.S. Synthetic aperture radar observation of the sea surface imprints of upstream atmospheric solitons generated by flow impeded by an island. *J. Geophys. Res.* **2004**, *109*, 235–250. [CrossRef]
28. Chunchuzov, I.; Vachon, P.W.; Li, X. Analysis and modeling of atmospheric gravity waves observed in Radarsat SAR images. *Remote Sens. Environ.* **2000**, *74*, 343–361. [CrossRef]
29. Li, X.; Zheng, W.; Zou, C.Z.; Pichel, W.G. A SAR observation and numerical study on ocean surface imprints of atmospheric vortex streets. *Sensors* **2008**, *8*, 3321–3334. [CrossRef] [PubMed]
30. Li, X.; Clemente-Colón, P.; Pichel, W.G.; Vachon, P.W. Atmospheric vortex streets on a Radarsat SAR image. *Geophys. Res. Lett.* **2000**, *27*, 1655–1658. [CrossRef]
31. Li, X.; Yang, X.; Zheng, W.; Zhang, J.A. Synergistic use of satellite observations and numerical weather model to study atmospheric occluded fronts. *IEEE Trans. Geosci. Remote Sens.* **2015**, *53*, 1–11. [CrossRef]
32. Mouche, A.; Collard, F.; Chapron, B.; Dagestad, K.; Guitton, G.; Johannessen, J.A.; Kerbaol, V.; Hansen, M.W. On the use of Doppler shift for sea surface wind retrieval from SAR. *IEEE Trans. Geosci. Remote Sens.* **2012**, *50*, 2901–2909. [CrossRef]

33. Wackerman, C.C.; Pichel, W.G.; Clemente-Colon, P. Automated estimation of wind vectors from SAR. In Proceedings of the 12th Annual Conference on Satellite Meteorology at the 83rd Annual Meteorological Association Meeting, Long Beach, CA, USA, 9–13 February 2003.
34. Koch, W. Directional analysis of SAR images aiming at wind direction. *IEEE Trans. Geosci. Remote Sens.* **2004**, *42*, 702–710. [CrossRef]
35. Zhang, G.; Perrie, W.; Li, X.; Zhang, J.A. A hurricane morphology and sea surface wind vector estimation model based on C-band cross-polarization SAR imagery. *IEEE Trans. Geosci. Remote Sens.* **2017**, *55*, 1–9. [CrossRef]
36. Zhang, B.; Perrie, W.; Vachon, P.W.; Li, X.; Pichel, W.G.; Guo, J.; He, Y. Ocean vector winds retrieval from C-band fully polarimetric SAR measurements. *IEEE Trans. Geosci. Remote Sens.* **2012**, *50*, 4252–4261. [CrossRef]
37. Liu, G.; Yang, X.; Li, X.; Zhang, B. A systematic comparison of the effect of polarization ratio models on sea surface wind retrieval from C-band synthetic aperture radar. *IEEE J. Sel. Top. Appl. Earth Obs. Remote Sens.* **2013**, *6*, 1100–1108. [CrossRef]
38. Wackerman, C.; Shuchman, R.; Fetterer, F. Estimation of wind speed and wind direction from ERS-1 imagery. In Proceedings of the Geoscience and Remote Sensing Symposium, Pasadena, CA, USA, 8–12 August 1994; pp. 1222–1224.
39. Yang, J.S.; Huang, W.G.; Zhou, C.B.; Bin, F.U.; Shi, A.Q.; Li, D.L. Coastal ocean surface wind retrieval from SAR imagery. *J. Remote Sens.* **2001**, *5*, 13–16.
40. Ulaby, F.T.; Moore, R.K.; Fung, A.K. *Microwave Remote Sensing: Active and Passive, Volume 2—Radar Remote Sensing and Surface Scattering and Emission Theory*; Addison-Wesley: Boston, MA, USA, 1982.
41. Wang, L.; Zhou, Y.; Liu, Y.; Zhang, M. Wind retrieval over the ocean using advanced synthetic aperture radar. In Proceedings of the International Conference on Geoinformatics, Shanghai, China, 24–26 June 2011; pp. 1–5.
42. Dee, D.P.; Uppala, S.M.; Simmons, A.J.; Berrisford, P.; Poli, P.; Kobayashi, S.; Andrae, U.; Balmaseda, M.A.; Balsamo, G.; Bauer, P. The ERA-interim reanalysis: Configuration and performance of the data assimilation system. *Q. J. R. Meteorol. Soc.* **2011**, *137*, 553–597. [CrossRef]

*Article*

# Assimilation of Sentinel-1 Derived Sea Surface Winds for Typhoon Forecasting

**Yi Yu [1,*], Xiaofeng Yang [2,3], Weimin Zhang [1], Boheng Duan [1], Xiaoqun Cao [1] and Hongze Leng [1]**

1 School of Computer Science, National University of Defense Technology, Changsha 410073, China; wmzhang104@139.com (W.Z.); bhduan@foxmail.com (B.D.); caoxiaoqun@nudt.edu.cn (X.C.); hzleng@nudt.edu.cn (H.L.)

2 State Key Laboratory of Remote Sensing Science, Institute of Remote Sensing and Digital Earth, Chinese Academy of Sciences, Beijing 100101, China; yangxf@radi.ac.cn

3 The Key Laboratory for Earth Observation of Hainan Province, Sanya 572029, China

* Correspondence: yuyi2019@nudt.edu.cn; Tel.: +86-0731-84573667

Academic Editors: Xiaofeng Li, Ferdinando Nunziata and Alexis Mouche
Received: 16 June 2017; Accepted: 10 August 2017; Published: 14 August 2017

**Abstract:** High-resolution synthetic aperture radar (SAR) wind observations provide fine structural information for tropical cycles and could be assimilated into numerical weather prediction (NWP) models. However, in the conventional method assimilating the $u$ and $v$ components for SAR wind observations (SAR_uv), the wind direction is not a state vector and its observational error is not considered during the assimilation calculation. In this paper, an improved method for wind observation directly assimilates the SAR wind observations in the form of speed and direction (SAR_sd). This method was implemented to assimilate the sea surface wind retrieved from Sentinel-1 synthetic aperture radar (SAR) in the basic three-dimensional variational system for the Weather Research and Forecasting Model (WRF 3DVAR). Furthermore, a new quality control scheme for wind observations is also presented. Typhoon Lionrock in August 2016 is chosen as a case study to investigate and compare both assimilation methods. The experimental results show that the SAR wind observations can increase the number of the effective observations in the area of a typhoon and have a positive impact on the assimilation analysis. The numerical forecast results for this case show better results for the SAR_sd method than for the SAR_uv method. The SAR_sd method looks very promising for winds assimilation under typhoon conditions, but more cases need to be considered to draw final conclusions.

**Keywords:** SAR; sea surface wind; assimilation; observational error; typhoon

## 1. Introduction

Sea surface wind is the primary power source for atmospheric movement over the ocean surface [1] and remarkably affects the air-sea exchange process. Tropical cyclones (TCs), storm surges, and many other severe ocean conditions are driven by the sea surface wind. However, the usual measurements of sea surface wind observations from the buoys and ships are scarce and distributed irregularly. The observations for bad weather are often far away from ships on limited routes, and especially, there are few observations of high precision and resolution in the area of TCs. The lack of observations regarding the initial analysis in the TC model would greatly reduce the accuracy of the TC forecast.

In recent years, microwave remote sensing instruments such as microwave scatterometers and synthetic aperture radar (SAR) have been used to retrieve sea surface wind fields [2–4]. Although the scatterometer observes the ocean surface with wide coverage [5], the spatial resolution of scatterometers is 12.5~25 km and the accuracy of scatterometers' measurements of winds is low in coastal regions [6].

However, the SAR observations are of high spatial resolution [7–9], and these data can provide very detailed information on the structure of tropical cyclones [10]. The accuracy of sea surface wind data retrieved from SAR is also comparable with scatterometer data [11,12], and these wind fields can be used with a data assimilation system to provide the initial conditions for the numerical weather prediction (NWP) model [13]. Some researchers have tried to adopt the SAR observations in the assimilation system. Danielson designed a plan to assimilate SAR wind information in Environment Canada's high-resolution three-dimensional variational (3DVAR) analysis system [14]. Perrie et al. assimilated the SAR derived wind, which captured Hurricane Isanbe'eye, and found that the analysis from the experiment provided new information about Isabel's central region [15]. Choisnard et al. assessed the quality of a marine wind vector retrieved from the variational data assimilation of SAR backscatter observation and inferred that wind direction information from wind streaks could be of interest to add some wind direction sensitivity [16].

The SAR derived wind products are in the form of the wind speed (*spd*) and the wind direction (*dir*). In current data assimilation systems, including the Weather Research and Forecasting Model Data Assimilation system (WRFDA), the input wind products are transformed to a longitudinal component ($u$ wind) and a latitudinal component ($v$ wind). Then, $u$ and $v$ are used in the assimilation calculation as the vectors for wind observations [17]. In this conventional method, *dir* is not a state vector and can not influence the analysis directly. Furthermore, the impact of the *dir* observational error on $u$ and $v$ is ignored. This will cause large errors in the $u$ and $v$ assimilation, especially when the wind observations are largely different from the background wind, e.g., during typhoon events [18]. In fact, for SAR wind retrieval, the wind speed is derived from the radar sea surface backscattering signal, while the wind direction is usually extracted from wind-aligned patterns on SAR imagery [10]. Consequently, the *spd* and *dir* errors of SAR observations are independent, and the *dir* observational error should not be ignored in the wind data assimilation.

Recently, a novel method to assimilate the wind observations has been proposed by Huang et al., which directly assimilates *spd* and *dir* based on the transformation of the state vectors $u$ and $v$ to *spd* and *dir* by the observation operator [18]. Gao et al. further tested this method in the experiments for the satellite-derived Atmospheric Motion Vectors (AMV) and the surface dataset in the Meteorological Assimilation Data Ingest System (MADIS) [19]. However, the quality control process is ignored for the ideal observations in these studies, and the derived values of $u$ and $v$ were not checked together in the quality control stage, even though they were derived from a single wind vector. In this study, an improved method is proposed based on Huang et al., which can assimilate the SAR wind data under typhoon conditions. A new quality control scheme is also presented. The proposed methods are tested and compared with the traditional methods within the WRF/3DVAR framework for the prediction of Typhoon Lionrock (2016). The impact of assimilation of SAR sea surface winds on the typhoon track and intensity will be examined using different methods.

The rest of the paper is organized as follows. In Section 2, we introduce the sea surface wind retrieval method, the assimilation methods, and the quality control scheme. The numerical model and experiment setup are described in Section 3. The experimental results are compared and discussed in Section 4. The conclusions are presented in Section 5.

## 2. Data and Methodology

### 2.1. SAR Wind Retrieval

On 3 April 2014, the European Space Agency (ESA) launched the C-band Sentinel-1A satellite. They acquired the first Sentinel-1 typhoon image in the northwest Pacific on 4 October 2014 [20]. Sentinel-1 provides free and open SAR data for ocean, land changes, and emergency response applications, and the data have been utilized in research for hurricane/typhoon studies [21–23]. Sentinel-1 has four operational modes, i.e., Strip Map Mode, Interferometric Wide Swath Mode, Extra-wide Swath Mode, and Wave Mode. Sentinel-1 has selectable single polarization (VV, vertical transmit and vertical receive or HH, horizontal transmit and horizontal receive) for the Wave Mode and selectable dual polatization (VV + VH, vertical transmit and horizontal receive or HH + HV, horizontal transmit and vertical receive) for all other modes. In this study, an Extra-wide Swath Mode dual polarization (VV + VH) SAR image is used to derive the sea surface wind map of Typhoon Lionrock (as shown in Figure 1). Its swath is 400 km, and its spatial resolution is 25 × 100 m.

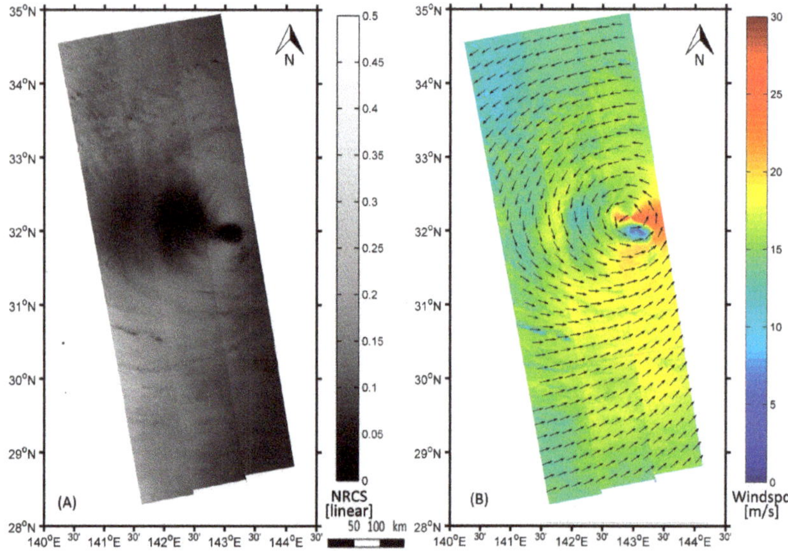


**Figure 1.** Windfield retrieved from Sentinel-1A synthetic aperture radar (SAR) Extra-wide Swath Mode data on 29 August 2016 at 08:32 UTC. (**A**) Sea surface radar backscattering map; (**B**) the SAR derived sea surface wind map.

The raw data of Sentinel-1 observations are first calibrated to give the Normalized Radar Cross-section (NRCS), which expresses the radar return signal per unit area and depends on the instant wind stress over the ocean surface and the radar viewing geometry. The Geophysic model function (GMF) is used for SAR wind retrieval and here a C-band geophysical model function (CMOD-5) [24] is used to convert the calibrated NRCS measurement to sea surface winds. The general form of the GMF is:

$$\sigma_0 = M(U, \varphi, \theta; p, f) \tag{1}$$

where $\sigma_0$ is NRCS; $U$ is the wind speed at 10 m; $\varphi$ is the wind direction with respect to the radar look direction; $\theta$ is local incidence angle; $p$ is the method of polarization; and $f$ is the frequency of the incident radar wave. As shown in Equation (1), the wind speed retrieval from SAR requires a priori information about the wind direction as the SAR is only capable of observing each point on the ocean surface from a single look angle. This wind direction information is usually obtained from in situ measurements, numerical model outputs, or SAR imagery. For the typhoon wind field, there are wind streaks in SAR imagery, and the wind direction can be extracted from this image pattern. In this study, a method called Discrete Wavelet Transform (DWT) is chosen to derive the sea surface wind direction. The details of the DWT methods can be found in Du et al., 2002 [25] and Zhou et al., 2013 [26]. In the SAR wind retrieval, the heavy rain has been filtered out.

### 2.2. Data Assimilationof SAR Sea Surface Winds

Data assimilation provides initial conditions to atmospheric models by concentrating on searching for a solution that minimizes simultaneously the distance between observations and the background and the distance between the initial guess variables and the analysis variables [27]. In the three dimensional variational data assimilation (3DVAR) framework, the initial conditions are the best estimations obtained through the minimization of a cost function based on Bayes theory [28], defined as:

$$J(x) = J_b + J_o = \frac{1}{2}(x - x_b)^T B^{-1}(x - x_b) + \frac{1}{2}[y - H(x)]^T R^{-1}[y - H(x)] \tag{2}$$

Here, $J_b$ is the background term and describes the misfit between the model state variable x and the background state $x_b$, which is derived from short-range forecast. $J_o$ is the observation term, describing the misfit between the observation vector $y$ and the vector equivalent to the model state variable, which is projected to the observation space by the observation operator $\boldsymbol{H}$ [29,30]. $B$ and $R$ are the background and observational error covariance matrices, respectively, and both are assumed Gaussian distributions. The superscripts $T$ and $-1$ denote the inverse and adjoint, respectively.

Since the state vectors for wind observations in the assimilation system are $u$ and $v$, the conventional method (referred as SAR_uv) assimilates the $u$ and $v$ wind derived from the initial observation of *spd* and *dir*. The impact of *dir* errors on the uncertainly of $u$ and $v$ is not considered during the assimilation process, and the *dir* errors have no independent influence on the assimilation results. A new method (referred as SAR_sd) proposed by Huang et al. [18] can directly assimilate *spd* and *dir*. In this method, the state vectors $u$ and $v$ need to be converted into *spd* and *dir* by the observation operator $H$, then the observation vector $y$ contains the variables $sp^o$ and $dir^o$, as below:

$$y_{sd} = (\cdots, sp^o, dir^o, \cdots)^T \tag{3}$$

Assuming the observation vector only contains the wind observation, the observation term $J_o$ can be written as:

$$J_o = \frac{1}{2}\left(\frac{sp^o - sp^b}{\sigma_{sp}^o}\right)^2 + \frac{1}{2}\left(\frac{dir^o - dir^b}{\sigma_{dir}^o}\right)^2 \tag{4}$$

Here, $sp^b$ and $dir^b$ are the wind speed and direction from the background, respectively, and $\sigma_{sp}^o$ and $\sigma_{dir}^o$ are the observational errors for wind speed and direction, respectively. By this method, the observational error of wind direction is considered for the wind observations. The details of this method were described in Huang et al. [18].

### 2.3. Observation Quality Control Scheme

Wind observations may carry many kinds of errors such as the error from the instruments, the error from the retrieval method, and so on. Adequate quality control can filter out spurious observations from the data assimilation system and obtain a more accurate field. The observation quality control process is usually verified by the observation innovation and the observation errors [31]. In the WRFDA system, when the observation innovation is greater than five times the observation error, the observation is rejected [32], as shown in the Equation (5):

$$|d| > 5\sigma \tag{5}$$

Here, $d$ is the observation innovation and $\sigma$ is the observational error. The observation innovation is derived by subtracting the background by the observation, as shown in the following equation:

$$d = y - H(x^b) \tag{6}$$

For *spd*, if the observation error is 2 m/s, then when the *spd* innovation is larger than 10 m/s, the observation is rejected. The same is for $u$ and $v$. For *dir*, make a definition that the observation difference is *inv*, and:

$$inv = \left| dir^o - dir^b \right| \quad (7)$$

Here, $dir^o$ is the *dir* observation and $dir^b$ is the *dir* background. Then the *dir* observation innovation is defined as below:

$$\begin{cases} d = inv & if \;\; inv \leq 180^\circ \\ d = 360^\circ - inv & if \;\; inv > 180^\circ \end{cases} \quad (8)$$

For example, if $dir^b$ = 0°, $dir_1^o = 30^\circ$, and $dir_2^o = 330^\circ$, then the observation innovation is 30° for both $dir_1^o$ and $dir_2^o$. If the *dir* observation error is 20° and the *dir* innovation is greater than 100°, then the *dir* observation is rejected. In this study, a *dir* observation error of 20° is applied, and a wind speed error of 2 m/s is applied for $u$, $v$, and *spd*.

There are two different quality control methods for the two components in the single wind vector, both for $u$ and $v$ in the SAR_uv assimilation or for *spd* and *dir* in the SAR_sd assimilation. As in most assimilation systems, including WRFDA, the assumptions are widely used that the errors for different state variables are independent and that the quality controls for different state variables are independent [33]. Here we refer to the method that indicates that the quality control for the two components in one single wind vector is independent as Quality Controlled alone (QC_al). However, the two components $u$ and $v$ are not observed independently and are calculated from one wind vector. For the *spd* and *dir* observations, sometimes they are independent, like the *spd* measured by the rotating cup anemometer in most operational 10-m synoptic observation stations while the *dir* is observed by a vane, and sometimes they are dependent, like the Atmospheric Motion Vectors (AMV) derived from satellite imagery. If the two components are not observed independently, they should be quality controlled together. Here we refer the method that indicates that two components in one single wind vector are checked by each other during the quality control as Quality Controlled corporately (QC_co).To discuss the two methods of quality control, one wind vector from the background (BKG) and four observation wind vectors are presented in Figure 2, and the values for all five vectors are detailed in the Table 1. In this figure, BKG is represented by a blue arrow, and the red arrows represent four different examples of observations. For SAR_sd assimilation, if we use the QC_co method for quality control, only observation 1 (OBS1) and observation 2 (OBS2) of the four observations are accepted in *dir* quality control, as their wind direction innovations are less than 100°, like all the observations distributed in the right hand angle between the green boundary lines Boundary1 and Boundary2, but OBS2 is rejected in *spd* quality control as the *spd* innovation is larger than 10 m/s. Finally, only OBS1 remains in the quality control. If we use the QC_al method, OBS2 can keep the *dir* observation, while the *spd* observation is rejected, and observation 3 (OBS3) and observation 4 (OBS4) can keep the *spd* observations, but the *dir* observations are rejected. For SAR_uv assimilation, the direction error is not considered. If the QC_co method is applied, OBS1 and OBS4 are kept, OBS2 is rejected as the $v$ innovation is larger than 10m/s, and the OBS3 is rejected with $u$ innovation larger than 10 m/s. If the QC_al method is applied, OBS2 can keep $u$ wind and OBS3 can keep $v$ wind. The results are concluded in Table 2.

From the discussion above, we conclude that, in SAR_uv assimilation, no matter which quality control method is adopted, observations like OBS4 filling in the left angle of the two green boundary lines could be assimilated, which may result bad analysis. If using the QC_al method, SAR_uv assimilation will reject all the observations with high speed such as $v$ of OBS2 and $u$ of OBS3, only keeping the $u$ or $v$ speed close to the background.

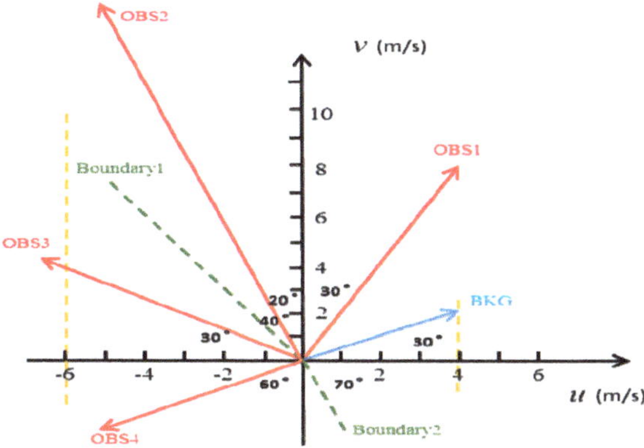


**Figure 2.** Diagram of background wind vector BKG and the observation wind vectors (OBS1, OBS2, OBS3, and OBS4) used to present the difference between thequality control procedures, QC_co and QC_al, of SAR_uv (standard assimilation method in theWeather Research and Forecasting Model Data Assimilation system (WRFDA), assimilating SAR wind observation in the form of $u$ and $v$ components) and SAR_sd (new assimilated method, assimilating SAR wind observation in the forms of wind speed and direction).

**Table 1.** The detail value of the background and four kinds of observations.

| Wind Vectors | *Spd* (m/s) | *Dir* (°) | *U* (m/s) | *V* (m/s) |
|---|---|---|---|---|
| BKG | 4.61 | 30 | 4.00 | 2.31 |
| OBS1 | 8.00 | 60 | 4.00 | 6.93 |
| OBS2 | 14.62 | 110 | −4.99 | 13.74 |
| OBS3 | 6.80 | 150 | −5.89 | 3.40 |
| OBS4 | 5.60 | 210 | −4.85 | −2.80 |

**Table 2.** The remaining components after quality control for the two assimilation methods.

| Observation Types | SAR_sd | | SAR_uv | |
|---|---|---|---|---|
| | QC_co | QC_al | QC_co | QC_al |
| OBS1 | spd, dir | spd, dir | u, v | u, v |
| OBS2 | - | dir | - | u |
| OBS3 | - | spd | - | v |
| OBS4 | - | spd | u, v | u, v |

## 3. Case Study—Typhoon Lionrock of 2016

### 3.1. Description of Typhoon Lionrock

Typhoon Lionrock was the tenth named storm in 2016. It was a powerful, long-lived, severe tropical cyclone and caused remarkable flooding and casualties in North Korea and Japan [34]. After forming as a hybrid disturbance located about 585 km to the west of Wake Island on 15 August, it moved southwestward and intensified into a typhoon, then moved northeastward and became a super typhoon. On 29 August, Lionrock weakened and moved on an unprecedented path towards the northeastern region of Japan. Right before weakening into a severe tropical storm at 0900 UTC on 30 August, Lionrock made landfall near Ōfunato, a city in Iwate Prefecture. After landing, the center of the Lionrock cluster intensified, moving further to the northwest as a Pacific storm at the rare 70 to 80 km per hour. After just a few hours, Lionrock swept northeast of Japan and moved again into the Sea of Japan at noon on the day. This makes Lionrock the first tropical cyclone to make landfall over the Pacific coast of the Tōhoku region of Japan since the Japan Meteorological Agency began record-keeping in 1951. Lionrock transformed to be a tropical storm at the night of 31 August and landed again in the vicinity of Vladivostok, Russia. It continued moving westward, turned into a temperate cyclone on 1 September, and was last noted on 2 September in the Jilin Province, China.

### 3.2. Experimental Design

Three groups of experiments are designed to investigate the impact of SAR wind data in three-dimensional variational system for the Weather Research and Forecasting Model (WRF 3DVAR). These runs are referred to as the basic control experiment (CNTL) without observations, the SAR_uv experiment assimilating $u$ and $v$, and the SAR_sd experiment assimilating the SAR wind speed and direct retrievals. The background fields were derived from a WRF simulation, which was integrated from 0000 UTC 28 August 2016 for 24 h, and the National Centers for Environmental Prediction (NCEP) final (FNL) analysis data was adopted in the WRF model. All the groups of experiments used the same background field. The analysis time for data assimilation was 0900 UTC 29 August 2016. In the SAR_uv and SAR_sd experiments, the assimilation window was 6 h, and the SAR observations were sounded at about 0830 UTC 29 August. After the assimilation, the analysis field was inputted to the WRF forecast model and run from the analysis time for 33 h. The input wind speed of the observations was checked to be less than 25 m/s and thinned to be 15 km. In both cases, 23,207 wind vectors were inputted. After quality control, only 327 wind vectors were accepted; 22,880 wind vectors were rejected in the SAR_sd case, and, in the SAR_uv case, 2896 wind components ($u$ and $v$ wind) were accepted and 20,311 wind vectors were rejected. The accepted observations for the two cases are plotted in Figure 3. The details of the three experiments are provided in Table 3.

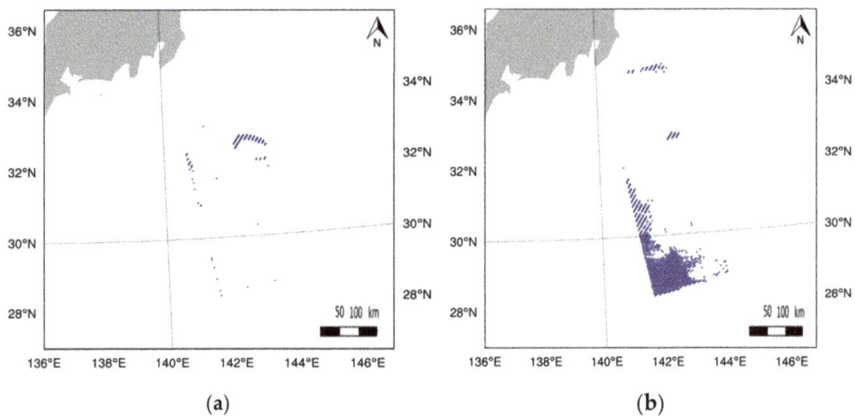


**Figure 3.** The accepted observtion wind vectors in the (**a**) SAR_sd and in the (**b**) SAR_uv experiments. The blue points are the wind vectors.

**Table 3.** Details of the three experiments.

| Experiment | Data | Operator | Quanlity Controll | Accept Obs. | Reject Obs. |
|---|---|---|---|---|---|
| CNTL | - | - | - | - | - |
| SAR_uv | $u$ and $v$ componets | UV operator | QC_co | 2896 | 20,311 |
| SAR_sd | *spd* and *dir* | SD operator | QC_co | 327 | 22,880 |

### 3.3. Model Description

The WRF model developed by the United States National Centers for Environmental Prediction was used in this study, and the 3DVAR version 3.5 was used as the basic assimilation system for the

SAR observations [29]. The UV operator and the SD operator were implemented in the 3DVAR system for the assimilation of the SAR UV wind and SD wind retrievals, respectively. The WRF 3.8 version was run as the forecast model.

Various physical parameterization schemes are developed in the non-hydrostatic WRF model for the researchers to choose properly in order to simulate the atmospheric structures. The major physical options in our experiments include the WRF Single-Moment 3-class (WSM3) microphysics scheme [35], the Kain-Fritsch (KF) convective parameterization [36], and the Yonsei University (YSU) boundary layer scheme [37]. The domain of the WRF 3DVAR analysis and the WRF model simulation has 260 × 250 grid points, and the center of the domain is located at 30.0°N, 135°E (as shown in Figure 4). The horizontal resolution is 15km, and the vertical levels are 51 in the WRF model framework; the no-nesting method is utilized, and the Terrian-following coordinate ($\sigma$-coordinate) [38] at the top of the model atmosphere was located at 10 hPa. The SAR observed the area covering the center of the typhoon, and the structure of the wind field around the typhoon center can be seen clearly in Figure 1. Like the wind retrieval from the scatterometer observations, the SAR wind retrieval with speeds over 25 m/s also exhibits larger errors and is considered to be less reliable; the SAR wind retrieval with speeds less than 25 m/s is used in the assimilation.

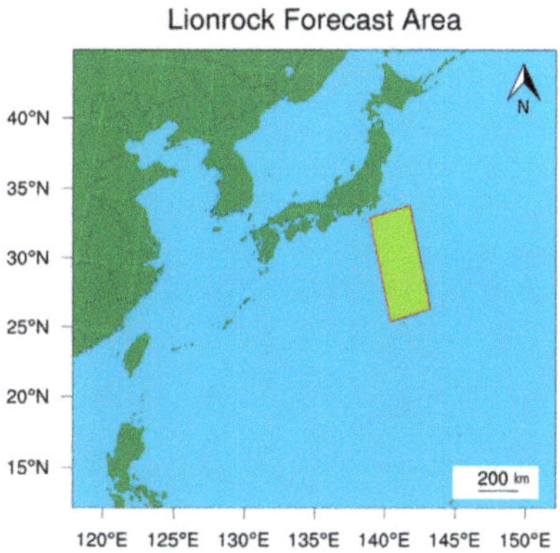


**Figure 4.** The domain area for theWeather Research and Forecasting (WRF) model simulation. The corverage of SAR observations is shownin the red frame.

## 4. Experimental Results

### 4.1. Wind Analysis at 10 m

To investigate the impact of the assimilation of SAR sea surface wind observations on the wind analysis at 10 m by the two different assimilation methods, the NCEP FNL data at 0900 UTC 29 August is used as the true wind field as the FNL data at the time of analysis time is of high accuracy [39].

Figure 5 shows the wind fields at 10 m at 0900 UTC 29 August 2016. The Figure 5a,b, show the wind fields at 10 m from the NCEP FNL data and from the background field, respectively. The center of the typhoon in the FNL data is at 142°E, 32°N, while the typhoon center in the background is further north than the location in the FNL data. It can be seen from Figure 5a that the typhoon center of the real wind field is basically symmetrical and the wind field in the typhoon revolves around the typhoon center in the reverse clock direction. Compared with the real wind field, the center of the typhoon in the background field from Figure 5b is not obviously symmetrical. The direction of the wind field around the typhoon center is also counter-clockwise, but the strength simulation was significantly weaker than the real wind field. Moreover, the vortex structures and the pressure are very different in the two panels.

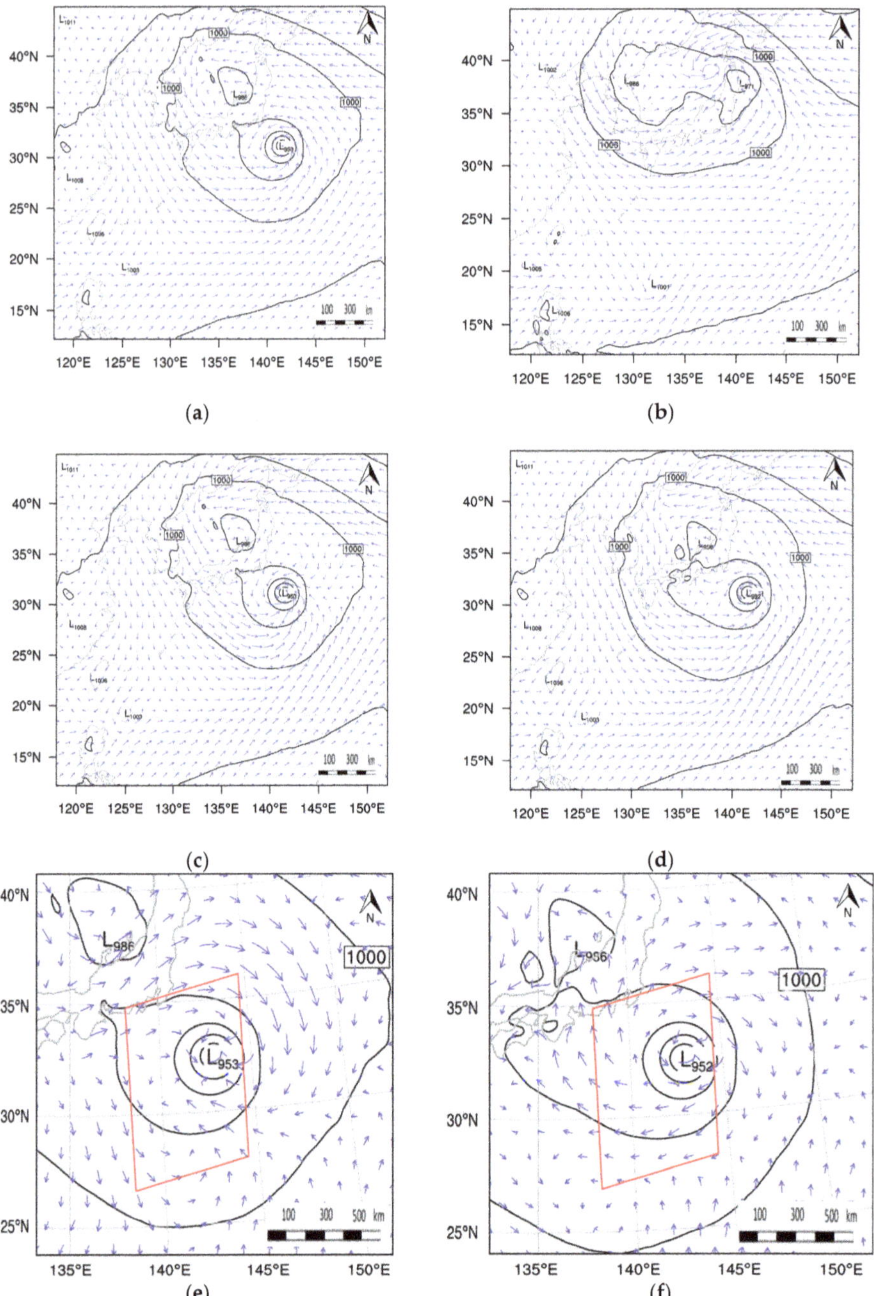


**Figure 5.** The 10m wind field at 0900 UTC 29 August 2016 from (**a**) the NCEP FNL data in (**b**) the background field; the wind analysis field at 10m from (**c**) SAR_sd and (**d**) SAR_uv. The bottom two panelsare the analysis bias derived from the analysis subtracting the FNL data for (**e**) SAR_sd and (**f**) SAR_uv. The red frames mark the area of the SAR wind observations. Note the different spatial scales of (**e**,**f**) as compared to (**a**–**d**).

The Figure 5c,d, are the wind analysis at 10 m from the SAR_sd and SAR_uv, respectively. The tropical centers in both analysis fields are very close to the location in the FNL data. The wind fields from two analysis experiments are close to the real wind field. The wind field structure and the intensity of the typhoon in both assimilation analysis fields are basically consistent with those of the typhoon wind field. The observation data of the assimilated SAR can effectively adjust the background field and provide an accurate wind field at 10 m for the forecasting model. However, the pressure intensity in the typhoon center from the SAR_uv is slightly lower than that for the true field. The typhoon structure in the SAR_sd analysis is more symmetrical than the SAR_uv analysis. The spatial pressure distribution in SAR_sd is similar to the true wind field (Figure 5a), and the wind field in SAR_sd is also close to the true wind field. These results indicate that the analysis field from the SAR_sd experiment is closer to the FNL truth.

The Figure 5e,f are the analytical deviation at 10 m in the center of typhoon from the SAR_sd and the SAR_uv experiments, respectively. The red diamond region represents the area covered by the SAR observations. The analysis deviation is derived from the true field subtracted by the analysis. The smaller the analysis deviation, the more accurate the analysis field is. In the red frames, most of the arrows in Figure 5e are shorter than the arrows in Figure 5f; this indicates that the magnitude of the wind speed deviation of SAR_sd is smaller than that of SAR_uv. We can infer that the wind speed of the SAR_sd analysis is closer to the real wind speed; even the direction deviation is sometimes the same as the real wind field and sometimes contrary to the real wind field. A clockwise wind field orientation in Figure 5e,f indicates that the wind speed is underestimated, whereas a counter-clockwise wind field orientation indicates that the wind speed is overestimated. Almost all the wind direction of the deviation of SAR_uv is opposite to the real wind field, indicating that the wind speed is underestimated. However, comparing this with the observation bias outside the observation coverage area, it can be found that the impact of the analysis bias of SAR_sd is spread beyond the observation range and the deviation in both the left and right sides of the observation area is relatively large. The assimilation of SAR observations by the method to assimilate *spd* and *dir* can have a good influence in the observation area, and it may have some negative effects out of the periphery of the scanning area. This also can be seen in the northeastern side of the typhoon center on the right side of the SAR observation area in Figure 5c, where the wind speed and wind direction are significantly different from the real wind field. The impact of the analysis of SAR_uv is small both inside and outside of the periphery of the observation area, but it is not negative out of the periphery of the observed coverage area.

### *4.2. Analysis Bias at Different Height*

To assess the impact of the SAR sea surface wind observations on the analysis in of the vertical height of a typhoon, the analysis biases of wind speed for SAR_sd and SAR_uv at 10 m at 850 hPa, 700 hPa, 500 hPa, and 300 hPa are investigated. The analysis bias is derived from the true field subtracted by the analysis, and the NCEP FNL analysis at the analysis time is used as the true field.

The top two panels are the analysis bias of the wind speed at 10 m. There is an area of bias of −6 to −2 m/s in SAR_sd, and the area of bias is significant smaller than the area of SAR_uv out of the periphery of the TC. Generally the minimum bias of −2 to 2 m/sis the dominance area of the two panels, and this kind of area in SAR_sd is larger than in SAR_uv. These results indicate that the analysis by SAR_sd at 10 m is more accurate than that of SAR_uv. The middle two panels show the analysis bias of wind speed at 850 hPa. Figure 6c shows that, although there is an area of analysis bias of SAR_sd lower than −10 m/s in the northeast of the TC center, the bias is small in the southwest of the TC center, where many SAR sea surface wind observations exist. There is a large area of bias of −6 to −2 m/s in the west of the periphery of the TC from SAR_sd; this indicates that, in this area, the wind speed of SAR_sd is slower than the true wind field. Figure 6d shows there is a large area of analysis bias of −10 to −6 m/s just around the TC center from SAR_uv. There is a large area of analysis bias of 2 to 6 m/s in the periphery of the TC from SAR_uv, indicating that the wind speed is larger than the true wind field in this area for SAR_uv. Figure 6e,f show the analysis bias of wind speed at the 700 hPa for SAR_sd and SAR_uv, respectively. The two panels are similar, the main bias is −2 to 2 m/s, the lower bias is more from SAR_sd, and the higher bias is more from SAR_uv. The results at 500 hPa and 300 hPa are almost the same; the figures are not shown.

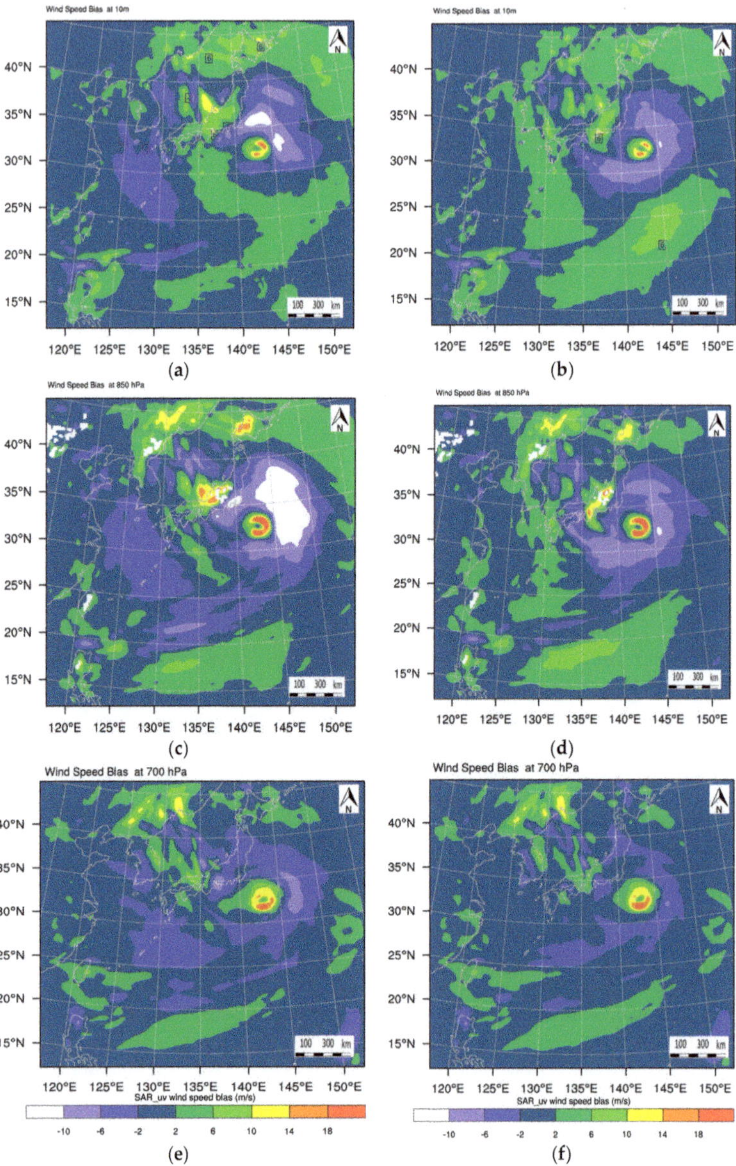


**Figure 6.** The analysis bias the SAR_sd (left panels) and the SAR_uv (right panels) at (**a**,**b**) 10 m at (**c**,**d**) 850 hPa, and at (**e**,**f**) 700 hPa.

### 4.3. Analysis Increment for Different Analysis Parameters

The results described above show that the analysis of SAR_sd is closer to the NCEP FNL data; it is more accurate than the analysis of SAR_uv. We can investigate the analysis increment to derive some verification for the two types of SAR wind observations. The larger the analysis increment is, the more improvement brought by the observations, as both assimilation experiments were based on the same background and literal boundary conditions.

Figure 7 presents the average root mean square error (RMSE) of the analysis increment for both the SAR_sd and SAR_uv experiments. The analysis of $u$ wind, $v$ wind, temperature, and relative humidity is described in Figure 7a–d, respectively. The results from Figure 7a,b are almost the same. It can be seen that the SAR wind observations improved $u$ and $v$ analysis mainly from the surface to 450 hPa. Above 450 hPa, the analysis increment is nearly 0 and the information on $u$ and $v$ comes from the background. For the analysis of both $u$ and $v$, the RMSE increment of the two experiments reached the maximum at 950 hPa. For the Figure 7c,d, it can be seen that the analysis increment exists in the whole vertical layer for both temperature and relative humidity. The maximum RMSE exists between 850 hPa and 800 hPa, and the RMSE of temperature is smaller than that of relative humidity. Generally, from the four panels, the results show that the RMSE from SAR_sd is larger than the RMSE from SAR_uv; this indicates the assimilation of *spd* and *dir* has larger improvement than the assimilation of $u$ and $v$ for the same SAR sea surface wind.

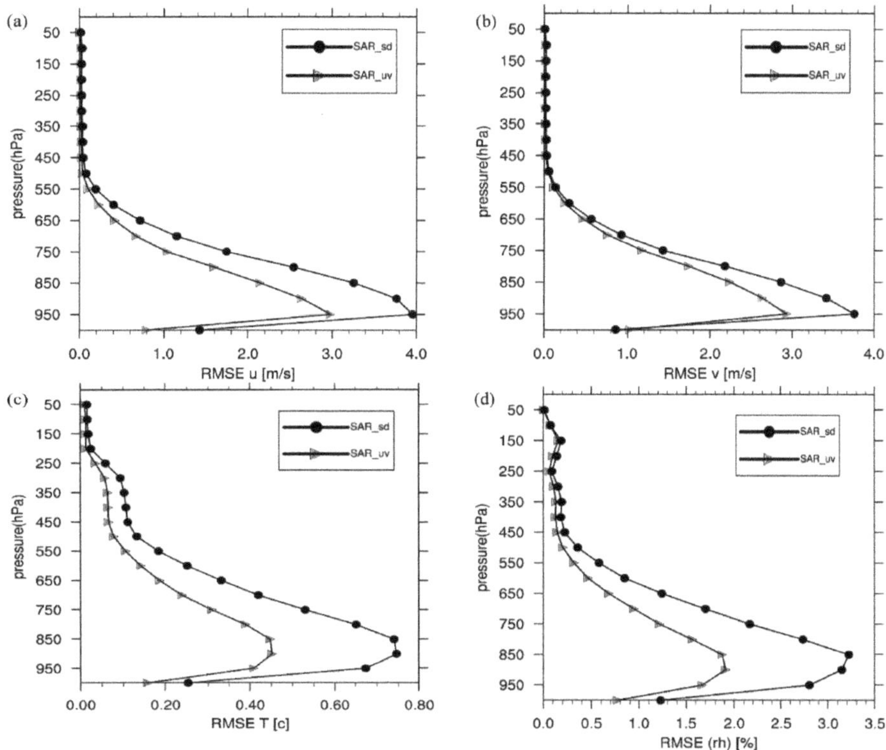


**Figure 7.** Root mean square error (RMSE) profiles of the analysis increments from the SAR_sd experiment (round point) and the SAR_uv experiment (triagle piont) for (**a**) $u$ wind, (**b**) $v$ wind, (**c**) temperature, and (**d**) relatively humidity.

### 4.4. Forecast Results

To verify how the assimilation of SAR sea surface wind affects the TC forecasts, the forecast skills of the track, the track error, the minimum sea level pressure (MSLP) error, and the absolute maximum wind speed error were assessed by comparing the forecasts from the SAR_sd and SAR_uv experiments with the CNTL experiment in Figure 8. The 'best track', minimum sea level pressure, and absolute maximum wind speed data for Typhoon Lionrock are obtained from the China Meteorological Administration (CMA).

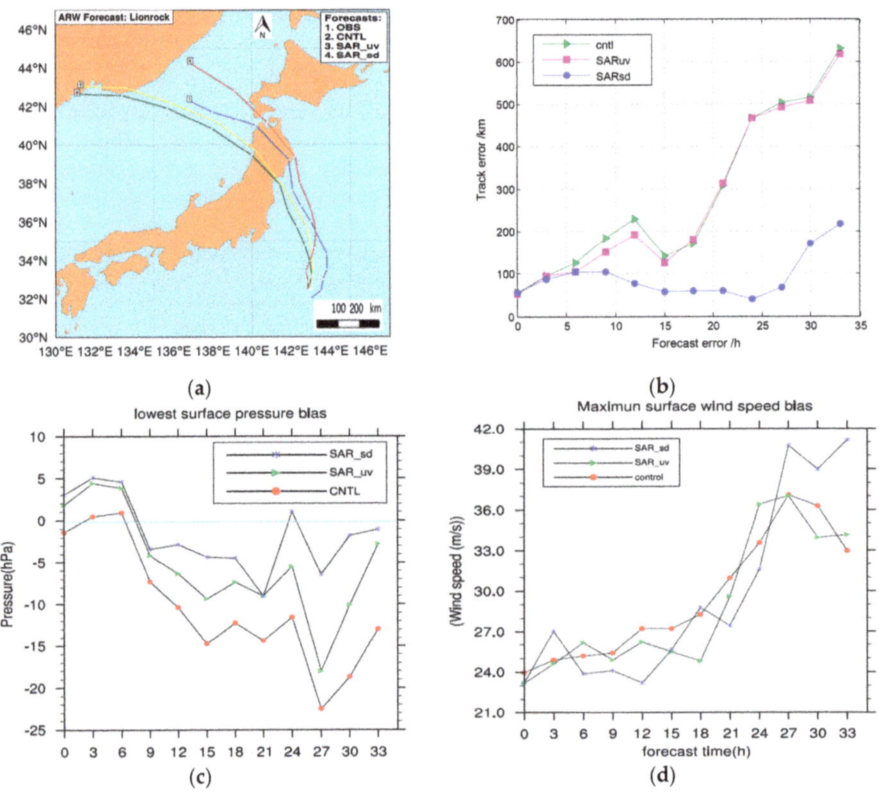


**Figure 8.** For Typhoon Lionrock: (**a**) 33h track forecast initialized at 0900 UTC August 2016; (**b**) mean absolute track errors; (**c**) mean absolute maximum wind speed errors; (**d**) minimum sea level pressure as a function of forecast lead time.

Figure 8a shows the 33 h track forecasts of Lionrock initialized at 0900 UTC 29 August 2016. The best track positions from CMA are the blue line. The forecast tracks from the two SAR observation experiments agree better with the best track than that of the CNTL experiment, especially the track forecast from SAR_sd experiment after the 9 h forecast. The SAR sea surface wind prevented the track forecast from moving clearly westward too fast, even though SAR_sd moved westward in the first 9 h. All the predicted tracks moved faster than the best track. Generally, the track forecast from SAR_sd was closer to the best track than the track forecast from SAR_uv, and the landfall position in Japan from SAR_sd was closest to the position from the best track.

Mean absolute track error, MSLP error, and mini and absolute maximum wind speed error as a function of forecast range for Lionrock are shown in Figure 8b–d. We found that the track errors of

SAR_uv and CNTL are almost the same, and they consistently increase, excepting the 15 h and 18 h forecast. The track error of SAR_sd is smaller those that of SAR_uv and CNTL, excepting the first 3 h, and is basically within 100km. The time after the 24 h forecast is the landing time for Lionrock; the track error from SAR_sd is about 45 km and is the smallest in the whole forecast time. The positive impact on the track forecast lasted for 27 h in SAR_sd, but the track error increases after 27 h. The MSLP error from SAR_sd is almost less than 5hPa for the whole forecast time, excepting the time after the 21 h forecast, and reached the minimum after the 9 h forecast even though it is the largest in the first 9 h compared with CNTL and SAR_uv. The maximum wind speed bias of SAR_sd is smaller than that of SAR_uv and CNTL during the forecast time 5~15 h and 20~25h. However, the bias from all three experiments is larger than 22 m/s, which may be related to the lack of perfection of the model itself [40].

## 5. Discussion and Conclusions

This study investigates the impact of SAR sea surface wind observations assimilated in the WRFDA system under typhoon conditions. The method to assimilate the wind speed (*spd*) and the direction (*dir*) was implemented in the WRFDA system and was compared with the conventional method, which assimilates the $u$ and $v$ wind components. The sea surface wind observations from the satellite-derived C-band Sentinel-1SAR were assimilated, and observational errors in the quality control were discussed for the two forms of wind state vectors. NCEP FNL analysis and reanalysis data distributed by the CMA were used to test and validate both assimilation methods.

Compared with the CNTL experiment, only the SAR sea surface wind observations in the center of the typhoon were assimilated in this study, but the results show that the SAR wind observations improved the analysis of Typhoon Lionrock in terms of the whole vertical height, and the improvement is significant below the height of 450 hPa, especially near 850 hPa, where the maximum improvement was reached. The assimilation of the SAR sea surface wind not only improved the analysis of wind but also improved the analysis of temperature and relative humidity. The SAR sea surface wind observations with a high resolution improved the depiction of the dynamic and thermodynamic vortex structure, especially by the method of the assimilation of *spd* and *dir*, which resulted in a reasonable observational error of *dir*, rather than the method of the assimilation of $u$ and $v$ wind without the consideration of the impact of *dir*. In the experiments, SAR_sd brought the TC environment fields closer to the NCEP FNL and observations and produced better analyses and forecasts for Typhoon Lionrock, compared with SAR_uv, using the same background and lateral boundary.

In this study, the SAR sea surface observations increase the number of effective observations in the typhoon area. However, many observations were still excluded after the thinning of the observations. This is because NWP systems run at low spatial resolutions, which is typically at 10 to 50 km resolution, and the high-resolution SAR data should be thinned to be compatible with the data assimilation system. The next generation data assimilation system with high resolution designed for limited area models could benefit from assimilating nearly full-resolution SAR data and could improve the typhoon forecast.

The strict QC_co method was applied for both kinds of assimilation methods to assimilate more accurate observations. The QC_co method was applied to the methods of assimilating $u$ and $v$ wind, but many wind vectors like the OBS4 in Figure 2 were still accepted after the quality control, and these observations may reduce the impact of the analysis. In addition, *dir* was derived independently from *spd* for satellite-derived SAR sea surface wind observations. It is likely that the QC_co method is too strict for the method of assimilating *spd* and *dir* to accept the most useful observations. For example, in the Typhoon Lionrock case, the wind vectors with huge but appropriate *spd* were rejected because the *dir* could not pass the quality control. The wind speed and wind direction are separately derived from SAR observations, and their errors are independent. Thus QC_al could be suitable for the method of assimilating *spd* and *dir*, and larger thresholds with more than five times the innovation could be applied for the wind *dir* for this procedure. Further investigation of the quality control schemes for

SAR-derived winds should be studied in the future. The numerical forecast results for this case show better results for the SAR_sd method than for the SAR_uv method. The SAR_sd method looks very promising for wind assimilation under typhoon conditions, but more cases need to be considered to draw final conclusions.

**Acknowledgments:** The authors acknowledge the support of the National Natural Science Foundation (41605070) and the Key Research and Development Program of Hainan Province (ZDYF2017167).

**Author Contributions:** Yi Yu decided on the direction of the study with help from Xiaofeng Yang, implemented the code for comparison, and completed the writing and implementation of the co-authors' comments. Xiaofeng Yang retrieved sea surface winds from the Sentinel-1 synthetic aperture radar (SAR). Weimin Zhang supported the method and the interpretation of the results with comments and discussions. Boheng Duan supported the implementation of the code for the data assimilation of wind speed and wind direction. Xiaoqun Cao and Hongze Leng provided guidance on the meteorological side of the method. All co-authors have participated in revising the draft from the first collection of ideas to the final version of the paper.

**Conflicts of Interest:** The authors declare no conflicts of interest.

## References

1. Zhang, Y.; Chen, Y.; Zhu, M. Research on ocean surface wind field retrievals from SAR. *Electron. Meas. Technol.* **2007**, *30*, 36–38.
2. Dagestad, K.F.; Horstmann, J.; Mouche, A.; Perrie, W.; Shen, H.; Zhang, B.; Li, X.; Monaldo, F.; Pichel, W.; Lehner, S.; et al. Wind retrieval from synthetic aperture radar-an overview. In Proceedings of the 4th SAR Oceanography Workshop (SEASAR 2012), Tromsø, Norway, 18–22 June 2012.
3. Chang, R.; Zhu, R.; Badger, M.; Hasager, C.B.; Xing, X. Offshore wind resources assessment from multiple satellite data and WRF modeling over South China Sea. *Remote Sens.* **2015**, *7*, 467–487. [CrossRef]
4. Hasager, C.; Badger, M.; Peña, A.; Larsén, X.; Bingöl, F. SAR-based wind resource statistics in the Baltic Sea. *Remote Sens.* **2011**, *3*, 117–144. [CrossRef]
5. Yu, Y.; Zhang, W.; Wu, Z.; Yang, X.; Cao, X.; Zhu, M. Assimilation of HY-2A scatterometer sea surface wind data in a 3DVAR data assimilation system—A case study of Typhoon Bolaven. *Front. Earth Sci.* **2015**, *92*, 192–201. [CrossRef]
6. Yang, X.; Li, X.; Zheng, Q.; Gu, X.; Pichel, W.G.; Li, Z. Comparison of Ocean-Surface Winds Retrieved From QuikSCAT Scatterometer and Radarsat-1 SAR in Offshore Waters of the U.S. West Coast. *IEEE Geosci. Remote Sens. Lett.* **2011**, *8*, 163–167. [CrossRef]
7. Danielson, R.; Dowd, M.; Ritchie, H. Marine Wind Analysis with the Benefit of Radarsat-1 Synthetic Aperture. In Proceedings of the OceanSAR 2006—Third Workshop on Coastal and Marine Applications of SAR, St. John's, NL, Canada, 23–25 October 2006.
8. Geldsetzer, T.; Pogson, L.; Scott, A.; Buehner, M.; Carrieres, T.; Ross, M.; Caya, A. Retrieval of sea ice and open water from SAR imagery for data assimilation. In Proceedings of the 7th IICWG Workshop on Sea Ice Data Assimilation and Verification, Frascati, Italy, 5–7 April 2016.
9. Lin, H.; Xu, Q.; Zheng, Q. An overview on SAR measurements of sea surface wind. *Proc. Nat. Sci.* **2008**, *18*, 913–919. [CrossRef]
10. Li, X.; Zhang, J.A.; Yang, X.; Pichel, W.G.; DeMaria, M.; Long, D.; Li, Z. Tropical cyclone morphology from spaceborne synthetic aperture radar. *Bull. Am. Meteorol. Soc.* **2013**, *94*, 215–230. [CrossRef]
11. Perrie, W.; Zhang, W.; Bourassa, M.; Shen, H.; Vachon, P.W. Impact of satellite winds on marine wind simulations. *Weather Forecast.* **2008**, *23*, 290–303. [CrossRef]
12. Yang, X.; Li, X.; Pichel, W.G.; Li, Z. Comparison of ocean surface winds from ENVISAT ASAR, MetOp ASCAT scatterometer, buoy measurements, and NOGAPS model. *IEEE Trans. Geosci. Remote Sens.* **2011**, *49*, 4743–4750. [CrossRef]
13. Ahsbahs, T.; Badger, M.; Karagali, I.; Larsén, X. Validation of Sentinel-1A SAR Coastal Wind Speeds against Scanning LiDAR. *Remote Sens.* **2017**, *9*, 552. [CrossRef]
14. Danielson, R.; Fillion, L.; Ritchie, H.; Dowd, M. Assimilation of SAR Wind Information IN Environment Canada's High Resolution 3D-Var Analysis System. In Proceedings of the Third International Workshop, Frascati, Italy, 25–29 January 2010.

15. Perrie, W.; Zhang, W.; Bourassa, M.; Shen, H.; Vachon, P.W. SAR-derived Winds from Hurricanes: Assimilative Blending with Weather Forecast Winds. In Proceedings of the Proceedings Ocean SAR 2006, St. John's, NL, Canada, 1–3 October, 2006.
16. Choisnard, J.; Laroche, S. Properties of variational data assimilation for synthetic aperture radar wind retrieval. *J. Geophys. Res.* **2008**, *113*, C050061-13. [CrossRef]
17. Barker, D.; Huang, X.Y.; Liu, Z.; Auligné, T.; Zhang, X.; Rugg, S.; Ajjaji, R.; Bourgeois, A.; Bray, J.; Chen, Y.; et al. The weather research and forecasting model's community variational/ensemble data assimilation system: WRFDA. *Bull. Am. Meteorol. Soc.* **2012**, *93*, 831–843. [CrossRef]
18. Huang, X.Y.; Gao, F.; Jacobs, N.A.; Wang, H. Assimilation of wind speed and direction observations: A new formulation and results from idealized experiments. *Tellus A* **2013**, *65*, 19936. [CrossRef]
19. Gao, F.; Huang, X.Y.; Jacobs, N.A.; Wang, H. Assimilation of wind speed and direction observations: Results from real observation experiments. *Tellus A* **2015**, *67*, 27132. [CrossRef]
20. Li, X. The first Sentinel-1 SAR image of a typhoon. *Acta Oceanol. Sin.* **2015**, *34*, 1–2. [CrossRef]
21. Friedman, K.; Li, X. Storm patterns over the ocean with wide swath SAR. *Johns. Hopkins Univ. Appl. Phys. Lab. APL Tech. Dig.* **2000**, *21*, 80–85.
22. Li, X.; Pichel, W.G.; He, M.; Wu, S.Y.; Friedman, K.S.; Clemente-Colón, P.; Zhao, C. Observation of hurricane-generated ocean swell refraction at the Gulf Stream north wall with the RADARSAT-1 synthetic aperture radar. *IEEE Trans. Geosci. Remote Sens.* **2002**, *40*, 2131–2142.
23. Zhang, G.; Li, X.; Perrie, W.; Hwang, P.A.; Zhang, B.; Yang, X. A Hurricane Wind Speed Retrieval Model for C-Band RADARSAT-2 Cross-Polarization ScanSAR Images. *IEEE Trans. Geosci. Remote Sens.* **2017**, *55*, 4766–4774. [CrossRef]
24. Hersbach, H.; Stoffelen, A.; De Haan, S. An improved C-band scatterometer ocean geophysical model function: CMOD5. *J. Geophys. Res.* **2007**, *112*, C030061-18. [CrossRef]
25. Du, Y.; Vachon, P.W. Characterization of hurricane eyes in RADARSAT-1 images with wavelet analysis. *Can. J. Remote Sens.* **2003**, *29*, 491–498. [CrossRef]
26. Zhou, X.; Yang, X.; Li, Z.; Yu, Y.; Bi, H.; Ma, S.; Li, X. Estimation of tropical cyclone parameters and wind fields from SAR images. *Sci. China Earth Sci.* **2013**, *56*, 1977–1987. [CrossRef]
27. Le Dimet, F.X.; Talagrand, O. Variational algorithms for analysis and assimilation of meteorological observations: Theoretical aspects. *Tellus A* **1986**, *38*, 97–110. [CrossRef]
28. Zou, X.; Navon, I.M.; Sela, J. Control of gravitational oscillations in variational data assimilation. *Mon. Weather Rev.* **1993**, *121*, 272–289. [CrossRef]
29. Barker, D.M.; Huang, W.; Guo, Y.R.; Bourgeois, A.J.; Xiao, Q.N. A three-dimensional variational data assimilation system for MM5: Implementation and initial results. *Mon. Weather Rev.* **2004**, *132*, 897–914. [CrossRef]
30. Huang, X.Y.; Xiao, Q.; Barker, D.M.; Zhang, X.; Michalakes, J.; Huang, W.; Henderson, T.; Bray, J.; Chen, Y.; Ma, Z.; et al. Four-dimensional variational data assimilation for WRF: Formulation and preliminary results. *Mon. Weather Rev.* **2009**, *137*, 299–314. [CrossRef]
31. Velden, C.S.; Hayden, C.M.; Paul Menzel, W.; Franklin, J.L.; Lynch, J.S. The impact of satellite-derived winds on numerical hurricane track forecasting. *Weather Forecast.* **1992**, *7*, 107–118. [CrossRef]
32. Holmlund, K.; Velden, C.S.; Rohn, M. Enhanced automated quality control applied to high-density satellite-derived winds. *Mon. Weather Rev.* **2001**, *129*, 517–529. [CrossRef]
33. Hollingsworth, A.; Lönnberg, P. The statistical structure of short-range forecast errors as determined from radiosonde data. Part I: The wind field. *Tellus A* **1986**, *38*, 111–136. [CrossRef]
34. Raymond, T. Report on TC's Key Activities and Main Events in the Region 2016. In Proceedings of the 49 Session ESCAP/WMO Typhoon Committee, Yokohama, Japan, 21–24 February, 2017.
35. Hong, S.Y.; Dudhia, J.; Chen, S.H. A revised approach to ice microphysical processes for the bulk parameterization of clouds and precipitation. *Mon. Weather Rev.* **2004**, *132*, 103–120. [CrossRef]
36. Kain, J.S. The Kain–Fritsch convective parameterization: An update. *J. Appl. Meteorol.* **2004**, *43*, 170–181. [CrossRef]
37. Shin, H.H.; Hong, S.Y. Intercomparison of planetary boundary-layer parametrizations in the WRF model for a single day from CASES-99. *Bound. Layer Meteorol.* **2011**, *139*, 261–281. [CrossRef]

38. Skamarock, W.C.; Klemp, J.B.; Dudhia, J. Prototypes for the WRF (Weather Research and Forecasting) model. Preprints. In Proceedings of the Ninth Conference Mesoscale Processes, American Meteorological Society, Fort Lauderdale, FL, USA, 11–15 July 2001.
39. Mejia, J.F.; Murillo, J.; Galvez, J.M.; Douglas, M.W. Accuracy of the NCAR global tropospheric analysis (FNL) over Central South America based upon upper air observations collected during the SALLJEX. In Proceedings of the 8th International Conference on Southern Hemisphere Meteorology and Oceanography (ICSHMO), Foz do Iguaçu, Brazil, 24–28 April 2006.
40. Xu, D.; Liu, Z.; Huang, X.Y.; Min, J.; Wang, H. Impact of assimilating IASI radiance observations on forecasts of two tropical cyclones. *Meteorol. Atmos. Phys.* **2013**, *122*, 1–18. [CrossRef]

*Article*

# Assimilation of Typhoon Wind Field Retrieved from Scatterometer and SAR Based on the Huber Norm Quality Control

**Boheng Duan [1,*], Weimin Zhang [1], Xiaofeng Yang [2,3], Haijin Dai [1] and Yi Yu [1]**

1 Academy of Ocean Science and Engineering, National University of Defense Technology, Changsha 410073, China; wmzhang104@139.com (W.Z.); hj_dai@nudt.edu.cn (H.D.); yuyi2019@nudt.edu.cn (Y.Y.)
2 State Key Laboratory of Remote Sensing Science, Chinese Academy of Sciences, Beijing 100101, China; yangxf@radi.ac.cn
3 The Key Laboratory for Earth Observation of Hainan Province, Sanya 572029, China
* Correspondence: bhduan@foxmail.com; Tel.: +86-156-7482-7043

Received: 17 June 2017 ; Accepted: 20 September 2017 ; Published: 23 September 2017

**Abstract:** Observations of sea surface wind field are critical for typhoon prediction. The scatterometer observation is one of the most important sources of sea surface winds, which provides both wind speed and wind direction information. However, the spatial resolution of scatterometer wind is low. Synthetic Aperture Radar (SAR) can provide a more detailed wind structure of the tropical cyclone. In addition, the cross-polarization observation of SAR can provide more detailed information of high speed wind ($>25\ \mathrm{m \cdot s^{-1}}$) than the scatterometer. Nevertheless, due to the narrow swath of SAR, the number of retrieved sea surface wind data used in the data assimilation is limited, and another limitation of SAR wind observation is that it does not provide true wind direction information. In this paper, the joint assimilation of the Advanced Scatterometer (ASCAT) wind and Sentinel-1 SAR wind was investigated. Another limitation in the current operational typhoon prediction is the inefficient quality control (QC) method used in the data assimilation since a large number of high speed wind observations was rejected by the traditional Gaussian distribution QC. We introduce the Huber norm distribution quality control (QC) into the data assimilation successfully. A numerical simulation experiment of typhoon by Lionrock (2016) is conducted to test the proposed method. The experimental results showed that the new quality control scheme not only greatly increases the availability of wind data in the area of the typhoon center, but also improves the typhoon track prediction, as well as the intensity prediction. The joint assimilation of scatterometer and SAR winds does have a positive impact on the typhoon prediction.

**Keywords:** tropical cyclone; scatterometer wind; SAR wind; data assimilation; quality control; Huber norm

---

## 1. Introduction

The demand for more accurate predictions of tropical typhoons is increasing in order to minimize losses and destruction. One primary objective is to enhance the observation targeting and observability of cyclones. Satellite observations can effectively compensate for the shortcomings of traditional methods of sea surface measurement and provide all-weather observation over the sea surface, which is of great significance to improve the numerical prediction of strong convective weather in the marine area. The spaceborne scatterometer observes the backscattering caused by the sea surface roughness, and then, the sea surface wind can be retrieved. ASCAT is one of the instruments carried on-board the Meteorological Operational (Metop) polar satellites launched by the European Space Agency (ESA) and operated by the European organization for the exploitation of Meteorological

Satellites (EUMETSAT) [1]. Its operating frequency is C-band (5.255 GHz), so the effects of clouds and precipitation in the observation are small. ASCAT has two swaths, and each has a scanning width of 550 km. It can achieve a daily quasi-global coverage.

Scatterometer data were first used in a numerical weather forecasting operational system in 1996, when the European Center for Medium-Range Weather Forecasts (ECMWF) incorporated ERS-1 (European remote sensing satellite) scatterometer data into its global three-dimensional variational system [2]. Previous works have shown that scatterometer data have significant impacts on weather forecasting and climate monitoring [3–10]. Especially, it has been demonstrated useful in the prediction of tropical cyclones [5] and extratropical cyclones [4]. ASCAT surface wind data have been used in many forecasting operational organizations such as the ECMWF, the United Kingdom's National Weather Service (Met Office), the National Weather Service of France (Meteo-France) and Environment Canada. In July 2009, the Japan Meteorological Agency (JMA) began to use ASCAT data for the global spectrum model (GSM) and found that the ASCAT wind can capture the development of the low pressure system and improve the prediction precision. Hersbach (2010) pointed out that the neutral wind retrieved by ASCAT had a positive effect on the ECMWF forecasting system [11]. In 2011, Li evaluated the role of the ASCAT wind in the global data assimilation system of the NCEP (National Centers for Environmental Prediction), and the results showed that ocean surface wind of ASCAT has a positive effect on the forecast of wind and temperature [1].

Spaceborne SAR systems are an important data source for sea surface monitoring. In 1978, the United States of America launched the first synthetic aperture radar satellite SEASAT. Since then, many countries have begun to carry out the study of the spaceborne SAR system vigorously. Most of the systems used the single-band, single polarization imaging radar systems in the early stage, such as ESA's ERS-1/2, Canada's Radarsat-1, and so on. At the beginning of the 20th Century, the spaceborne SAR system was developed into multi-band, multi-polarization and multi-mode. The Envisat satellite launched by ESA in 2002, with the multi-polarization interferometric imaging model, is widely used in natural disaster monitoring and resource and environmental survey. The launch of the Japanese ALOS (Advanced Land Observing Satellite) in 2006 aimed to provide full polarimetric SAR data. The launch of the COSMO-SkyMed (Constellationof small Satellites for the Mediterranean basin Observation) satellite in 2006 by Italy made the satellite resolution increase to 3m or even 1m. In 2007, Germany launched the TerraSAR-X satellite with a revisit cycle of 11 days, which greatly improved the coherence of interference data. Canada's Radarsat-2 satellite has been able to provide full polarimetric image with high-resolution since 2007. The Sentinel-1 was launched by ESA in April 2014 to provide data services for more users with its wide range of multi-mode, multi-application features. The Sentinel-1 carries a single C-band synthetic aperture radar instrument operating at a center frequency of 5.405 GHz. Its extra-wide (EW) swath mode data can cover a wide area of 400 km at a medium resolution of 20 m by 40 m on the ground. It also has the capability of dual polarization, a short revisit cycle and rapid productization. Using pre-programmed, conflict-free operation mode, the Sentinel-1 can track and monitor a typhoon center dynamically.

A variety of meteorological hydrological elements can be retrieved by SAR observation, and it has been widely used in data assimilation in recent years. The first attempt to sequentially assimilate ESA's ERS SAR estimations of surface soil moisture was conducted in 2003 [12]. Matgen, P. (2010) presented a new concept for the sequential assimilation of SAR-derived water stages into coupled hydrologic-hydraulic models [13]. Scott, K. A. (2015) investigated the assimilation of binary observations calculated from SAR images of sea ice [14]. Phan, X. V. (2014) introduced a variational data assimilation scheme coupling TerraSAR-X radiometric data with the snowpack evolution model Crocus, and the results indicated that X-band SAR data can be taken into account to modify the evolution of snowpack simulated by Crocus [15]. Pichelli, E. (2015) developed a technique to retrieve integrated water vapor from interferometric synthetic aperture radar (InSAR) data, and the computation of statistical indices shows that the InSAR assimilation improves the forecast of weak to moderate precipitation [16]. Advanced Synthetic Aperture Radar (ASAR) wide swath data were used to measure soil moisture by [17], and they have sufficient resolution to allow soil moisture variations due to local topography to be detected, which helps to take into account the spatial heterogeneity of hydrological processes. Another important use of SAR is the monitoring of typhoons and the study of typhoon structure [18–24]. Many studies have also been carried out about the retrieval of ocean winds from SAR image [25–28], the and estimation of the retrieved wind shows that the SAR wind has an ideal accuracy [29–32]. However, the use of SAR-retrieved wind in numerical weather forecasts is a relatively new area, especially for typhoon prediction.

Quality control of observation is an indispensable process for data assimilation [33]. It ensures that the wrong observations are removed before assimilation, which would otherwise result in inaccurate analysis [34]. In general, the deviation of background (b) and observation (o) is used as a basis for evaluating the quality of the data. It is generally believed that the observation error distribution satisfies the Gaussian distribution, assuming that the background error is a Gaussian distribution, and then, the distribution of deviation should also be satisfied with the Gaussian distribution. However, according to the statistical results, the distribution of deviation for many observations does not strictly follow the Gaussian distribution. The traditional Gaussian distribution QC method carries out strict threshold control by the magnitude of the deviation. However, large deviation between the observation and background does not mean that the observation is wrong. In extreme weather conditions, observation and background tend to have a larger deviation, and the use of traditional Gaussian distribution QC will result in rejection of a large number of effective observations. In fact, the deviation often complies with a Huber norm distribution [35,36]. In 2009, the Huber norm-based QC method was applied to both the deterministic and ensemble forecasting system at ECMWF, but only for conventional observations. Unlike traditional QC methods, this method takes appropriate weights based on the magnitude of deviation from the observation and background, making it possible to utilize more observations. Based on the Huber norm QC scheme, this paper calculates the transition point of the Huber norm distribution and adjusts the weight of the observation error for the wind data.

In this paper, a data assimilation scheme is proposed to jointly use SAR and scatterometer retrieved winds in the Weather Research and Forecasting (WRF) model. The improved Huber norm QC method is also introduced. Following this Introduction, a brief description of the ASCAT scatterometer wind and the Sentinel-1 SAR wind is given in Section 2. Section 3 introduces the detailed scheme of the Huber norm QC. In Section 4, we use a numerical simulation experiment of the typhoon Lionrock case to test the proposed method. Finally, the conclusion is given in Section 5.

## 2. Retrieved Wind of Scatterometer and SAR

### 2.1. ASCAT Wind

Two sets of three antennas are used in the ASCAT to generate radar beams looking 45 degrees forward, sideways and 45 degrees backwards with respect to the satellite's flight direction, on both sides of the satellite ground track. For each wind vector cell (WVC), ASCAT obtains three independent backscatter measurements using the three different viewing directions, separated by a short time delay. Then, the surface wind speed and direction can be obtained by using these 'triplets' within a geophysical model functions (GMF). The wind product we used in the paper is obtained through the processing of scatterometer data originating from the ASCAT instrument of EUMETSAT's Metop-B satellite with a resolution of 12.5 km. Figure 1 shows the ASCAT wind field of the center of the typhoon "Lionrock" at 9 a.m. on 29 August 2016. However, the wind field does not cover a complete typhoon eye due to the limitation of the ASCAT swath.

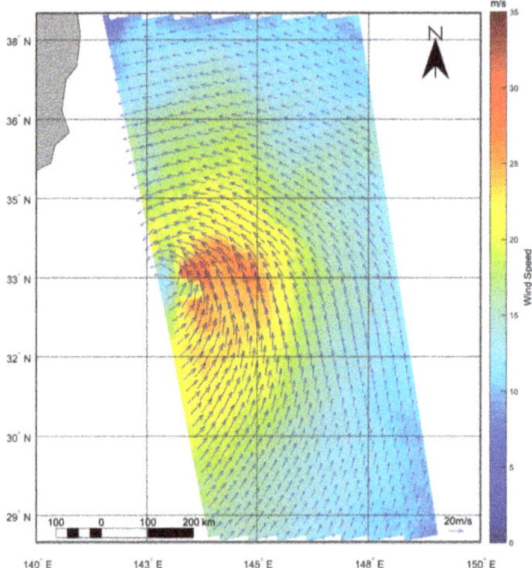


**Figure 1.** ASCAT wind field of typhoon "LionRock" with a resolution of 12.5 km.

### 2.2. SAR Wind of Satellite Sentinel-1

There exists an undisclosed problem for the retrieval of wind through SAR observation, that is one SAR observation corresponds to numerous wind speed and wind direction solutions through the geophysical model function (GMF). To solve the problem of SAR wind inversion, new information, which can be obtained from the SAR image itself or from the numerical forecast model or buoy observation data, must be added. In this paper, the wind direction information used for the wind inversion of Sentinel-1 SAR data comes from interpolation of the numerical forecast model (NCEP/GFS), and then, the wind speed can be retrieved by using the C-2PO (C-band Cross-Polarization Ocean) GMF [37].

In this study, a Sentinel-1A EWswath mode dual-polarization (VV/VH) SAR image is used to retrieve sea surface wind speed (see Figure 2). Its overpass time was 20:36:44 UTC on 29 August 2016. The observation using VV polarization can get a good signal to noise ratio (SNR) in the low speed wind conditions, and the accuracy of the retrieved wind is high. However, either the scatterometer nor the SAR observation using VV polarization are sensitive to the physical feedback of the sea surface under high speed wind conditions (wind speed greater than 25 $\text{m}\cdot\text{s}^{-1}$), and the detection signal would reach a level of saturation. In other words, the effective range of the wind speed inversion for scatterometer and SAR observation with VV polarization is limited to 25 $\text{m}\cdot\text{s}^{-1}$. The SAR observation using VH polarization is sensitive to high speed wind conditions, and it can retrieve the wind speed greater than 25 $\text{m}\cdot\text{s}^{-1}$, but the SNR is low under low speed wind conditions. As shown in Figure 2b, VH polarization observation of the typhoon region can get more high speed wind information, but there is a large amount of noise in the non-high speed wind region.

In order to combine the advantages of two kinds of polarization modes, we use a simple linear weighted method to composite these two retrieved wind fields. The main idea is that more weight is given to the VV polarization retrieved wind when the wind speed (we use the $v_{VV}$ as the reference wind speed) is less than 25 $\text{m}\cdot\text{s}^{-1}$, on the other hand, more weight for the VH polarization retrieved wind when the wind speed is greater than 25 $\text{m}\cdot\text{s}^{-1}$. The specific expression is as follows:

$$v_S = \begin{cases} (1-\lambda_1)\cdot v_{VV} + \lambda_1 \cdot v_{VH} & \text{if } v_{VV} \leq 25\ \text{m}\cdot\text{s}^{-1} \\ (1-\lambda_2)\cdot v_{VV} + \lambda_2 \cdot v_{VH} & \text{if } v_{VV} > 25\ \text{m}\cdot\text{s}^{-1}, \text{ and } 25\ \text{m}\cdot\text{s}^{-1} < v_{VH} < 35\ \text{m}\cdot\text{s}^{-1} \\ v_{VH} & \text{if } v_{VV} > 25\ \text{m}\cdot\text{s}^{-1}, \text{ and } v_{VH} \geq 35\ \text{m}\cdot\text{s}^{-1} \end{cases} \tag{1}$$

where $v_{VV}$,$v_{VH}$ and $v_S$ represent VV polarization wind speed, VH polarization wind speed and synthetic wind speed, respectively; $\lambda_1$ and $\lambda_2$ are the linear functions of the wind speed, where $\lambda_1 \in [0, 0.5]$, $\lambda_2 \in [0.5, 1]$. The setup of $\lambda_1$ and $\lambda_2$ makes sure that the VV polarization retrieved wind gets more weight when $v_{VV} \leq 25$ m·s$^{-1}$ and the VH polarization retrieved wind gets more weight when $v_{VV} > 25$ m·s$^{-1}$. As shown in Figure 2c, the synthesized wind field not only preserves the low speed wind information of VV polarization observation, but also introduces the high speed wind information of VH polarization observation.

Compared to the ASCAT wind field, the SAR wind field has a higher resolution, so that a fine typhoon structure can be demonstrated. It can also be seen from the figure that there are still some discontinuities among different radar beams in the SAR wind field. However, we can see later that these discontinuities can be removed by thinning the observations.

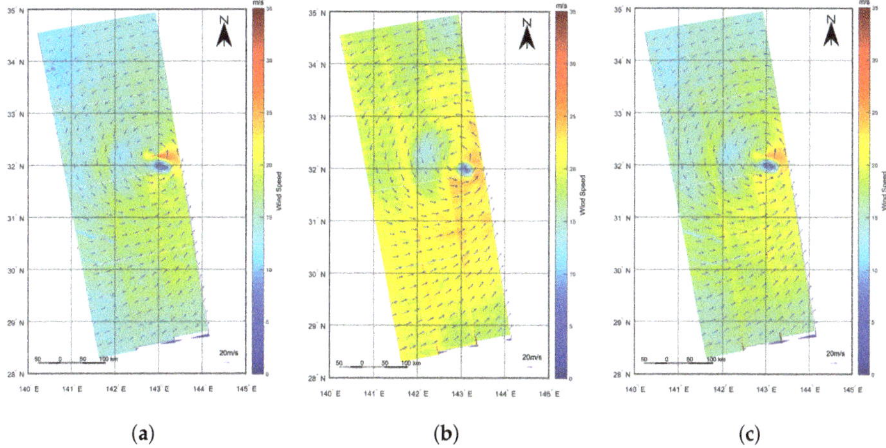

(**a**) (**b**) (**c**)

**Figure 2.** SAR wind field of satellite Sentinel-1 in the typhoon "Lionrock" region with a resolution of 500 m (the arrows here are thinned for clarity): (**a**) retrieved wind using VV polarization observation; (**b**) retrieved wind using VH polarization observation; (**c**) synthetic wind combined with VV and VH polarization retrieved wind.

## 2.3. *Joint Wind Field*

In order to obtain a complete typhoon wind field, we combine the ASCAT wind field with the Sentinel-1 SAR synthetic wind field. First, we thin the two wind fields to 25 km by sampling to accommodate the needs of the assimilation system. For the ASCAT wind field with a resolution of 12.5 km, we just need to sample every two winds from both zonal and meridional directions. Additionally, we sample every 50 winds for SAR synthetic wind field with a resolution of 500 m. Then, the overlapped winds of the SAR synthesis wind field with the ASCAT wind field are removed. Since the swath of the SAR synthetic wind field is smaller than that of the ASCAT and has no real wind direction information, it is mainly used as a complement to the ASCAT wind field. The joint wind field is illustrated in Figure 3, which forms a complete wind observation of typhoon "Lionrock". It can also be seen from the figure that the discontinuities among different radar beams in the SAR wind field are greatly eliminated by the thinning process.

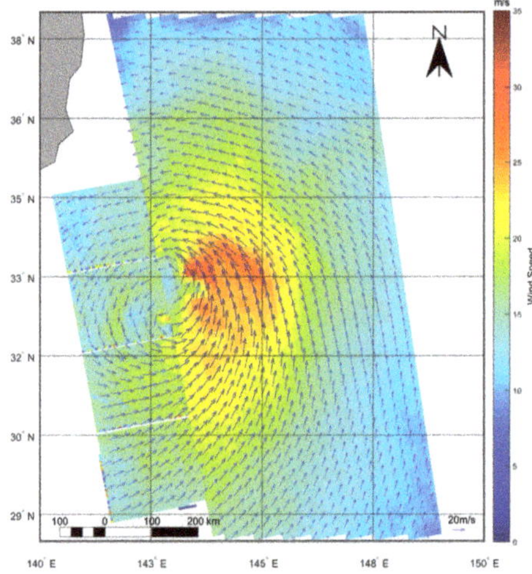


**Figure 3.** Joint wind field of typhoon "Lionrock" with a resolution of 25 km.

## 3. Huber Norm Distribution QC

### 3.1. Gaussian Distribution QC Scheme

The traditional Gaussian distribution QC assumes that the deviation between observation and background satisfies the Gaussian distribution. Since observation with a big deviation will cause the instability of the assimilation process, the observation that goes into the assimilation process should be strictly controlled. In the Gaussian distribution QC, observation that satisfies $(o-b)^2 < \alpha^2(\sigma_o^2+\sigma_b^2)$ can be received by the assimilation system, where $\sigma_o$ and $\sigma_b$ represent the standard deviation of the error of observation and background error, respectively (it should be noted that observation errors inside the assimilation system are different from the errors of observation here). In the assimilation system we used in this study, $\alpha$ is set to five, which means observation with deviation more than five-times the observation error was rejected. However, in extreme weather conditions, although observation and background tend to have a large deviation, it does not mean that the observation has a gross error, since the background sometimes also has large errors. Most of the high speed wind observation using the traditional Gaussian distribution QC were rejected before the assimilation, as shown in Figure 4, which shows the joint typhoon wind field after the traditional Gaussian distribution QC. In this paper, the magnitude of the observation error is defined as 2 m·s$^{-1}$ (namely, $\sqrt{(\sigma_o^2+\sigma_b^2)} = 2$ m·s$^{-1}$). It can be seen that QC using the Gaussian distribution can cause a large number of observations to be unusable in typhoon center areas, while these observations often contain key information of the typhoon structure.

The actual situation is that the deviation between observation and background does not strictly follow the Gaussian distribution, as shown in Figure 5, which demonstrates the deviation distribution of the $u, v$ component of the joint wind field and its best Gaussian fitting curve. In the figure, the true distribution of the $u$ component deviation shows a very obvious asymmetry, and the Gaussian curve does not fit the distribution well. In addition, it can be seen from the figure that there is a certain bias between the joint wind field and the background, so it is necessary to undertake the bias correction process before assimilation (the cause of the bias correction was not analyzed in this study).

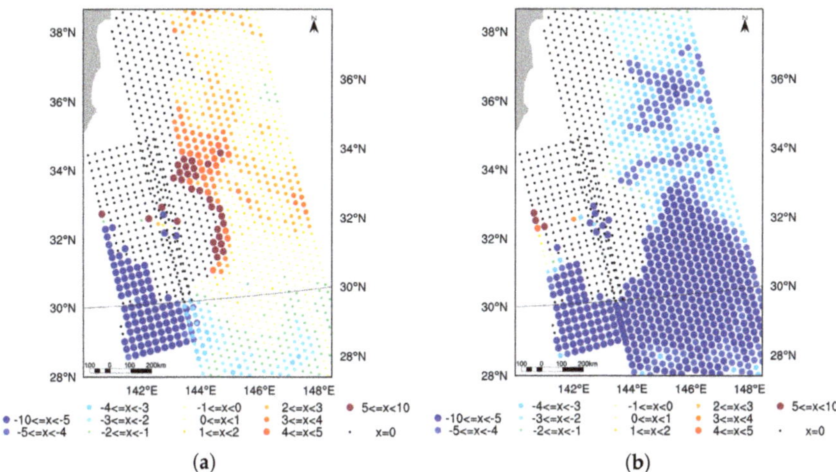


(a) (b)

**Figure 4.** Gaussian distribution QC of the joint wind field during assimilation, where $x$ represents the magnitude of the deviation ($x = o - b$), and the black dots indicate the rejected observations: (**a**) $u$ component; (**b**) $v$ component.

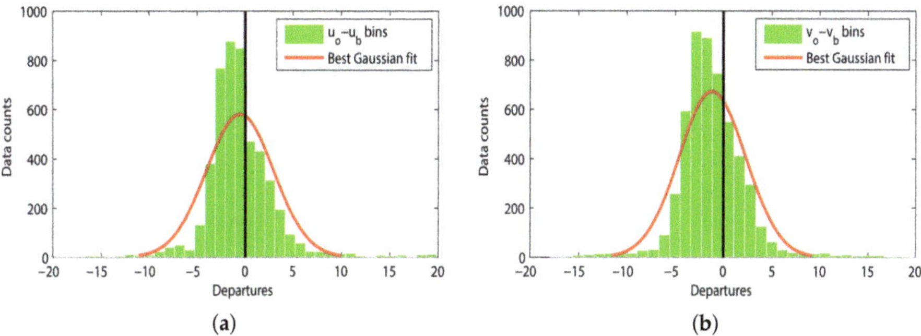


(a) (b)

**Figure 5.** Best Gaussian fit to the observation deviation distribution: (**a**) $u$ component; (**b**) $v$ component.

### 3.2. Huber Norm Distribution QC Scheme

#### 3.2.1. Definition of Huber Norm

Research shows that the Huber norm distribution is more consistent with the deviation distribution of the actual observations (Tavolato and Isaksen, 2014). The Huber norm uses a combination of the Gaussian distribution with an exponential distribution, where the Gaussian fit is used for the middle part of the distribution, while the exponential fit is used on both sides, as shown in the equation:

$$f(x) = \frac{1}{\sigma_o\sqrt{2\pi}} \cdot e^{-\frac{\rho(x)}{2}} \tag{2}$$

where:

$$\rho(x) = \begin{cases} \dfrac{x^2}{\sigma_o^2} & \text{if } |x| \le c \\ \dfrac{2c|x| - c^2}{\sigma_o^2} & \text{if } |x| > c \end{cases} \tag{3}$$

In the assimilation system, $x = y - H(x_b)$, where $y$ represents the observation, $x_b$ is the background field, $H$ is the observation operator and $\sigma_o$ represents the observation error. $c$ is the transition point where the Gaussian fitting curve connects with the exponentially fitting curve, that is where the Gaussian fitting curve ends and the exponential fitting curve begins. This definition ensures that the derivative of the function $f$ and $f$ itself are continuous. Since the distribution of the deviation is not strictly symmetrical, the left transition point $c_L$ of the function can be different from the right transition point $c_R$, and different parameter values are selected according to the type of observation.

### 3.2.2. Calculation of the Optimal Transition Point

Before calculating the transition point of the Huber norm fitting curve for the joint wind field, it is necessary to correct the bias of the observation (this is not discussed in detail here). The Huber norm fitting curve is mainly determined by the parameters $\sigma_o, c_L, c_R$, where $\sigma_o$ is known, and $c_L, c_R$ can be calculated separately. For the calculation of the transition point, it is usually done by searching the value between interval [0.0, 5.0] with a given step size of 0.1 [38], where the optimal value of the transition point is the one that has the minimum misfit error between the data distribution and the given Huber norm fitting curve. For each $(c_L, c_R)$ pair, the misfit between the Huber norm curve and the data distribution is defined as:

$$M(c_L, c_R) = \sum_{i=1}^{n} (\frac{N_i}{N_{sum}} - \frac{S_i}{S_f})^2 \tag{4}$$

where $N_i$ is the population in range bin $i$ and $N_{sum}$ is the number of all data, $S_i$ is the integration in range bin $i$ and $S_f$ is the integration of function $f(x)$ with the specific Huber distribution (since $f(x)$ is a density function, so its integration $S_f$ equals one).

For the $u, v$ component of the joint wind field, we calculate the transition points of the Huber norm distribution for them separately. The optimal Huber norm fitting curve is shown in Figure 6, where the optimal left and right transition point values of the $u$ component are (1.4, 1.2), and the optimal left and right transition points of the $v$ component are (0.6, 0.8). It can be seen from the figure that the Huber norm curve can better fit the distribution of observation deviation compared to the Gaussian fitting curve.

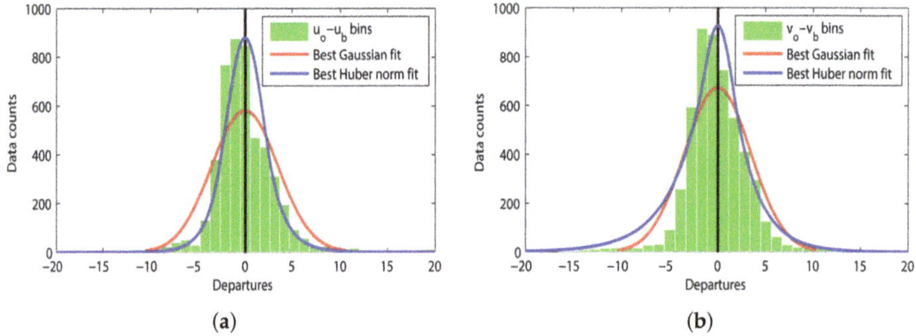


**Figure 6.** Best Huber norm fit to the observation deviation distribution: (**a**) $u$ component; (**b**) $v$ component.

### 3.2.3. Definition of Observation Weight

The traditional Gaussian distribution QC scheme gives full weight to the observation satisfying the condition $(o - b)^2 < \alpha^2(\sigma_o^2 + \sigma_b^2)$. For the Huber norm distribution QC scheme, the main purpose is to give a more reasonable weight to the observation. Observation with a smaller deviation to the background is given a larger weight, while observation with a bigger deviation is given a smaller weight. By this mean, it makes sure that observation with a big deviation can still affect the final analysis, while ensuring the stability of the assimilation process.

The cost function of the QC for a single observation is [39]:

$$J_O^{QC} = -\frac{1}{2}\ln(f(x)) = \rho(x) + const \tag{5}$$

For the weight attached to a single observation, the value of the weight is given by the ratio of $J_o^{QC}$ to its cost function under the Gaussian assumption, namely:

$$W = \frac{J_o^{QC}}{J_o^{Gaussian}} \tag{6}$$

When $|x| < c$ (or $-c_L \leq x \leq c_R$),

$$W = \frac{2\ln(\sigma_o\sqrt{2\pi}) + (\frac{y - h(x_b)}{\sigma_o})^2}{2\ln(\sigma_o\sqrt{2\pi}) + (\frac{y - h(x_b)}{\sigma_o})^2} = 1 \tag{7}$$

When $|x| > c$ (or $x < -c_L or x > c_R$),

$$W = \frac{2\ln(\sigma_o\sqrt{2\pi}) - \frac{c^2}{\sigma_o^2} + \frac{2c|y - h(x_b)|}{\sigma_o^2}}{2\ln(\sigma_o\sqrt{2\pi}) + (\frac{y - h(x_b)}{\sigma_o})^2} \tag{8}$$

Since $|x| > c$,

$$(c - |x|)^2 = c^2 - 2c|y - h(x_b)| + (y - h(x_b))^2 > 0 \tag{9}$$

which is:

$$(y - h(x_b))^2 > -c^2 + 2c|y - h(x_b)| \tag{10}$$

we can conclude that $W < 1$, that is to say, it reduces the weight of observation when the deviation drops out of the transition point.

Figure 7 shows the weight assigned to observation using the Gaussian distribution QC scheme and the Huber norm distribution QC scheme. As shown in the figure, the observation whose deviation drops between the two transition points is given equal weight, while the weight of others is reduced by the ratio of $J_o^{QC}$ to its cost function under the Gaussian assumption.

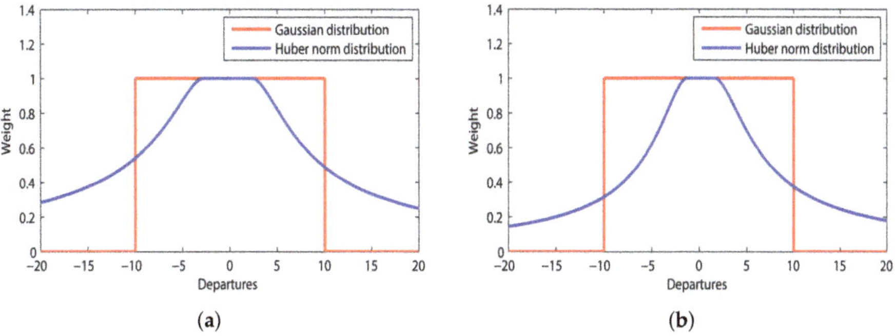


**Figure 7.** The corresponding weights after applying the variational QC. Red line: Gaussian distribution; blue line: Huber norm distribution: (**a**) $u$ component; (**b**) $v$ component.

## 4. Assimilation Experiments

### 4.1. Experimental Setup

In this study, we select typhoon Lionrock (2016) as a numerical example, which was generated in the Northwest Pacific at 2000 UTC 19 August 2016, in the vicinity of (33.1° N, 141.4° E). The center's maximum wind speed was about 18 m·s$^{-1}$, and its pressure was about 994 hPa. Lionrock weakened to an extratropical cyclone on 31 August. The the Weather Research and Forecasting model data assimilation system (WRFDA) developed by the National Center for Atmospheric Research (NCAR) [40] is adopted in this study. The WRFDA system is a widely-used operational system that can produce a multivariate incremental analysis in the WRF model space [41]. The grid size of the assimilation region is 260 × 250; the horizontal resolution is 15 km; and the vertical discretization is 30 layers. The time of assimilation is based on the time window of the joint wind field of the typhoon region, which was 0900 UTC 29 August 2016. The NCEP FNL (Final) Operational Global Analysis data are used as the initial field and boundary conditions. This product is on one-degree by one-degree grids prepared operationally every six hours. It is from the Global Data Assimilation System (GDAS), which continuously collects observational data from the Global Telecommunications System (GTS) and other sources, for many analyses. We take the 21-h forecast adjustment from 1200 UTC 28 August 2016 to 0900 UTC 29 August 2016 as the background field of the assimilation system. After the assimilation, a 30-h forecast is made, which is a forecast to 1500 UTC 30 August 2016.

In this study, a set of assimilation and comparison experiments is carried out. The assimilation experimental design is shown in Table 1, in terms of the QC scheme (traditional QC and Huber norm QC) and observation types (SAR wind, ASCAT wind and joint wind). The control experiment is just a forecast of the background field without assimilation. The purpose of Experiments 1, 2 and 4 is to show if the assimilation of joint wind improves the analysis compared to the single wind type scheme, while the aim of Experiments 3 and 4 is to show if the Huber norm method gives a better result than the traditional QC scheme.

**Table 1.** Data assimilation experimental design.

| Experiment Name | QC Scheme | Observation Type |
|---|---|---|
| 1 | Huber norm distribution QC | SAR wind |
| 2 | Huber norm distribution QC | ASCAT wind |
| 3 | Gaussian distribution QC | SAR + ASCAT wind |
| 4 | Huber norm distribution QC | SAR + ASCAT wind |

### 4.2. Experimental Results and Discussion

We take the FNL data at the analysis time as the reference (close to the truth, but not the real truth) and to see the analysis errors (analysis minus reference) of different experiments. Figure 8 gives the analysis errors of the pressure field of different experiments at the 10-m height of the center area of typhoon Lionrock at the analysis time. We still see some big analysis errors of different experiments in the typhoon center; however, the joint wind assimilation using the Huber norm distribution QC has the minimum analysis errors of pressure field compared to the others. An accurate initial field (the analysis is used as the initial field for the forecast) is crucial to a good numerical forecast, as can be seen in the following forecast results.

Figure 9a shows the observed typhoon path and forecasted typhoon paths of different experiments. It is apparent that the location of the typhoon center based on different QC schemes is very close to the control experiment (with no assimilation of the scatterometer wind) at the time of the assimilation, and this may due the defect of the position algorithm of the typhoon center. However, the forecasted typhoon paths of wind assimilation all show some improvement compared to that of the control experiment. The time indicators in Figure 9a demonstrate that the joint wind assimilation using the Huber norm distribution QC has the best forecasted path. In order to better compare the accuracy of the assimilation experiments with the typhoon moving path, Figure 9b gives the error of the forecasted typhoon path. As Figure 9 shows, with an increase in forecast time, the typhoon path error of most of the assimilation experiments becomes obviously smaller than that of the control experiment. The experiment with SAR wind, however, shows some large errors after a forecast of 27 h, mainly because that SAR wind alone contains no new direction information of wind, and the amount of data is quite small after thinning to a resolution of 25 km. We get the benefit from the joint assimilation of SAR wind and ASCAT wind comparing to the assimilation of a single type of wind, and it improves the typhoon forecasted path using the Huber norm distribution QC compared to the Gaussian distribution QC scheme.

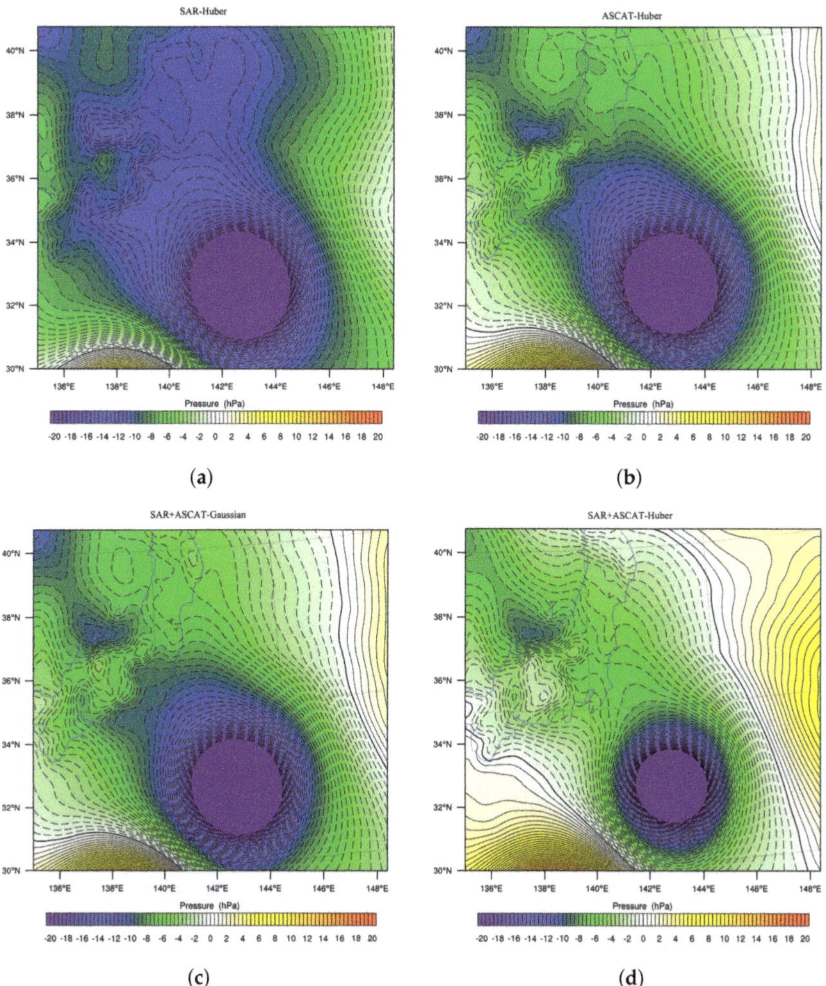


**Figure 8.** Pressure error of the center area of typhoon Lionrock at the analysis time: (**a**) assimilation of SAR wind using the Huber norm distribution QC; (**b**) assimilation of ASCAT wind using the Huber norm distribution QC; (**c**) assimilation of joint wind using the Gaussian distribution QC; (**d**) assimilation of joint wind using the Huber norm distribution QC.

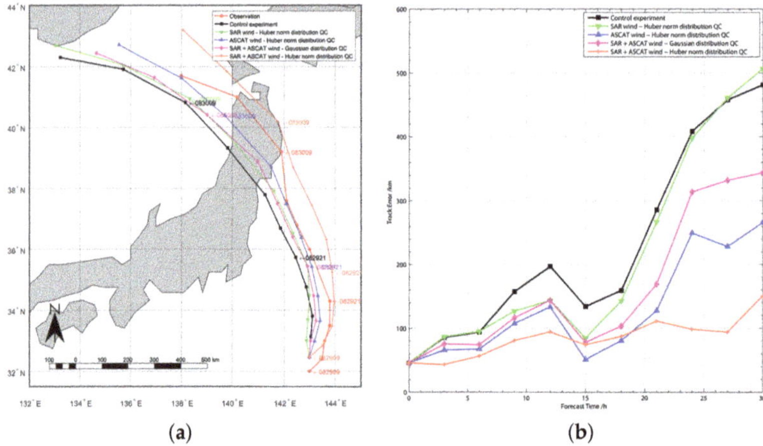


**Figure 9.** Forecasted path and track errors of typhoon Lionrock: (**a**) forecasted path of different experiments and the observed path. The time indicator is given every 12 h. (**b**) Track errors of different experiments compared to the observed path.

The intensity forecasts of the typhoon based on the different schemes are compared in Figure 10, and Table 2 gives the quantitative analysis of the data assimilation experiments. As shown in the figure, the assimilation of wind data improves the intensity forecast of the typhoon. The joint wind assimilation using the Huber norm, however, in some way weakens the minimum pressure and strengthens the maximum wind speed of the typhoon eye at the time of the assimilation. Although there is a bigger misfit between the intensity forecast and the observed truth than other experiments at the analysis time, it has a better analysis field than the others (see Figure 8). As can be seen from Table 2, the joint assimilation of SAR and ASCAT wind using the Huber norm distribution QC has the minimum average errors on track forecast (85.24 km) and intensity forecast ($-5.61$ m$\cdot$s$^{-1}$ for speed and $-1.84$ hPa for pressure, respectively). Therefore, the joint wind assimilation using the Huber norm distribution QC has also improved the intensity forecast of the typhoon.

Figures 11 and 12 show the $(O - A)$ (observation minus analysis, namely residual) diagram and $O/A$ (observation / analysis ) comparison of the bias, root-mean-square value and standard derivation of the $u$ and $v$ components of the four experiments. Assimilation experiment with the SAR wind field using Huber norm distribution QC shows a big deviation from the analysis (Figure 11a,e), thus having a big $rms$ ($rms_u = 5.46$, $rms_v = 4.50$) for the analysis (Figure 12a,e). This is mainly because most of the wind observations of SAR were located in the center area of the typhoon (as can be seen in Figure 2), and about one third of the observations have a deviation more than five-times the observation error to the background ($n_{x>5\sigma} = 93$, $n_{x\leq5\sigma} = 188$). Observations with a bigger deviation from the background have a smaller weight using the Huber norm distribution QC scheme. Smaller weight in the observation means the analysis gets closer to the background or a bigger residual, as shown in the Figure 11a,e. This also can be seen from Figure 11b,d,f,h, where the residual is much bigger in the center area of the typhoon. While the joint wind assimilation using the Gaussian distribution QC has the minimum $rms$ ($rms_u = 1.19$, $rms_v = 1.06$) compared to the others (Figure 12c,g), since observations with a big deviation from the background have been decreased in the QC step (Figure 11c,g), and the remaining observations have an equal weight.

**Table 2.** Quantitative analysis of data assimilation experiments.

| Observation Type/QC Scheme | $n_{x \leq 5\sigma}$ * | $n_{x > 5\sigma}$ | $rms_u$ ** | $rms_v$ | $\overline{e}_{track}$ *** | $\overline{e}_{speed}$ | $\overline{e}_{pressure}$ |
|---|---|---|---|---|---|---|---|
| SAR/Huber norm | 188 | 93 | 5.46 | 4.50 | 214.44 | −6.43 | 7.05 |
| ASCAT/Huber norm | 4844 | 75 | 1.48 | 1.27 | 129.50 | −6.52 | 7.14 |
| SAR + ASCAT/Gaussian | 4717 | 0 | 1.19 | 1.06 | 163.26 | −6.84 | 6.48 |
| SAR + ASCAT/Huber norm | 4776 | 113 | 3.96 | 1.85 | 85.24 | −5.61 | −1.84 |

* $n$ is the number of observations used in the assimilation, $x = o - b, \sigma = \sqrt{\sigma_o^2 + \sigma_b^2}$;
** $rms$ is the root mean square of $o - a$; the unit is $\text{m}\cdot\text{s}^{-1}$;
*** $\overline{e}$ is the average forecast error; the units are km, $\text{m}\cdot\text{s}^{-1}$ and hPa, respectively.

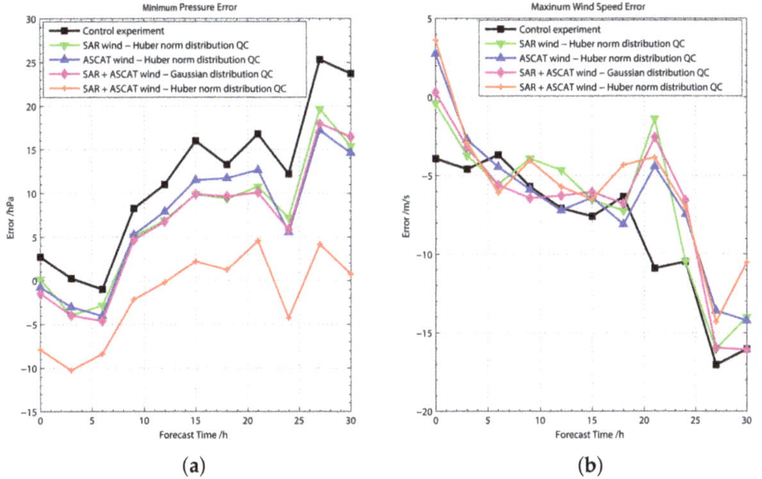


**Figure 10.** Intensity forecast errors of the typhoon eye: (**a**) minimum pressure forecast errors. (**b**) maximum wind speed forecast errors.

It can be seen from Table 2 that the *rms* mainly depends on the ratio of the observations with a big deviation to the background ($x > 5\sigma$) in the wind field. The higher the ratio, the bigger the *rms*. However, it also can be seen from the table that the effect on the assimilation is not determined by the *rms*, but the distribution of the observation. The experiment with ASCAT wind and the experiment with joint wind using the Huber norm QC have almost the same ratio as the observations with a big deviation, but the latter has a more complete observation of the typhoon center assimilated in the assimilation step (see Figure 11b,d,f,h), thus a more positive impact on the typhoon prediction. This is also the truth for the comparison of two different QC schemes (see Figure 11c,d,g,h).

It can be seen from Equation (5) that with the Huber norm distribution applied, the $J_o^{QC}$ is an $L^2$ norm in the center of the distribution and an $L^1$ norm in the tails (Tavolato and Isaksen, 2014). This makes the Huber norm QC a robust method that allows the use of observations with a large deviation from the background. It also makes it safe to use observations with a few erroneous outliers, since observation with a very large deviation only has a small weight and affects the analysis very little. Figure 13 shows that the QC scheme with the Huber norm has a fast convergence to the minima in the minimization process, as well as the Gaussian distribution QC scheme, which proves that the Huber norm QC is a robust method.

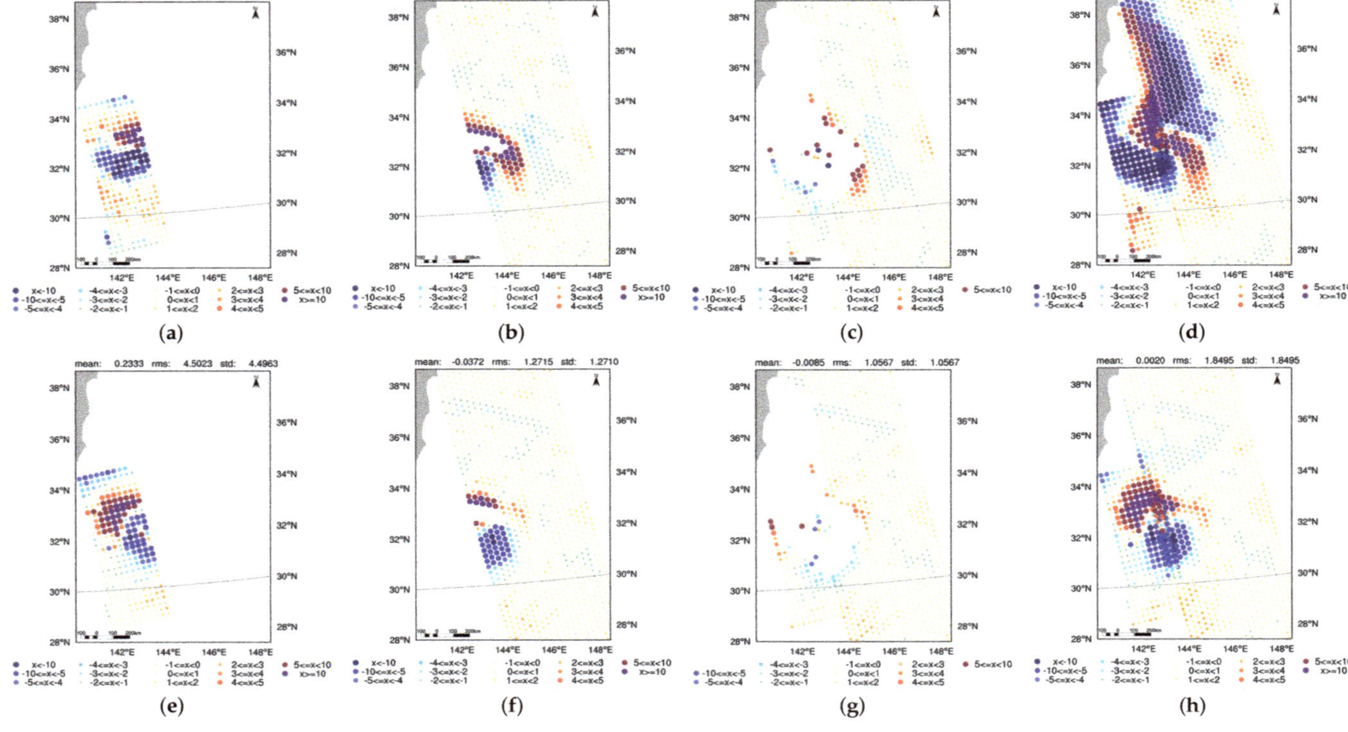

**Figure 11.** Observation residual after assimilation, where $x$ represents the magnitude of the residual ($x = o - a$): (**a**) residual of the $u$ component of SAR wind using the Huber norm distribution QC; (**b**) residual of the $u$ component of ASCAT wind using the Huber norm distribution QC; (**c**) residual of the $u$ component of joint wind using the Gaussian distribution QC; (**d**) residual of the $u$ component of joint wind using the Huber norm distribution QC; (**e**) residual of the $v$ component of SAR wind using the Huber norm distribution QC; (**f**) residual of the $v$ component of ASCAT wind using the Huber norm distribution QC; (**g**) residual of the $v$ component of joint wind using the Gaussian distribution QC; (**h**) residual of the $v$ component of joint wind using the Huber norm distribution QC.

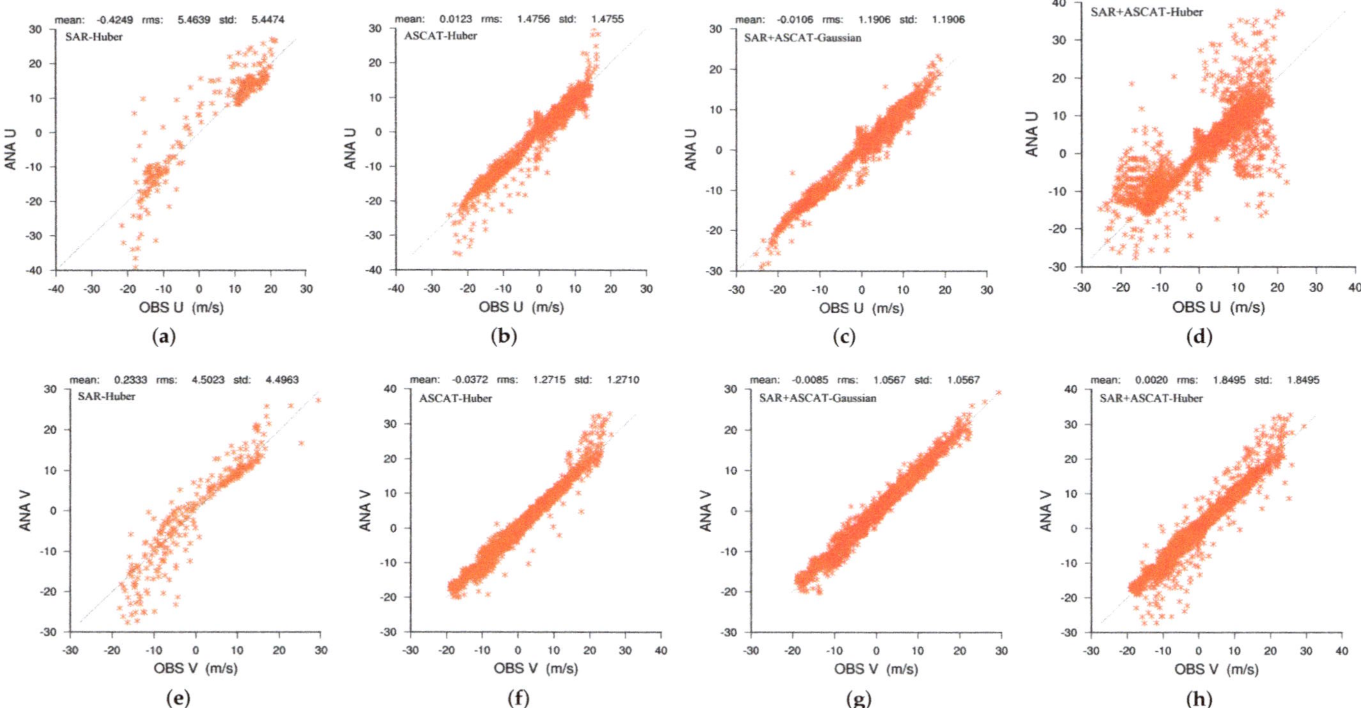


**Figure 12.** $O/A$ (observation / analysis) comparison of the bias, root-mean-square value and standard derivation of the $u$ and $v$ wind components: (**a**) $u$ component of SAR wind using the Huber norm distribution QC; (**b**) $u$ component of ASCAT wind using the Huber norm distribution QC; (**c**) $u$ component of joint wind using the Gaussian distribution QC; (**d**) $u$ component of joint wind using the Huber norm distribution QC; (**e**) $v$ component of SAR wind using the Huber norm distribution QC; (**f**) $v$ component of ASCAT wind using the Huber norm distribution QC; (**g**) $v$ component of joint wind using the Gaussian distribution QC; (**h**) $v$ component of joint wind using the Huber norm distribution QC.

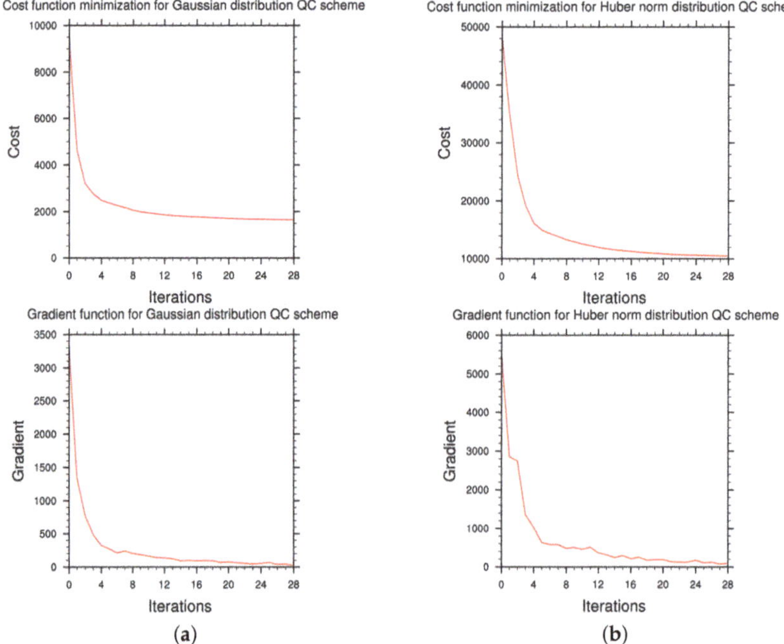


**Figure 13.** Minimization process of the cost function and the gradient for assimilation of the joint wind field: (**a**) Gaussian distribution QC scheme; (**b**) Huber norm distribution QC scheme.

However, although the joint wind field provides a detailed wind field of the typhoon, the information of the wind in the center area of the typhoon is not fully used with the Huber norm QC, since observation only have a small weight. This is mainly due to the bad background, which usually cannot provide the high speed wind structure of the typhoon.

We can see from the results that when a detailed wind field is available for the typhoon, a more accurate analyze was obtained when the Huber norm QC was applied, thus a more accurate prediction for the typhoon.

## 5. Conclusions

Based on NCAR's WRFDA system, we conducted a joint assimilation experiment of the ASCAT wind and SAR wind for the typhoon "Lionrock (2016)" and introduced the Huber norm QC scheme. Combining the SAR wind field with the ASCAT wind field, we take advantage of both observations, which not only make up the small swath limit of the SAR wind, but also fill the vacancy of the ASCAT wind field in the typhoon area. The results of assimilation experiments show that the joint assimilation improved the typhoon track forecast results. In addition, the Huber norm distribution QC scheme is adopted to increase the usage of the observation in the typhoon center area and to assign more reasonable weight to the observation, thus improving the analysis.

Although VH polarization observation of SAR can retrieve high speed wind, the impact on typhoon forecast is limited due to the quality of the background and parameterization scheme of the typhoon model. On the one hand, the background wind field of the typhoon is generally smoother than the real typhoon field, resulting in big deviation for the high speed wind observation, which led to the small weight of observation in the process of assimilation. On the other hand, since the parameterization of the typhoon model is not optimal, the mechanism of high speed wind for the development of the typhoon is not clear. Therefore, the improvement of the parameterization of the typhoon model is the next step we should consider.

**Acknowledgments:** This work was supported by the Key Research and Development Program of Hainan Province (ZDYF2017167) and the projects of the National Natural Science Foundation (NSFC40775064 and NSFC41675097).

**Author Contributions:** Boheng Duan, Weimin Zhang, Xiaofeng Yang and Yi Yu initiated the research. Under supervision of Weimin Zhang and Xiaofeng Yang, Boheng Duan performed the experiments and analysis. Boheng Duan, Weimin Zhang, Xiaofeng Yang and Haijin Dai wrote and revised the manuscript. All authors read and approved the final version of the manuscript.

**Conflicts of Interest:** The authors declare no conflict of interest.

## References

1. Li, B.; Jung, J.A.; Morgan, M.C. Assessment of Assimilating ASCAT Surface Wind Retrievals in the NCEP Global Data Assimilation System. *Mon. Weather Rev.* **2011**, *139*, 3405–3421.
2. Isaksen, L.; Janssen, P.A. Impact of ERS scatterometer winds in ECMWF's assimilation system. *Q. J. R. Meteorol. Soc.* **2004**, *130*, 1793–1814.
3. Stoffelen, A.; Cats, J.C. The impact of Seasat-A scatterometer data on high-resolution analyses and forecasts: The development of the QEII storm. *Mon. Weather Rev.* **1991**, *119*, 2794–2802.
4. Stoffelen, A.; Beukering, P.V. The Impact of Improved Scatterometer Winds on HIRLAM Analyses and Forecasts. HIRLAM Technical Report 31. HIRLAM. Available online: http://hirlam.org/ (accessed on 1 January 2017).
5. Isaksen, L.; Stoffelen, A. ERS scatterometer wind data impact on ECMWF's tropical cyclone forecasts. *IEEE Trans. Geosci. Remote Sens.* **2000**, *38*, 1885–1892.
6. Atlas, R.; Hoffman, R.N. *The Use of Satellite Surface Wind Data to Improve Weather Analysis and Forecasting at the NASA Data Assimilation Office*; Elsevier Oceanography Series; Elsevier: New York, NY, USA, 2000; pp. 57–78.
7. Candy, B. *The Assimilation of Ambiguous Scatterometer Winds Using a Variational Technique: Method and Forecast Impact*; Met Office, NWP Division: Berkshire, UK, 2001.
8. Atlas, R.; Hoffman, R.N.; Leidner, S.M.; Sienkiewicz, J. The effects of marine winds from scatterometer data on weather analysis and forecasting. *Bull. Am. Meteorol. Soc.* **2001**, *82*, 1965–1990.
9. Andersson, E.; Haseler, J.; Undén, P. The ECMWF implementation of three-dimensional variational assimilation (3D-Var). III: Experimental results. *Q. J. R. Meteorol. Soc.* **1998**, *124*, 1831–1860.
10. Prasad, S.V.; Gupta, A.; Rajagopal, E.N. Impact of OSCAT surface wind data on T574L64 assimilation and forecasting system—A study involving tropical cyclone Thane. *Curr. Sci.* **2013**, *104*, 627–631.
11. Hersbach, H. *Assimilation of Scatterometer Data as Equivalent-Neutral Wind*; ECMWF Publications: Reading, UK, 2010.
12. Francois, C.; Quesney, A.; Ottle, C. Sequential Assimilation of ERS-1 SAR data into a doupled land surface-hydrological model using an extended kalman filter. *J. Hydrometeorol.* **2003**, *4*, 473–487.
13. Matgen, P.; Montanari, M.; Hostache, R.; Pfister, L.; Hoffmann, L.; Plaza, D.; Pauwels, V.R.N.; De Lannoy, G.J.M.; Keyser, R.D.; Savenije, H.H.G. Towards the sequential assimilation of SAR-derived water stages into hydraulic models using the particle filter: Proof of Concept. *Hydrol. Earth Syst. Sci.* **2010**, *14*, 1773–1785.
14. Scott, A.K.; Ashouri, Z.; Buehner, M.; Pogson, L.; Carrieres, T. Assimilation of ice and water observations from SAR imagery to improve estimates of sea ice concentration. *Tellus A* **2015**, *67*, doi:10.3402/tellusa.v67.27218.
15. Phan, V.X.; Ferro-Famil, L.; Gay, M.; Durand, Y.; Dumont, M.; Morin, S.; Allain, S.; D'Urso, G.; Girard, A. 1D-Var multilayer assimilation of X-band SAR data into a detailed snowpack model. *Cryosphere* **2014**, *8*, 1975–1987.
16. Pichelli, E.; Ferretti, R.; Cimini, D.; Panegrossi, G.; Perissin, D.; Pierdicca, N.; Rocca, F.; Rommen, B. InSAR water vapor data assimilation into mesoscale model MM5: Technique and pilot study. *IEEE J. Sel. Top. Appl. Earth Obs. Remote Sens.* **2015**, *8*, 3859–3875.
17. Mason, C.D.; Garcia-Pintado, J.; Cloke, H.L.; Dance, S.L. Evidence of a topographic signal in surface soil moisture derived from ENVISAT ASAR wide swath data. *Int. J. Appl. Earth Obs. Geoinf.* **2016**, *45*, 178–186.

18. Li, X.F.; Pichel, W.G.; He, M.X.; Wu, S.Y. Observation of hurricane-generated ocean swell refraction at the Gulf Stream north wall with the RADARSAT-1 synthetic aperture. *IEEE Trans. Geosci. Remote Sens.* **2002**, *40*, 2131–2142.
19. Li, X.F.; Zhang, J.A.; Yang, X.F.; Pichel, W.G. Tropical Cyclone Morphology from Spaceborne Synthetic Aperture Radar. *Bull. Am. Meteorol. Soc.* **2013**, *94*, 215–230.
20. Jin, S.H.; Wang, S.; Li, X.F. Typhoon eye extraction with an automatic SAR image segmentation method. *Int. J. Remote Sens.* **2014**, *35*, 11–12.
21. Li, X.F. The first Sentinel-1 SAR image of a typhoon. *Acta Oceanol. Sin.* **2015**, *34*, 1–2.
22. Monaldo, F.M.; Jackson, C.R.; Pichel, W.G.; Li, X.F. A Weather Eye on Coastal Winds. *Eos* **2015**, *96*, 18–19.
23. Lee, I.K.; Shamsoddini, A.; Li, X.F.; Trinder, J.C. Extracting hurricane eye morphology from spaceborne SAR images using morphological analysis. *ISPRS J. Photogramm. Remote Sens.* **2016**, *117*, 115–125.
24. Zheng, G.; Yang, J.S.; Liu, A.K.; Li, X.F. Comparison of Typhoon Centers from SAR and IR Images and Those from Best Track Data Sets. *IEEE Trans. Geosci. Remote Sens.* **2016**, *54*, 1000–1012.
25. Zhou, X.; Yang, X.F.; Li, Z.W.; Yang, Y. Estimation of tropical cyclone parameters and wind fields from SAR images. *Sci. China Earth Sci.* **2013**, *56*, 1977–1987.
26. Monaldo, F.M.; Li, X.F.; Pichel, W.G.; Jackson, C.R. Ocean Wind Speed Climatology from Spaceborne SAR Imagery. *Bull. Am. Meteorol. Soc.* **2014**, *95*, 565–569.
27. Zhang, B.; Li, X.F.; Perrie, W.; He, Y.J. Synergistic measurements of ocean winds and waves from SAR. *J. Geophys. Res. Oceans* **2015**, *120*, 6164–6184.
28. Hwang, P.A.; Li, X.F.; Zhang, B. Retrieving Hurricane Wind Speed from Dominant Wave Parameters. *IEEE J. Sel. Top. Appl. Earth Obs. Remote Sens.* **2017**, *10*, 2589–2598.
29. Yang, X.F.; Li, X.F.; Pichel, W.G.; Li, Z.W. Comparison of Ocean Surface Winds from ENVISAT ASAR, MetOp ASCAT Scatterometer, Buoy Measurements, and NOGAPS Model. *IEEE Trans. Geosci. Remote Sens.* **2011**, *49*, 4743–4750.
30. Yang, X.F.; Li, X.F.; Zheng, Q.N.; Gu, X.F. Comparison of Ocean-Surface Winds Retrieved from QuikSCAT Scatterometer and Radarsat-1 SAR in Offshore Waters of the U.S. West Coast. *IEEE Geosci. Remote Sens. Lett.* **2011**, *8*, 163–167.
31. Monaldo, F.M.; Jackson, C.; Li, X.F.; Pichel, W.G. Preliminary Evaluation of Sentinel-1A Wind Speed Retrievals. *IEEE J. Sel. Top. Appl. Earth Obs. Remote Sens.* **2016**, *9*, 2638–2642.
32. Zhang, G.S.; Perrie, W.; Li, X.F.; Zhang, J.A. A Hurricane Morphology and Sea Surface Wind Vector Estimation Model Based on C-Band Cross-Polarization SAR Imagery. *IEEE Trans. Geosci. Remote Sens.* **2017**, *55*, 1743–1751.
33. Lorenc, C.A.; Hammon, O. Objective quality control of observationsusing Bayesian methods—Theory, and practical implementation. *Q. J. R. Meteorol. Soc.* **1988**, *114*, 515–543.
34. Lorenc, C.A. Analysis methods for the quality control of observations. In Proceedings of the ECMWF Workshop on the Use and Quality Control of Meteorological Observations for Numerical Weather Prediction, Reading, UK, 6–9 November 1984; pp. 397–428.
35. Huber, J.P. Robust estimates of a location parameter. *Ann. Math. Statist.* **1964**, *35*, 73–101.
36. Huber, J.P. The 1972 Wald Lecture Robust statistics: A review. *Ann. Math. Stat.* **1972**, *43*, 1041–1067.
37. Zhang, B.; Perrie, W. Cross-Polarized Synthetic Aperture Radar: A New Potential Measurement Technique for Hurricanes. *Bull. Am. Meteorol. Soc.* **2012**, *93*, 531–541.
38. Tavolato, C.; Isaksen, L. On the Use of a Huber Norm for Observation Quality Control in the ECMWF 4D-Var. *Q. J. R. Meteorol. Soc.* **2015**, *141*, 1514–1527.
39. Lorenc, A.C. Analysis methods for numerical weather prediction. *Q. J. R. Meteorol. Soc.* **1986**, *112*, 1177–1194.
40. Barker, M.D.; Huang, W.; Guo, Y.R. A three-dimensional variational data assimilation system for MM5: Implementation and initial results. *Mon. Weather Rev.* **2004**, *132*, 897–914.
41. Zhang, Q.F.; Weng, Y.H.; Sippel, J.A. Cloud-resolving hurricane initialization and prediction through assimilation of Doppler radar observations with an ensemble Kalman filter. *Mon. Weather Rev.* **2009**, *137*, 2105–2125.

*Article*

# An Empirical Algorithm for Wave Retrieval from Co-Polarization X-Band SAR Imagery

**Weizeng Shao [1], Jing Wang [1], Xiaofeng Li [1,3,*] and Jian Sun [2]**

[1] Marine Science and Technology College, Zhejiang Ocean University, Zhoushan 316000, China; shaoweizeng@zjou.edu.cn (W.S.); momozjhn@163.com (J.W.)

[2] Physical Oceanography Laboratory, Ocean University of China, Qingdao 266100, China; sunjian77@ouc.edu.cn

[3] Global Science and Technology, National Oceanic and Atmospheric Administration (NOAA)-National Environmental Satellite, Data, and Information Service (NESDIS), College Park, MD 20740, USA

* Correspondence: xiaofeng.li@noaa.gov; Tel.: +1-301-683-3144

Academic Editors: Xiaofeng Yang, Ferdinando Nunziata and Alexis Mouche

Received: 17 April 2017; Accepted: 6 July 2017; Published: 11 July 2017

**Abstract:** In this study, we proposed an empirical algorithm for significant wave height (SWH) retrieval from TerraSAR-X/TanDEM (TS-X/TD-X) X-band synthetic aperture radar (SAR) co-polarization (vertical-vertical (VV) and horizontal-horizontal (HH)) images. As the existing empirical algorithm at X-band, i.e., XWAVE, is applied for wave retrieval from HH-polarization TS-X/TD-X image, polarization ratio (PR) has to be used for inverting wind speed, which is treated as an input in XWAVE. Wind speed encounters saturation in tropical cyclone. In our work, wind speed is replaced by normalized radar cross section (NRCS) to avoiding using SAR-derived wind speed, which does not work in high winds, and the empirical algorithm can be conveniently implemented without converting NRCS in HH-polarization to NRCS in VV-polarization by using X-band PR. A total of 120 TS-X/TD-X images, 60 in VV-polarization and 60 in HH-polarization, with homogenous wave patterns, and the coincide significant wave height data from European Centre for Medium-Range Weather Forecasts (ECMWF) reanalysis field at a 0.125° grid were collected as a dataset for tuning the algorithm. The range of SWH is from 0 to 7 m. We then applied the algorithm to 24 VV and 21 HH additional SAR images to extract SWH at locations of 30 National Oceanic and Atmospheric Administration (NOAA) National Data Buoy Center (NDBC) buoys. It is found that the algorithm performs well with a SWH stander deviation (STD) of about 0.5 m for both VV and HH polarization TS-X/TD-X images. For large wave validation (SWH 6–7 m), we applied the empirical algorithm to a tropical cyclone Sandy TD-X image acquired in 2012, and obtained good result with a SWH STD of 0.3 m. We concluded that the proposed empirical algorithm works for wave retrieval from TS-X/TD-X image in co-polarization without external sea surface wind information.

**Keywords:** SAR; significant wave height; co-polarization; TerraSAR-X/TanDEM-X

---

## 1. Introduction

It is well known that space-borne synthetic aperture radar (SAR) is an efficiently instrument for wind and wave observation in a large coverage with high spatial resolution at seas. Most satellite SAR operates at X-band (TerraSAR-X (TS-X), TanDEM-X (TD-X), and Cosmo-SkyMed), C-band (Radarsat-1/2, ERS-1/2, Envisat-ASAR, Sentinel-1A/-1B and Chinese Gaofen-3), and L-band (Japanese ALOS-1/ALOS-2). TS-X and its twin TD-X have 514 km orbit height above earth and a 100-min orbit period with fine spatial resolution of image up to 1 m. TS-X and TD-X SAR are officially operated by Germen Aerospace Center (DLR). In the past few years, several algorithms for winds [1–5] and waves [6–9] retrieval from TS-X/TD-X image have been developed. Geophysical model function

(GMF) XMOD2 [4] and polarization ratio XPR2 [5] are the latest achievements for wind retrieval from VV-polarization and HH-polarization TS-X/TD-X image, respectively. After employing SAR-derived wind speed, waves can be estimated from TS-X/TD-X image by using the theoretic-based algorithm "Parameterized First-guess Spectrum Method" (PFSM) [9] or the empirical algorithms [6,7]. Recently, empirical wave retrieval algorithm is adapted for coastal application [8], considering the ship and wave breaking etc. in offshore region.

Algorithm PFSM [10–12] was originally exploited for wave retrieval from C-band SAR, which is based on the wave mapping mechanism on SAR, including tilt modulation, hydrodynamic modulation [13] and velocity bunching [14]. PFSM is developed similar to the "Max-Planck Institute" algorithm (MPI) [15–17], "Semi Parametric Retrieval Algorithm" scheme (SPRA) [18] and "Partition Rescaling and Shift Algorithm" (PARSA) [19]. All these algorithms need a first-guess wave spectrum and the "true" wave spectrum is inverted through a set of iterations by minimizing a cost function [15]. Algorithm MPI and PARSA take the outputs from numeric wave model [20] as the first-guess spectrum, and both of these require a long computing time. Algorithm SPRA employs wind speed from scatterometer to produce the first-guess wind-sea spectrum by using empirical parametric wave function and information on swell is regarded as the difference between the retrieval results mapping spectrum and the original SAR spectrum. In other words, the error in the wind-sea retrieval process is delivered into the swell retrieval process in SPRA scheme. PFSM separates the non-linear wind-sea and the linear-mapping swell spectrum by calculating the separation threshold of the wave number. Moreover, it searches for the best parameters, e.g., dominate wave phase velocity and peak propagation direction, together with SAR-derived wind speed so as to produce the best fit first-guess wind-sea wave spectrum by using empirical parametric wave function, e.g., Jonswap [21]. The composite wave spectrum is obtained, after the different wave spectrum portions are inverted from corresponding SAR intensity spectrum portions. In addition, there are several unconstrained algorithms [22,23], which can also be applied for waves retrieval in a particular sea state, e.g., a long wave dominant regime, however, the retrieval result usually contains information on swell due to the portion produced by shorter waves in a SAR spectrum is missing in the inversion schemes. In fact, the basic scattering physics is independent on radar frequency and imaging polarization. In our previous study [9], it was already proven that algorithm PFSM can be applied to invert wave spectrum from TS-X/TD-X image and then wave parameters are derived from the inverted wave spectrum. Validation against the third-generation wave model WaveWatch-III outputs through 16 HH-polarization TS-X/TD-X images show a 0.43 m Root-Mean-Square Error (RMSE) of significant wave height (SWH).

Due to the complex nature of modulation transfer functions (MTF) in these theoretic-based algorithms, researchers also exploited the empirical algorithms such as CWAVEs (CWAVE_ERS [24] and CWAVE_ENV [25]) algorithms for C-band SAR and XWAVE [6–8] algorithms for X-band SAR. In parallel, a few researches recently have made effort to build empirical algorithms for retrieving SWH through azimuthal cutoff wavelength on SAR [26–29]. CWAVEs describe a relationship among wave and several other variables, e.g., wind speed, radar cross section and a set of orthonormal decompositions in a two-dimensional SAR spectrum derived from SAR intensity image. However, CWAVEs were designed to retrieve wave information from particular C-band SAR mode image, e.g., wave mode that has a fixed incidence angle around 23°. XWAVE inherits the idea behind CWAVEs, which is exploited through a number of TS-X/TD-X images at full incidence angle ranged from 20° to 50°. In the development of XWAVE [6,7], the algorithm coefficients were primarily tuned using VV-polarization TS-X/TD-X data acquired over National Oceanic and Atmospheric Administration (NOAA) moored buoys in the open ocean. The SWH retrieval results show good agreement with the outputs from numerical wave model data provided by DWD. SAR-derived wind speed is necessary in algorithm XWAVE. XMOD2 [4] are tuned by using an amount of VV-polarization TS-X/TD-X images and collocated winds from DWD. However, no reliable wind retrieval above 20 m/s is achieved from TS-X/TD-X image by using XMOD2 due to no available data at such wind speeds in the tuning process. Thus far, two existing algorithms, PFSM and XWAVE, have not been implemented under tropical cyclone conditions yet. This is because XMOD is tuned and validated through VV-polarization TS-X/TD-X images and the DWD with the wind speeds up to 25 m/s. Moreover, the signal saturation problem also exists for SAR in tropical cyclone [30,31].

Recently, a new empirical approach was reported in [32], by which it is possible to directly retrieve SWH in tropical cyclones from normalized radar cross section (NRCS) of C-band wide ScanSAR image, e.g., Envisat-ASAR and Radarsat-1/2. Interestingly, that empirical model can be conveniently applied similarly to the SAR wind retrieval methodology. Although the validation against outputs from the third–generation wave models, including WaveWatch-III and SWAN, has exhibited encouraging results, there are still some weaknesses existed in the model as mentioned by the authors.

In this study, we propose an empirical algorithm for SWH retrieval from X-band SAR through improving existing XWAVE model. In particular, this developed model can be directly applied for HH-polarization TS-X/TD-X image without converting NRCS in HH-polarization into NRCS in VV-polarization. The proposed algorithm avoids using SAR-derived wind speed, which is known having large retrieval errors in tropical cyclone. Data collected at high sea state (SWH > 5 m) are also included in the tuning dataset and the algorithm performs well under tropical cyclone condition.

The paper is organized as follows. SAR images and collocated NOAA in situ buoys dataset are introduced in Section 2. In Section 3, methodology of the proposed empirical model for SWH retrieval is presented and the coefficients of the proposed empirical function are tuned by the dataset. The comparison of SWH retrieved from SAR imagery and those measured by buoys is shown in Section 4. A case study for wave retrieval using two TS-X/TD-X images acquired during in tropical cyclone Sandy in 2012 is also presented. Conclusions are summarized in Section 5.

## 2. Data Description

SAR data used in this study includes 60 VV-polarization and 60 HH-polarization SAR images acquired between 2008 and 2015. As examples, a HH-polarization TS-X image in StripMap mode acquired in Gulf of Alaska at 03:05 UTC on 3 November 2011 is shown in Figure 1a while another VV-polarization ScanSAR mode TD-X image acquired near Southeast Newfoundland Coast at 21:17 UTC on 4 October 2013 is shown in Figure 1b.

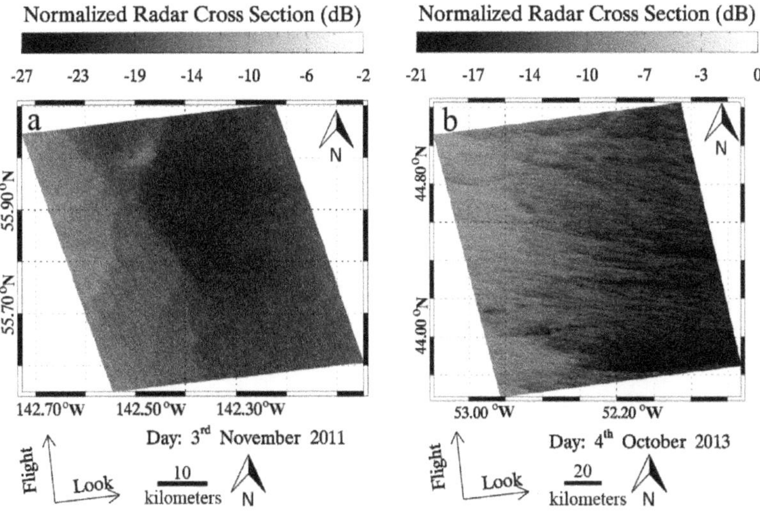


**Figure 1.** (**a**) A HH-polarization StripMap mode TerraSAR(TS-X) image acquired in Gulf of Alaska at 03:05 UTC on 3 November 2011; and (**b**) a VV-polarization ScanSAR mode TanDEM-X (TD-X) image acquired to the Southeast of Newfoundland at 21:17 UTC on 4 October 2013.

In this study, the European Centre for Medium-Range Weather Forecasts (ECMWF) global atmospheric-marine reanalyzed data is matched up against the SAR measurements. Here, we used ECMWF reanalysis SWH data at a 0.125° grid (approximate 12.5 km) at an interval of six hours. To perform the matchup, every TS-X/TD-X imagery was divided into sub-scenes with a spatial coverage of 1.5 × 1.5 km for StripMap mode and 4 × 4 km for ScanSAR mode images in azimuth and range direction, respectively. Then, ECMWF SWH data in every sub-scene were calculated by both bilinear interpolation in space and time. To eliminate inhomogeneous sub-scenes, we compute the image variance and only keep those with values smaller than 1.05 [26]. Moreover, the SAR spectrum is smoothed to reduce the distortions of other marine phenomena. Figure 2a,b shows the ECMWF SWH data that correspond to the two images shown in Figure 1a,b, respectively.

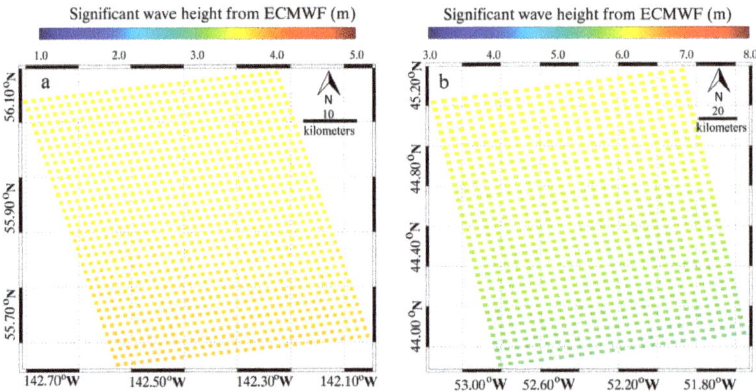


**Figure 2.** (**a**) Space and time interpolated European Centre for Medium-Range Weather Forecasts (ECMWF) SWH data corresponding to the SAR image in Figure 1a; and (**b**) same as (**a**) but for the SAR image in Figure 1b.

In total, our dataset consists of more than one thousand SAR-derived and ECMWF reanalysis SWH matchup points for algorithm tuning. Histograms of SWH matchups are shown in Figure 3, in which the SWH ranges from 0 to 7 m at interval of 0.3 m.

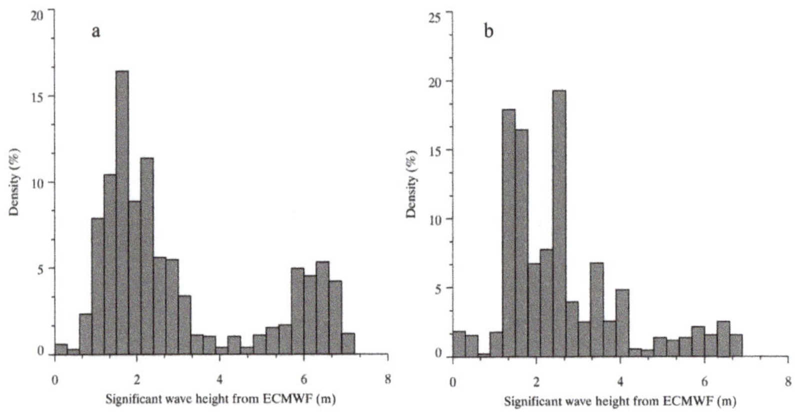


**Figure 3.** Histograms of SWH matchups at interval of 0.3 m. SWH ranges from 0 to 7 m: (**a**) VV-polarization; and (**b**) HH-polarization.

## 3. Development of Empirical Algorithm for Wave Retrieval at X-Band

### 3.1. Existing X-Band SAR Wind and Wave Algorithms

The wind retrieval from SAR is a matured technology. The initial X-band GMF, XMOD1, simply related VV-polarization X-band radar normalized radar cross section (NRCS) from TS-X/TD-X images to wind speed in a pre-launch study [1]. Then, similar to the development of C-band GMF CMOD5 [33] that was derived from ERS-1 SAR images and ECMWF reanalysis wind data, XMOD2 has been exploited in [4] by using collocated VV-polarization TS-X/TD-X images and National Data Buoy Center (NDBC) buoy measurements and it was found that a 1.44 m/s RMSE of wind speed was achieved against NOAA in situ buoys. Besides, another X-band GMF, called SIRX-MOD, was proposed in [3] by retuning the coefficients in the C-band GMF CMOD-IFR2 [34] with the VV-polarization Spaceborne Imaging Radar (SIR) X-band SAR NRCS data and ECMWF reanalysis wind data. XMOD2 and SIRX-MOD take the general form of:

$$\sigma_0 = B_0(1 + B_1 \cos\phi + B_2 \cos 2\phi) \quad (1)$$

where $\sigma_0$ is the NRCS in linear unit, and $\phi$ represents the angle between the radar look direction and the wind direction. The coefficients $B_0$, $B_1$ and $B_2$ are functions of the radar incidence angle $\theta$ and sea surface wind speed $U_{10}$ at 10 m height above sea surface. Figure 4 shows the XMOD2 and SIRX-MOD curves at $\theta$ of 30° and $\phi$ of 45°, showing X-band NRCS is linearly related to wind speed. This behavior is consistent with the observations of microwave backscattering signatures of the ocean at X-band during the experiment using an airborne microwave scatterometer-radiometer system [35]. As for wind retrieval from HH-polarization TS-X/TD-X image, polarization ratio (PR) model is used to convert NRCS values from VV to HH. X-band PR (XPR) models for TS-X/TD-X are given for [2,5]. It was reported in [5] that the comparison of wind speed by using the combination method, that is XMOD2 together with XPR2, shows a RMSE of 1.79 m/s against winds measured by NOAA buoys. However, these algorithms are only valid for wind speeds up to 25 m/s, because they are exploited through low-to-moderate wind speeds.

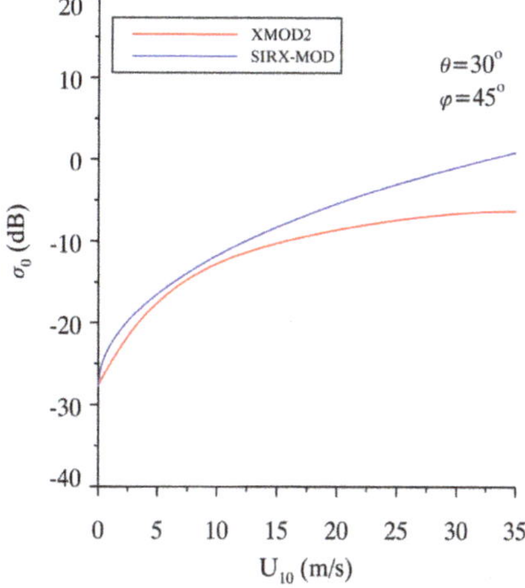


**Figure 4.** The simulation of XMOD2 and SIRX-MOD at $\theta$ of 30° and $\phi$ of 45°.

Based on SAR-derived wind speed, two wave retrieval algorithms, a theoretic-based PFSM [9] algorithm and an empirical XWAVE model [6], have been developed for TS-X/TD-X image. XWAVE takes the form of:

$$H_s = A_1\sqrt{E_s \tan\theta} + A_2U_{10} + A_3 + A_4 \cos\phi \quad (2)$$

where $H_s$ is the SWH, $\theta$ is the radar incidence angle, $U_{10}$ is the wind speed at 10 m above sea surface, $\phi$ is the wave peak direction relative to azimuth direction ranged from 0 to 90°, $E_s$ $(= \int_0^{2\pi}\int_{k_{min}}^{k_{max}} \bar{S}(k,\theta)dkd\theta)$ is the integrated value of the normalized SAR intensity spectrum $\bar{S}(k,\theta)$in wavelength domain $L_{min}$ (= $2\pi/k_{min}$) of 30 to $L_{max}$ (=$2\pi/k_{max}$) of 600 m, and the coefficients $A_1$ to $A_4$ are the constants tuned by VV-polarization TS-X/TD-X images together with SWH from DWD and NOAA buoys in [6,7]. XWAVE is conveniently applied for waves retrieval from TS-X/TD-X images without transferring SAR intensity spectrum into wave spectrum. Although SAR-derived wind speed from VV-polarization and HH-polarization TS-X/TD-X image has known accuracy at within 2 m/s RMSE of wind speed [4,5], XWAVE is restrictedly used during operational application, due to prior wind direction is necessary in the process of wind retrieval by using XMODs.

### 3.2. Empirical Algorithm for Wave Retrieval in Both VV- and HH-Polarization

X-band GMF XMOD2 and PR model XPR2 are valid for winds up to 25 m/s due to no available higher winds in the tuning dataset. In this study, we develop an empirical wave retrieval algorithm by replacing wind speed with NRCS in Equation (2). The purpose of this kind of development is that the empirical algorithm can be conveniently implemented without calculating the sea surface wind speed. The proposed empirical model takes the form:

$$H_s = C_1\sqrt{E_s \tan\theta} + C_2\sigma^0 + C_3 + C_4 \cos\alpha \quad (3)$$

where, $\alpha$ represents the peak direction relative to azimuth direction in a SAR spectrum instead of wave peak direction $\phi$ in Equation (2) for convenient application. The collocated dataset, including ECMWF SWH data and the three other variables derived from SAR intensity spectrum, is used for tuning the coefficients $C_1$ to $C_4$ in VV-polarization and HH-polarization. The values of matrix C in Equation (3) for VV-polarization and HH-polarization are shown in Tables 1 and 2, respectively.

The statistical analysis between the ECMWF reanalysis SWH and the simulated SWH by using proposed algorithm is exhibited in Figure 5 for 10° of incidence angle bins between 20° and 50° and 1 m of SWH bins ranged from 0 to 7 m. The result shows the correlation is about 0.8. Under this circumstance, it is indicated that the proposed algorithm is suitable for $H_s$ retrieval from VV and HH polarization TS-/TD-X image. However, it is necessary to figure out if the proposed algorithm relies on good-quality power spectra of SAR image.

**Table 1.** Tuned coefficients in Equation (3) for VV-polarization.

| | |
|---|---|
| $C_1$ | 2.90 |
| $C_2$ | 3.31 |
| $C_3$ | 0.47 |
| $C_4$ | 0.58 |

**Table 2.** Tuned coefficients in Equation (3) for HH-polarization.

| | |
|---|---|
| $C_1$ | 2.11 |
| $C_2$ | 2.21 |
| $C_3$ | 0.91 |
| $C_4$ | 0.64 |

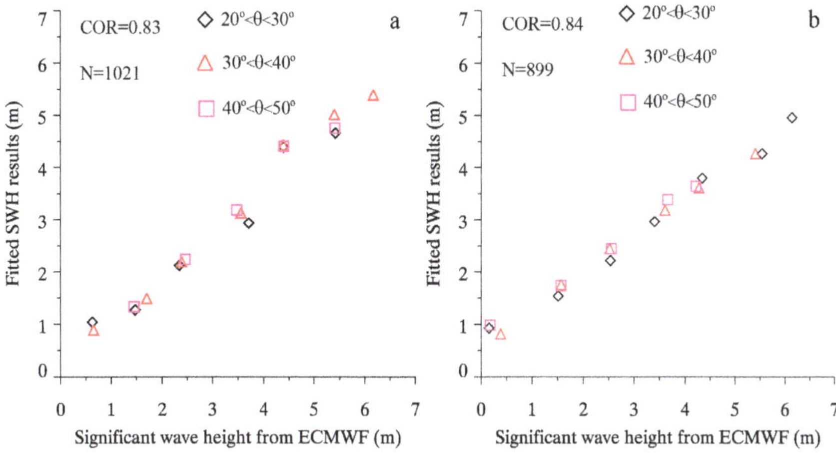


**Figure 5.** ECMWF SWH data versus fitted SWH results, for 10° incidence angle bins between 20° and 50° and 1 m/s SWH bins ranging from 0 to 7 m/s: (**a**) VV-polarization; and (**b**) HH-polarization.

## 4. Validation

### *4.1. Validation Against Buoys*

As a case study, the image of HH-polarization TS-X image in StripMap mode acquired at 16:19 UTC on 1 February 2012 at 13:59 UTC is shown in Figure 6, which covers the NOAA in situ buoy (ID: 46047). A sub-scene of 2048 × 2048 pixels with a 1.25 × 1.25 pixel size has been extracted from TS-X image, which covers the location of NDBC buoy. The sub-scene is normalized and then the sub-scene is divided into 2 × 2 small scenes. The corresponding four two-dimensional SAR spectra are calculated by using the two-dimensional Fast Fourier Transform (FFT-2) method. The smooth two-dimensional spectrum, which is obtained by averaging the four two-dimensional SAR spectra, is used here.

The image of sub-scene and the corresponding two-dimensional wave spectrum in term of length $\lambda$, is shown in Figure 7a,b. The SAR-derived SWH in area A centered at the buoy location is 2.01 m and the buoy-measured SWH is 2.48 m. As for this case study, the difference between retrieve SWH and observed SWH is 0.47 m.

We apply the X-band wave retrieval algorithm to extract SWH values from 24 VV-polarization and 21 HH-polarization TS-X/TD-X images. In these SAR images, they also contain 30 NOAA buoy locations. The information of SAR images and corresponding NDBC buoys is shown in Appendix A. The SAR-derived wave information is matched up against co-located NOAA buoy measurements.

As shown in Figure 8, the RMSE of SWH is 0.5 m with a 27% scatter index (SI) for VV-polarization images and the RMSE of SWH is 0.52 m with a 36% SI for HH-polarization images. The stander deviation (STD) of SWH is 0.5 m between retrieval results from co-polarization TS-X/TD-X images and buoy measurements. We found that SAR-derived SWH by using the proposed algorithm has a similar accuracy to the analysis results by using the existing wave retrieval algorithms, which has a SWH STD of around 0.5 m as validated against observations from moored buoys or altimeters [18,24,25]. Again, it should be noted that the proposed empirical XWAVE model can be directly applicable without knowing the information on wind speed and PR model is not required as it is applied for HH-polarization TS-X/TD-X image.

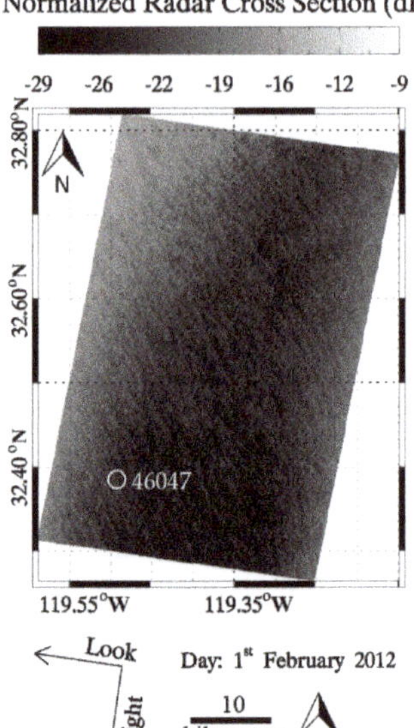


**Figure 6.** The image of HH-polarization TS-X image in StripMap mode acquired at 16:19 UTC on 1 February 2012, covering the NOAA in situ buoy (ID: 46047).

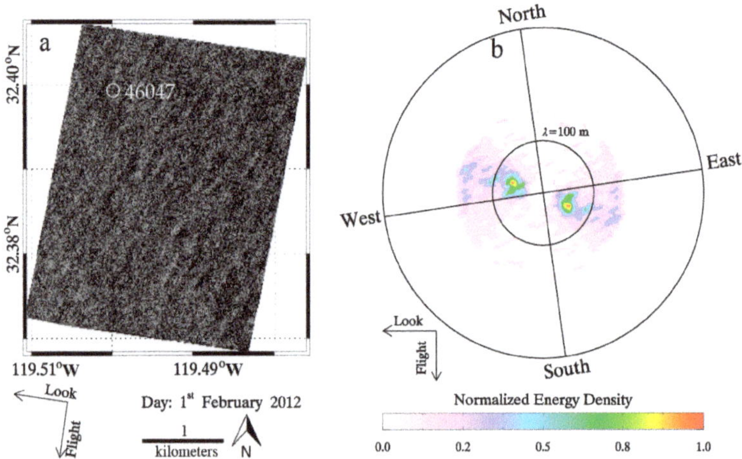


**Figure 7.** (**a**) Intensity image of sub-scene covering the NOAA in situ buoy (ID: 46047); and (**b**) the two-dimensional SAR spectrum in term of length λ corresponding to the sub-scene.

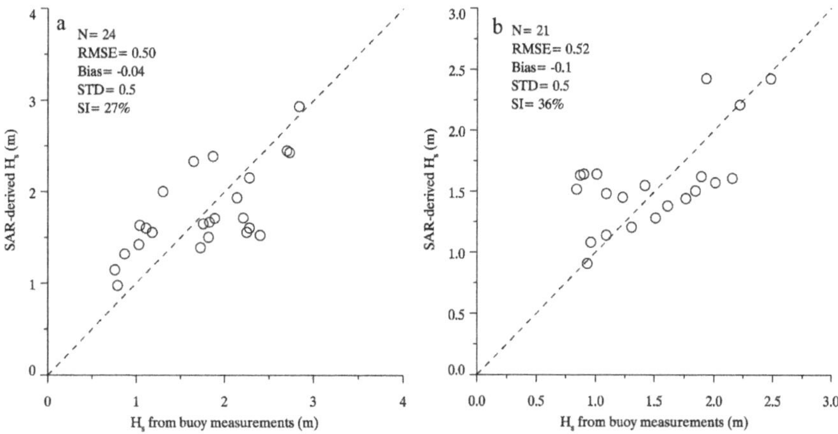


**Figure 8.** SWH retrieval results from TS-X/TD-X images are compared with buoy measurements: (**a**) VV-polarization images; and (**b**) HH-polarization images.

### *4.2. Application in Trpocial Cyclone*

Further, we also validate the algorithm for one TD-X image taken during tropical cyclone Sandy in 2012. SAR has the capacity of all-weather field monitoring, especially in tropical cyclones. Through several tropical cyclones captured by SAR, some achievements have been exhibited in [36–40], e.g., morphology of cyclones [36–38], hurricane-generated ocean swell refraction [39] and a new method of high wind speed retrieval [40]. The comparison of wind retrieval using VV-polarization C-band SAR backscattering in hurricanes was reported in [41]. The results show RMSE of wind speed is 6.2–6.5 m/s against measurements from Stepped Frequency Microwave Radiometer (SFMR), due to winds encounter saturation problem as winds growing under tropical cyclone condition [30,31]. Therefore, SAR-derived wind speeds have a large deviation with reality in tropical cyclones. To eliminate this source of errors, we replaced the wind speed with the NRCS in the existing XWAVE formula in this study. The advantage of this development is that the application of proposed algorithm avoids using SAR-derived wind speed, which is not working at high winds.

The multi-look ground range detected (MGD) VV-polarization TD-X SAR image in ScanSAR mode acquired over tropical cyclone Sandy at 22:49 UTC on 28 October 2012 is shown in Figure 9. The TD-X image has an 8.25 × 8.25 m pixel size in both azimuth and range directions and then it was divided into sub-scenes of 512 × 512 pixels, which correspond to a spatial coverage of about 4 × 4 km. The sub-scenes were processed to retrieve SWH by using the developed algorithm. However, about 15% of sub-scenes are contaminated by the rain. These data were excluded in this study.

There are no NOAA buoys within the TD-X image's coverage. Therefore, we only perform the comparison against ECMWF results. The commonly used WaveWatch-III model output has a spatial resolution of 0.5° grid, which is too coarser than the ECMWF model results. Figure 10 shows that the SAR-derived SWH from the TD-X image in tropical cyclone Sandy and ECMWF reanalysis SWH at a 0.125° grid, in which the black rectangle represents the coverage of TD-X image. In particular, the time between the TD-X imaging time and ECMWF reanalysis SWH data is comparatively close, i.e. within 2 h. In general, the SAR-derived SWH is agreeable to the ECMWF reanalysis SWH data. Then the SAR-derived SWH points matched up closest to ECMWF grid points are selected. Figure 11 shows a 0.35 m RMSE of the SWH comparison. The unique ECMWF reanalysis SWH data were used for tuning and validating the proposed empirical algorithm, causing a better 0.3 m STD than a 0.5 m STD of SWH. Although ECMWF reanalysis SWH data deviate from reality, the statistical analysis of the case study still reveals the proposed empirical wave retrieval algorithm has a creditable performance under tropical cyclone condition.

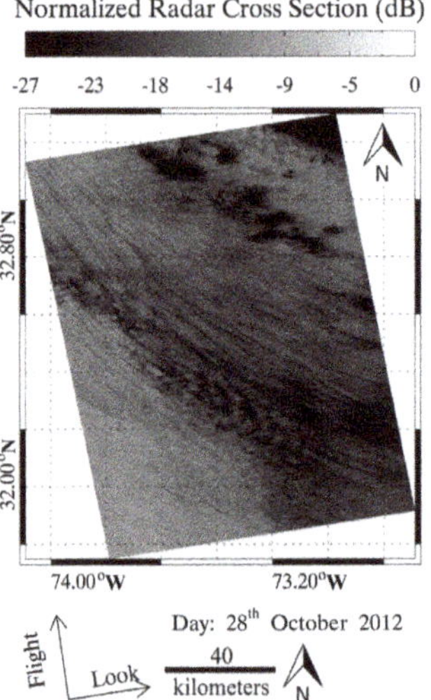


**Figure 9.** The image of VV-polarization TD-X SAR image in ScanSAR mode acquired over tropical cyclone Sandy at 22:49 UTC on 28 October 2012.

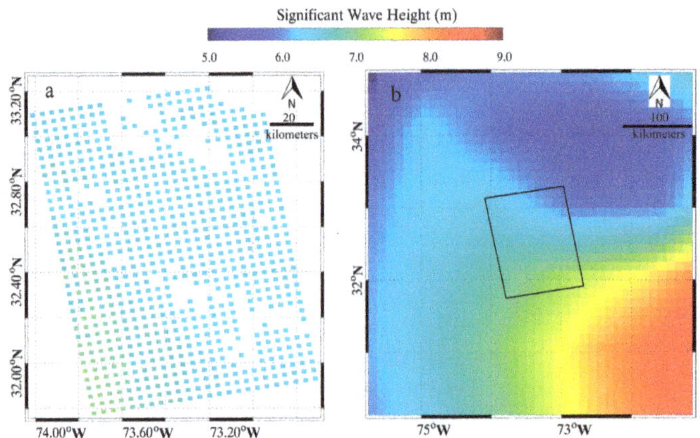


**Figure 10.** (**a**) The SAR-derived SWH from TD-X image over tropical cyclone Sandy at 22:49 UTC on 28 October 2012. (**b**) ECMWF reanalysis SWH data at 00:00 UTC on 29 October 2012 at a 0.125° gird, in which the rectangle represents the coverage of TD-X image.

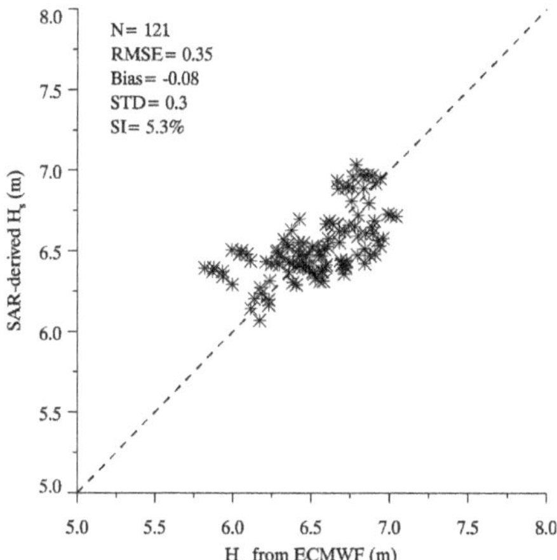


**Figure 11.** The comparison between SAR-derived SWH by using the proposed empirical algorithm and ECMWF reanalysis SWH data.

## 5. Conclusions

XWAVE's design was aimed to wave retrieval from VV-polarization TS-X/TD-X image, which relies on SAR-derived wind speeds. Although several algorithms have been recently exploited for wind retrieval from co-polarization TS-X/TD-X image, such as GMF SIRX-MOD [3], GMF XMOD2 [4] and polarization ratio XPRs [2,5], these algorithms are only valid for wind speeds up to 25 m/s. When XWAVE is applied for wave retrieval from HH-polarization TS-X/TD-X image, XPR has to be used for converting NRCS in HH-polarization to NRCS in VV-polarization to retrieve wind speed. It is well known that SAR NRCS has a strong relation with wind speed. This is true for C-band [33] and X-band SAR [35]. In this study, we proposed an empirical algorithm by replacing NRCS instead of wind speed in the existing XWAVE model. Therefore, this development benefits the operation of waves retrieval from X-band SAR due to its application without using SAR-derived wind speeds.

In our work, 60 TS-X/TD-X images in VV-polarization and 60 TS-X/TD-X images in HH-polarization were collected over whole seas. All these images were divided into numbers of sub-scenes, which were collocated with ECMWF SWH data at a 0.125° grid. We have more than one thousand matchups to tune the proposed empirical algorithm. An additional 24 images in VV-polarization and 21 images in HH-polarization were implemented using the proposed empirical algorithm and the retrieval results were validated against the observations from 30 NOAA in situ buoys, showing a 0.5 STD of SWH. XWAVE needs convert NRCS in HH-polarization to NRCS in VV-polarization by using XPRs and it relies on SAR-derived wind speed which has a deviation with reality. The proposed algorithm directly works for both VV-polarization and HH-polarization without using XPRs. The correlation between the ECMWF reanalysis SWH and the simulated SWH is about 0.8. Therefore, we think the proposed algorithm is suitable for wave retrieval from co-polarization TS-X/TD-X image.

The validation of wind speed retrieved from TS-X/TD-X image using XMOD has not been investigated yet under tropical cyclone condition. Therefore, the advantage of the proposed empirical algorithm is that wind speed is no longer needed for wave retrieval in tropical cyclone. One VV-polarization TD-X image in tropical cyclone Sandy in 2012 was used to confirm the applicability of the proposed algorithm. Because no moored buoys were available in the TS-X coverage, and wave data from WaveWatch-III model have a 0.5° grid, which is too coarse for validation, we use ECMWF reanalysis SWH to preliminary evaluate the performance of the proposed algorithm. The comparison between SAR-derived SWH and ECMWF reanalysis SWH data shows a 0.3 m STD of SWH meaning the proposed empirical algorithm also works under tropical cyclone condition. A new method of wind retrieval in tropical cyclone was proposed in [40], in which wind speeds up to 65.4 m/s were retrieved from the information on waves using the fetch-limited wind wave growth function. The validation shows a good agreement with hurricane hunter measurements and there is no indication of saturation problem in the wind retrieval. In the near future, we plan to validate the proposed algorithm through more X-band SAR images in tropical cyclones, covering the moored buoys. Then winds can be retrieved from X-band SAR image through SAR-derived SWH.

**Acknowledgments:** TS-X/TD-X images were acquired via the science AOs of OCE3150 and OCE3356 provided from German Aerospace Center (DLR). Buoy data are downloaded via http://www.ndbc.noaa.gov/. ECMWF reanalysis wave data were openly accessed via http://www.ecmwf.int. The research is partly supported by the National Key Research and Development Program of China under Grant No. 2017YFA0604901, National Natural Science Foundation of China under Grant No. 41376010 and Foundation of Zhejiang Ocean University (2015). The views, opinions, and findings contained in this report are those of the authors and should not be construed as an official NOAA or U.S. Government position, policy or decision.

**Author Contributions:** Weizeng Shao, Jian Sun and Xiaofeng Li came up the original idea and designed the experiments. Weizeng Shao and Jing Wang collected and analyzed the data. All authors contributed to the writing and revising of the manuscript.

**Conflicts of Interest:** The authors declare no conflict of interest.

## Appendix A

**Table A1.** The information of TS-X/TD-X images and corresponding NDBC buoys used in our study.

| Buoy ID | TS-X/TD-X Acquisition Time (YYYY-MM-DD) | Imaging Mode | Buoy ID | TS-X/TD-X Acquisition Time (YYYY-MM-DD) | Imaging Mode |
|---|---|---|---|---|---|
| 46013 | 2008-02-22 02:08 | StripMap | 46011 | 2011-03-13 14:07 | ScanSAR |
| 46026 | 2008-03-02 14:15 | StripMap | 46054 | 2011-03-15 01:59 | StripMap |
| 46029 | 2009-03-19 02:02 | ScanSAR | 46050 | 2011-10-23 14:30 | StripMap |
| 41048 | 2009-05-13 22:31 | StripMap | 41047 | 2011-10-25 10:51 | StripMap |
| 46025 | 2009-06-10 01:50 | ScanSAR | 46050 | 2011-10-26 02:11 | StripMap |
| 46221 | 2009-07-19 01:42 | ScanSAR | 46229 | 2011-10-26 02:10 | StripMap |
| 46221 | 2009-09-12 01:42 | ScanSAR | 46015 | 2011-11-06 02:10 | StripMap |
| 46028 | 2010-04-08 01:59 | ScanSAR | 41047 | 2011-11-06 22:39 | StripMap |
| 46015 | 2010-06-13 02:01 | ScanSAR | 46015 | 2011-11-28 02:10 | StripMap |
| 46013 | 2010-07-05 02:00 | ScanSAR | 46050 | 2011-12-04 02:02 | StripMap |
| 46029 | 2010-07-05 02:02 | StripMap | 46015 | 2011-12-04 02:01 | StripMap |
| 42036 | 2010-07-06 11:42 | StripMap | 41043 | 2012-01-23 22:20 | StripMap |
| 46028 | 2010-07-16 01:59 | StripMap | 46047 | 2012-02-01 13:59 | StripMap |
| 46011 | 2010-11-12 14:07 | StripMap | 51000 | 2012-02-26 16:19 | StripMap |
| 46022 | 2010-11-16 14:31 | StripMap | 46222 | 2012-03-27 13:59 | StripMap |
| 46011 | 2010-11-17 14:15 | StripMap | 46011 | 2012-05-23 01:51 | StripMap |
| 44008 | 2010-11-20 22:25 | StripMap | 46025 | 2013-02-03 14:07 | ScanSAR |
| 51000 | 2010-12-13 16:19 | ScanSAR | 41048 | 2013-02-05 22:32 | StripMap |
| 46053 | 2010-12-15 14:07 | ScanSAR | 41002 | 2013-04-04 11:07 | ScanSAR |
| 46012 | 2010-12-17 02:00 | ScanSAR | 46011 | 2013-06-28 01:59 | ScanSAR |
| 51000 | 2010-12-20 04:14 | ScanSAR | 46053 | 2013-09-22 14:07 | ScanSAR |
| 51000 | 2010-12-24 16:19 | ScanSAR | 52200 | 2015-04-13 08:22 | StripMap |
| 46050 | 2011-03-04 02:02 | ScanSAR | | | |

## References

1. Ren, Y.Z.; He, M.X.; Lehner, S. An algorithm for the retrieval of sea surface wind fields using X-band TerraSAR-X data. *Int. J. Remote Sens.* **2012**, *33*, 7310–7336. [CrossRef]
2. Shao, W.Z.; Li, X.M.; Lehner, S.; Guan, C.L. Development of polarization ratio model for sea surface wind field retrieval from TerraSAR-X HH polarization data. *Int. J. Remote Sens.* **2014**, *35*, 4046–4063. [CrossRef]

3. Ren, Y.Z.; Li, X.M.; Zhou, G. Sea surface wind retrievals from SIR-C/X-SAR data: a revisit. *Remote Sens.* **2015**, *7*, 3548–3564. [CrossRef]
4. Li, X.M.; Lehner, S. Algorithm for sea surface wind retrieval from TerraSAR-X and TanDEM-X Data. *IEEE Trans. Geosci. Remote Sens.* **2014**, *52*, 2928–2939. [CrossRef]
5. Shao, W.Z.; Zhang, Z.; Li, X.M.; Wang, W.L. Sea surface wind speed retrieval from TerraSAR-X HH-polarization data using an improved polarization ratio model. *IEEE J. Sel. Top. Appl. Earth Obs. Remote Sens.* **2016**, *9*, 4991–4997. [CrossRef]
6. Bruck, M.; Lehner, S. Coastal wave field extraction using TerraSAR-X data. *J. Appl. Remote Sens.* **2013**, *107*, 333. [CrossRef]
7. Bruck, M. Sea state measurements using the XWAVE algorithm. *Int. J. Remote Sens.* **2015**, *36*, 3890–3912. [CrossRef]
8. Pleskachevsky, A.L.; Rosenthal, W.; Lehner, S. Meteo-Marine parameters for highly variable environment in coastal regions from satellite radar images. *ISPRS J. Photogramm.* **2016**, *119*, 464–484. [CrossRef]
9. Shao, W.Z.; Li, X.F.; Sun, J. Ocean wave parameters retrieval from TerraSAR-X images validated against buoy measurements and model results. *Remote Sens.* **2015**, *7*, 12815–12828. [CrossRef]
10. Sun, J.; Guan, C.L. Parameterized first-guess spectrum method for retrieving directional spectrum of swell-dominated waves and huge waves from SAR images. *Chin. J. Oceanol. Limn.* **2006**, *24*, 12–20.
11. Sun, J.; Kawamura, H. Retrieval of surface wave parameters from SAR images and their validation in the coastal seas around Japan. *J. Oceanogr.* **2009**, *65*, 567–577. [CrossRef]
12. Lin, B.; Shao, W.Z.; Li, X.F.; Li, H.; Du, X.Q.; Ji, Q.Y.; Cai, L.N. Development and validation of an ocean wave retrieval algorithm for VV-polarization Sentinel-1 SAR data. *Acta Oceanol. Sin.* **2017**, *36*, 95–101. [CrossRef]
13. Alpers, W.; Ross, D.B.; Rufenach, C.L. On the detectability of ocean surface waves by real and synthetic radar. *J. Geophys. Res.* **1981**, *86*, 10529–10546. [CrossRef]
14. Alpers, W.; Bruning, C. On the relative importance of motion-related contributions to SAR imaging mechanism of ocean surface waves. *IEEE Trans. Geosci. Remote Sens.* **1986**, *24*, 873–885. [CrossRef]
15. Hasselmann, K.; Hasselmann, S. On the nonlinear mapping of an ocean wave spectrum into a synthetic aperture radar image spectrum. *J. Geophys. Res.* **1991**, *96*, 10713–10729. [CrossRef]
16. Hasselmann, S.; Bruning, C.; Hasselmann, K. An improved algorithm for the retrieval of ocean wave spectra from synthetic aperture radar image spectra. *J. Geophys. Res.* **1996**, *101*, 6615–6629. [CrossRef]
17. Collard, F.; Ardhuin, F.; Chapron, B. Extraction of coastal ocean wave fields from SAR images. *IEEE J. Ocean. Eng.* **2005**, *30*, 526–533. [CrossRef]
18. Mastenbroek, C.; de Valk, C.F. A semi-parametric algorithm to retrieve ocean wave spectra from synthetic aperture radar. *J. Geophys. Res.* **2000**, *105*, 3497–3516. [CrossRef]
19. Schulz-Stellenfleth, J.; Lehner, S.; Hoja, D. A parametric scheme for the retrieval of two-dimensional ocean wave spectra from synthetic aperture radar look cross spectra. *J. Geophys. Res.* **2005**, *110*, 297–314. [CrossRef]
20. WAMDIG. The WAM model—A third generation ocean wave prediction model. *J. Phys. Oceanogr.* **1988**, *18*, 1775–1810.
21. Hasselmann, K. Measurements of wind-wave growth and swell decay during the Joint North Sea Wave Project (JONSWAP). *Erganzung zur Deut. Hydrogr. Z.* **1973**, *12*, 1–95.
22. Johnsen, H.; Engen, G.; Hogda, K.A.; Chapron, B.; Desnos, Y.L. Validation of ENVISAT-ASAR wave mode Level 1 and 2 products using ERS SAR data. *IEEE Int. Geosci. Remote Sens. Sym. Pro.* **2000**, *4*, 1498–1500.
23. Lyzenga, D.R. Unconstrained inversion of wave height spectra from SAR images. *IEEE Trans. Geosci. Remote Sens.* **2002**, *40*, 261–270. [CrossRef]
24. Schulz-Stellenfleth, J.; Konig, T.; Lehner, S. An empirical approach for the retrieval of integral ocean wave parameters from synthetic aperture radar data. *J. Geophys. Res.* **2007**, *112*, 1–14. [CrossRef]
25. Li, X.M.; Lehner, S.; Bruns, T. Ocean wave integral parameter measurements using Envisat ASAR wave mode data. *IEEE Trans. Geosci. Remote Sens.* **2011**, *49*, 155–174. [CrossRef]
26. Stopa, J.E.; Ardhuin, F.; Collard, F.; Chapron, B. Estimating wave orbital velocities through the azimuth cut-off from space borne satellites. *J. Geophys. Res.* **2016**, *120*, 7616–7634. [CrossRef]
27. Shao, W.Z.; Zhang, Z.; Li, X.F.; Li, H. Ocean wave parameters retrieval from Sentinel-1 SAR imagery. *Remote Sens.* **2016**, *8*, 707–721. [CrossRef]

28. Grieco, G.; Lin, W.; Migliaccio, M.; Nirchio, F.; Portabella, M. Dependency of the Sentinel-1 azimuth wavelength cut-off on significant wave height and wind speed. *Int. J. Remote Sens.* **2016**, *37*, 5086–5104. [CrossRef]
29. Stopa, J.E.; Mouche, A. Significant wave heights from Sentinel-1 SAR: validation and applications. *J. Geophys. Res.* **2017**. [CrossRef]
30. Fernandez, D.E.; Carswell, J.R.; Frasier, S.; Chang, P.S.; Black, P.G.; Marks, F.D. Dual-polarized C- and Ku-band ocean backscatter response to hurricane-force winds. *J. Geophys. Res.* **2006**, *111*, 275–303. [CrossRef]
31. Hwang, P.A.; Fois, F. Surface roughness and breaking wave properties retrieved from polarimetric microwave radar backscattering. *J. Geophys. Res.* **2015**, *120*, 3640–3657. [CrossRef]
32. Romeiser, R.; Graber, H.C.; Caruso, M.J.; Jensen, R.E.; Walker, D.T.; Cox, A.T. A new approach to ocean wave parameter estimates from C-band ScanSAR images. *IEEE Trans. Geosci. Remote Sens.* **2015**, *53*, 1320–1345. [CrossRef]
33. Hersbach, H.; Stoffelen, A.; Haan, S.D. An improved C-band scatterometer ocean geophysical model function: CMOD5. *J. Geophys. Res.* **2007**, *112*, C03006. [CrossRef]
34. Quilfen, Y.; Bentamy, A.; Elfouhaily, T.; Katsaros, K.; Tournadre, J. Observation of tropical cyclones by high-resolution scatterometry. *J. Geophys. Res.* **1998**, *103*, 7767–7786. [CrossRef]
35. Masuko, H.; Okamoto, K.; Shimada, M.; Niwa, S. Measurement of microwave backscattering signatures of the ocean surface using X band and Ka band airborne scatterometers. *J. Geophys. Res.* **1986**, *91*, 13065–13083. [CrossRef]
36. Li, X.F. The first Sentinel-1 SAR image of a typhoon. *Acta Oceanol. Sin.* **2015**, *34*, 1–2. [CrossRef]
37. Li, X.F.; Zhang, J.A.; Yang, X.F.; Pichel, W.G.; DeMaria, M.; Long, D.; Li, Z.W. Tropical cyclone morphology from spaceborne synthetic aperture radar. *Bull. Am. Meteorol. Soc.* **2013**, *94*, 215. [CrossRef]
38. Friedman, K.; Li, X.F. Storm patterns over the ocean with wide swath SAR. *Johns Hopkins APL Tech. Dig.* **2000**, *21*, 80–85.
39. Li, X.F.; Pichel, W.G.; He, M.X.; Wu, S.; Friedman, K.; Clemente-Colon, P.; Zhao, C. Observation of hurricane-generated ocean swell refraction at the gulf stream north wall with the RADARSAT-1 synthetic aperture radar. *IEEE Trans. Geosci. Remote Sens.* **2002**, *40*, 2131–2142.
40. Hwang, P.; Li, X.F.; Zhang, B. Retrieving hurricane wind speed from dominant wave parameters. *IEEE J. Sel. Top. Appl. Earth Obs. Remote Sens.* **2017**, 1–10. [CrossRef]
41. Zhang, B.; Perrie, W.G. Cross-polarized synthetic aperture radar: A new potential technique for hurricanes. *Bull. Am. Meteorol. Soc.* **2012**, *93*, 531–541. [CrossRef]

*Article*

# Modulation Model of High Frequency Band Radar Backscatter by the Internal Wave Based on the Third-Order Statistics

**Pengzhen Chen [1,2,3], Lei Liu [4], Xiaoqing Wang [5], Jinsong Chong [1,2,*], Xin Zhang [1,2] and Xiangzhen Yu [6]**

1 National Key Laboratory of Science and Technology on Microwave Imaging, Beijing 100190, China; cpz0820@sina.com (P.C.); zhxinw@vip.163.com (X.Z.)
2 Institute of Electronics, Chinese Academy of Sciences, Beijing 100190, China
3 School of Electronics, Electrical and Communication Engineering, University of Chinese Academy of Sciences, Beijing 100190, China
4 Institute of Spacecraft System Engineering, China Academy of Space Technology, Beijing 100094, China; liulei211@126.com
5 Institute of Microelectronics of Chinese Academy of Sciences, Beijing 100029, China; huadaqq@126.com
6 Shanghai Radio Equipment Institute, Shanghai 200090, China; yxz8302@163.com
* Correspondence: iecas_chong@163.com; Tel.: +86-135-0100-8982

Academic Editors: Xiaofeng Yang, Xiaofeng Li, Ferdinando Nunziata, Alexis Mouche and Prasad S. Thenkabail
Received: 31 March 2017; Accepted: 17 May 2017; Published: 19 May 2017

**Abstract:** Modulation model of radar backscatters is an important topic in the remote sensing of oceanic internal wave by synthetic aperture radar (SAR). Previous studies related with the modulation models were analyzed mainly based on the hypothesis that ocean surface waves are Gaussian distributed. However, this is not always true for the complicated ocean environment. Research has showed that the measurements are usually larger than the values predicted by modulation models for the high frequency radars (X-band and above). In this paper, a new modulation model was proposed which takes the third-order statistics of the ocean surface into account. It takes the situation into consideration that the surface waves are Non-Gaussian distributed under some conditions. The model can explain the discrepancy between the measurements and the values calculated by the traditional models in theory. Furthermore, it can accurately predict the modulation for the higher frequency band. The model was verified by the experimental measurements recorded in a wind wave tank. Further discussion was made about applicability of this model that it performs better in the prediction of radar backscatter modulation compared with the traditional modulation model for the high frequency band radar or under lager wind speeds.

**Keywords:** radar backscatter; modulation model; internal wave; third-order statistics; high frequency band radar

---

## 1. Introduction

Internal waves usually result from the sharp density change occurring along the interface of the stratified density structure of the two fluids and travel with the interior of a fluid [1]. In the process of SAR imaging of internal waves, the internal wave firstly induce the variable current. Then, the current will directly interact with the surface waves, which results in the modulation of the radar backscatters [2]. Therefore, modulation model building is very crucial for the study of interaction between the radar backscatter and internal wave.

Many joint experiments, such as SAXON-FPN [3] (the Synthetic Aperture Radar and X Band Ocean Nonlinearities-Forschungs-platform Nordsee), JOWIP [4] (Joint Canada-U.S. Ocean Wave

Investigation Project), SARSEX [5] (SAR Internal Wave Signature Experiment), CoastWatch-95 [6], and SCSE [7] (South China Sea Experiment) were carried out and in situ measurements [8–12] were made to investigate the modulation mechanism of radar backscatter. Weak hydrodynamic interaction theory [13–15] was used to describe the distribution of Bragg wave spectrum modulated by internal waves [2]. The imaging of internal waves is attributed to variations in the spectral energy density of Bragg waves induced by weak current variations associated with internal waves, similarly as the analysis of the imaging of bottom topography. A two-scale composite surface model derived from a modified Kirchhoff model is used to calculate the L-/X-band radar backscatter modulation [16]. A full-spectrum model of the modulation of internal wave is established taking account of the wave spectral perturbations over the entire spectrum of waves [17]. Existing models are mainly based on the assumption that fluctuation of heights on the water surface is a random Gaussian distribution.

However, the comparison between the theoretical model and experimental results showed that the measured modulation in SAR images is underestimated [17], especially for high frequency band (higher than X-band) radar signals. Some investigators pointed out that the contribution of the backscatter from breaking waves should not be ignored, especially for higher-band radar. RIM (Radar Imaging Model) [18] adds the energy source of breaking waves into the formation of a wave-current model. RIM simulates the wave modulation induced by convergent current taking account of breaking waves and finds that the spectral modulation of the shorter wave (between 10 and 1000 rad/m) is larger than the modulation calculated by the wave-current model without waves breaking. The radar signatures of internal wave are more visible for HH polarization than VV polarization because of the impact of breaking waves, as reported in [7]. In substance, RIM adopts the improved hydrodynamic model, the composite surface model and Phillips's semi-empirical model [19] of breaking waves scattering to describe the scattering processing and explain the discrepancy of the modulation.

A modulation model of internal wave based on the third-order statistics of surface backscattering is proposed in this paper. It can effectively explain the discrepancy mentioned above by taking the non-Gaussian distribution of ocean surface slope into consideration. The IEM [20] (Integral Equation Model) was introduced to calculate radar backscatter coefficients. Compared with traditional models, the modulation model proposed in this paper combined the small perturbation method (SPM) [21] and the physics optical method (POM) [22], and it does not need to divide the ocean surface into different scales. As a result, the modulation of radar backscatter by internal wave could be calculated more precisely. The model explains the contradiction between the radar backscatter and the values predicted by traditional models. Experimental measurements were analyzed to verify the model. Information recorded by a CCD (Charge-coupled Device), which has high spatial and temporal resolution, was used to calculate the theoretical modulation attributed to second-order and third-order statistics. Results were compared with the data obtained by X and Ka band radar showing good agreement with the measured data by considering the third-order statistics. Moreover, these theoretical analyses and experimental observations demonstrate that the contribution of ocean surface third-order statistics to the modulation is significant for high frequency band radar. In other words, for high frequency band radar, it is necessary to add the contribution of ocean surface third-order statistics to the modulation by a variable surface current.

This paper is organized as follows: the modulation model of radar backscatter by internal wave based on the third-order statistics was derived in Section 2. In Section 3, an experiment was briefly described, as well as the data processing. In Section 4, results of experimental data were analyzed and discussed to validate the proposed model. Finally, main conclusions were given in Section 5.

## 2. Modulation Model of Radar Backscatters by Internal Wave Based on Third-Order Statistics

### 2.1. Radar Backscatters of Ocean Surface Based on Third-Order Statistics

An ocean surface scattering model, which is related to the ocean surface roughness spectrum, aims to quantify the relationship between the radar backscatter intensity and ocean surface statistics.

The surface roughness spectrum is defined as the Fourier transform of the correlation of ocean surface wave [23], that is,

$$W(k_x, k_y) = \frac{1}{4\pi^2}\int\int \sigma^2\rho(\xi,\zeta)\exp(-jk_x\xi - jk_y\zeta)d\xi d\zeta, \quad (1)$$

where $\sigma^2\rho(\xi,\zeta)$ is the second-order statistics of the surface wave $z(x,y)$ and can be calculated by Equation (2), $\rho(\xi,\zeta)$ is the correlation function of surface wave, and $\sigma^2$ is the variance of the surface wave height:

$$\sigma^2\rho(\xi,\zeta) = \langle z(x,y)z(x+\xi,y+\zeta)\rangle, \quad (2)$$

where the $\langle\cdot\rangle$ stands for the ensemble average. Incorporating a non-Gaussian distribution of the surface wave height results in a difference between radar backscatters recorded downwind and upwind [24]. The skewness function $s(\xi,\zeta;\tau,\varsigma)$ represents the distribution of the surface skewness coefficient [23]. It is commonly used to measure the departure from symmetry and can be calculated by

$$\langle z(x,y)z(x+\xi,y+\zeta)z(x+\tau,y+\varsigma)\rangle = \sigma^3 s(\xi,\zeta,\tau,\varsigma). \quad (3)$$

The Fourier transform of the bicorrelation function $\sigma^3 s(\xi,\zeta,\tau,\varsigma)$ is the bispectrum, that is,

$$B\left(k_x, k_y; \bar{k}_x, \bar{k}_y\right) = \frac{1}{16\pi^4}\int \sigma^3 s(\xi,\zeta;\tau,\varsigma)\exp\left(-jk_x\xi - jk_y\zeta - j\bar{k}_x\tau - j\bar{k}_y\varsigma\right)d\xi d\zeta d\tau d\varsigma. \quad (4)$$

It is a function of four variables. Two special cases were considered in the following calculation of the model. When $\tau = \xi, \varsigma = \zeta$, we can get $\langle z(x,y)z^2(x+\xi,y+\zeta)\rangle = \sigma^3 s(\xi,\zeta)$. For the case $\tau = \varsigma = 0$, we can get $\langle z^2(x,y)z(x+\xi,y+\zeta)\rangle = \sigma^3 s(-\xi,-\zeta)$. We can decompose the skewness function into two parts, the symmetric part $s_s(\xi,\zeta)$ and the asymmetric part $s_a(\xi,\zeta)$, as

$$s_s(\xi,\zeta) = \frac{s(\xi,\zeta) + s(-\xi,-\zeta)}{2}, \quad (5)$$

$$s_a(\xi,\zeta) = \frac{s(\xi,\zeta) - s(-\xi,-\zeta)}{2}. \quad (6)$$

The bispectrum is the Fourier transform of $s(\xi,\zeta)$ and can be written as

$$\begin{aligned} B(k_x,k_y) &= B_s(k_x,k_y) + jB_a(k_x,k_y) \\ &= \tfrac{1}{2\pi}\int \sigma^3 s(\xi,\zeta)e^{-jk_x\xi - jk_y\zeta}d\xi d\zeta \\ &= \tfrac{1}{2\pi}\int \sigma^3 \tfrac{s_a(\xi,\zeta)+s_a(\xi,\zeta)}{2}e^{-jk_x\xi - jk_y\zeta}d\xi d\zeta \end{aligned}, \quad (7)$$

where

$$\begin{aligned} B_s(k_x,k_y) &= \tfrac{1}{2\pi}\int \sigma^3 s_s(\xi,\zeta)\exp(-jk_x\xi - jk_y\zeta)d\xi d\zeta \\ jB_a(k_x,k_y) &= \tfrac{1}{2\pi}\int \sigma^3 s_a(\xi,\zeta)\exp(-jk_x\xi - jk_y\zeta)d\xi d\zeta \end{aligned}. \quad (8)$$

They present the symmetric and asymmetric property of the random ocean surface waves, respectively. Radar backscatters of the ocean surface can be further calculated (see Appendix A) by the theory of electromagnetic scattering [23]. That is,

$$\sigma^0_{pp} = \frac{k^2}{4\pi}|\Gamma_{pp}|^2 e^{-4k_z^2\sigma^2}\iint\left\{\exp\left[4k_z^2\sigma^2\rho(\xi,\zeta) + j8k_z^3\sigma^3 s_a(\xi,\zeta)\right] - 1\right\}e^{-2jk_x\xi}d\xi d\zeta, \quad (9)$$

where $k$ is the wavenumber of the radar, $\Gamma_{pp}$ is the coefficient defined as [23], $k_z = 2k\cos\theta$, and $k_x = 2k\sin\theta$. From Equation (9), we can see that the radar backscatter $\sigma^0_{pp}$ is a weighted value of the contribution of the second-order statistics $\sigma^2\rho(\xi,\zeta)$ and the third-order statistics $\sigma^3 s_a(\xi,\zeta)$. The coefficient of the contribution is related to the $k_z$ and $\sigma$.

### 2.2. Modulation Transfer Function of Radar Backscatter by Internal Wave

To simplify the following analysis, we name the modulation model that only considers the contribution of second-order statistics of ocean surface IEM2 model. The model with consideration of third-order statistics is called the IEM3 model.

Defining $\eta(\xi,\zeta) = \exp\left[4k_z^2\sigma^2\rho(\xi,\zeta) + j8k_z^3\sigma^3 s_a(\xi,\zeta)\right] - 1$, we can get its Fourier transform as

$$Q(k_1,k_2) = \iint \left\{\exp\left(4k_z^2\sigma^2\rho(\xi,\zeta) + j8k_z^3\sigma^3 s_a(\xi,\zeta)\right) - 1\right\}\exp(-jk_1\xi - jk_2\zeta)d\xi d\zeta. \quad (10)$$

Therefore, Equation (9) can be rewritten as

$$\sigma_{pp}^0 = \frac{k^2}{4\pi}\left|\Gamma_{pp}\right|^2 \exp\left(-4k_z^2\sigma^2\right) Q(2k_x,0). \quad (11)$$

We can get the modulation transfer function of radar backscatter by internal wave, that is,

$$\mathrm{M_{IEM3}}(k) = \frac{\widetilde{\sigma}_{pp}^0}{\sigma_{pp}^0} = \frac{\widetilde{Q}(2k_x,0)}{Q_0(2k_x,0)}, \quad (12)$$

where $\widetilde{\sigma}_{pp}^0$ and $\widetilde{Q}(2k_x,0)$ are the modulated radar backscatter and the spectrum, respectively. $Q_0(2k_x,0)$ is the background spectrum without modulation of internal waves.

For the case that the ocean surface is a Gaussian distribution, there will be no third-order component existing, $s_a(\xi,\zeta) = 0$, and the radar backscatter is only attributed to the second-order statistics, Equation (9) can be rewritten as

$$\sigma_{pp}^0 = \frac{k^2}{4\pi}\left|\Gamma_{pp}\right|^2 \exp\left(-4k_z^2\sigma^2\right) W(2k_x,0), \quad (13)$$

where $W(k_1,k_2) = \iint \left\{\exp(4k_z^2\sigma^2\rho(\xi,\zeta)) - 1\right\}\exp(-jk_1\xi - jk_2\zeta)d\xi d\zeta$. We can find that Equation (13) is the same with the expression in [25]. Similarly, the modulation of radar backscatter by internal wave can be obtained:

$$\mathrm{M_{IEM2}}(k) = \frac{\widetilde{\sigma}_{pp}^0}{\sigma_{pp}^0} = \frac{\widetilde{W}(2k_x,0)}{W_0(2k_x,0)}, \quad (14)$$

where $\widetilde{\sigma}_{pp}^0$ and $\widetilde{W}(2k_x,0)$ are the modulated radar backscatter and the spectrum, respectively. $W_0(2k_x,0)$ is the background spectrum without modulation of internal waves.

We can see that Equations (12) and (14) are exactly the same when the ocean surface is Gaussian distributed. The contribution of third-order statistics can be ignored as long as $j8k_z^3\sigma^3 s_a(\xi,\zeta) \ll 4k_z^2\sigma^2\rho(\xi,\zeta)$ according to Equations (10)–(12). However, the value of $k_z$ usually becomes larger for the high frequency band radar that has a large wave number $k$, and $s_a(\xi,\zeta) \neq 0$ for the case of high wind speeds, which is likely resulting in the asymmetric distribution of the ocean surface. In this situation, the modulation of radar backscatter should include the contribution of third-order statistics. Therefore, it can explain the discrepancy between the measured radar backscatter and values calculated by traditional modulation models that only take the second-order statistics into consideration.

## 3. Experimental Validation of the Model

We used the data of wind-wave tank experiment to validate the model proposed in Section 2. Experiments were carried out in a large wind-wave tank filled with stratified water. The tank is 12 m × 1.2 m × 1.2 m shown as Figure 1. Wind waves and internal wave were generated to simulate the condition of ocean surface.

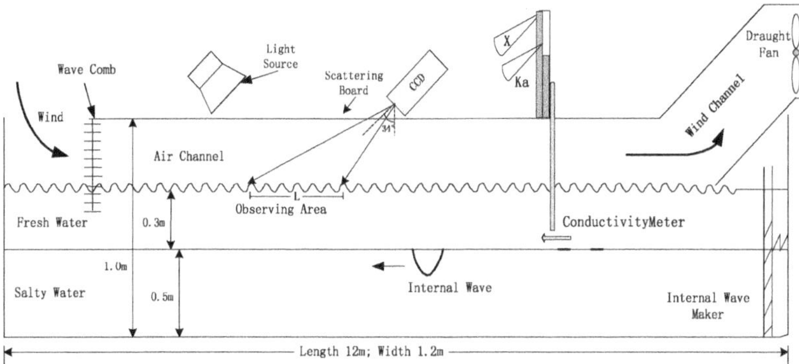


**Figure 1.** Schematic side view of the experimental wind-wave tank.

### 3.1. Experiment Description

X-/Ka-band radars and CCD were employed in experiments to record the modulation of reflected microwave signals by internal waves. Specifications of radar system are listed in Table 1.

**Table 1.** Specifications of radar system.

| Specifications | Values | |
|---|---|---|
| Band | X | Ka |
| Frequency | 9.4 GHz | 35 GHz |
| Beam Width | $9^\circ \times 9^\circ$ | $6^\circ \times 6^\circ$ |
| Incidence Angle | $50^\circ$ | $57^\circ$ |

The CCD array has high spatial and temporal resolution. It was used to record the information of surface waves in the tank. As the optical system, it can obtain the wave slope by retrieving the intensity of reflected light from the water surface. Specifications of the CCD are listed in Table 2.

**Table 2.** Specifications of CCD.

| Specifications | Values |
|---|---|
| Swath Width | 36 cm |
| Resolution (Geometrical) | 0.3 mm |
| Frame Repetition | 300 Hz |
| Analog-to-Digital Convert Frequency | 300 KHz |

Radar system and CCD array are shown in Figure 2.

Figure 2. Experimental devices: (a) radar system; (b) CCD array.

Experiments were carried out under different experimental conditions. Table 3 shows the wind, internal wave and fetch parameters used during the experiments. $U_w$ is the wind speed measured in the air channel, and its corresponding 10 m wind speed is $U_{10}$. $D$ is the depth of the water. $F$ is the fetch where the surface waves were recorded. $D_i$ is the depth of the internal wave.

Table 3. Description of experiments.

| | $U_w$ (m/s) | $U_{10}$ (m/s) | $D$ (m/s) | $F$ (m) | $D_i$ (m) |
|---|---|---|---|---|---|
| No. 1 | 3.2 | 4.1 | 0.8 | 5 | 0.3 |
| No. 2 | 4 | 5.2 | 0.8 | 5 | 0.3 |
| No. 3 | 5 | 6.9 | 0.8 | 5 | 0.3 |
| No. 4 | 6 | 8.6 | 0.8 | 5 | 0.3 |

## 3.2. Experimental Data Processing

According to Bragg scattering theory [21] and parameters listed in Table 1, the frequencies of surface Bragg waves should be 11.3 Hz (X-band) and 61.8 Hz in theory. Moreover, the surface current and the orbital velocity of long surface wave which can be estimated by $V_c = 0.6u_*$ Ref. [26] also result in Doppler frequency shifts of 5.2 Hz and 20.3 Hz for the X-band and Ka-band radar, respectively. Figure 3 shows the Doppler spectrum of radar measured in the experiment at wind speed 4 m/s. It is reasonable that the center of the Doppler frequencies are mainly concentrated around 17 Hz(X) and 78 Hz (Ka) before the internal wave generated (200 s–400 s).

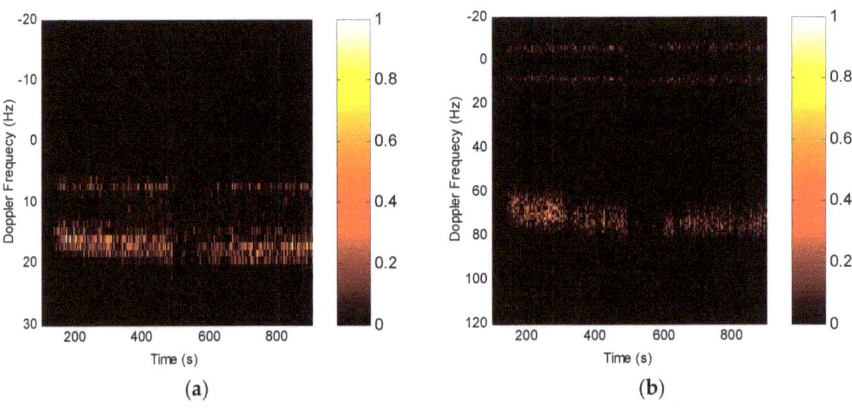


Figure 3. Doppler spectrum at the wind speed 4 m/s (a) X band; (b) Ka band.

We can also see that an interference frequency about 7 Hz was shown in Figure 3. It is caused by the echo data of water surface related to the antenna sidelobe. Therefore, we made the filtering in frequency domain in the following data processing.

The surface wave height can be obtained by integrating the surface slope recorded by CCD array. Figure 4a shows the water surface wave height recorded in 1 s. Figure 4b is the wave spectrum measured at different wind speeds. We can see that the spectrum increases with the increasing wind speed.

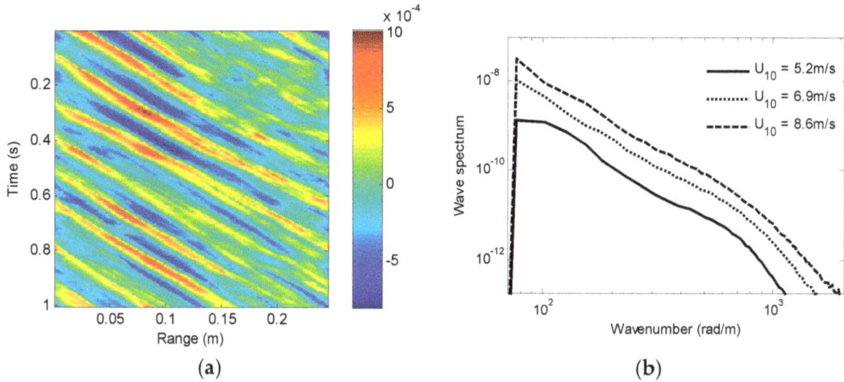


**Figure 4.** Surface wave recorded by CCD: (**a**) wave height; (**b**) wave spectra at different wind speeds.

## 4. Results of Experiments and Discussion

### *4.1. Comparison between Radar Measurements and Values Calculated by Models*

Given the high spatial and temporal resolution of CCD array, we took CCD data as the input of modulation models to calculate the theoretic value. Radar systems used in the experiments were not calibrated. Therefore, we cannot measure the absolute value of radar backscatters. In the further data processing, we only calculate the change of the radar backscatter modulated by the internal wave.

Results of the IEM2 model, IEM3 model and contribution of the bispectrum were compared with the radar data as shown in Figure 5. The 10 m wind speed is 5.2 m/s. Internal solitary wave passed by the observing area at 500 s.

From Figure 5a, we can see that IEM2 modulation model can predict the modulation well for X-band radar. However, the value predicted by IEM2 modulation model for Ka-band radar shown in Figure 5b is smaller than the measured value about 5 dB. As has been analyzed before, IEM2 modulation model only takes the second-order statistics of the ocean surface into consideration. It seems that the IEM2 model is not suitable for high frequency band radar. Moreover, the modulation of Ka-band radar backscatter shown in Figure 5f is larger than the X-band radar backscatter as shown in Figure 5e according to the measured radar data and the values calculated by the IEM3 model. This may result from the contribution of breaking waves or bound waves that commonly have a small wave length. They usually exist in the front of the long waves, resulting in the asymmetric distribution of the ocean surface. Therefore, it is reasonable that modulation of Ka-band radar calculated by IEM3 is larger than the result of IEM2, even larger than the modulation of X-band.

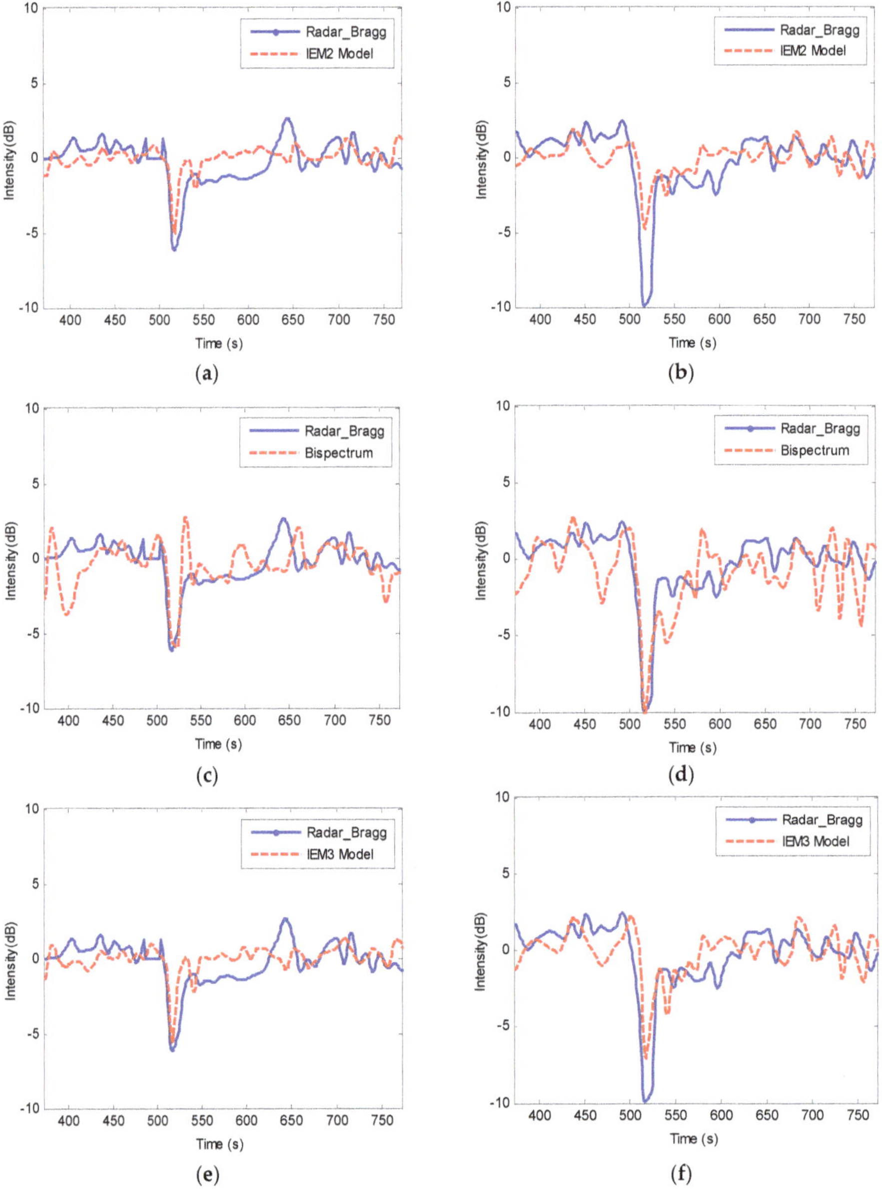


**Figure 5.** Comparison between measurements of radar and values calculated by models in theory. 10 m wind speed 5.2 m/s: (**a**) IEM2 Model vs. Radar: X-band; (**b**) IEM2 Model vs. Radar: Ka-band; (**c**) Bispectrum vs. Radar: X-band; (**d**) Bispectrum vs. Radar: Ka-band; (**e**) IEM3 Model vs. Radar: X-band; (**f**) IEM3 Model vs. Radar: Ka-band.

The IEM2 model describes the contribution of surface roughness spectrum of the ocean. On the contrast, modulation calculated by IEM3 model is a weighted sum of second-order statistics (surface roughness spectrum) and third-order statistics (bispectrum) according to Equation (10). Contribution of bispectrum was compared with the radar data in Figure 6c,d. We can see that the contribution of bispectrum is larger than the surface roughness spectrum for Ka-band radar from Figure 6b,d.

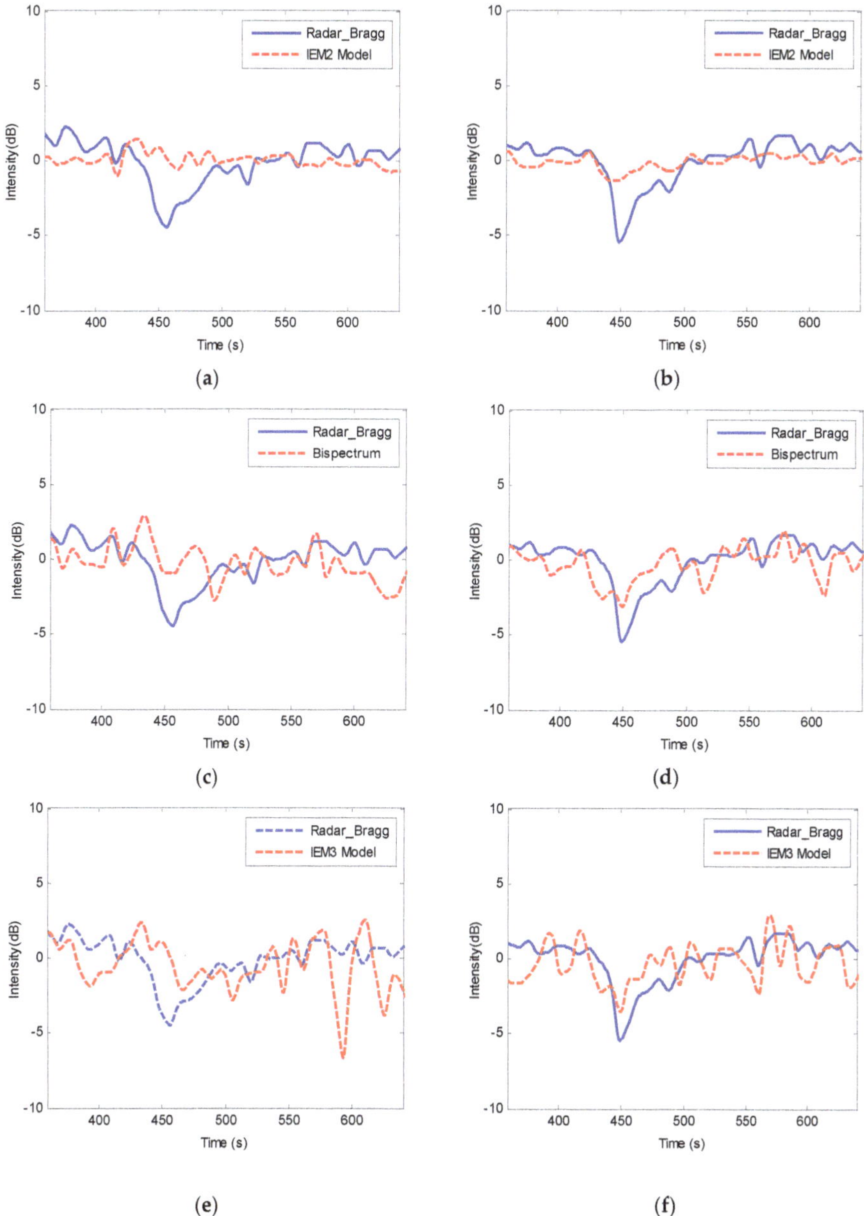


**Figure 6.** Comparison between measurements of radar and values calculated by models in theory, 10 m wind speed 6.9 m/s; (**a**) IEM2 Model vs. Radar: X-band; (**b**) IEM2 Model vs. Radar: Ka-band; (**c**) Bispectrum vs. Radar: X-band; (**d**) Bispectrum vs. Radar: Ka-band; (**e**) IEM3 Model vs. Radar: X-band; (**f**) IEM3 Model vs. Radar: Ka-band.

Figure 6 shows the results at wind speed 6.9 m/s. Compared with Figure 5, the relaxation rates of the surface waves increase with the increasing wind speed. Correspondingly, the modulation of the intensity of the radar decreases about 3 dB for X-band radar and 5 dB for Ka-band radar. The modulation estimated by IEM2 model can hardly be recognized as shown in Figure 6a,b. It was submerged under the signals of background and, as a consequence, the internal wave cannot be detected. On the contrast, predicted values of IEM3 model as shown in Figure 6e,f are closer to the radar data, which implies that the third-order statistics become increasingly important with the wind speed increasing.

### 4.2. Relation between Modulation Depth of Radar Backscatter and Wind Speeds

In this section, we will further discuss the modulation of high frequency band radar by internal waves at different wind speeds. To quantify the modulation, we introduce a new parameter modulation depth $\Delta M$ defined as

$$\Delta M = \frac{\sigma_{\max} - \sigma_{\min}}{\sigma_0}, \tag{15}$$

where $\sigma_0$ is the backscatter coefficient of background, $\sigma_{\max}$ and $\sigma_{\min}$ are the maximum and minimum of the radar backscatter coefficient modulated by internal wave, respectively. We used the data recorded in the experiments at different wind speeds ranging from 4.1 m/s to 8.6 m/s and made the statistics. Results were compared with the predicted value of IEM2 and IEM3 models shown as in Figure 7.

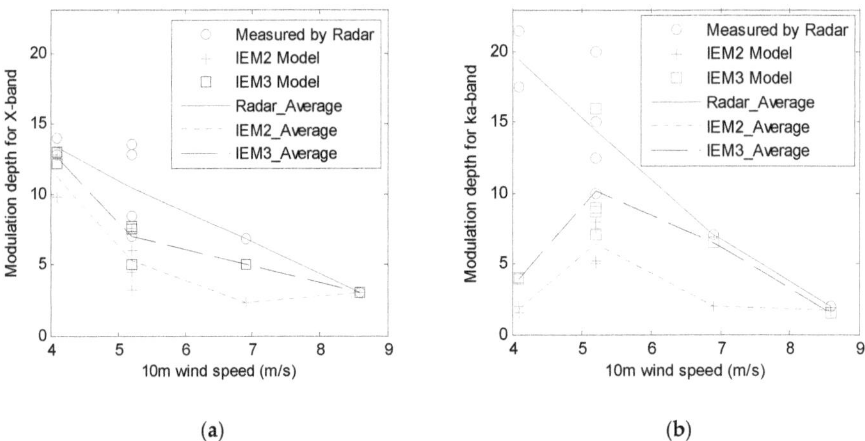


**Figure 7.** Modulation depth of radar backscatters as a function of wind speed: (**a**) X-band; (**b**) Ka-band.

We can see that all of the modulation depth measured by X-/Ka-band radar shows a decrease with increasing wind speed. It is similar for the predicted values calculated by modulated models except for Ka-band at 4.1 m/s. This is reasonable for the increasing wind speed corresponding to small relaxation rates [27,28] of the surface Bragg waves. As a result, the effect of internal wave on surface wave spectrum decreases and results in small modulation depth.

Modulation depth calculated by the IEM3 model is superior to the results of the IEM2 model compared with the experimental measurements. The difference between them is more obviously for Ka-band than X-band. It is reasonable that Ka-band corresponds to a large $k_z$ and the contribution of third-order statistics has a large impact on the total scatters. As for the value calculated at wind speed 4.1 m/s for Ka-band radar, this might be accounted for the few Bragg waves existing at low wind speed and the scatters mainly attributed to other sources, that is, multi-scattering or wedge scattering [29], which is not considered in the IEM3 model.

The predicted values by IEM2 of X-band and Ka-band are very close at wind speed larger than 4.1 m/s. However, the modulation depth of X-band is smaller than the Ka-band predicted by IEM3. With the increasing wind speed, the contribution of bispectrum increases especially for the higher band radar.

## 5. Conclusions

In this paper, the discrepancy between traditional modulation model and the measurements of high frequency band radars are addressed. Based on the third-order statistics of ocean surface, a modulation model of high frequency band radar backscatters by internal wave was proposed. It takes the non-Gaussian distribution of the ocean surface into consideration.

Data of experiments conducted in a wind-wave tank was employed to evaluate performance of the proposed model. Modulation depth of radar backscatter coefficients were calculated based on the IEM model and compared with the measured results by X-/Ka-band radar. The IEM3 model that considers the third-order statistics shows a better consistency with the radar data than the IEM2 model. Further processing and analysis to the model were made and showed that the third-order statistics of ocean surface are more important to the high frequency band radar. The relation between modulation depth and wind speed are also given. The larger radar frequency as well as the wind speed corresponds to a greater weight to third-order statistics in the radar backscatters modulated by internal waves. For the Ka-band radar, there are some other scattering mechanisms at low wind speed, which will be explored in future studies.

This proposed model can be applied to high frequency band SAR imaging of internal waves. It can enhance the image quality and show more information. Furthermore, it can be used in other SAR's marine applications such as the imaging of sea bottom topography and eddies, since they consist of similar imaging mechanisms.

**Acknowledgments:** This work was supported by the National Natural Science Foundation of China (No. 61302166 and No. 41406206).

**Author Contributions:** Pengzhen Chen, Lei Liu and Xiaoqing Wang conceived and performed the experiments; Xiaoqing Wang and Jinsong Chong supervised and designed the research and made contribution to the article's organization; Xin Zhang Xiangzhen Yu and provided help for the manuscript revision. Pengzhen Chen and Lei Liu drafted the manuscript, which was revised by all authors. All authors have read and approved the final manuscript.

**Conflicts of Interest:** The authors declare no conflict of interest.

## Appendix A. Radar Backscatter Coefficient Function

Radar backscatter coefficient is a function of average receiving power $p_{pp}$. It can be simplified by using IEM. Details are shown as follows:

$$\begin{aligned} P_{pp} &= \left\langle E_{pp}^{s} E_{pp}^{s\,*} \right\rangle - \left\langle E_{pp}^{s} \right\rangle \left\langle E_{pp}^{s\,*} \right\rangle \\ &= \left\langle E_{pp}^{k} E_{pp}^{k\,*} \right\rangle - \left\langle E_{pp}^{k} \right\rangle \left\langle E_{pp}^{k\,*} \right\rangle + 2\mathrm{Re}\left[ \left\langle E_{pp}^{c} E_{pp}^{k\,*} \right\rangle - \left\langle E_{pp}^{c} \right\rangle \left\langle E_{pp}^{k\,*} \right\rangle \right] + \left\langle E_{pp}^{c} E_{pp}^{c\,*} \right\rangle - \left\langle E_{pp}^{c} \right\rangle \left\langle E_{pp}^{c\,*} \right\rangle \end{aligned}, \tag{A1}$$

where $E_{pp}^{s}$ is the scattering electromagnetic field given by [20]. The right of Equation (A1) consists of three parts, and they are

$$\begin{aligned} P_{pp}^{k} &= \left\langle E_{pp}^{k} E_{pp}^{k\,*} \right\rangle - \left\langle E_{pp}^{k} \right\rangle \left\langle E_{pp}^{k\,*} \right\rangle \\ P_{pp}^{kc} &= 2\mathrm{Re}\left[ \left\langle E_{pp}^{c} E_{pp}^{k\,*} \right\rangle - \left\langle E_{pp}^{c} \right\rangle \left\langle E_{pp}^{k\,*} \right\rangle \right] \\ P_{pp}^{c} &= \left\langle E_{pp}^{c} E_{pp}^{c\,*} \right\rangle - \left\langle E_{pp}^{c} \right\rangle \left\langle E_{pp}^{c\,*} \right\rangle \end{aligned}. \tag{A2}$$

The scattering field $E_{pp}^{s}$ is

$$E_{pp}^{s} = E_{pp}^{k} + E_{pp}^{c}, \tag{A3}$$

where

$$E_{pp}^{k} = CE_0 \int f_{pp} \exp\left(-j2\vec{k}_i \cdot \vec{r}\right) dxdy, \quad \text{(A4)}$$

$$E_{pp}^{c} = \frac{CE_0}{8\pi^2} \int F_{pp} \exp\left[-j\vec{k}_i \cdot \vec{r} - j\vec{k}_i \cdot \vec{r}' + ju(x-x') + jv(y-y')\right] dxdydx'dy'dudv, \quad \text{(A5)}$$

where $E_0$ is the average transmitting power, $C = -jke^{-jkR}/4\pi R$,. Therefore, the first item of Equation (A2) can be expressed as

$$\begin{aligned} P_{\mathrm{pp}}^{k} &= |CE_0 f_{pp}|^2 \left\langle \iint e^{-j2\vec{k}_i \cdot \vec{r} + j2\vec{k}_i \cdot \vec{r}} dxdydx'dy' \right\rangle - |CE_0 f_{pp}|^2 \left\langle \int e^{-j2\vec{k}_i \cdot \vec{r}} dxdy \right\rangle \left\langle \int e^{j2\vec{k}_i \cdot \vec{r}'} dx'dy' \right\rangle \\ &= |CE_0 f_{pp}|^2 \iint \left\langle e^{-j2k_z(z-z')} \right\rangle e^{-j2k_x(x-x') - j2k_y(y-y')} dxdydx'dy' - \\ &\quad |CE_0 f_{pp}|^2 \left[\int \left\langle e^{-j2k_z z} \right\rangle e^{-j2k_x x - j2k_y y} dxdy\right] \left[\int \left\langle e^{j2k_z z} \right\rangle e^{j2k_x x' + j2k_y y'} dx'dy'\right] \end{aligned} . \quad \text{(A6)}$$

Since the averages of the quantities referred in Equation (A6) can be placed by [20] ,

$$\begin{cases} \left\langle e^{-j2k_z(z-z')} \right\rangle = \exp\{4k_z^2\sigma^2[\rho(\xi,\zeta) - 1] + j8k_z^3\sigma^3 s_a(\xi,\zeta)\} \\ \left\langle e^{-j2k_z z} \right\rangle = \exp(-2k_z^2\sigma^2 + j4k_z^3\sigma^3/3) \\ \left\langle e^{j2k_z z} \right\rangle = \exp(-2k_z^2\sigma^2 - j4k_z^3\sigma^3/3) \end{cases} . \quad \text{(A7)}$$

Defining $x - x' = \xi$ and $y - y' = \zeta$, (A6) can then be further rewritten as

$$\begin{aligned} P_{\mathrm{pp}}^{k} &= |CE_0 f_{pp}|^2 \int \exp\{4k_z^2\sigma^2[\rho(\xi,\zeta) - 1] + j8k_z^3\sigma^3 s_a(\xi,\zeta)\} e^{-j2k_x\xi - j2k_y\zeta} d\xi d\zeta - \\ &\quad |CE_0 f_{pp}|^2 \int \exp(-2k_z^2\sigma^2 + j4k_z^3\sigma^3/3) \exp(-2k_z^2\sigma^2 - j4k_z^3\sigma^3/3) e^{-j2k_x\xi - j2k_y\zeta} d\xi d\zeta \\ &= |CE_0 f_{pp}|^2 A_0 e^{-4k_z^2\sigma^2} \int \{\exp[4k_z^2\sigma^2\rho(\xi,\zeta) + j8k_z^3\sigma^3 s_a(\xi,\zeta)] - 1\} e^{-j2k_x\xi - j2k_y\zeta} d\xi d\zeta \end{aligned} . \quad \text{(A8)}$$

For the second item in Equation (A2),

$$\begin{aligned} P_{\mathrm{pp}}^{kc} = 2\mathrm{Re}\Bigg\{ & \frac{|CE_0|^2}{8\pi^2} \left\langle \iint F_{pp} f_{pp}^* e^{-j\vec{k}_i \cdot \vec{r} - j\vec{k}_i \cdot \vec{r}' + j2\vec{k}_i \cdot \vec{r}''} e^{ju(x-x') + jv(y-y')} dxdydx'dy'dx''dy''dudv \right\rangle - \\ & \frac{|CE_0|^2}{8\pi^2} \left\langle \int F_{pp} e^{-j\vec{k}_i \cdot \vec{r} - j\vec{k}_i \cdot \vec{r}'} e^{ju(x-x') + jv(y-y')} dxdydx'dy'dudv \right\rangle \left\langle \int f_{pp}^* e^{j2\vec{k}_i \cdot \vec{r}''} dx''dy'' \right\rangle \Bigg\} \end{aligned} . \quad \text{(A9)}$$

We take the situation that only single scattering occurs during the radar radiation, that is, $\vec{r} = \vec{r}'$. Therefore, $x = x'$, $y = y'$, $x - x'' = x' - x'' = \xi$, and $y - y'' = y' - y'' = \zeta$. Equation (A9) can be further expressed as

$$\begin{aligned} P_{\mathrm{pp}}^{kc} &= 2\mathrm{Re}\left\{ \frac{|CE_0|^2}{8\pi^2} \left\langle \iint F_{pp} f_{pp}^* e^{-2j\vec{k}_i \cdot \vec{r} + j2\vec{k}_i \cdot \vec{r}''} dxdydx''dy'' \right\rangle - \frac{|CE_0|^2}{8\pi^2} \left\langle \int F_{pp} e^{-2j\vec{k}_i \cdot \vec{r}} dxdy \right\rangle \left\langle \int f_{pp}^* e^{j2\vec{k}_i \cdot \vec{r}''} dx''dy'' \right\rangle \right\} \\ &= 2\mathrm{Re}\Bigg\{ \frac{|CE_0|^2}{8\pi^2} \iint F_{pp} f_{pp}^* \left\langle e^{-j2k_z(z-z'')} \right\rangle e^{-j2k_x(x-x'') - j2k_y(y-y'')} dxdydx''dy'' - \\ &\quad \frac{|CE_0|^2}{8\pi^2} \left[\int F_{pp} \left\langle e^{-j2k_z z} \right\rangle e^{-j2k_x x - j2k_y y} dxdy\right] \left[\int f_{pp}^* \left\langle e^{j2k_z z''} \right\rangle e^{-j2k_x x'' - j2k_y y''} dx''dy''\right] \Bigg\} \end{aligned} . \quad \text{(A10)}$$

The third item in Equation (A2) is

$$\begin{aligned} P_{\mathrm{pp}}^{c} = & \left|\frac{CE_0 F_{pp}}{8\pi^2}\right|^2 \left\langle \iint \iint e^{-j\vec{k}_i \cdot \vec{r} - j\vec{k}_i \cdot \vec{r}' + j\vec{k}_i \cdot \vec{r}'' + j\vec{k}_i \cdot \vec{r}'''} e^{ju(x-x') + jv(y-y') - ju(x''-x''') - jv(y''-y''')} dxdydx'dy'dx''dy''dx'''dy'''dudv \right\rangle - \\ & \left|\frac{CE_0}{8\pi^2}\right|^2 \left\langle \iint F_{pp} e^{-j\vec{k}_i \cdot \vec{r} - j\vec{k}_i \cdot \vec{r}'} e^{ju(x-x') + jv(y-y')} dxdydx'dy'dudv \right\rangle \left\langle \iint F_{pp}^* e^{j\vec{k}_i \cdot \vec{r}'' + j\vec{k}_i \cdot \vec{r}'''} e^{-ju(x''-x''') - jv(y''-y''')} dx''dy''dx'''dy'''dudv \right\rangle . \end{aligned} \quad \text{(A11)}$$

For the same condition assumed above, we can get $\vec{r} = \vec{r}', \vec{r}'' = \vec{r}'''$, $x = x', y = y'$, and $x'' = x''', y'' = y'''$. Therefore, we can further obtain

$$\begin{aligned} P_{\mathrm{pp}}^{c} &= \left|\frac{CE_0F_{pp}}{8\pi^2}\right|^2\left\langle \iint e^{-j2\vec{k}_i\cdot\vec{r}+2j\vec{k}_i\cdot\vec{r}''}dxdydx''dy''\right\rangle - \left|\frac{CE_0}{8\pi^2}\right|^2\left\langle \int F_{pp}e^{-j2\vec{k}_i\cdot\vec{r}}dxdy\right\rangle\left\langle \iint F_{pp}^{*}e^{j2\vec{k}_i\cdot\vec{r}''}dx''dy''\right\rangle \\ &= \left|\frac{CE_0F_{pp}}{8\pi^2}\right|^2\iint\left\langle e^{-j2k_z(z-z'')}\right\rangle e^{-j2k_x(x-x'')-j2k_y(y-y'')}dxdydx''dy'' - \\ &\left|\frac{CE_0}{8\pi^2}\right|^2\left[\int F_{pp}\left\langle e^{-j2k_zz}\right\rangle e^{-j2k_xx-j2k_yy}dxdy\right]\left[\int F_{pp}^{*}\left\langle e^{j2k_zz''}\right\rangle e^{-j2k_xx''-j2k_yy''}dx''dy''\right] \end{aligned} \quad \text{(A12)}$$

Defining $x - x'' = \xi, y - y'' = \zeta$, we can rewrite (A12) as

$$\begin{aligned} P_{\mathrm{pp}}^{c} &= \left|\frac{CE_0F_{pp}}{8\pi^2}\right|^2\left[\int\left\langle e^{-j2k_z(z-z'')}\right\rangle e^{-j2k_x\xi-j2k_y\zeta}d\xi d\zeta - \int\left\langle e^{-j2k_zz}\right\rangle\left\langle e^{j2k_zz''}\right\rangle e^{-j2k_x\xi-j2k_y\zeta}d\xi d\zeta\right] \\ &= \left|\frac{CE_0F_{pp}}{8\pi^2}\right|^2 A_0e^{-4k_z^2\sigma^2}\int\left\{\exp\left[4k_z^2\sigma^2\rho(\xi,\zeta) + j8k_z^3\sigma^3s_a(\xi,\zeta)\right] - 1\right\}e^{-j2k_x\xi-j2k_y\zeta}d\xi d\zeta \end{aligned} \quad \text{(A13)}$$

By substituting Equations (A2), (A8), (A10) (A13) into Equation (A1), we obtain

$$\begin{aligned} P_{\mathrm{pp}} &= P_{\mathrm{pp}}^{k} + P_{\mathrm{pp}}^{kc} + P_{\mathrm{pp}}^{c} \\ &= |CE_0|^2\left[\left|f_{pp}\right|^2 - \frac{\mathrm{Re}\left(F_{pp}f_{pp}^{*}\right)}{4\pi^2} + \frac{\left|F_{pp}\right|^2}{64\pi^4}\right] \\ &e^{-4k_z^2\sigma^2}\int\left\{\exp\left[4k_z^2\sigma^2\rho(\xi,\zeta) + j8k_z^3\sigma^3s_a(\xi,\zeta)\right] - 1\right\}e^{-j2k_x\xi-j2k_y\zeta}d\xi d\zeta \end{aligned}, \quad \text{(A14)}$$

where $k_x = 2k_0\sin\theta$, $k_z = 2k_0\cos\theta$, $f_{pp}$ and $F_{pp}$ are coefficient defined by [20]. The radar backscatter coefficient $\sigma_{pp}^0$ is a function of $P_{pp}$, that is,

$$\sigma_{pp}^0 = 4\pi R^2\frac{P_{pp}}{E_0^2A_0}, \quad \text{(A15)}$$

where $A_0$ is the area of antenna. Substitute Equation (A14) into Equation (A15) and use the method introduced in [25], we can get

$$\begin{aligned} \sigma_{pp}^0 &= \frac{k^2}{4\pi}\left[\left|f_{pp}\right|^2 + \frac{1}{4\pi^2}\mathrm{Re}\left(F_{pp}f_{pp}^{*}\right) + \frac{1}{64\pi^4}\left|F_{pp}\right|^2\right]e^{-4k_z^2\sigma^2} \\ &\iint\left\{\exp\left[4k_z^2\sigma^2\rho(\xi,\zeta) + j8k_z^3\sigma^3s_a(\xi,\zeta)\right] - 1\right\}e^{-2jk_x\xi}d\xi d\zeta \end{aligned}. \quad \text{(A16)}$$

Since $F_{pp} = \left[F_{pp}(k_x,0) + F_{pp}(-k_x,0)\right]/2$, we can obtain

$$\mathrm{Re}\left(F_{pp}f_{pp}^{*}\right) = \frac{F_{pp}(k_x,0) + F_{pp}(-k_x,0)}{4}f_{pp}^{*} + \frac{F_{pp}^{*}(k_x,0) + F_{pp}^{*}(-k_x,0)}{4}f_{pp}. \quad \text{(A17)}$$

By substituting Equation (A17) into Equation (A16), we can further get

$$\sigma_{pp}^0 = \frac{k^2}{4\pi}\left|\Gamma_{pp}\right|^2e^{-4k_z^2\sigma^2}\iint\left\{\exp\left[4k_z^2\sigma^2\rho(\xi,\zeta) + j8k_z^3\sigma^3s_a(\xi,\zeta)\right] - 1\right\}e^{-2jk_x\xi}d\xi d\zeta, \quad \text{(A18)}$$

where

$$\Gamma_{pp} = f_{pp} + \frac{1}{4\pi^2}\left(\frac{F_{pp}(k_x,0) + F_{pp}(-k_x,0)}{4}\right) \quad \text{(A19)}$$

## References


1. Jackson, C.R.; Apel, J.R. *Syntheic Aperture Radar Marine User's Manual*; NOAA/NESDIS: Washington, DC, USA, 2004.
2. Alpers, W. Theory of radar imaging of internal waves. *Nature* **1985**, *314*, 245–247. [CrossRef]

3. Plant, W.J.; Alpers, W. The Saxon-FPN Experiment. In Proceedings of the IEEE Geoscience and Remote Sensing Symposium, Remote Sensing: Global Monitoring for Earth Management, Espoo, Finland, 3–6 June 1991; IEEE: New York, NY, USA, 1991.; pp. 1983–1987.
4. Hughes, B.A.; Dawson, T.W. Joint Canada-U.S. Ocean wave investigation project: An overview of the Georgia Strait experiment. *J. Geophys. Res. Oceans* **1988**, *93*, 12219–12234. [CrossRef]
5. Gasparovic, R.F.; Apel, J.R.; Kasischke, E.S. An overview of the sar internal wave signature experiment. *J. Geophys. Res. Oceans* **1988**, *93*, 12304–12316. [CrossRef]
6. Johanessen, J.A.; Korsbakken, E.; Samuel, P.; Jenkins, A.; Espedal, H. Coast watch: Using SAR imagery in an operational system for monitoring coastal currents, wind, surfactants and oil spills. In *Operational Oceanography: The Challenge for European Co-Operation*; Stel, J.H., Ed.; Elsevier: Amsterdam, The Netherlands, 1997.
7. Plant, W.J.; Keller, W.C.; Hayes, K.; Chatham, G.; Lederer, N. Normalized radar cross section of the sea for backscatter: 2. Modulation by internal waves. *J. Geophys. Res. Oceans* **2010**, *115*. [CrossRef]
8. Bai, X.; Liu, Z.; Li, X.; Chen, Z.; Hu, J.; Sun, Z.; Zhu, J. Observations of high-frequency internal waves in the Southern Taiwan Strait. *J. Coast. Res.* **2013**, *29*, 1413–1419. [CrossRef]
9. Li, X.; Jackson, C.R.; Pichel, W.G. Internal solitary wave refraction at Dongsha Atoll, South China Sea. *Geophys. Res. Lett.* **2013**, *40*, 3128–3132. [CrossRef]
10. Bai, X.; Liu, Z.; Li, X.; Hu, J. Generation sites of internal solitary waves in the Southern Taiwan Strait revealed by MODIS true-colour image observations. *Int. J. Remote Sens.* **2014**, *35*, 4086–4098. [CrossRef]
11. Liu, B.; Yang, H.; Zhao, Z.; Li, X. Internal solitary wave propagation observed by tandem satellites. *Geophys. Res. Lett.* **2014**, *41*, 2077–2085. [CrossRef]
12. Dong, D.; Yang, X.; Li, X.; Li, Z. SAR observation of eddy-induced mode-2 internal solitary waves in the South China Sea. *IEEE Trans. Geosci. Remote Sens.* **2016**, *54*, 6674–6686. [CrossRef]
13. Longuet-Higgins, M.S.; Stewart, R. Radiation stresses in water waves; a physical discussion, with applications. In *Deep Sea Research and Oceanographic Abstracts*; Elsevier: Amsterdam, The Netherlands, 1964; pp. 529–562.
14. Whitham, G. A general approach to linear and non-linear dispersive waves using a lagrangian. *J. Fluid Mech.* **1965**, *22*, 273–283. [CrossRef]
15. Bretherton, F.P. A note on hamilton's principle for perfect fluids. *J. Fluid Mech.* **1970**, *44*, 19–31. [CrossRef]
16. Thompson, D.R. Calculation of radar backscatter modulations from internal waves. *J. Geophys. Res. Oceans* **1988**, *93*, 12371–12380. [CrossRef]
17. Lyzenga, D.R.; Bennett, J.R. Full-spectrum modelling of synthetic aperture radar internal wave signatures. *J. Geophys. Res. Oceans* **1988**, *93*, 12345–12354. [CrossRef]
18. Kudryavtsev, V.; Akimov, D.; Johanessen, J.A.; Chapron, B. On radar imaging of current features: 1. Model and comparison with observations. *J. Geophys. Res. Oceans* **2005**, *110*. [CrossRef]
19. Phillips, O.M. *The Dynamics of the Upper Ocean*, 2nd ed.; Cambridge University Press: Cambridge, UK, 1977; pp. 37–81.
20. Fung, A.K. *Microwave Scattering and Emission Models and Their Applications*; Artech House: London, UK, 1994.
21. Wright, J. A new model for sea clutter. *IEEE Trans. Antennas Propag.* **1968**, *16*, 217–223. [CrossRef]
22. Valenzuela, G.R. Theories for the interaction of electromagnetic and ocean waves—A review. *Bound. Layer Meteorol.* **1978**, *13*, 61–85. [CrossRef]
23. Chen, K.S.; Fung, A.K.; Weissman, D.A. A backscattering model for ocean surface. *IEEE Trans. Geosci. Remote Sens.* **1992**, *30*, 811–817. [CrossRef]
24. Plant, W.J. A new interpretation of sea-surface slope probability density functions. *J. Geophys. Res. Oceans* **2003**, *108*. [CrossRef]
25. Plant, W.J. A stochastic multiscale model of microwave backscatter from the ocean. *J. Geophys. Res. Oceans* **2002**, *107*, 3120. [CrossRef]
26. Plant, W.J.; Wright, J.W. Phase speeds of upwind and downwind traveling short gravity waves. *J. Geophys. Res. Oceans* **1980**, *85*, 3304–3310. [CrossRef]
27. Hughes, B.A. The effect of internal waves on surface wind waves 2. Theoretical analysis. *J. Geophys. Res. Oceans* **1978**, *83*, 455–465. [CrossRef]

28. Plant, W.J. A relationship between wind stress and wave slope. *J. Geophys. Res. Oceans* **1982**, *87*, 1961–1967. [CrossRef]
29. Plant, W.J.; Keller, W.C.; Hayes, K.; Chatham, G. Normalized radar cross section of the sea for backscatter: 1. Mean levels. *J. Geophys. Res. Oceans* **2010**, *115*. [CrossRef]

*Article*

# Satellite Survey of Internal Waves in the Black and Caspian Seas

**Olga Lavrova * and Marina Mityagina**

Space Research Institute of Russian Academy of Sciences, Moscow 117997, Russia; mityag@iki.rssi.ru
* Correspondence: olavrova@iki.rssi.ru; Tel.: +7-495-333-4256

Received: 30 June 2017; Accepted: 25 August 2017; Published: 28 August 2017

**Abstract:** The paper discusses the results of a study of short-period internal waves (IWs) in the Black and Caspian Seas from their surface manifestations in satellite imagery. Since tides are negligible in these seas, they can be considered non-tidal. Consequently, the main generation mechanism of IWs in the ocean—interaction of barotropic tides with bathymetry—is irrelevant. A statistically significant survey of IW occurrences in various regions of the two seas is presented. Detailed maps of spatial distribution of surface manifestations of internal waves (SMIWs) are compiled. Factors facilitating generation of IWs are determined, and a comprehensive discussion of IW generation mechanisms is presented. In the eastern and western coastal zones of the Black Sea, where large rivers disembogue, intrusions of fresh water create hydrological fronts that are able to generate IWs. At the continental shelf edge, on the west and northwest of the Black Sea and near the Crimean Peninsula, IWs are generated primarily due to relaxation of coastal upwelling and inertial oscillations associated with hydrological fronts. In addition, IWs can be formed at sea fronts associated with the passage of cold eddies. In the Caspian Sea, seiches are the main source of the observed IWs.

**Keywords:** Caspian Sea; Black Sea; internal waves; satellite remote sensing; sea surface; SAR imagery

## 1. Introduction

Internal waves (IWs) exist in the stably stratified ocean which corresponds to water density increase in the direction of gravity force. They are wave-like oscillations of water particles around a stable equilibrium position under the restoring action of Archimedes (buoyancy) forces [1]. IWs are characteristic of the dynamics of all density-stratified water bodies—oceans, seas, lakes and reservoirs. The differences in layer densities may be caused by differences in water temperature or salinity. In fact, such processes also occur at the air/water interface. Unlike surface waves, IWs entrain fluid particles that undergo vertical displacements with maximum amplitude located deep in water rather than at the surface. At the same time, maximum amplitudes of horizontal displacements (orbital motion in IW) are found in the immediate proximity to the ocean surface where these amplitudes can reach almost 1 m/s and produce perceptible changes in small-scale wind wave spectrum [2].

IWs are known to strongly affect processes in the ocean. Motions induced by IWs transpierce the whole body of ocean water and also play an important role in processes at its surface. They can propagate over several hundred kilometers and transport both mass and momentum. IW propagation is accompanied by a considerable velocity shear that can lead to turbulence and mixing. That is why IWs, and mechanisms of their generation, development, propagation and decay are always a research focus.

With the employment of synthetic aperture radars (SARs), satellite remote sensing techniques provided a new outlook for the study of oceanic IWs [3]. Satellite observations of surface manifestations of internal waves (SMIWs) enable determining their spatial parameters and regions of regular occurrence, as well as analyzing possible generation mechanisms and evolution. Today, using SAR data in IWs studies is standard practice around the world. By the initiative of John Apel, an atlas of surface manifestations of IWs has been created and is maintained [4], presently comprising over 300 instances of IW patterns from 54 regions of the world. Hundreds of publications have been devoted to remote sensing of IWs and the flow of works does not seem to run out. For example, the development of remote-sensing techniques in the study of IWs as well as the internal wave studies in the South China Sea were summarized in [5]. Single or a series of SAR images makes it possible to investigate the two dimensional structure of IWs by determining such characteristics as: half-width, crest length, number of waves, propagation direction, distance between neighboring trains, distance between neighboring IWs, and wave velocity. By using these IW parameters in models, oceanographers have been able to derive important physical information, such as oceanic mixed layer depth [5–15]. However, most of the observations concern IWs generated by tidal currents in shelf zones. It is widely accepted that IW generation is closely related to tidal activity. Specifically, tidal IWs occur as a result of tidal currents flowing normal to the local bathymetry features. Such processes are statistically reproducible, given the same season, the same phase of both daily and fortnightly tides, and the same bathymetry. That is why most remote sensing techniques are tailored for studying IWs generated by tidal currents and internal tides in shelf zones.

By contrast, IWs of non-tidal origin have received much less attention. In non-tidal seas, IWs can be induced by intensive dynamic processes such as coastal upwelling, eddies, wind-driven water movements, hydrological front oscillations, etc. There are some works based on contact methods focused on observation and numerical modeling of generation and propagation of non-tidal IWs [16,17]. However, publications on satellite observation of IWs in non-tidal seas are scarce and focus mainly on IWs in large lakes [18–20].

Up to date, there are no reports on studies of IWs in the Black and Caspian Seas based on their manifestations in satellite images, with the exception of publications by the researchers of Space Radar Laboratory of the Space Research Institute RAS, where the authors belong. Our works, initially performed for the northeastern Black Sea and Middle Caspian, revealed a variety of IW types which could possibly be explained not only by bottom topography features, but also a diversity of generation mechanisms. Preliminary results were presented in international conferences and published in Russian language scientific journals [21–25]. Soon it became clear that IWs in non-tidal seas differed substantially from those generated by barotropic tides on oceanic shelf and their investigation was a challenge. Reconstruction of a satisfactory picture of generation, propagation and interaction of IWs in the seas without tides and validation of IW generation hypotheses demanded accumulation of a large bank of data. In this paper, we summarize the results obtained over a 10-year survey of the Black and Caspian Seas using satellite data. The large amount of examined data permitted some generalizations and statistically significant results on spatial and temporal variability of various SMIWs in satellite images. Factors facilitating generation of non-tidal IWs were determined based on joint analysis of available satellite remote sensing data of the sea surface in microwave, visible (VIS) and near-infrared (NIR) ranges complemented by results of in-situ measurements.

During our long-term monitoring of the Black and Caspian Seas [21], we came across a significant number of SMIWs in satellite imagery. Tides are so negligible in these seas that they cannot induce IWs [26].

The main goal of our work reported in this paper was to broaden our insights into the physics and geography of IWs in non-tidal seas based on satellite data.

This goal implied the following primary tasks: (1) identify regions of non-tidal seas where SMIWs are regularly detected; (2) investigate features of SMIWs based on satellite data in various ranges of the electromagnetic spectrum (microwaves, VIS); (3) determine the reasons for considerable interannual, seasonal and spatial variations in SMIWs both at the sea surface and in radar images of the surface; (4) map SMIWs in the Black and Caspian Seas; and (5) determine probable generation mechanisms of IWs from joint analysis of satellite observations, hydrometeorological data, and in-situ measurements.

Note that over the past year, new satellite data have emerged ensuring a much higher degree of quality of remote observation of IWs. We refer to the continual flow of various high resolution satellite data from the sensors installed on board new Sentinel satellites launched by European Space Agency. Hence, the possibilities are open to: (1) consider smaller scales since data with resolution of units of meters are now available; (2) compare radar and optical data obtained almost simultaneously in same region and at same high resolution; and (3) investigate IW development and propagation in time due to smaller intervals between the observations.

## 2. Study Areas

### 2.1. The Caspian Sea

The Caspian Sea has no natural connection with the World Ocean and is the largest inland body of water in the world. It contains about 3.5 times more water, by volume, than all five of North America's Great Lakes combined. The Caspian is a lake by geographical definition but it is a saltwater lake. Average salinity of the Caspian Sea water is 12.85‰ and that is approximately three times lower than salinity of ocean. The Caspian has characteristics common to both seas and lakes, but its size, hydrometeorological conditions, currents, water level oscillations, flora and fauna are more characteristic of seas [27]. It is basically considered a non-tidal sea since tide heights do not exceed 12 cm at the coastline and 2 cm in open sea. Nevertheless, IWs do occur here and their generation mechanisms can be explained by some specific characteristics of this water basin (Figure 1):

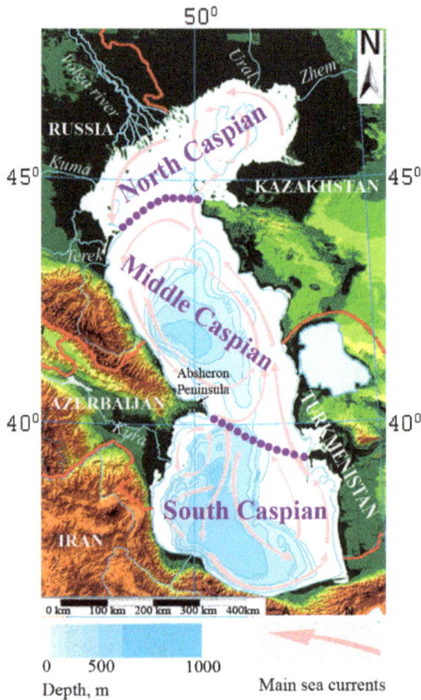


**Figure 1.** Map of the Caspian Sea. Dots indicate the division into North, Middle and South Caspian.

- The Caspian Sea is divided into three, approximately equal, parts: North, Middle and South.
- The most characteristic features of bottom topography are big shallow water areas in North, and deep areas in Middle and South Caspian.

- The temperature regime of the Caspian Sea is rather unusual. On the one hand, it is characterized by considerable differences in wintertime temperature between North and South, and, on the other hand, by equalizing the temperatures in summertime. Seasonal oscillations of water temperatures are more pronounced in Middle Caspian. In North Caspian, there is no water temperature stratification due to its shallowness.
- The most typical hydrometeorological feature of the Caspian Sea is the winds that induce 8–10 m high and up to 100–150 m long strong waves throughout November to March in Middle Caspian.
- Well-pronounced atmospheric-forced motions are common. Seiches-like oscillations (amplitudes up to 35 cm, periods from 8–10 min to several hours) occur often [28]. The currents are primarily wind-generated. Horizontal dynamics is characterized by predominating cyclonic circulation at the sea center and by generation of separate local eddies [29].

### 2.2. The Black Sea

The Black Sea is an inner enclosed sea on the southeast of Europe. As part of the Atlantic Basin, it connects to the ocean on the south via the Mediterranean Sea through the Bosporus, Dardanelles and Gibraltar (Figure 2). On the northeast, it connects to the Azov Sea through the Kerch Strait. The total area of the sea is about 422 thousand km$^2$, coastline is 3400 km long, average depth is 1240 m, and maximum depth is 2210 m. Like the Caspian Sea, the Black Sea is considered non-tidal since tide levels do not exceed 0.1 m because Mediterranean tides subside in the straits and the Black Sea is not big enough to develop considerable tides of its own. The level of the sea changes primarily due to wind-driven and seiche oscillations, as well as river outflows. Seiches are common in the Black Sea with periods of several minutes to two hours and level oscillations of 0.4–0.5 m, sometimes up to 1 m. The seiche lifetime varies in a broad range: from rapid subsiding to persisting for several days [30].

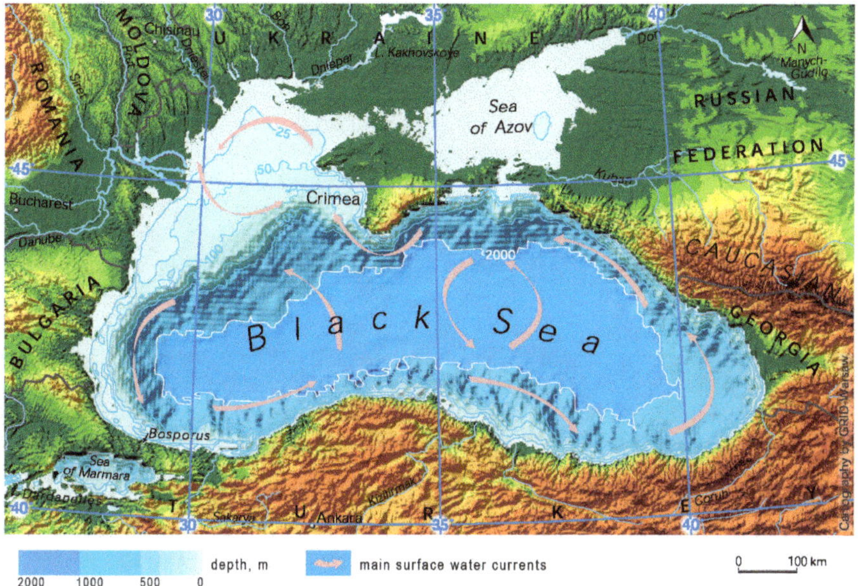


**Figure 2.** Map of the Black Sea.

The continental shelf of the Black Sea has a slight inclination (to the depths of 100–150 m) over the distance of a few kilometers off mountainous coasts. It ends with an abrupt (up to 20°–30°) slope to the depth of more than 1000 m. The only exception is the northwest of the sea where the shelf is up to 190 km wide.

The basic circulation in the Black Sea is characterized by a strong cyclonic basin-wide current along the coastline which is referred to as the Rim Current [31,32]. It is highly hydrodynamically unstable and consists of a system of moving mesoscale rings and eddies [21,33]. Eddies of various scales are typical of the whole coastal zone of the Black Sea [2,23,34,35]. The hydrological regime of the sea is determined by water exchange with the Marmora and Azov Seas, inflow of coastal fresh waters and climatic conditions. Coastal waters significantly freshen the near-surface water layer. The major part of river water (up to 80%) flows in on the northwest, where the largest rivers, Danube, Dnepr and Dniester, disembogue. On the east, rivers Inguri, Rioni, and Chorokh as well as many smaller ones carry their waters into the sea. Along the rest of the coastline, the inflow of coastal waters is insignificant [30].

All year round, the temperature of the near-surface layer of water tends to grow in the direction from northwest to southeast which is conditioned by the regional climate. In coastal zones, water temperature is strongly influenced by storm surges and upwellings. Negative storm surges or upwellings occurring near the Crimean Peninsula (Crimea) and the east coast in summer were reported to drop the near-surface water temperature from 25 to 7 °C [30].

## 3. Data and Methods

### 3.1. Data and Data Processing

Data were processed and analyzed using the toolkit of the satellite information system developed at Space Research Institute of the Russian Academy of Sciences (IKI RAS). The system is named "See the Sea" (STS). STS functionality, goals and current status are detailed in [36,37]. STS provides not only instruments for fast and easy access to satellite data and products, but also various tools for specialized data analysis. In addition, it enables joint analysis of data different in nature, spatial resolution, unit of measurement and acquisition time. The archive of STS is constantly automatically updated by its data receiving unit communicating with a number of data source centers. Today, it holds over 100,000 individual scenes. STS enables easy search of the distributed image archive using sensor type, time interval and location as search criteria. The selected image is visualized in the map area of the interface along with its geographic basis and related cartographic data. All information (regardless of sensor type or product) is presented for viewing in the same cartographic projection for a given geographic area. This facilitates data selection for analysis. Combining data of different nature (active/passive microwave, VIS and NIR), spatial resolution and swath width allowed us to better understand the complex picture of meteorological and hydrodynamic processes in the study regions, reveal factors favoring generation of non-tidal IWs, and, consequently, come up with a hypothesis on generation mechanisms.

For purposes of complex analysis, STS has a specialized cartographic web-interface allowing researchers to search, process and analyze various remote sensing data and related (for instance, meteorological) information.

Our investigation of IWs in the Black and Caspian Seas began with selecting satellite data required for the reconstruction of IWs generation and propagation in the regions of interest. After expert visual identification of SMIWs in satellite images, the images were processes and analyzed. This could include image contrasting, conversion into different units, comparison of images of different acquisition times, creating color composites using different bands and data types (including data acquired at different times), as well as parameter estimation of different objects, pixel-by-pixel analysis for various products, etc.

]Using the STS interface, the main characteristics of IWs were determined in an interactive semi-automatic mode. The result was a comprehensive description of the phenomena, including: the train's center coordinates, water depth at the manifestation location, number of waves in the train, its width, maximal wave length in the train, direction of propagation, presence of nonlinear interactions, near-surface wind speed and direction. The description was stored in the database. The database tools allow storing and visualization of graphical and attributive information, hierarchical classification of observed processes, search by spatial, time and typological criteria, and mapping IWs distribution in the studied seas.

### 3.2. Basics of SMIWs in Sea Surface Radar Imagery

Case studies performed during numerous field experiments have revealed main characteristics of IWs in oceans and seas. They propagate in trains, usually rank-ordered, with the largest, the fastest, the greatest-wavelength, and the longest-crested oscillations appearing at the train front. The maximal amplitudes of fluid particles' vertical displacements in IW occur at the pycnocline (the layer of rapid vertical changes in water column density), while the maximal amplitudes of horizontal displacements and horizontal orbital velocities are observed at the sea surface and may reach tens of cm/s, substantially modulating small-scale wind waves [38]. SMIWs usually look as alternating parallel bands of enhanced and attenuated surface roughness. The resulting surface roughness variations are translated into contrast variations in radar data [39]. SMIWs are made visible in SAR data due to: (i) associated with IW orbital currents which modulate, in one way or another, the short surface wave spectra; and (ii) surface films that cause changes in surface tension.

The pertinent theory used in analyzing and interpreting SAR data includes a few mechanisms of IW effects on the disturbed sea surface due to modulation of short wind gravity-capillary waves. Historically, the first one was a kinematic model based on the wave action balance equation [2,40]. It described the significant features of IW-induced modulations of ocean surface observed in early satellite data and explained image contrasts in the case of relatively long electromagnetic waves (tens of centimeters and longer). Two hypotheses based on the Bragg resonant backscatter were suggested to account for the contrasts observable in the range of centimeter waves: the cascade mechanism of ripple modulation based on the known fact that a nonlinear IW can generate "parasitic" capillary ripples near its crest in a resonant manner [41,42], and the mechanism of modulation of the momentum flux towards the surface [43]. Further non-Bragg mechanism of microwave scattering on the SMIWs was studied in [44–47]. Respective quantitative contributions of micro-breaking of steep surface waves in the convergence zones to electromagnetic wave scattering were evaluated in these studies and it was shown that, under certain conditions, they can be very important.

Modulation of centimeter- and decimeter-scale surface waves caused by redistribution of surface films by currents within IWs can also be the dominant factor behind SMIWs imaged by SAR operating in short centimeter wavelengths. Elongated areas with increased film concentrations are present on the sea surface at IW depressions resulting in significant attenuation of ripples and formation of slicks. Since increased surfactant concentration is common for semi-enclosed and inland seas, this mechanism is expected to predominate in such regions [48,49].

Under moderate viewing angles implemented in satellite SARs, SMIWs can be distinctly visualized in co-polarized SAR images on both VV- and HH-polarizations. Co-polarized SAR images are extensively employed in this study.

Figure 3 presents images taken by Advanced Synthetic Aperture Radar (ASAR) on board Envisat satellite and illustrates two different mechanisms of SMIWs in radar imagery. Figure 3a features parallel bands of enhanced backscatter alternating with wide zones of slightly attenuated signal and shows no signs of surfactant film. By contrast, Figure 3b depicts long narrow parallel bands of slicks due to concentrated film trapped by a SMIW. The inset graphs show variations in radar signal along the cross sections of the trains (white lines).

Combined analysis of satellite radar and VIS/NIR data is effective for resolving ambiguities in interpretation of radar data, detecting SMIWs, as well as revealing sources and mechanisms of IW generation. VIS/NIR data, especially obtained with sun glint, can provide additional information on the processes and phenomena on the sea surface at low winds. In these conditions, the returned radar backscatter is low, and the surface may appear featureless and uniformly dark across the radar image. Sun glint is observed by optical (VIS/NIR) sensors when sunlight incidence angle is equal to the angle of reflection. Surface roughness excites tilting of numerous small wave facets reflecting sunlight at a variety of different angles. Size and shape of a sun glint pattern depend on the probability distribution of the slopes of the facets, incident sunlight direction and sensor viewing angle. Thus, differences in sea surface roughness variance in a sun glint area are manifested as variations in image brightness. An example of SMIW imaged by an optical multispectral sensor is shown in Figure 4c.

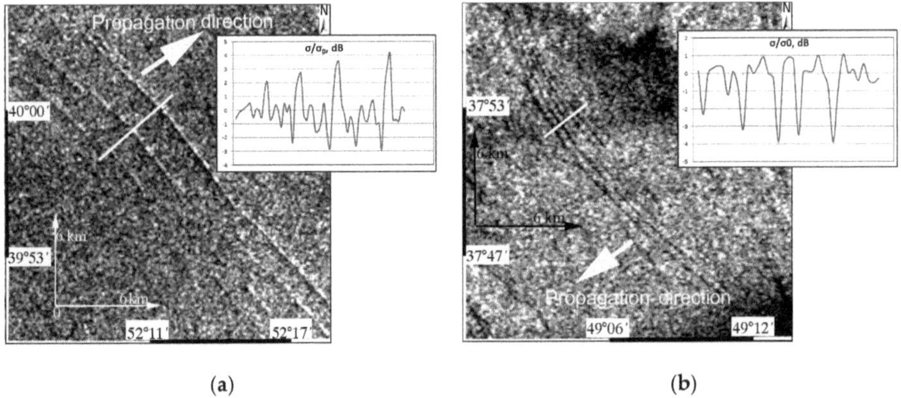


(a) (b)

**Figure 3.** Mechanisms of Surface manifestations of internal waves imaging by SAR in the Caspian Sea: (**a**) restructuring of ripple spectrum, Envisat ASAR, HH-polarization, 16 July 2010, 06:44 UTC; and (**b**) redistribution of surface films, Envisat ASAR, HH-polarization, 26 May 2011, 06:59 UTC.

## 4. Results

### *4.1. Radar Observations of IWs in the Caspian Sea*

#### 4.1.1. Parameters of IWs Derived from Satellite Data

Numerous SMIWs were identified in satellite data obtained over the Caspian Sea [50], some examples are shown in Figure 4. IWs may appear as isolated solitons, such as the one in Figure 4a. It has a wave length of 900 m and a wave crest length of 56 km and propagates southeast. However, the majority of IWs look like classical soliton trains (Figure 4b). The inset depicts variation of the radar signal due to the SMIW along the marked cross section (white line). The leading wave crest is about 68 km long and the maximal wave length in the train is 1500 m. Sometimes, one image manifests several IW trains that propagate at different angles to each other and interact nonlinearly. In a false-color Landsat-8 OLI image (Figure 4c), five IW trains propagating over the depths of 100–150 m are clearly visible (A, B, C, D, and E). The most distinct trains, A, B and C, propagate northeast, the distance between A and B is 9 km, the distance between B and C is 8 km. Each of them contains 4–5 individual waves with average lengths of 350 m, rising intensities from rear to front of a train and leading wave crest lengths of about 55 km. Train D propagates southeast, intersects with C and B, interacting with them in a nonlinear way. Train E propagates north, and consists of four waves with leading wave crest length of 15 km and 400 m average wave length. The SMIW patterns mimic bottom topography features. Specific curvatures of the leading waves allow for the identification of IW origination areas. They are usually topography inhomogeneities, sandbanks, and steep elevations that are typical of the eastern region of Middle Caspian.

Some statistics on SMIWs in Middle Caspian were derived from the satellite data. Distributions of spatial dimensions of IW trains detected in radar images are shown in Figure 5. The widths of trains varied from 1 to 6 km depending on the number of waves in train. The length of the longest wave in trains ranged from 150 to 2000 m (Figure 5a). The lengths of leading wave crests ranged tens of kilometers, between 8 and 80 km (Figure 5b).

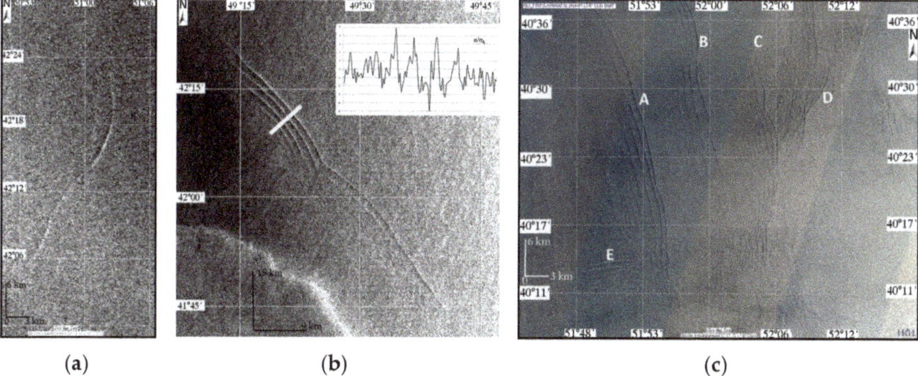


(a) (b) (c)

**Figure 4.** Surface manifestations of internal waves in the Caspian Sea: (**a**) Solitary IW, Envisat ASAR, HH-polarization, 25 July 2010, 18:20 UTC; (**b**) IW train, Envisat ASAR, HH-polarization, 11 September 2010, 06:52 UTC; and (**c**) Multiple trains of IWs propagating in different directions, Landsat-8 OLI, 13 July 2016, 07:13 UTC.

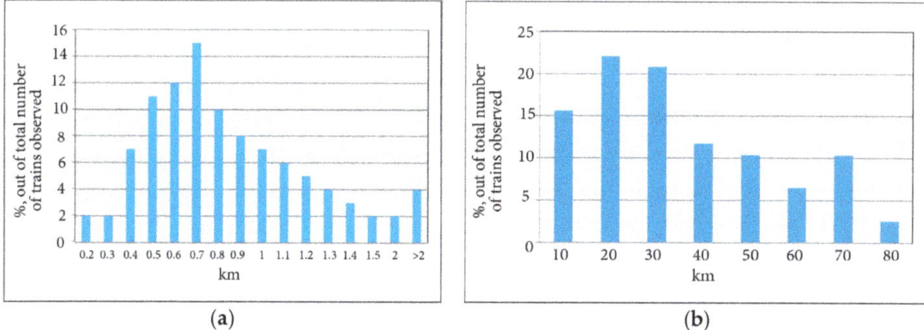


(a) (b)

**Figure 5.** Distributions of spatial dimensions of IW trains detected in satellite images of the Caspian Sea: (**a**) maximal wave length in train; and (**b**) leading wave crest length.

### 4.1.2. Spatial and Seasonal Variability of SMIWs in the Caspian Sea

Spatial distribution of SMIWs in Middle and South Caspian derived from radar data is presented in Figure 6. The data cover only the period from May to October because no IWs were observed throughout the other months.

As can be inferred from Figure 6, the SMIWs are distributed unevenly, concentrating in three main regions. Their occurrence is season-dependent. In the end of May and in the first half of June, SMIWs are found only in the western part of South Caspian (I); in the second part of June and in July, all SMIWs are observed on the east, near the Absheron Sill (II); and in August, they shift northeast (III). By far the largest number of SMIWs is observed in July and August, all located in the eastern Caspian, north of Absheron Sill. In the other parts of the sea, SMIWs occur less often.

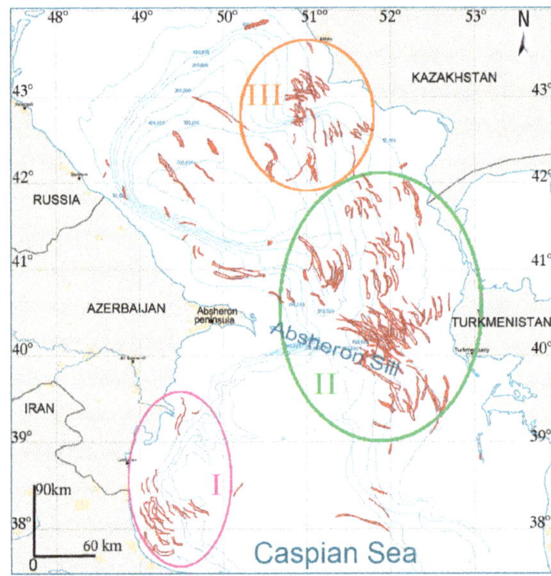


**Figure 6.** Surface manifestations of internal waves detected in satellite imagery of the Caspian Sea.

### *4.2. Satellite Observations of IWs in the Black Sea*

A large archive of radar and VIS/NIR satellite data allowed us to estimate contributions of various factors into generation mechanisms of SMIWs and their spatial distribution in different areas of the Black Sea. There are four main areas in the Black Sea where SMIWs are mostly observed: near the Danube Delta and Crimea on the west, near Novorossiysk on the northeast, and on the continental shelf near Georgia on the east (Figure 7).

#### 4.2.1. SMIWs near the Danube Delta

The greatest number of SMIWs is observed on the Romanian shelf, in the area of the Danube Delta. One high resolution satellite image, for example an image acquired by Landsat-8 OLI, with a resolution of 15 m for the panchromatic band 8 and swath of 185 km, can manifest up to twenty IW trains. Statistical analysis of satellite data shows that SMIWs are mostly observed from May to July in the region. The largest amount of IWs propagating both toward and off the coast was observed in 2010. Most likely, this fact can be explained by specific hydrological conditions of that year. However, no in situ measurements are available to confirm this. Figure 8 shows mapped locations of the leading fronts of IWs revealed in 2010 radar (Envisat ASAR, ERS-2 SAR) and VIS (Landsat-5 TM, Landsat-7 ETM+) images. Orientation peculiarities of the SMIW allow suggesting two main sites the IWs originate from: one is attached to the Danube water plumes (dashed area), while the other is farther off the coast, over the isobaths of 50–75 m.

IWs from the first site propagate in various directions, they contain from 10 to 20 waves with lengths up to 150 m, and their leading fronts are rather distorted. To our view, the source of IWs here is the low-salinity water plumes of the Danube River.

Generation of IWs associated with river outflow is quite common. It is theoretically examined, for example, in [51,52], and satellite observations are described in [53]. Comprehensive measurements of intense IWs induced by an intrusion of low-salinity surface waters were conducted from a marine platform on the northwestern shelf of the Black Sea [54], and numerical modeling based on those data was performed in [55]. These works consider IW generation by a solitary front of estuarine type when wave crests are congruent with the front and propagate to the open sea.

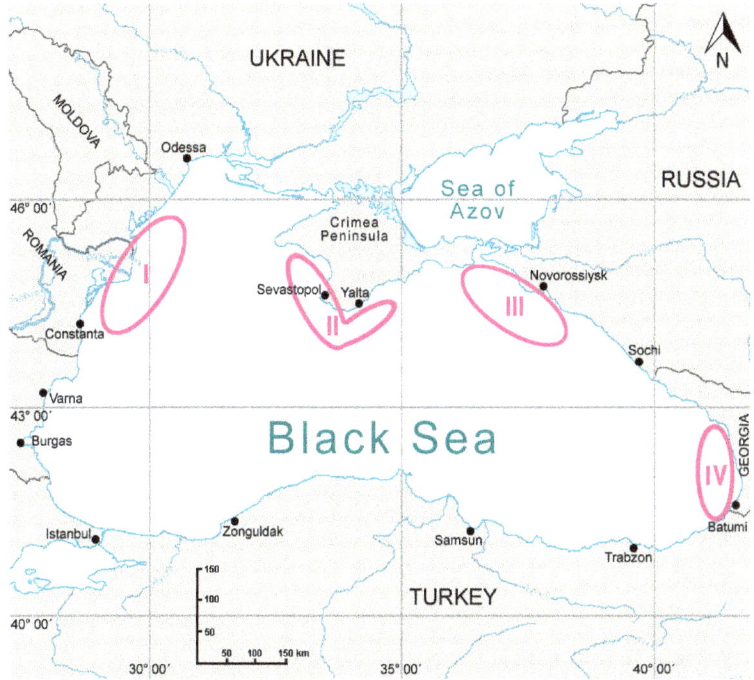


**Figure 7.** Main areas of SMIWs in the Black Sea, close to: (I) Danube Delta; (II) Crimean Peninsula; (III) city of Novorossiysk; and (IV) shelf of Georgia.

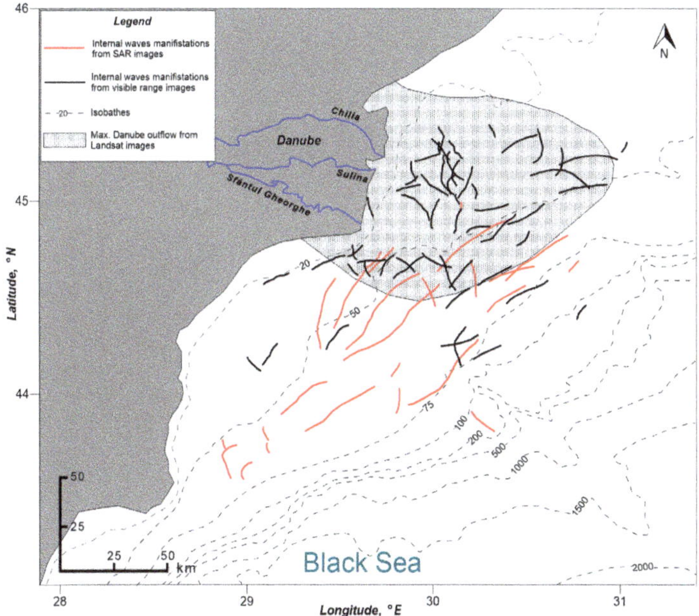


**Figure 8.** Locations of IW fronts revealed in 2010 satellite data of the Danube Delta region, western Black Sea.

The Danube waters flow into the surface layer of the Black Sea through many delta arms. The main volume comes via three distributaries: Chilia, Sulina, and Sfântul Gheorghe (Saint George) (Figure 8). In the distributaries, waters have slightly different parameters. Their interaction with each other and the seawater produces multiple relatively small-scale non-stationary fronts, eddies, apparently generating IWs in varying directions. As a rule, the IWs are parallel to the fronts outlining river water plumes of different turbidities.

A color composite of Landsat-5 TM (bands 3, 2, 1, Figure 9) clearly displays turbid fresh Danube waters and multiple (over 20) IWs propagating in different directions. Such direction variability is explained by high non-uniformity of the river outflow as well as bottom topography features. Many IWs sit directly on turbid plume leading fronts so that they propagate together. Multiple IWs cross each other, their fronts become distorted, and new ones occur.

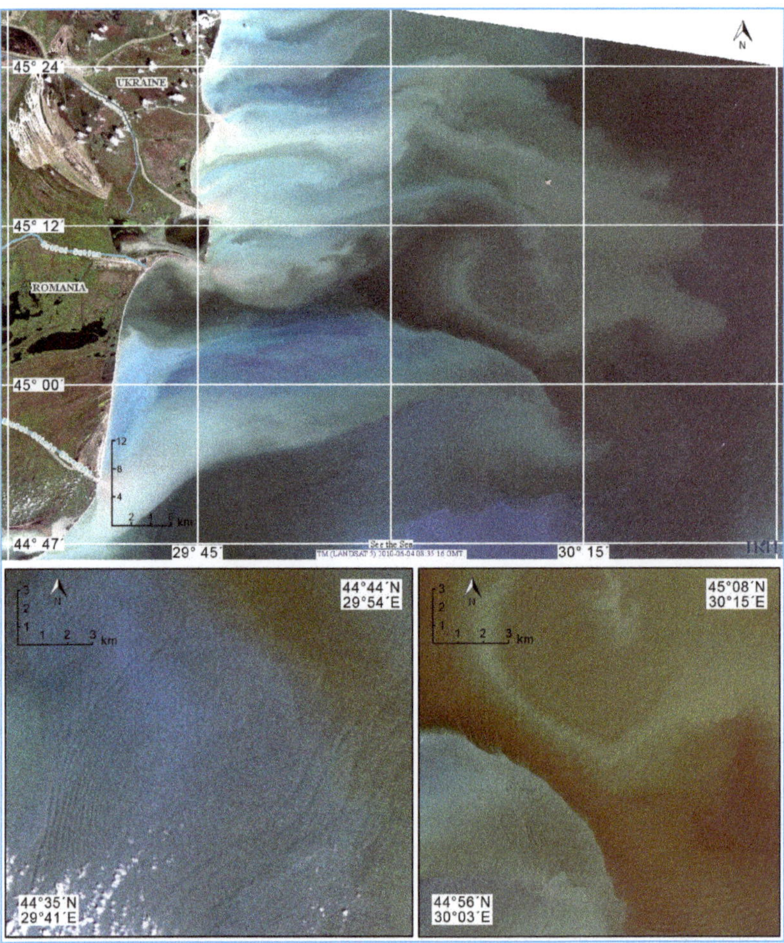


**Figure 9.** Multiple SMIWs near the Danube Delta. Intense color variation pertains to high turbidity of the Danube plume. Landsat-5 TM (bands 3, 2, and 1), 4 June 2010.

On the second site (farther off the coast), SMIWs almost always indicate propagation toward the coast with IW fronts parallel to local isobaths. The number of waves in a train usually does not exceed 10, wave length can reach 300 m. The IWs propagate shoreward, being induced, to our view, by inertial period IWs entering the shelf zone, as it occurs in open ocean mainly under the impact of tides. In the Black Sea, the same generation mechanism of intense IWs was also observed near the southern coast of Crimea. Examples of SMIWs imaged by SAR in the second site are shown in Figure 10. To prove our IW generation hypothesis, let us make some estimates in the way shown in [9,56]. For the Black Sea, the inertial period is about 17.2 h. From the SAR image in Figure 10, we estimate the distance between the IW trains at 13 km and, consequently, IW velocity at 0.21 m/s. The latter value is the characteristic phase velocity of short-period IWs in the Black Sea, which is confirmed by in situ measurements [16]. The periods of waves range 10–20 min by in situ measurements [16,24]. These values are the characteristic periods of intense short period IWs. Thus, the wave length is estimated at 126–252 m, which is in agreement with satellite data. There are no reasons for the IW trains propagating with the front of internal inertial wave to have phase velocity differing from the velocity of the long (inertial) IW.

With known vertical density (Brunt–Väisälä frequency) profile, it is easy to solve the equation for internal inertial waves and obtain estimates of phase velocity and lengths for 17.2-h period and short period waves. For arbitrary profile, the equation can be solved by numerical methods. IW parameters were obtained by a numerical solution of the equation for hydrological conditions of the Danube shelf over the depths of 70–80 m (assuming flat sea floor, which is acceptable for this region). From temperature and salinity profiles measured by Romanian researchers, seawater density was determined and then input into calculations of IW parameters and dispersion characteristics. To simplify calculations, the waves were assumed linear. For 10-min period waves, phase velocity was estimated at 0.42 m/s, for 20-min period waves, it was 0.46 m/s. Wave lengths were 252 m and 552 m, respectively. Note that, while the lengths of short period waves are of the order of several kilometers in open ocean [9], they are much less in the Black Sea. A good example of a wave train generated by inertial IW on the Black Sea shelf, with wave parameters estimated, is presented in [16].

Nonlinear interactions manifested in Figure 10 were used to estimate the degree of nonlinearity of IWs in the Danube Delta region. The Ursell number ($U_r$)—wave nonlinearity to variance ratio—was calculated with the IWs modeled using the Korteweg–de Vries equation [57]:

$$U_r = (\alpha/\beta) \times c_0 \times \eta \times {\Delta_s}^2 \quad (1)$$

where $\eta$ is wave height, ${\Delta_s}^2$ is wave half-width, and $\alpha$ and $\beta$ are nonlinearity and variance coefficients defined in the case of two-layer fluid ($H = h_1 + h_2$) as:

$$\alpha = (3/2) \times (h_1 - h_2)/(h_2 \times h_1);\ \beta = c_0 \times h_1 \times h_2/6 \quad (2)$$

To estimate $\alpha/\beta$ we used hydrological data obtained by Romanian researchers during ship expeditions. The measurements were performed in a time period close to the satellite observations at depths of 25, 55 and 70 m. Water temperature and salinity data were taken at standard depth levels yielding rather coarse vertical profiles of the parameters. Nevertheless, they allowed for a two-layer approximation of the water column [58]. For example for $H$ = 70 M, $h_1$ = 15 m, $h_2$ = 55 m.

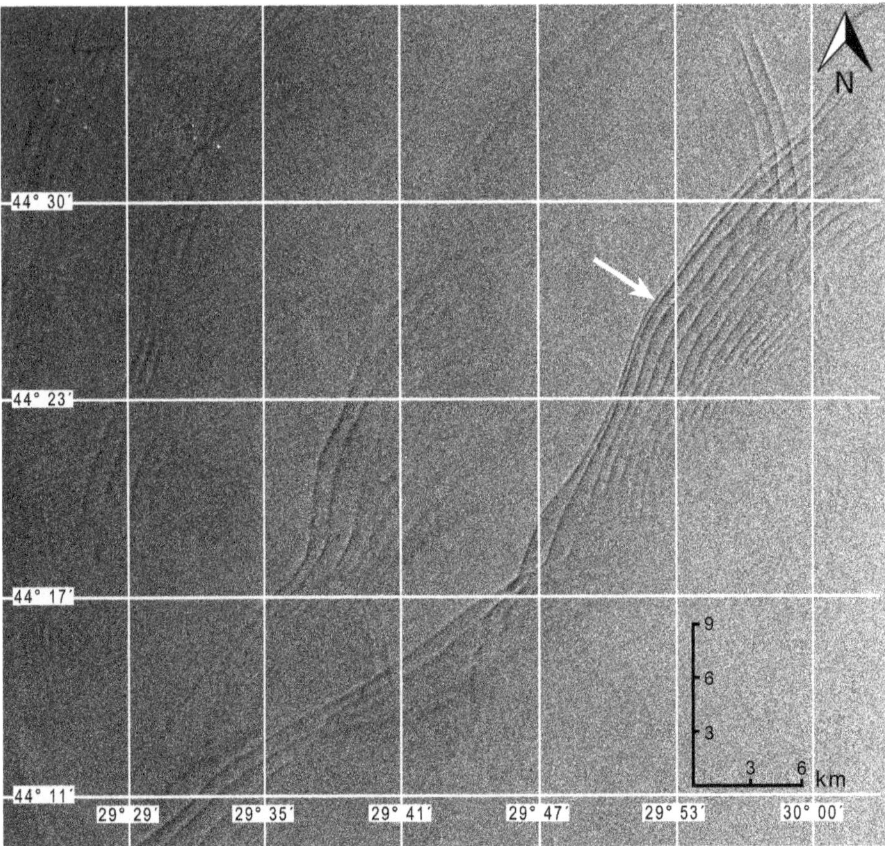


**Figure 10.** SMIWs near the Danube Delta indicating coastward propagation, Envisat ASAR, VV-polarization, 8 June 2010. Arrow points to the IW train for which the Ursell number was estimated.

Some SMIW spatial characteristics were derived from the image (Figure 10). The one indicated by white arrow has the following characteristics. The train propagates northwest over a wide shelf with a depth about 75 m. There are 14 waves in the train. The front wave is the narrowest one, its length, estimated as the distance between the two successive low-signal bands in the image, is 325 m. The second wave is twice as long—650 m. Lengths of the following waves are close to this value. At present, it is not possible to estimate IW height from satellite images. Nevertheless, our multi-year experience of measurements in the Black Sea [24] allows us to suggest an estimate of 5 m. For this wave height, $U_r$ is assessed at 67.7. If we assume the leading wave height to be 1 m, then $U_r$=13, which is close to soliton-type value equal to 12 [59]. This number characterizes the degree of IW field nonlinearity. The higher $U_r$, the stronger are nonlinear effects [60]. Considering the obtained $U_r$ estimates (13 ... 67.7), we can conclude that nonlinearity prevails over variance in IWs propagating towards the coast from open sea in the region of the Danube Delta.

#### 4.2.2. IWs near the Crimean Peninsula

IWs in the region of the Crimean Peninsula have been closely studied and described in detail by researchers of Marine Hydrophysical Institute (Sevastopol) and Acoustics Institute (Moscow). Many results were obtained due to regular observations from a marine scientific platform in Katsiveli (Yalta Region) conducted for the past 30 years; the most recent are presented in [24,61].

We have conducted monitoring of the region since 2009. Until April 2012, we used primarily Envisat ASAR and ERS-2 SAR data, then Landsat data. Starting from October 2014, we have used Sentinel-1 SAR-C data, and since August 2015, Sentinel-2 MSI as well. According to our satellite observations, SMIWs most frequently appear in two areas: near the south and west coasts of Crimea. A typical SMIW near the west coast is shown in Figure 11. The most distinct IW train (another two less pronounced IWs are closer to the coast) propagates east over the depths of 80–90 m, contains 6 waves with an average length of 200 m and a leading wave crest length about 17 km. All detected IW trains near Crimea propagated toward the coast. Fronts in the trains frequently had the shape of concentric arcs, which made it possible to approximately locate their sources (Figure 12). In this region, at shelf edge, velocity of the Rim Current, the principal element of water circulation in the sea, can reach 1 m/s and more. Its meanders and inertial oscillations can induce current components normal to the coastline with velocities up to 50 cm/s, comparable with current velocities in tidal seas. Another IW generation mechanism involves local hydrological fronts moving over the shelf. For instance, they can be generated by storm surges near south Crimea [16]. Anyways, SMIWs are quite rarely detected in SAR data of the region. During 2009–2011, as few as eight SMIWs were observed, and they occurred not only during June–August, but also in October (for example, 1 October 2010) [34].

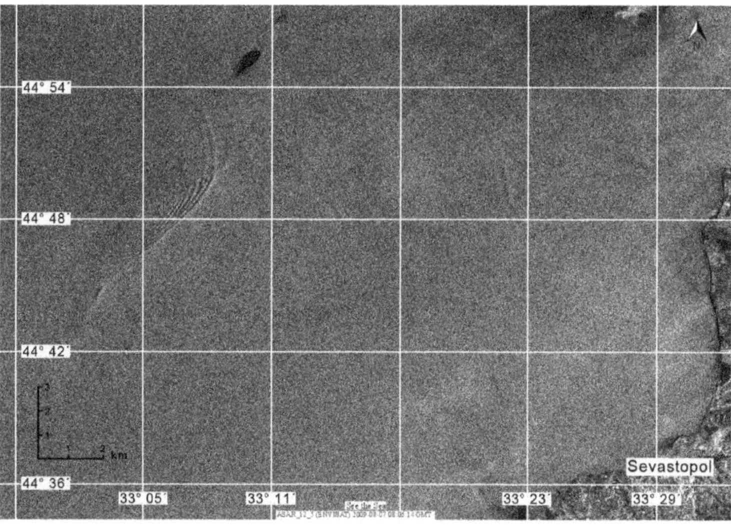


**Figure 11.** Typical SMIW near the west coast of Crimea, Envisat ASAR, VV-polarization, 7 August 2009.

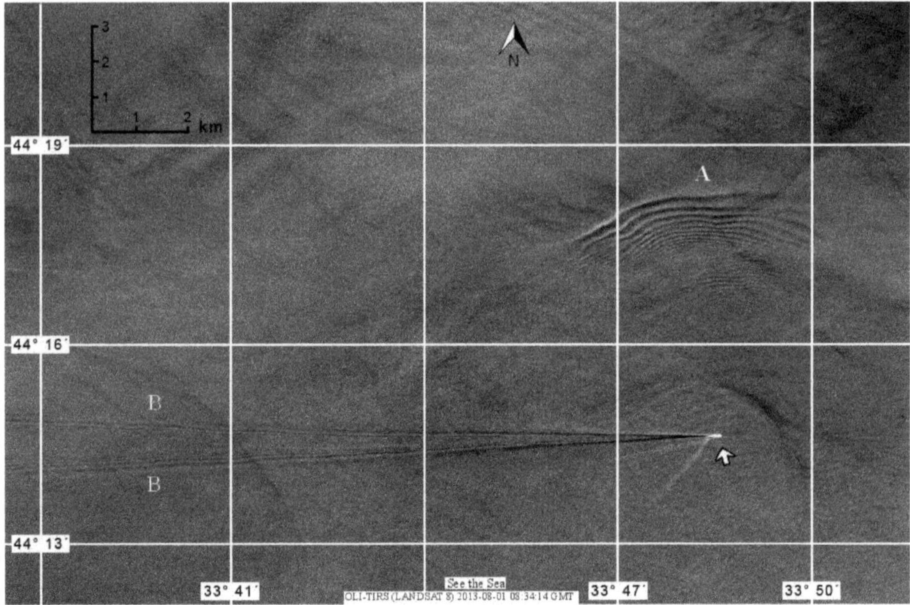


**Figure 12.** Different types of IWs propagating near the south coast of Crimea: (**A**) natural IW; and (**B**) IW generated by a moving ship, arrow points to the ship; Landsat-8 OLI (panchromatic band, resolution 15 m), 1 August 2013.

In satellite image of 1 August 2013 (Figure 12) obtained off south Crimea, beside SMIW marked A, we can see two SMIWs marked B. These are IWs generated by a moving ship. Such IWs can be clearly detected in high resolution satellite images obtained by Landsat-8 OLI at a resolution of 15 m (band 8) or Sentinel 1 and 2 at a resolution of 10 m. Natural (A); and ship-wake (B) IWs in the region have similar wave lengths, about 80–100 m. They obviously differ in crest length, as can be inferred only from satellite data analysis. Natural IW crest length scarcely exceeds 5–6 km, while ship wakes containing IWs can span tens of kilometers. Thus, one may encounter IWs of different origins when interpreting data from regions with ship routes.

#### 4.2.3. IWs in the Eastern Black Sea

As mentioned above (Section 2.2), the eastern Black Sea is characterized by a narrow shelf (the 200 m isobath is at the distance of only 5 km from the coast) and many rivers carrying waters into the sea. There are important differences between SMIWs on the west (Danube Delta) and east (shelf of Georgia). In contrast to the wide western shelf, no IWs generated at shelf edge were observed on the east. All SMIWs were associated with river plumes. Analysis shows that submesoscale IWs occur due to inertial currents of (always) unstable sharp plume boundary front. IWs of this type are manifested only in high resolution data and are best pronounced in VIS images. Examples are presented in Figure 13. Wave length in the trains does not exceed 50 m, the length of leading wave crest depends on the shape of river plume. The seasonal character of these SMIWs is associated with variations in river outflow volumes rather than thermocline behavior.

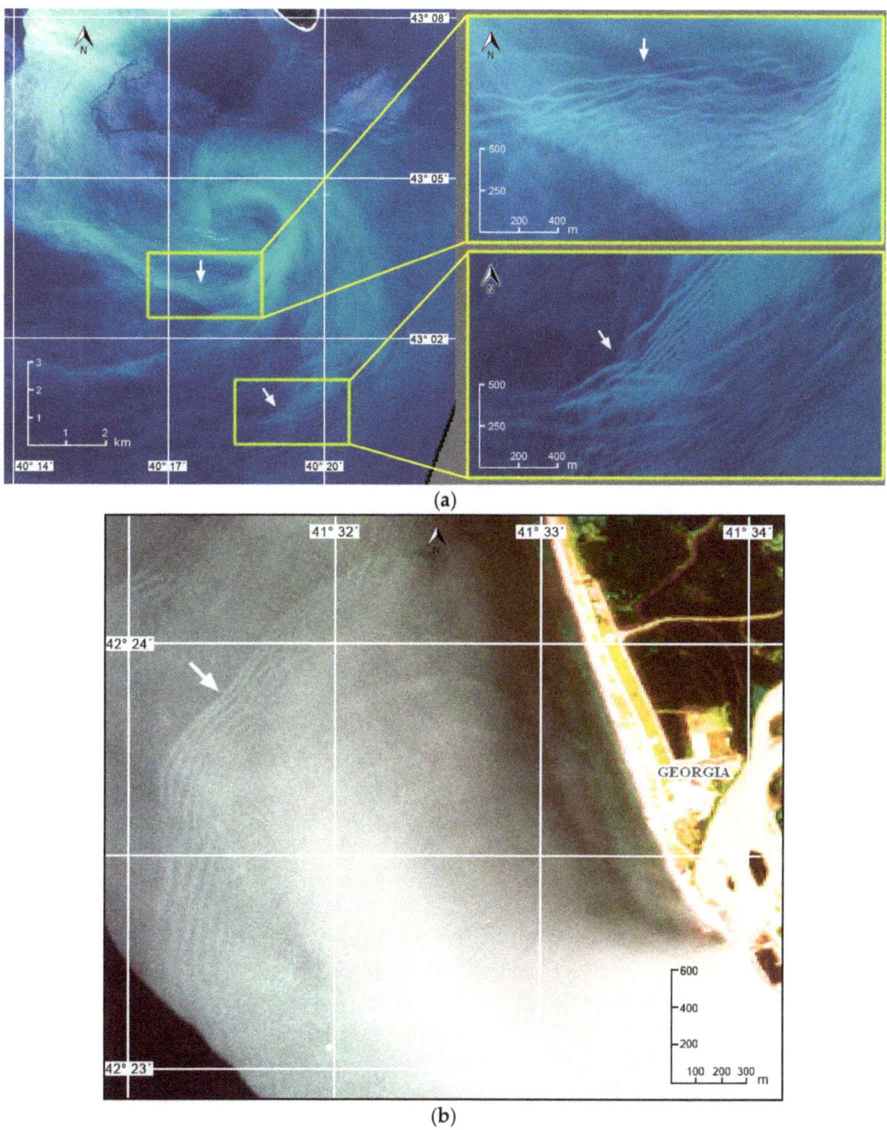


**Figure 13.** SMIWs in the eastern Black Sea due to river plumes of: (**a**) the Bzyp River, Sentinel-2A MSI, 1 May 2017; and (**b**) the Enguri River, Sentinel-2A MSI, 1 August 2016.

### 4.2.4. IWs in the Northeastern Black Sea

SMIWs in radar imagery of the northeastern Black Sea are scarce. We observed only fifteen instances in satellite surveys of 2005–2016. Six cases were registered in 2006, of them four in June and two in July. One SMIW was registered in August 2007, one case in June 2009, and none in years 2008 and 2010. Four SMIWs were revealed in July 2011. The 2012–2014 data gap was due to the interruption of the continuous SAR data flow (Envisat out of operation, Sentinel not launched yet). After the launch of the Sentinel satellites, we continued radar survey of the area and detected two SMIWs in June 2015 and one in June 2016. The reasons for such a considerable interannual variability in SMIWs in radar images are discussed in Section 5.1.

Spatial distribution of SMIWs identified in radar data in the northeastern Black Sea is depicted in Figure 14.

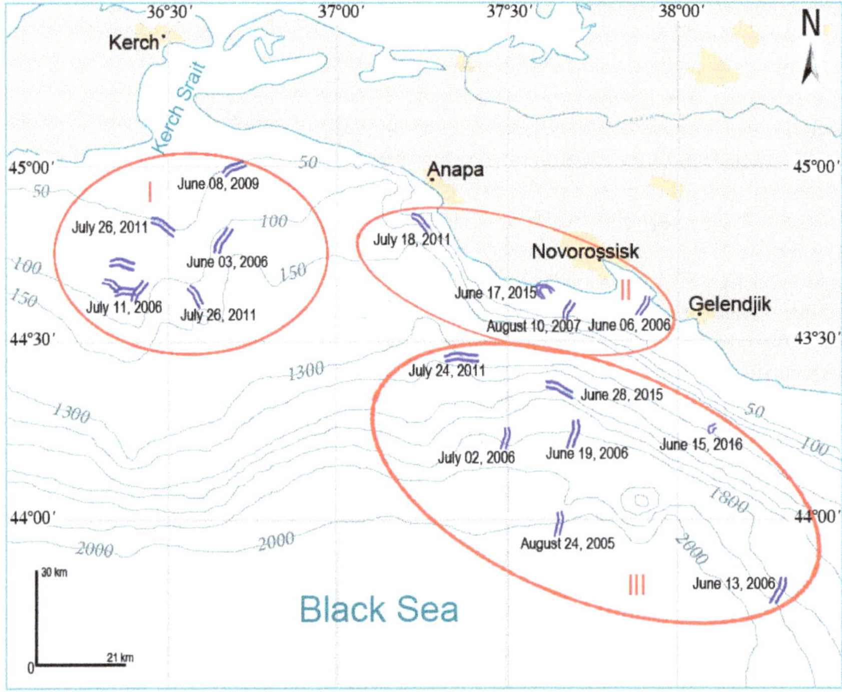


**Figure 14.** A map of SMIWs detected in SAR images of the northeastern Black Sea.

According to Figure 14, we can outline three regions where IWs are observed:

I. Area near the Kerch Strait between isobaths 50 and 150 m.
II. Area near the Anapa-Gelendjik coastal line.
III. Area 45–50 km southwest off the Novorossiisk-Tuapse coastline between isobaths 1300 and 2000 m.

In the northeastern Black Sea, IWs are manifested as trains containing up to 20 waves with maximal wave length in a train not exceeding 500 m. Leading wave crest is typically substantially curved and its length varies between 7 and 18 km. Train width varies between 850 and 3500 m depending on the number of waves in train. The distribution of IWs spatial characteristics derived in radar images of this region is presented in Figure 15.

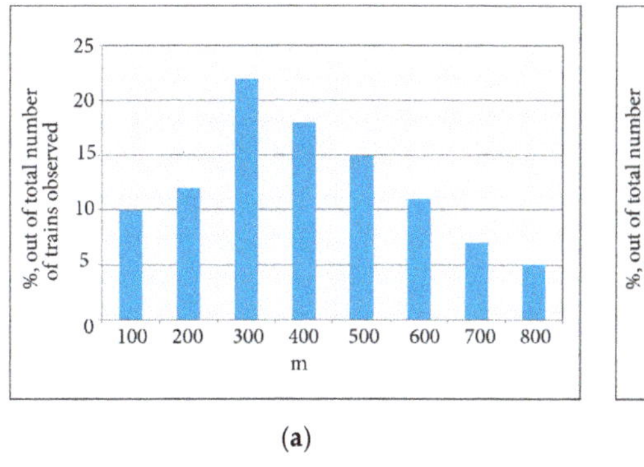


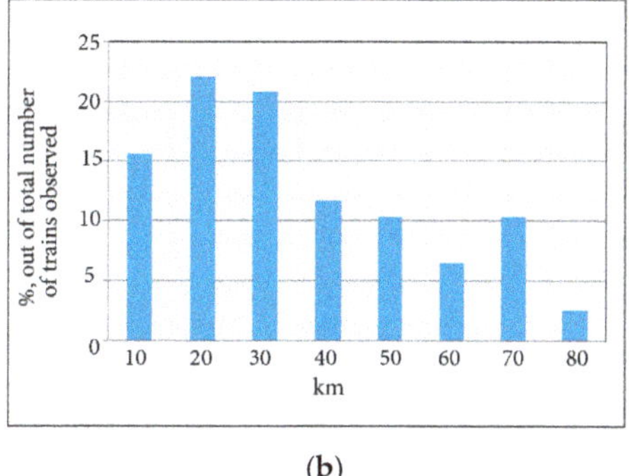


(**a**) (**b**)

**Figure 15.** Distribution of spatial dimensions of IW trains detected in satellite images of the northeastern Black Sea: (**a**) maximal wave length; and (**b**) leading wave crest length.

## 5. Discussion

### 5.1. Effect of Sea Water Stratification on Detectability of IWs in Radar Imagery

The above results are evidence of a significant interannual, seasonal and spatial variability of SMIWs both at the sea surface and in radar images of the surface. Below we discuss possible reasons and show that the key factors are shape and depth of the pycnocline.

We observed a considerable interannual variability in the number of SMIWs in the northeastern Black Sea (Figure 15). Most of the SMIWs in this region were documented in 2006. To find out the reason, we analyzed the data of conductivity-temperature-depth (CTD) water column profiling performed during summer cruises of R/V Akvanavt by researchers of the South Branch of Institute of Oceanography RAS in 2006–2008. Hydrological measurements were carried out along a transection perpendicular to the coastline [62].

Figure 16 shows profiles of sea buoyancy frequency derived from CTD measurements over the depths of 1800 and 50 m. Buoyancy frequency describes the oscillation of a water parcel about its equilibrium depth and is used as a parameter to express the strength of the seawater stratification. It is inherently related to internal gravity waves and determines an upper bound on the existence of free internal inertia-gravity waves in the sea.

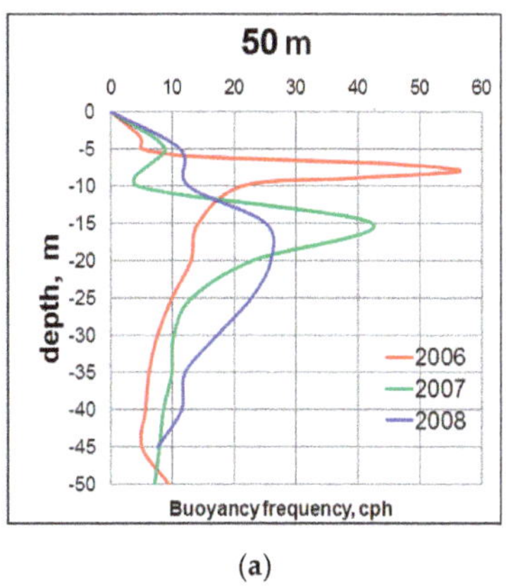


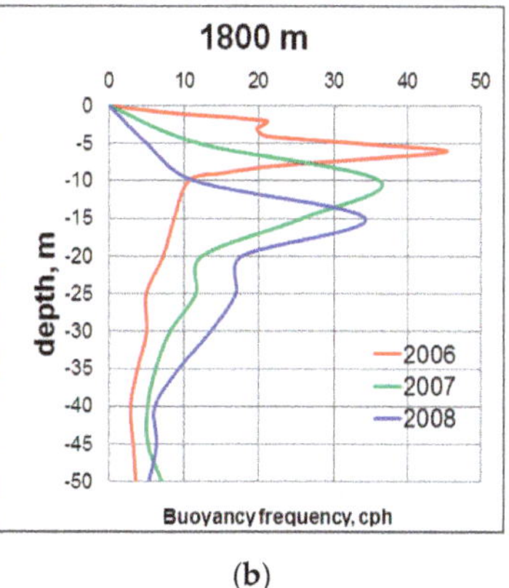


(**a**) (**b**)

**Figure 16.** Vertical profiles of buoyancy frequency (cycles per hour, cph) based on in situ measurements in areas with depth of: (**a**) 50 m; and (**b**) 1800 m (northeastern Black Sea).

The pycnocline shape of June 2006 (when most of the SMIWs were observed) has significant differences as compared to the other years, and these differences persist for different depths (Figure 16). The buoyancy frequency peak is very sharp and located at a depth of 5–7 m.

Numerous SMIWs were identified in satellite data obtained over the Caspian Sea. All were observed within the period from end of May to end of August. Three main regions of their localization depending on time were outlined. In May, SMIWs were observed on the west of South Caspian, in June and July, all SMIWs were located on the east, near the Absheron Sill, then in August, SMIWs moved north. Comparison of satellite data and in situ measurements has demonstrated (Figure 6) that the observed seasonal variability of SMIWs can also be explained by variations in pycnocline properties. Periods of numerous SMIWs coincide with seasonal thermocline formation periods. Most SMIWs registered in Middle Caspian are associated with sharp and shallow pycnocline. In summer, salinity is almost constant across all depths in Middle and South Caspian, so density gradient primarily depends on temperature gradient. As upper water layer warms up starting the second half of July, and thermocline deepens down to bottom, SMIW localization areas shift northeast, where a noticeable stratification exists.

Based on the results obtained for the northeastern Black Sea and Caspian Sea, we can state that the maximal number of SMIWs is associated with the most sharp and shallow pycnocline. Such conditions favor both generation of IWs and their manifestation at the sea surface by inducing strong orbital motions in the near-surface water layer which modulate the wind wave spectrum to produce SMIWs in radar data.

It should be noted that all SMIWs were detected in satellite data taken over Middle and South Caspian. North Caspian is the shallowest area of the sea with average depth of only 6 m. Its depth does not exceed 10 m and almost 20% of the region is less than 1 m deep. There is no thermal stratification in North Caspian due to its shallowness. This fact explains the absence of SMIWs in this area.

### 5.2. Predominating Mechanisms of IW Generation in Inner Seas

The origins of IWs are diverse: tides, inertial oscillations, variations of atmospheric pressure and wind, earthquakes beneath the sea, currents flowing over sea floor irregularities, anthropogenic effects, etc. [61,63]. In coastal regions of oceans and tidal seas, IWs induced by interaction of tides with shelf edge prevail. By contrast, IWs in non-tidal basins, though much less intense, vary significantly by their generation mechanisms.

The main feature of IWs in non-tidal sea is their moderate amplitudes as compared to IWs in the ocean. Widely spread nonlinear effects are another characteristic of IWs on the shelf of non-tidal sea, the same as in seas with tides. Nonlinearity (vertical and horizontal profile asymmetry) of IWs is manifested in train forming intense waves, generated under certain conditions. These are a combination of processes responsible for generation of intense IWs in non-tidal sea. Previous results were based on in-situ observations from stationary platforms in the Caspian and Black Seas [59,64]. They enabled estimating some IW characteristics, their amplitudes and periods in the first place, and determine the corresponding hydrological conditions. However, local in situ measurements, even using a network of distributed probes, could not provide a comprehensive spatial picture and spatial characteristics of IWs in the Black and Caspian Seas, therefore limiting our outlook with regard to their origin. A complex analysis of multi-year satellite data complemented by meteorological and in-situ data, both reported in literature [16,54] and our own [24], yielded statistically valid conclusions on IW generation mechanisms in the studied seas. The results maintain that an effective generation mechanism of intense SMIWs in non-tidal sea is the intrusion of long IWs—internal seiches or quasi-inertial IWs—onto shelf that usually occurs after storms. Entering the shelf zone, long IWs play the same role in non-tidal sea as tides on ocean shelf (Section 4.2.1). On the way toward the coast, they first transform from sinusoidal to nonlinear, and then generate IW trains. In the case of a narrow shelf and abrupt continental slope (like in the Black Sea on the northeast and off the southern tip of Crimea), IWs generation is associated with surge-induced local fronts in the coastal area that are observed in periods of wind relaxation and restoration of water stratification disturbed by the surge (Section 4.2.2). In the case of a wide shelf (the Danube Delta region), intense IWs can be generated by a developing surface intrusion of fresh coastal waters (Sections 4.2.1 and 4.2.3).

Our long-term satellite survey revealed that practically all IW trains identified in satellite imagery of the northeastern Black Sea (Figure 14) are located close to a cold eddy structure or a cold hydrological front. This evidences in favor of the frontal generation mechanism, when IWs are radiated by a non-stationary (moving and/or inertially oscillating) front or edge of a mesoscale eddy.

I. IWs in the Black Sea area adjacent to the Kerch Strait are often generated by moving hydrological fronts. The Kerch Strait water exchange with the Black Sea is determined by the wind flows over the strait. The water level slopes from the Black to the Azov Sea under the impact of the winds blowing from the south. Under south winds, a front of salty and cold Black Sea waters can form and move towards the Kerch Strait. In satellite images, we often see SMIWs moving ahead of the front. A typical example of SMIWs generated by a cold seawater front and moving ahead of it is shown in Figure 17. This figure presents an SST chart derived from NOAA AVHRR data taken at a relatively close time to the moment of ASAR image acquisition. A thermal front separating two water masses is clearly seen. The red rectangle marks the location of IW trains propagating ahead of the front above the depths of 50–70 m. Each IW train in the Envisat ASAR image contains more than 10 waves with an average wave length of 300–400 m. The IWs trains interact nonlinearly. The SMIWs pattern is similar to tidal seas. However, these IWs are smaller in dimensions and considerably less intense in comparison to IWs in oceans and tidal seas.

II. In near-coastal waters (Region II), IW generation can be attributed to a storm surge or relaxation of upwelling.

III. Combined analysis of radar and VIS/NIR data obtained within a small time interval shows that all SMIWs revealed over deep waters (Region III) are located near the edge of a mesoscale eddy or an eddy dipole.

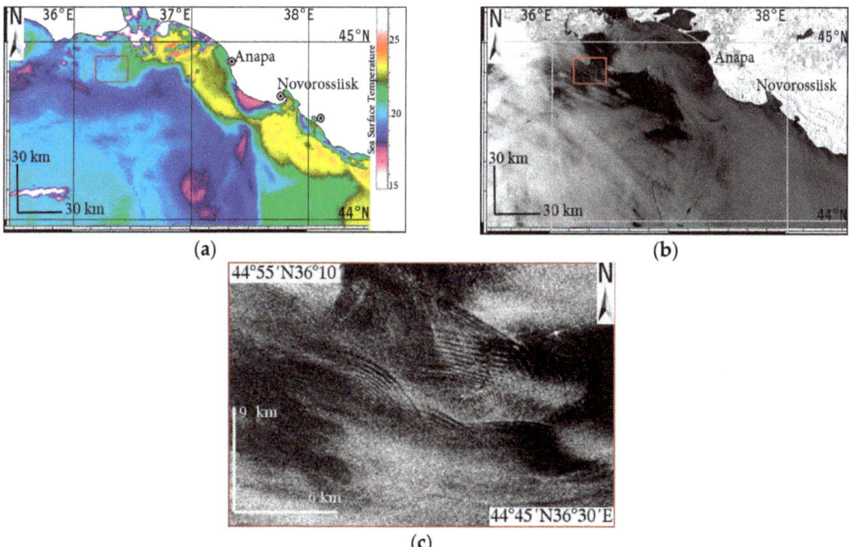


**Figure 17.** (**a**) Sea surface temperature chart from NOAA-15 data of 11 July 2006, 03:21 UTC; (**b**) Envisat ASAR, VV-polarization, image of 11 July 2006, 19:19 UTC; and (**c**) signatures of non-tidal IWs (zoom into red box in (**b**)).

Figure 18 presents an Envisat ASAR image acquired on 19 June 2006 at 19:10 UTC together with a water-leaving radiance chart derived from Aqua MODIS data obtained at a close time. Dark bands distinctly visible in the Envisat ASAR image correspond to the outer boundary of an eddy dipole visualized in the water-leaving radiance chart. In the Envisat ASAR image, the boundary is visible due to filamentary biogenic films accumulated in convergence zones of the surface water layer. Maximum contrasts delineate the eddy boundary in the water-leaving radiance chart. The SMIW detectable in the Envisat ASAR image is located close to the area of high contrast. The train consisting of over 20 waves with fronts parallel to the dipole jet propagates seaward at an angle to the shore. Most probably, the IW train is radiated by the oscillating dipole jet.

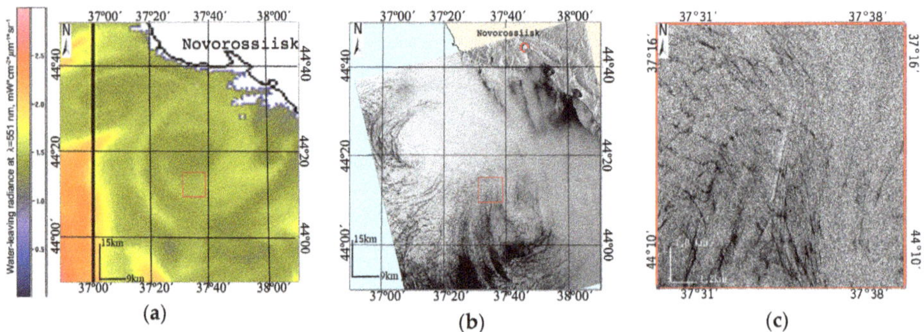


**Figure 18.** (**a**) Water-leaving radiance chart from Aqua MODIS data, wavelength 551 nm, 20 June 2006, 11:10 UTC; (**b**) mesoscale eddy dipole, Envisat ASAR, VV-polarization, 19 June 2006, 19:10 UTC; and (**c**) signature of non-tidal IWs located in close proximity to the central part of the eddy dipole (zoom into red box in (**b**). There are over 20 waves in the train. Leading wave crest length is 5600 m. Maximal wave length is 220 m. IWs propagate seaward at an angle to the coast with fronts parallel to the dipole jet.

As to the Caspian Sea, we can point out two of the most common sources of IW generation: (i) summer upwelling often observed near the east coast; and (ii) one-node seiches with nodes situated near Absheron Cape.

Summer upwelling plays a very important role for the Caspian Sea, radically changing water dynamics. It is observed every year along the east coast of Middle and parts of South Caspian. The rise of cold deep waters occurs when strong southeast winds set to push warm surface waters away from the coast. As a result, surface temperature decreases, horizontal temperature gradients reach 2.3 °C at the surface and 4.2 °C at a depth of 20 m. The center of the upwelling gradually moves from 41°–42°N in June to 43°–45°N in September.

Seiches occur as a result of rapid changes in atmospheric or wind pressure over the sea surface. Free seiche oscillations arise in Middle and South Caspian with predominating periods of 8.5–8.7 and 4.2–4.6 h. The nodal line of a longitudinal uninodal seiche extends approximately along the Absheron Sill separating Middle and South Caspian. Strong seiche currents bumping against the steep slope generate thermocline oscillations inducing IWs. The sill-formed IWs radiate to deeper sea, less affected by the bathymetry.

### 5.3. IWs and Look-Alikes in Non-Tidal Seas

Study of IWs in non-tidal seas based on satellite data is a complicated task. One of non-trivial problems is discrimination between the signatures of IWs generated in water and in the atmosphere above the sea surface.

Atmospheric internal gravity waves (AIWs) propagate all the time and everywhere in a stratified atmosphere. They have various origins: air flows past different obstacles generating so-called lee waves, interaction of continental and marine air masses, movement of cold atmospheric fronts, etc. In radar images of sea surface, we can observe AIWs when wind variations modulate sea surface roughness so as it results in backscatter modulation [65].

Radar signatures of oceanic IWs (OIWs) and AIWs are frequently confused. Both are shaped in alternating bands of enhanced and attenuated radar backscatter, similar in intensity. It is easier to discriminate them in regions where OIWs are generated by tides at shelf edge. Such OIWs have typical signatures of trains, as a rule, propagating toward the coast. Their localization and seasonal variability are well known [4]. Radar signatures of OIWs have high contrasts; bands of enhanced backscatter are extremely narrow, not more than 2–3 pixels (pixel resolution 75 m). Usually, AIWs propagate away from the coast (orography waves), or on the lee side of islands, or are associated with passage of an atmospheric front. AIWs have greater spatial dimensions: front lengths reach 200–300 km and more, wave lengths are seldom less than 1–2 km. However, our satellite monitoring shows that in inner seas, such as the Black and Caspian Seas, AIWs are manifested in a variety of patterns: from gigantic narrow solitary waves with front length reaching 600 km to small-scale trains containing more than 10 waves. Sometimes, the latter are very hard to differentiate from OIWs. An example is shown in Figure 19. Radar signatures of OIWs and AIWs are very similar here, but some distinctions are quite noticeable. OIWs are manifested as bright bands (suloys) with curved fronts (Figure 19a), while AIWs as parallel dark bands (slicks) (Figure 19b). Given the same resolution of both images (75 m), the suloys are narrower than the slicks. The signatures of AIWs appear less sharp, which is confirmed by the inset graphs of normalized radar cross section (NRCS) along the black lines.

Obviously, to reliably discriminate between OIWs and AIWs in enclosed non-tidal seas, in addition to comparing radar contrasts and spatial properties, we need to consider all available hydrometeorological information, data on sea surface state, sea floor bathymetry and coastal orography. The best approach would be to employ atmospheric sounding and compute the Scorer parameter, as it was made in [66]. Unfortunately, radiosonde observations are hardly affordable for most researchers. At present, the final decision is up to the human expert who analyzes radar data and all complementary information. The problem of discriminating surface manifestations of OIWs and AIWs in radar imagery is certainly a topic of an individual comprehensive consideration and cannot be examined within the present work, it is only mentioned here for the sake of completeness of the presentation. A detailed discussion of discrimination between OIWs and AIWs in radar data can be found in [67]. Despite the fact that the paper considers seas with tides, the principal conclusions agree with our results: the main discrimination criteria are based on the shape and structure of radar signatures. The authors also present useful findings on polarization differences of the signatures. AIWs are better pronounced in HH polarization and OIWs in VV polarization. We plan to further investigate this aspect based on radar observations of the Caspian Sea, where, unlike the Black Sea, OIW are frequently manifested.

Another problem of SMIWs studies is their differentiating from fine current structures within submesoscale eddies. It is well known that, in radar images, submesoscale eddies are primarily visualized by slicks [25,68]. At eddy edges, however, slicks may be shaped in quasi periodic structures similar to IW trains.

We hope that, with the Sentinel program in place, the steady flow of high resolution radar data that the sensors provide will facilitate important advances in the study of such infrequent phenomenon as IWs in non-tidal seas.

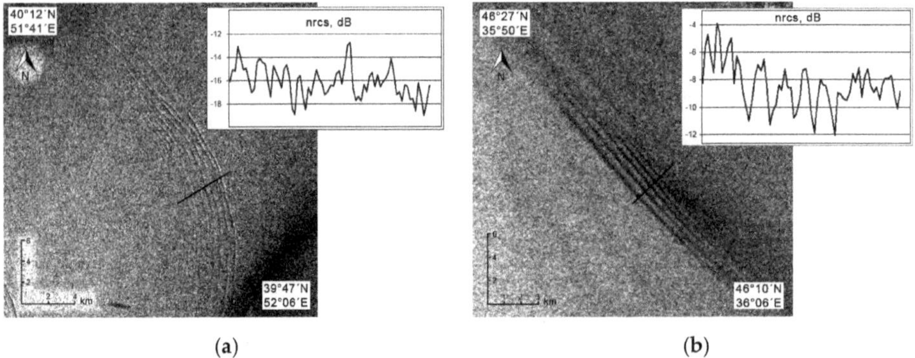


(a) (b)

**Figure 19.** Comparison of radar signatures of OIW and AIW: (**a**) OIW, Envisat ASAR, HH-polarization, 24 July 2009; and (**b**) AIW, Envisat ASAR, VV-polarization, 14 March 2009. The inset graphs show variations in radar signal along the cross sections of the trains (black lines).

## 6. Conclusions

We present results of our study of SMIWs in the Black and Caspian Seas. In these seas, no significant tides can ever develop, so they can be considered non-tidal. Consequently, the main mechanism of IW generation in ocean—interaction of tidal wave with bathymetry—is irrelevant.

Analysis of satellite data, primarily radar imagery (ERS-2 SAR, Envisat ASAR, and Sentinel-1 SAR-C) allowed us to identify regions of most regular SMIWs in these seas. In the Black Sea, there are four such regions: (I) near the Danube Delta on the northwest; (II) near the Crimean Peninsula; (III) near the city of Novorossiysk on the northeast; and (IV) the shelf of Georgia on the east. In the Caspian Sea, SMIWs have distinct seasonal variability. In the end of May and in the first half of June, SMIWs can be found only in the western part of South Caspian (I); in the second part of June and in July, all SMIWs are observed on the east, near the Absheron Sill (II); and, in August, they appear northeast of the Absheron Sill (III).

A comparative analysis of the observed non-tidal IWs was performed. A strong diversity of SMIW forms, propagation directions and types of surface wave modulation was revealed which implied different mechanisms of IW generation. For each region of regular SMIWs, the factors facilitating generation of non-tidal IWs were identified via joint analysis of radar, VIS and NIR satellite data and corresponding hydrometeorological information. For instance, in the eastern and western coastal zones of the Black Sea, where large rivers disembogue, intrusions of fresh water create hydrological fronts able to generate IWs. At continental shelf edge, on the west and northwest of the Black Sea and near Crimea, the main IW generation mechanisms are: relaxation of coastal upwelling, inertial oscillations associated with hydrological fronts, and storm surges. For the first time, we discovered IWs formed at the fronts associated with passage of cold eddies. Processes of the type we observed several times in the northwestern Black Sea. In the Caspian Sea, seiches were the main source of IWs that we observed in satellite images.

The relation between occurrence of SMIWs and position of the pycnocline peak was also established. Sharp and shallow pycnocline was shown to facilitate generation of IWs as well as enhance near-surface currents associated with IWs.

In conclusion, we should note that statistically significant estimates of spatial and temporal distribution of surface manifestations of non-tidal IWs in the Black and Caspian Seas have been obtained for the first time.

**Acknowledgments:** This research is supported by the Russian Science Foundation under the project # 14-17-00555. Basic functionality of the STS portal is implemented in frame of Theme "Monitoring", state register No. 01.20.0.2.00164 (FASO Russia). The authors are grateful to late Konstantin D. Sabinin for useful discussions and help in investigating SMIWs, to Andrey N. Serebryany for help in IW nonlinearity calculations (both from N.N. Andreev Acoustics Institute, Moscow, Russia) and to Tatiana Yu. Bocharova (Space Research Institute, Moscow, Russia) for the assistance in satellite data processing. We would like to thank two anonymous reviewers for their valuable comments and suggestions.

**Author Contributions:** Both authors contributed equally to the paper. Olga Lavrova and Marina Mityagina analyzed data, contributed with ideas and discussions, and wrote the manuscript.

**Conflicts of Interest:** The authors declare no conflict of interest.

## References

1. LeBlond, P.H.; Mysak, L.A. *Waves in the Ocean*; Elsevier: Amsterdam, The Netherlands, 1978; pp. 1–602.
2. Hughes, B.A. The effect of internal waves on surface wind waves. 2. Theoretical analysis. *J. Geophys. Res.* **1978**, *83*, 455–469. [CrossRef]
3. Alpers, W.; Salusti, E. Scylla and Charybdis observed from space. *J. Geophys. Res.* **1983**, *88*, 1800–1808. [CrossRef]
4. Jackson, C.; Apel, J. An Atlas of Internal Solitary-like Waves and their Properties. Second Edition. Available online: http://www.internalwaveatlas.com/Atlas2_index.html (accessed on 22 June 2017).
5. Zhao, Z.; Liu, B.; Li, X. Internal solitary waves in the China seas observed using remote-sensing techniques: A review and perspectives. *Int. J. Remote Sens.* **2014**, *35*, 3926–3946. [CrossRef]
6. Klemas, V. Remote sensing of ocean internal waves: An overview. *J. Coast. Res.* **2012**, *28*, 540–546. [CrossRef]
7. Prasad, K.; Rajasekhar, M. Space borne SAR observations of oceanic internal waves in North Bay of Bengal. *Nat. Hazards* **2011**, *57*, 657–667. [CrossRef]
8. Lorenzzetti, J.A.; Dias, F.G. Internal solitary waves in the Brazilian SE continental shelf: Observations by synthetic aperture radar. *Int. J. Oceanogr.* **2013**, *2013*, 403259. [CrossRef]
9. Da Silva, J.C.B.; New, A.L.; Magalhaes, J.M. On the structure and propagation of internal solitary waves generated at the Mascarene Plateau in the Indian Ocean. *Deep Sea Res. I* **2011**, *58*, 229–240. [CrossRef]
10. Da Silva, J.C.B.; Buijsman, M.C.; Magalhaes, J.M. Internal waves on the upstream side of a large sill of the Mascarene Ridge: A comprehensive view of their generation mechanisms and evolution. *Deep Sea Res. I* **2015**, *99*, 87–104. [CrossRef]
11. Hyder, P.; Jeans, D.R.G.; Cauquil, E.; Nerzic, R. Observations and predictability of internal solitons in the Northern Andaman Sea. *Appl. Ocean Res.* **2005**, *27*, 1–11. [CrossRef]
12. Alpers, W.; Brandt, P.; Rubino, A. Internal Waves Generated in the Straits of Gibraltar and Messina: Observations from Space. In *Remote Sensing of the European Seas*; Barale, V., Gade, M., Eds.; Springer: Heidelberg, Germany, 2008; pp. 319–330.
13. Susanto, R.D.; Mitnik, L.; Zheng, Q. Ocean internal waves observed in the Lombok Strait. *Oceanography* **2005**, *18*, 80–87. [CrossRef]
14. Liu, A.K.; Su, F.C.; Hsu, M.K.; Kuo, N.J.; Ho, C.R. Generation and evolution of mode-two internal waves in the South China Sea. *Cont. Shelf Res.* **2013**, *59*, 18–27. [CrossRef]
15. Li, X.; Jackson, C.R.; Pichel, W.G. Internal solitary wave refraction at Dongsha Atoll, South China Sea. *Geophys. Res. Lett.* **2013**, *40*, 3128–3132. [CrossRef]
16. Ivanov, V.A.; Serebryany, A.N. Short-period internal waves in the coastal zone of a non-tidal sea. *Izv. Atmos. Ocean. Phys.* **1985**, *21*, 648–656.
17. Vlasenko, V.I.; Ivanov, V.A.; Krasin, I.G.; Lisichenok, A.D. Generation of intense short-period internal waves in the Crimean shelf zone during an event of coastal upwelling. *Mar. Hydrophys. J.* **1997**, *3*, 3–16.
18. Bidokhti, A.A.; Shekarbaghani, A. The role of internal waves in the formation of layered structure at exchange flows between two closed basins (Middle and southern basins of the Caspian Sea). *Iran. J. Phys. Res.* **2006**, *5*, 99–104.
19. Filatov, N.; Terzevik, A.; Zdorovennov, R.; Vlasenko, V.; Stashchuk, N.; Hutter, K. Field Studies of Non-Linear Internal Waves in Lakes on the Globe. In *Nonlinear Internal Waves in Lakes*; Hutter, K., Ed.; Springer: Berlin, Germany, 2012; pp. 23–103.

20. Ivanov, A.Y.; Alpers, W.; Naumenko, M.A.; Karetnikov, S.G. Lake Ladoga surface features on the ERS—1 SAR imagery. In *ESA SP-414, Proceedings of the 3rd ERS Symposium, Florence, Italy, 14–21 March 1997*; ESA Publications Division: Noordwijk, The Netherlands, 1997; pp. 1035–1040.
21. Lavrova, O.Y.; Kostianoy, A.G.; Lebedev, S.A.; Mityagina, M.I.; Ginzburg, A.I.; Sheremet, N.A. *Complex Satellite Monitoring of the Russian Seas*; IKI RAN: Moscow, Russia, 2011. (In Russian)
22. Mityagina, M.; Lavrova, O. Surface manifestations of non-tidal internal waves in the North-Eastern Black Sea as viewed by satellite sensors. In Proceedings of the 2009 IEEE International Geoscience and Remote Sensing Symposium (IGARSS'09), Cape Town, South Africa, 12–17 July 2009.
23. Mityagina, M.I.; Lavrova, O.Y.; Karimova, S.S. Multi-sensor survey of seasonal variability in coastal eddy and internal wave signatures in the north-eastern Black Sea. *Int. J. Remote Sens.* **2010**, *31*, 4779–4790. [CrossRef]
24. Lavrova, O.Y.; Mityagina, M.I.; Serebryany, A.N.; Sabinin, K.D.; Kalashnikova, N.A.; Krayushkin, E.V.; Khymchenko, I. Internal waves in the Black Sea: Satellite observations and in-situ measurements. In Proceedings of the Remote Sensing of the Ocean, Sea Ice, Coastal Waters, and Large Water Regions 2014, Amsterdam, The Netherlands, 14 October 2014; Volume 9240.
25. Lavrova, O.Y.; Mityagina, M.I.; Sabinin, K.D.; Serebryany, A.N. Study of hydrodynamic processes in the shelf zone based on satellite data and subsatellite measurements. *Sovremennye Problemy Distantsionnogo Zondirovaniya Zemli iz Kosmosa* **2015**, *12*, 98–129. (In Russian)
26. Medvedev, I.P.; Rabinovich, A.B.; Kulikov, E.A. Tides in Three Enclosed Basins: The Baltic, Black, and Caspian Seas. *Front. Mar. Sci.* **2016**, *3*, 1–7. [CrossRef]
27. Kostianoy, A.G.; Kosarev, A.N. (Eds.) *The Caspian Sea Environment. The Handbook of Environmental Chemistry. Water Pollution, Part P*; Springer: Berlin/Heidelberg, Germany; New York, NY, USA, 2000; Volume 5, p. 271.
28. Medvedev, I.P.; Kulikov, E.A.; Rabinovich, A.B. Tidal Oscillations in the Caspian Sea. *Oceanology* **2017**, *57*, 360–375. [CrossRef]
29. Gunduz, M. Caspian Sea surface circulation variability inferred from satellite altimeter and sea surface temperature. *J. Geophys. Res.* **2014**, *119*, 1420–1430. [CrossRef]
30. Grinevetskiy, S.R.; Zonn, I.S.; Zhiltsov, S.S.; Kosarev, A.N.; Kostianoy, A.G. *The Black Sea Encyclopedia*; Springer: Berlin/Heidelberg, Germany; New York, NY, USA, 2015; p. 889.
31. Oguz, T.; Latun, V.S.; Latif, M.A.; Vladimirov, V.; Su, H.; Markov, A.; Ozsoy, E.; Kotovshchikov, B.; Eremeev, V.; Unluata, U. Circulation in surface and intermediate layers of the Black Sea. *Deep Sea Res. I* **1993**, *40*, 1597–1612. [CrossRef]
32. Stanev, E. Understanding Black Sea dynamics. An Overview of Recent Numerical Modeling. *Oceanography* **2005**, *18*, 56–75. [CrossRef]
33. Zatsepin, A.G.; Ginzburg, A.I.; Kostianoy, A.G.; Kremenetskiy, V.V.; Krivosheya, V.G.; Stanichny, S.V.; Poulain, P.M. Observations of Black Sea Mesoscale Eddies and Associated Horizontal Mixing. *J. Geophys. Res.* **2003**, *108*, 3246. [CrossRef]
34. Zatsepin, A.G.; Kremenetskiy, V.V.; Ostrovskii, A.G.; Baranov, V.I.; Kondrashov, A.A.; Korzh, A.O.; Soloviev, D.M. Submesoscale eddies at the Caucasus Black Sea shelf and the mechanisms of their generation. *Oceanology* **2011**, *51*, 554–567. [CrossRef]
35. Korotaev, G.K.; Oguz, T.; Nikiforov, A.; Koblinsky, C. Seasonal, interannual, and mesoscale variability of the Black Sea upper layer circulation derived from altimeter data. *J. Geophys. Res.* **2003**, *108*, 3122. [CrossRef]
36. Loupian, E.A.; Matveev, A.A.; Uvarov, I.A.; Bocharova, T.Y.; Lavrova, O.Y.; Mityagina, M.I. The satellite service See the Sea—A tool for the study of oceanic phenomena and processes. *Sovremennye Problemy Distantsionnogo Zondirovaniya Zemli iz Kosmosa* **2012**, *9*, 251–261. (In Russian)
37. Mityagina, M.I.; Lavrova, O.Y.; Uvarov, I.A. See the Sea: Multi-user information system for investigating processes and phenomena in coastal zones via satellite remotely sensed data, particularly hyperspectral data. In Proceedings of the SPIE 9240, Remote Sensing of the Ocean, Sea Ice, Coastal Waters, and Large Water Regions 2014, Amsterdam, The Netherlands, 22 September 2014.
38. Lighthill, M.J. *Waves in Fluids*, 1st ed.; Cambridge University Press: Cambridge, UK, 1978.
39. Alpers, W. Theory of radar imaging of internal waves. *Nature* **1985**, *314*, 245–247. [CrossRef]
40. Apel, J.R.; Gasparovic, R.F.; Thompson, D.R.; Gotwols, B.L. Signatures of Surface Wave/Internal Wave Interactions: Experiment and Theory. *Dyn. Atmos. Oceans* **1988**, *12*, 89–106. [CrossRef]
41. Gade, M.; Alpers, W.; Ermakov, S.A.; Hühnerfuss, H.; Lange, P.A. Wind-wave tank measurements of bound and freely propagating short gravity-capillary waves. *J. Geophys. Res.* **1998**, *103*, 21697–21710. [CrossRef]

42. Ebuchi, N.; Kawamura, H.; Toba, Y. Fine structure of laboratory wind-wave surfaces studied using an optical method. *Bound. Layer Meteorol.* **1987**, *39*, 133–151. [CrossRef]
43. Gorshkov, K.A.; Dolina, I.S.; Soustova, I.A.; Troitskaya, Y.I. Modulation of short wind waves in the presence of strong internal waves: The effect of growth-rate modulation. *Izv. Atmos. Ocean. Phys.* **2003**, *39*, 596–606.
44. Churyumov, A.N.; Kravtsov, Y.A.; Lavrova, O.Y.; Litovchenko, K.T.; Mityagina, M.I.; Sabinin, K.D. Signatures of resonant and non-resonant scattering mechanisms on radar images of internal waves. *Int. J. Remote Sens.* **2002**, *23*, 4341–4355. [CrossRef]
45. Kravtsov, Y.A.; Mityagina, M.I.; Churyumov, A.N. Electromagnetic waves backscattering by mesoscale breaking waves on the sea surface. *Uspekhi Fizicheskikh Nauk* **1999**, *63*, 1859–1865.
46. Voronovich, A.G.; Zavorotny, V.U. Theoretical model for scattering of radar signals in Ku- and C-bands from a rough sea surface with breaking waves. *Wave Random Media* **2001**, *11*, 247–269. [CrossRef]
47. Kudryavtsev, V.; Akimov, D.; Johannessen, J.; Chapron, B. On radar imaging of current features: 1. Model and comparison with observations. *J. Geophys. Res. Oceans* **2005**, *110*, 691–706. [CrossRef]
48. Ermakov, S.A.; da Silva, J.C.; Robinson, I.S. The role of surface films in SAR signatures of internal waves on the shelf. II. Internal tidal waves. *J. Geophys. Res.* **1998**, *103*, 8033–8043. [CrossRef]
49. Da Silva, J.C.B.; Ermakov, S.A.; Robinson, I.S. Role of surface films in ERS SAR signatures of internal waves on the shelf: 3.Mode transitions. *J. Geophys. Res.* **2000**, *105*, 24089–24104. [CrossRef]
50. Lavrova, O.U.; Mityagina, M.I.; Sabinin, K.D.; Serebryany, A.N. Satellite Observations of Surface Manifestations of Internal Waves in the Caspian Sea. *Izv. Atmos. Ocean. Phys.* **2011**, *47*, 1119–1126. [CrossRef]
51. Nash, J.D.; Moum, J.N. River plumes as a source of large-amplitude internal waves in the coastal ocean. *Nature* **2005**, *437*, 400–403. [CrossRef] [PubMed]
52. Stashchuk, N.; Vlasenko, V. Generation of internal waves by a supercritical stratified plume. *J. Geophys. Res. Oceans* **2009**, *114*, 362–370. [CrossRef]
53. Pan, J.; Jay, D.A.; Orton, P.M. Analyses of internal solitary waves generated at the Columbia River plume front using SAR imagery. *J. Geophys. Res. Oceans* **2007**, *112*, 1835–1855. [CrossRef]
54. Ivanov, V.A.; Serebryanyi, A.N. Vnutrennie volny na melkovodnom shel'fe besprilivnogo moray. *Izv. Akad. Nauk SSSR Fiz. Atmos. Okeana* **1983**, *19*, 661–665.
55. Pao, H.P.; Serebryany, A.N. Studies of intense internal gravity waves: Field measurements and numerical modeling. In *Advances in Engineering Mechanics Reflections and Outlooks*; In Honor of Theodore T-Y Wu; Chuang, A.T., Teng, M.H., Valentine, D.T., Eds.; World Scientific: New Jersey, NJ, USA, 2005; pp. 286–296.
56. Smyth, W.D.; Moum, J.N.; Nash, J.D. Narrowband oscillations in the upper equatorial ocean: Part II. Properties of shear instabilities. *J. Phys. Oceanogr.* **2011**, *41*, 412–428. [CrossRef]
57. Grimshaw, R. Korteweg de-Vries Equation. In *Nonlinear Waves in Fluids: Recent Advances and Modern Applications*; CISM International Centre for Mechanical Sciences: Udine, Italy, 2005; Volume 483, pp. 1–28.
58. Boegman, L. Currents in Stratified Water Bodies 2: Internal Waves. In *Encyclopedia of Inland Waters*; Likens, G.E., Ed.; Elsevier: Oxford, UK, 2009; Volume 1, pp. 539–558.
59. Serebryany, A.N. Nonlinearity effects in internal waves on the shelf. *Izv. Akad. Nauk SSSR Fiz. Atmos. Okeana* **1990**, *26*, 285–293.
60. Zheng, Q.; Vic Klemas, V.; Yan, X.-H.; Pan, J. Nonlinear evolution of ocean internal solitons propagating along an inhomogeneous thermocline. *J. Geophys. Res. Oceans* **2001**, *106*, 14083–14094. [CrossRef]
61. Sabinin, K.; Serebryany, A. Intense short period internal waves in the Ocean. *J. Mar. Res.* **2005**, *63*, 227–261. [CrossRef]
62. Mityagina, M.; Lavrova, O. Comparative analysis of internal waves surface manifestations observed by Envisat ASAR in different non-tidal seas. In *ESA Special Publication SP-686, Proceedings of ESA Living Planet Symposium, Bergen, Norway, 28 June–2 July 2010*; Lacoste-Francis, H., Ed.; ESA Publications Division: Noordwijk, The Netherlands, 2010.
63. Jackson, C.; da Silva, J.; Jeans, G. The generation of nonlinear internal waves. *Oceanography* **2012**, *25*, 108–123. [CrossRef]
64. Serebryany, A.N. Manifestation of the properties of solitons in the internal waves on the shelf. *Izv. Akad. Nauk Fiz. Atmos. Okeana* **1993**, *29*, 244–252.

65. Bulatov, M.G.; Kravtsov, Y.A.; Lavrova, O.Y.; Litovchenko, K.T.; Mityagina, M.I.; Raev, M.D.; Sabinin, K.D.; Trokhimovskii, Y.G.; Churyumov, A.N.; Shugan, I.V. Physical Mechanisms of Aerospace Radar Imaging of the Ocean. *Uspekhi Fizicheskikh Nauk* **2003**, *173*, 69–87. [CrossRef]
66. Da Silva, J.C.B.; Magalhaes, J.M. Satellites observations of large atmospheric gravity waves in the Mozambique Channel. *Int. J. Remote Sens.* **2009**, *30*, 1161–1182. [CrossRef]
67. Alpers, W.; Huang, W. On the discrimination of radar signatures of atmospheric gravity waves and oceanic internal waves on synthetic aperture radar images of the sea surface. *IEEE Trans. Geosci. Remote Sens.* **2011**, *49*, 1114–1126. [CrossRef]
68. Munk, W.; Armi, L.; Fisher, K.; Zachariasen, F. Spirals on the Sea. *Proc. R. Soc.* **2000**, *456*, 1217–1280. [CrossRef]

*Article*

# Contextual Region-Based Convolutional Neural Network with Multilayer Fusion for SAR Ship Detection

**Miao Kang, Kefeng Ji *, Xiangguang Leng and Zhao Lin**

School of Electronic Science and Engineering, National University of Defense Technology, Sanyi Avenue, Changsha 410073, China; kangmiao15@163.com (M.K.); luckight@163.com (X.L.); lzkmylz@163.com (Z.L.)
* Correspondence: jikefeng@nudt.edu.cn; Tel.: +86-731-8457-6384

Academic Editor: Xiaofeng Yang
Received: 21 July 2017; Accepted: 9 August 2017; Published: 20 August 2017

**Abstract:** Synthetic aperture radar (SAR) ship detection has been playing an increasingly essential role in marine monitoring in recent years. The lack of detailed information about ships in wide swath SAR imagery poses difficulty for traditional methods in exploring effective features for ship discrimination. Being capable of feature representation, deep neural networks have achieved dramatic progress in object detection recently. However, most of them suffer from the missing detection of small-sized targets, which means that few of them are able to be employed directly in SAR ship detection tasks. This paper discloses an elaborately designed deep hierarchical network, namely a contextual region-based convolutional neural network with multilayer fusion, for SAR ship detection, which is composed of a region proposal network (RPN) with high network resolution and an object detection network with contextual features. Instead of using low-resolution feature maps from a single layer for proposal generation in a RPN, the proposed method employs an intermediate layer combined with a downscaled shallow layer and an up-sampled deep layer to produce region proposals. In the object detection network, the region proposals are projected onto multiple layers with region of interest (ROI) pooling to extract the corresponding ROI features and contextual features around the ROI. After normalization and rescaling, they are subsequently concatenated into an integrated feature vector for final outputs. The proposed framework fuses the deep semantic and shallow high-resolution features, improving the detection performance for small-sized ships. The additional contextual features provide complementary information for classification and help to rule out false alarms. Experiments based on the Sentinel-1 dataset, which contains twenty-seven SAR images with 7986 labeled ships, verify that the proposed method achieves an excellent performance in SAR ship detection.

**Keywords:** context information; convolutional neural network (CNN); ship detection; synthetic aperture radar (SAR); Sentinel-1

## 1. Introduction

With the rapid development of spaceborne SAR, such as TerraSAR-X, RADARSAT-2 and Sentinel-1 [1–3], synthetic aperture radar (SAR) ship detection has been playing an increasingly essential role in marine monitoring and maritime traffic supervision [4–6]. Many investigations relating to ship detection in SAR imagery have been carried out. Traditional methods [7–9] detect targets after sea–land segmentation and utilize the hand-crafted features for discrimination, which has poor performance on nearshore areas and has difficulty ruling out false alarms, such as icebergs and small islands. Additionally, the existence of speckle noises and motion blurring in SAR images causes undesirable differences between ships, which creates difficulty for traditional SAR ship detection methods in extracting effective features for discrimination. Therefore, it is necessary to develop detectors with strong feature extraction capabilities to obtain better performances in SAR ship detection.

Deep neural networks are capable of feature representation and have been widely applied for object detection [10,11]. They provide a highly promising approach for end-to-end object detection. Since the breakthroughs made by the region-based convolutional neural network (R-CNN) [12] using the PASCAL VOC dataset, the process followed by a region-based proposal extractor with a detection network has been intensively investigated in recent years [13,14]. Ren et al. [15], introduced a Region Proposal Network (RPN) to replace the typical region proposal methods, which achieves end-to-end object detection and shares full-image convolutional features with a RPN and Fast R-CNN. Deep transfer learning algorithms [16–18], which tune the model with rich labeled source domains and small-scale target domains, are widely used to reduce the demand of labeled data and accelerate the convergence of networks.

Despite being capable of extracting discriminative representation, the sharing CNN has a tradeoff between the spatial resolution of the network and the semantic distinction of features. Specifically, the shallow layers of CNN have a higher spatial resolution but more coarse features. The feature maps of intermediate layers are complementary with a passable resolution. Moreover, with the depth of layer increasing, the feature map becomes highly semantic but abstract. Taking VGG16 [19] for example, a $32 \times 32$ pixel object will shrink to $2 \times 2$ when it comes to the last convolutional layer. In general, the mean area of the majority of ships on SAR images from Sentinel-1 is smaller than $32 \times 32$, which means that the ship detection on Sentinel-1 belongs to small-sized object detection. Therefore, when the bounding box predictions map to the last feature maps by ROI pooling, small-sized objects have little information for location refinement and classification, which naturally degrades the performance of detection.

In order to cover the shortage of small-sized object detection, experiments have been conducted by utilizing the different layers of CNN. SSD [14], MS-CNN [20] and FPN [21] predict objects on multiple layers and fuse the output in the end, which also consumes more time for training and testing. Tao Kong et al. [22] proposed a HyperNet to incorporate the intermediate layer with the downscaling shallow layer and up-sampled deep layers, and compress them into a uniform space, which obtained a comprehensive and relatively high resolution framework. MultiPath Network [23] and U-Net [24] utilize skip connection between different layers to provide better feature representation at the cost of a complex network structure.

Another way to improve the performance is to add contextual features for small-sized objects. Research shows that contextual information around the objects in input images can provide a valuable cue for object detection [25,26]. Especially for ship detection, the ocean surroundings can help detectors to better rule out false alarms on land. Thus, adding context information to deep object detection networks is a way to improve their distinction of small-sized ship detection. In ParseNet [27], global context features are appended to help clarify the local confusion. With the contextual information about the whole image, it has limited effects on object detection. Inside–Outside Net (ION) [28] integrated the contextual information outside the region of interest by using spatial recurrent neural networks with multiple layer feature maps. In order to obtain a better performance, the IRNN which is Recurrent Neural Networks with ReLU recurrent transitions, needs to be trained on extra semantic segmentation labels, which increases the difficulty of training. Chenchen Zhu et al. [29] presented a face object detection network named CMS-RCNN, which combined multi-scale information with body contextual information, for real-world face detection. However, this approach only builds fused feature maps, which have the same resolution as the deepest layer and the small-sized objects have little information for bounding box prediction.

This paper proposes a contextual convolutional neural network with multilayer fusion for SAR ship detection. Similar to R-CNN, the proposed network is composed of a RPN with high resolution and an object detection network with contextual features. Instead of using low-resolution feature maps from a single layer for proposal generation, the proposed method employs an intermediate layer combined with a downscaled shallow layer and an up-sampled deep layer to predict the bounding box. In this way, the fused feature maps integrate semantic, complementary, and high-resolution CNN features. The spatial resolution of a RPN is raised to the same level as the intermediate layer, which enlarges the response area of small-sized ships in feature maps. In the object detection network, region proposals are projected onto multiple layers with ROI pooling to extract the corresponding features. Contextual features around the ROI contain the environmental information of candidates, which can complement the computation of a confidence score and bounding box regression.

The rest of this paper is organized as follows. Section 2 introduces the details of the proposed method. Section 3 presents three experiments conducted on Sentinel-1 dataset to validate the effectiveness of the proposed framework. Section 4 discusses the results of the proposed method. Finally, Section 5 concludes this paper.

## 2. Proposed Method

In order to improve the performance of ship detection, the proposed network consists of a RPN with higher resolution and an object detection network with contextual features. As is shown in Figure 1, in a RPN, 13 convolutional layers of VGG16 [19] are employed for shared feature extraction. All convolutional layers adopt very small 3 × 3 filters in order to reduce the number of parameters and to decrease the demand of labeled data. In this paper, conv1_2, conv2_2, conv3_3, conv4_3 and conv5_3 of VGG16 are called conv1, conv2, conv3, conv4 and conv5 respectively. In order to improve the resolution of the network, a shallow layer and a deep layer ("conv1" and "conv3") are downscaled with max pooling and up-sampled with deconvolution respectively. Then, they are concatenated with the intermediate layer ("conv2") and compressed into a uniform space with $l_2$ normalization [26], which obtains the same resolution of the intermediate layer and more detailed information for region generation. The fused cube is reshaped to the same dimension as the intermediate layer and fed into bounding box regression for the sake of the region proposal. The predicted bounding boxes are mapped to different layers of VGG16 by ROI-pooling operations to obtain ROI features. Simultaneously, contextual features around each ROI are extracted. After normalization, concatenation and dimension reduction, the ROI features and contextual features are imported into two fully connected layers ("fc"). Finally, two flattened vectors are concatenated for classification and location refinement.

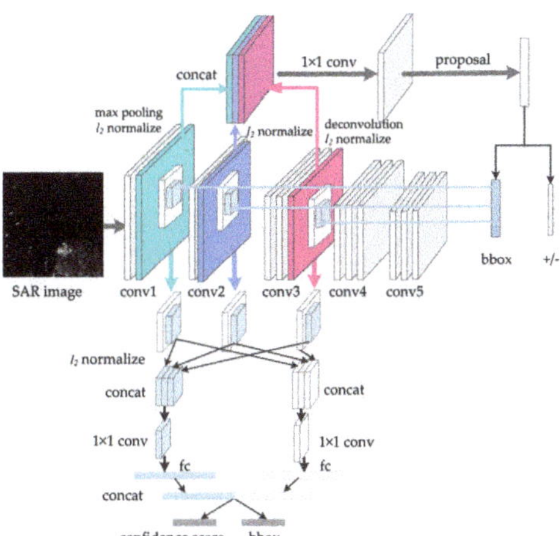


**Figure 1.** The architecture of the proposed network. SAR images are fed into VGG16 which has five sets of layers for feature extraction. The upper part is the RPN of the network. The white blocks and light blue blocks represent contextual features and ROI features respectively, which are processed concurrently in the object detection network.

The rest of this section introduces the details of the proposed method and explains the motivation of our design.

## 2.1. Concatenation of Multiple Layers

In order to reduce the number of parameters in the neural network, CNN always shrinks its feature maps by using the max pooling operation after convolution. That is, one pixel on the feature map corresponds to several pixels in the input image and the numerical correspondence is defined as the resolution of a network. Some feature maps are displayed in Figure 2, which shows that in VGG16 shallow layers keep more details of the input image. With the increase of resolution, the feature map becomes smaller and more abstract, and while small-sized objects hardly have responses on the deeper layers.

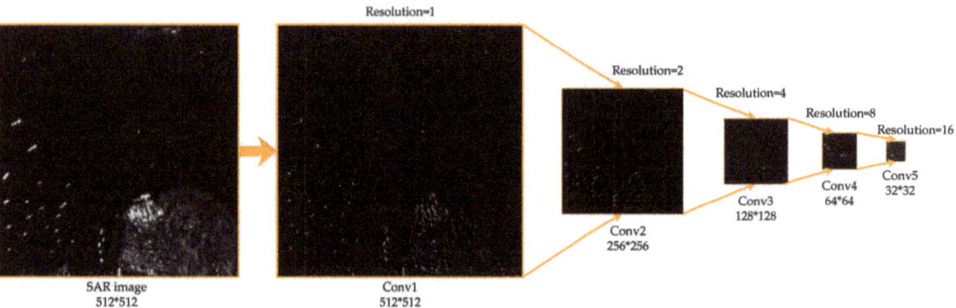


**Figure 2.** The feature maps of different layers in VGG16. With the increasing of resolution, the feature map becomes smaller and more semantic.

Due to the respective merits that different layers possess, multiple layers fusion is a popular way to enhance the performance of detection in the current top-performance detector. As CMS-RCNN [29] did, the first way is to integrate down-sampled earlier layers with the last layer of the sharing CNN. Despite the fact that the feature map information is increased, small-sized objects still only cause responses on a tiny area in a fused feature map. Another way is to increase the resolution of a network by up-sampling the deeper layer and connecting them with the shallower layer as proposed in this paper. With the integration of conv1, conv2 and conv3, the resolution of network changes from 16 to 2, which means a $32 \times 32$ sized object in the input image will have a $16 \times 16$ sized response on the fused feature map rather than a $2 \times 2$ sized response. The increase of resolution will naturally provide more detailed information for the following bounding box prediction.

## 2.2. Layer Up-Sampling with Deconvolution

Deconvolution, also known as transposed convolution, is extensively used in feature visualization, image generation and up-sampling [21,30,31]. Since a naive up-sampling inadvertently loses details, for feature map rescaling, a better option is to have a trainable up-sampling convolutional layer, namely a deconvolution layer [21,22], whose parameters will change during training. The implementation of deconvolution consists of two operations as shown in Figure 3. The first step is to insert zeros between the consecutive inputs according to the resolution requirements. After that, an operation similar to convolution is conducted, that is, defining a kernel of an appropriate size and sliding it with a stride to get a higher resolution output compared with the inputs. Since such an operation simply reverses the forward and backward passes of convolution, up-sampling with deconvolution is able to be performed in-network for end-to-end learning by backpropagation.

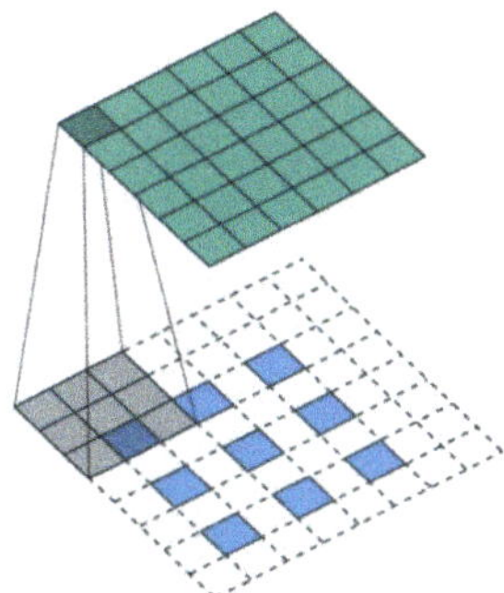

**Figure 3.** An illustration of deconvolution. It consists of inserting zeros and convolution operation [32].

### 2.3. $L_2$ Normalization

In general, with the depth of the network increasing, the scale and norm of feature maps always have a tendency to decrease. Concatenating multiple feature maps directly will lead to the dominance of shallow layers [27] and degrade the generalization ability of the model. Although the weights of layers are able to be tuned during the training, it takes a long time for the network to fill the dramatic gap in scale of value and it requires elaborate tricks to achieve a good performance. With the limited labeled data, overtraining will make the model learn the detail and noises in the training data and put the network at risk of overfitting. Therefore, a desirable approach is to employ $l_2$ normalization to constrain the scale of value of the different feature maps to the same level before integration.

$l_2$ normalization is applied to every pixel of the feature maps. For a layer that has $d$-channel feature maps sized with $(w, h)$, $l_2$-norm for a $d$-channel vector is represented with Equation (1)

$$\|\mathbf{x}\|_2 = \left(\sum_{i=1}^{d} |x_i|^2\right)^{1/2} \tag{1}$$

Per $d$-dimension pixels vector $\mathbf{x}$ of a layer is normalized as in Equation (2)

$$\hat{\mathbf{x}} = \frac{\mathbf{x}}{\|\mathbf{x}\|_2} \tag{2}$$

where $\hat{\mathbf{x}}$ is the $d$-dimension normalized pixels vector.

In order to accelerate the training, the scale value of layers is always rescaled with a factor $\mu$ for each channel $i$.

$$y_i = \mu x_i \tag{3}$$

The scale factor $\mu$ is able to be updated with the backpropagation and chain rule [29]. In this paper, a fixed scale factor, which makes the fused feature maps have the same mean level as the replaced layer in Faster RCNN, is adopted [28].

### 2.4. ROI Pooling to Multiple Layers

In Faster R-CNN, the prediction of the bounding box will be projected onto the last convolutional layer. Since region proposals are extracted from the fusion layer in this paper, projecting the region proposals to appropriate layers and fusing the region features as the fused layer will generate more accurate and comprehensive features for classification.

As shown in Figure 1, bounding box predictions are mapped to conv1, conv2 and conv3 respectively instead of a single layer. The corresponding regions on feature maps are normalized, rescaled and fused together.

### 2.5. Integrating Contextual Information

When searching for ships in a SAR image with the visual system of a human, context information is able to help us to increase the confidence of decision. For instance, an object located on land is highly unlikely to be considered a ship, while an object with bright intensity in the ocean area is prone to be affirmed as a positive object. In order to mimic the visual effect of a human being in a computer vision field, context information is always added into the deep neural network to recognize the small-sized objects [27,29,33].

As shown in Figure 4, the proposed method takes the surrounding pixels of the proposal as context information. In order to keep the same quantitative relation when the bounding boxes are projected to multiple layers to obtain contextual features, we keep $w_{context} = \lambda \times w_{proposal}$ and $h_{context} = \lambda \times h_{proposal}$, where $w$ and $h$ represent the width and height of the bounding box. After $l_2$ normalization and concatenation, the contextual features are flattened to a vector in the fully connected layers, which are combined with ROI features in a new fused vector for the final output.

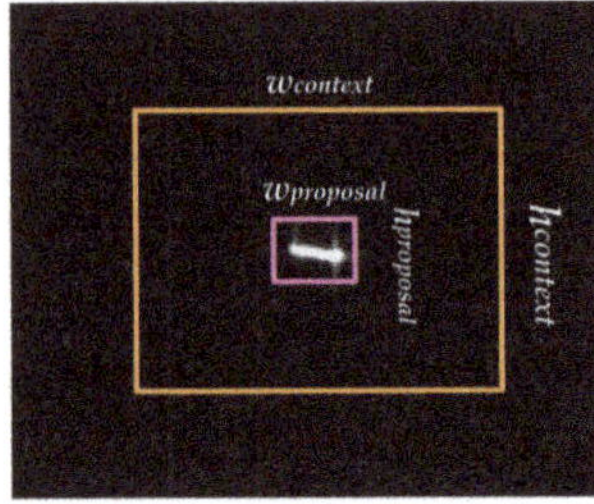


**Figure 4.** An illustration of context information. The bounding box in purple represents the region proposal of the network and the outer orange bounding box is the boundary of context information.

## 3. Experiments and Results

In this section, experiments are carried out to evaluate the performance of the proposed method. Two experiments are designed to explore the effect of different layer fusions and the influence of contextual features. Besides, the comparison with other methods indicates the outperformance of the proposed method.

### 3.1. Experiment Dataset and Settings

The dataset used in this paper is Sentinel-1, provided by the European Space Agency (ESA) on the Internet [34] for free, which was collected in Interferometric Wide swath (IW) mode. Compared with Extra-Wide swath (EW) mode, IW mode, as the main operational mode of Sentinel-1, is able to acquire more and higher resolution images. Full resolution Level-1 Ground Range Detected (GRD) products with 10 m pixel spacing were obtained. We labeled the location and the box of the ships on SAR images with ship detection software [35] and visual interpretation. Some of them were verified with Automatic Identification System (AIS) information [36]. Twenty-seven SAR images with 7986 labeled ships were utilized in this paper and seven of them, containing 1502 ships, were used for testing. Five-sixths and one-sixth of the remainders were used for training and validation sets respectively.

The histogram of the ship area is shown in Figure 5, according to the labeled ships that provided AIS information. More than 85% of ships have an area smaller than 8000 m$^2$, that is, around 80 pixels on a SAR image, which is less than the object size of the ImageNet dataset (more than 80% of objects have sizes between 40 and 140 pixels) [33]. Additionally, the ships which offer AIS information have an average length of 168.3 m. Furthermore, the average area is around 51 pixels which is far less than the area that is able to cause a response on the last convolutional layer of VGG16.

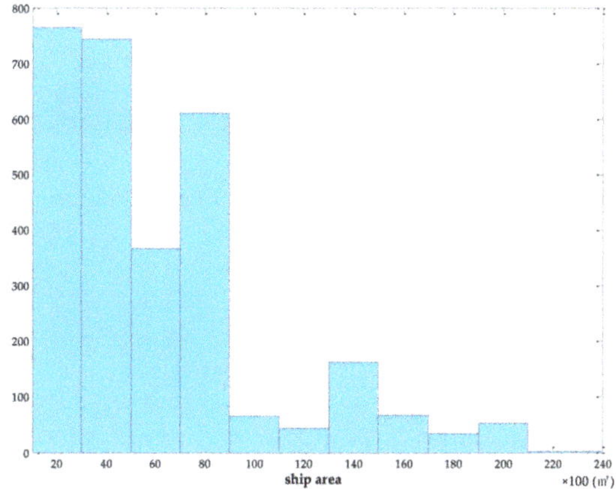


**Figure 5.** The distribution of the ship area in the Sentinel-1 dataset (only the ships have Automatic Identification System (AIS) information). More than 85% of ships have an area smaller than 80 pixels on SAR images.

The labeled SAR images were cut into 512 $\times$ 512 sized patches without overlap and the coordinates of the labeled bounding boxes were transformed into the location of the corresponding patch. Those patches with the labeled ship were selected to feed into the proposed network for training. The testing images are processed in the same way and are combined together for the detection result display.

All experiments are implemented in the Tensorflow deep learning framework [37] and are executed on a PC with an Intel single Core i7 CPU, NVIDIA GTX-1070 GPU (8 GB video memory), and 64 GB RAM. The PC operating system was Ubuntu 14.04.

As is common practice, the pre-trained model on the ImageNet dataset of VGG16 was used to initialize the model. According to the calculation of the mean norm of conv5 and pool5 of Faster RCNN, which is trained on the Sentinel-1 dataset, the scale factor $\mu$ for a RPN and object detection network is initialized to 20 and 40 respectively. The learning rate was set to $1 \times 10^{-4}$ initially and the maximal iteration was 10,000.

At the same time, we define the target detection probability as

$$p_d = \frac{N_{td}}{N_{ground\_truth}} \tag{4}$$

where $N_{td}$ is the number of detected targets and $N_{ground\_truth}$ denotes the total number of ground truth and in this paper we have $N_{ground\_truth}$ = 1502. Similarly, the estimation of the false alarm probability is defined as (5), where $N_{fd}$ denotes the number of false detected targets of all testing images and $N_{total\_target}$ denotes the total number of detected ships of all testing images.

$$p_f = \frac{N_{fd}}{N_{total\_target}} \tag{5}$$

In order to evaluate the overall performance of the detector, $F_1$ score which is defined as (6) is adopted in this paper. It reaches its best value at 1 and worst at 0.

$$F_1 = 2 \times \frac{p_d \times (1 - p_f)}{p_d + (1 - p_f)} \tag{6}$$

### *3.2. Influence of Different Layer Combination Strategies*

As mentioned before, feature maps from different layers differ in terms of spatial resolution and semantic distinction, giving them comparative advantages and disadvantages. Therefore, layer selection has a great impact on the performance of the detection system. In this section, four models with different layer combination strategies are trained for exploring the influence of different layer selections. Specifically, the first model combines the conv3, conv4 and conv5 of VGG16 together for region proposal. The second model integrates conv1, conv3 and conv5, and the final model selects conv1, conv2 and conv3. The baseline method is a model with a single layer conv5. All models have the same object detection network as the proposed method. The influence of different $\lambda$ for contextual features will be discussed in Section 3.3 and in this section we take $\lambda$ equal to 3 for all models to explore the effects of different layer fusion strategies.

As shown in Figure 6, in the open water areas, models have a comparative performance. Conv5 misses most of the ships around the tiny harbor where the denser ships berth. The situation improves greatly when conv3, conv4 and conv5 are combined. With the improvement of network resolution, more small-sized targets are picked up with the combination of conv1, conv3 and conv5. When the resolution of the network increases to the same level of conv2 in the model of conv 1+2+3, the best performance is achieved and only few tiny weak targets are missing. The comparison of the performance indicates that the detection performance of dense tiny ships improves dramatically with the fusion of layers and the increase in network resolution.

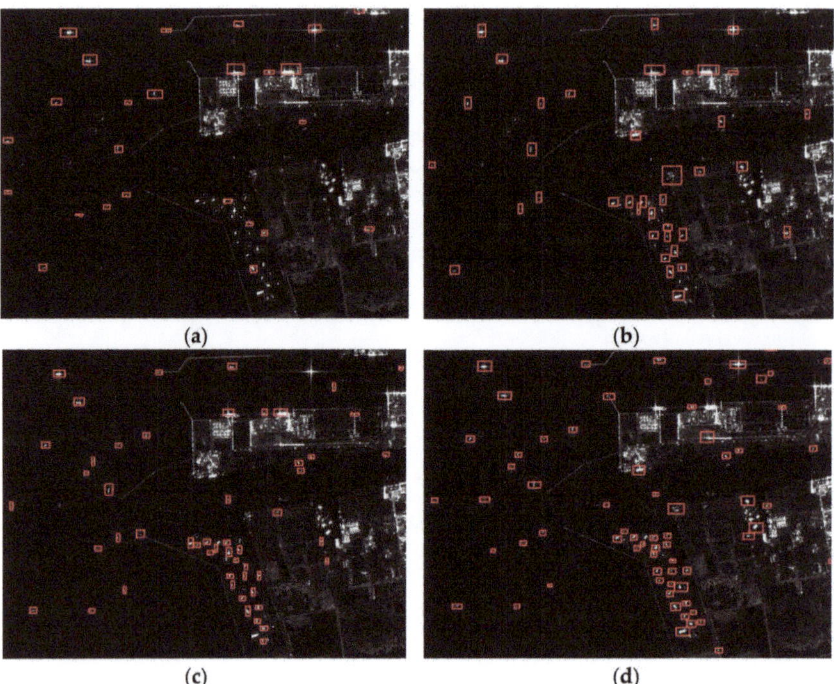

**Figure 6.** The comparisons of detection results with different layer combination strategies. The red and yellow rectangles represent the detection results and missing ships of detectors respectively. (**a**) conv5; (**b**) conv 3+4+5; (**c**) conv 1+3+5; (**d**) conv 1+2+3.

Table 1 displays the detection probability, false alarm probability and $F_1$ scores of different layer combination strategies. Compared with the performance on a single layer conv5, the networks with combined layers achieve higher detection probability and lower false alarm probability. With the combination of feature maps and a slight increase of resolution, the model with layer conv 3+4+5 detects more targets and obtains the lowest false alarm probability. The fusion of conv1, conv3 and conv5 promotes the detection probability to 80.43%. Conv 1+2+3 has the highest resolution compared with the other structures, which leads to a 12.71% increase in $P_d$ compared with a single layer. Compared with other fusion structures, conv 1+2+3 also has a slight but acceptable increase in false alarm probability, since the feature maps from shallow layers have lower semantic distinction. The highest $F_1$ score also indicates that the combination of conv1, conv2 and conv3 has the best performance in SAR ship detection.

**Table 1.** Detection performance with different layer combination strategies.

| Layers | $N_{total\_targets}$ | $N_{td}$ | $N_{fd}$ | $P_d$ (%) | $P_f$ (%) | $F_1$ |
|---|---|---|---|---|---|---|
| conv5 | 1463 | 1106 | 357 | 73.64 | 24.4 | 0.746 |
| conv 3+4+5 | 1355 | 1160 | 195 | 77.23 | **14.39** | 0.812 |
| conv 1+3+5 | 1418 | 1208 | 210 | 80.43 | 14.81 | 0.827 |
| conv 1+2+3 | 1540 | 1297 | 243 | **86.35** | 15.78 | **0.853** |

In summary, the increase of network resolution can dramatically improve the performance of detectors, especially in small-sized targets detection. Different layer combination strategies have a great impact on detection performance. As for SAR ship detection on Sentinel-1, since the sizes of most targets are smaller than 32 $\times$ 32 and the features of ships are relatively simple in intensity imagery, the combination of shallow layers from VGG16 is semantic enough to detect a ship in the background. In other words, resolution improvement plays a more important role than semantic feature for ships detection in SAR imagery.

### 3.3. Influence of Contextual Features

In order to identify the influence of contextual features, comparison experiments with different sizes of contextual features in the proposed network are conducted in this section. The network without contextual information means the object detection network only has one branch in the object detection network of Figure 1. In other models, $\lambda$ changes from 2 to 7 to obtain different sized contextual features. The combination strategy of conv 1+2+3 is adopted and all models have the same experiment settings.

Table 2 shows that when the bounding box of context information is relatively small, additional contextual information improves the overall performance to different degrees with higher $F_1$ scores. When a bounding box of contextual information five times larger than normal is appended, the best performance is obtained and the $F_1$ score changes to 0.873. Compared with the model without any contextual information, the model with fivefold contextual features increases by 4.53% in detection probability and decreases by 3.34% in false alarm rate. That is, extra contextual features provide more information for the model to pick up more targets. Meanwhile, the additional surrounding information of proposals also successfully assists to discriminate targets from false alarms. However, when the size of the bounding box enlarges to 6 or 7, the detection probability begins to decrease. One of the possible reasons is that most of the bounding boxes are oversized when $\lambda$ is too large, which leads to the dominance of contextual information in the concatenated features and aggravates the performance of the network. Thus, the size of contextual information should be moderated according to the detection task. Specifically, the proposed method possesses the best detection performance when fivefold contextual information is added and conv1, conv2 and conv3 are fused.

**Table 2.** Detection performance comparisons between different sized contextual features.

| Context Size | $N_{total\_targets}$ | $N_{td}$ | $N_{fd}$ | $P_d$ (%) | $P_f$ (%) | $F_1$ |
|---|---|---|---|---|---|---|
| without context | 1518 | 1259 | 259 | 83.82 | 17.06 | 0.834 |
| 2× context | 1534 | 1285 | 249 | 85.56 | 16.23 | 0.847 |
| 3× context | 1540 | 1297 | 243 | 86.35 | 15.78 | 0.853 |
| 4× context | 1564 | 1280 | 284 | 85.22 | 18.16 | 0.836 |
| 5× context | 1538 | 1327 | 211 | **88.35** | **13.72** | **0.873** |
| 6× context | 1519 | 1228 | 291 | 81.76 | 19.16 | 0.813 |
| 7× context | 1481 | 1231 | 250 | 81.95 | 16.88 | 0.825 |

### 3.4. Comparisons with Other Methods

In order to validate the effectiveness of the proposed method, Faster RCNN [15,38] and CMS-RCNN [29] are applied to Sentinel-1 dataset. CMS-RCNN, which has the same resolution as conv5, fuses conv3, con4 and conv5 by down-sampling. The other experiment settings of CMS-RCNN and Faster RCNN are the same as the proposed method.

Table 3 displays the performance of the three methods. Due to the increase of complexity in the network structure, the proposed method consumes more time in training. However, for a $512 \times 512$ sized image, the testing time of the proposed method remains at the same level as Faster RCNN and CMS-RCNN. With the layer fusion and the additional context information, the proposed network increases by 25.8% in detection probability and reduces the false alarm probability from 27.68 to 13.72% compared with Faster RCNN. Based on a higher network resolution than CMS-RCNN, the proposed method also promotes the detection performance significantly.

**Table 3.** Detection performance comparison between different methods.

| Method | $N_{total\_targets}$ | $N_{td}$ | $N_{fd}$ | $P_d$ (%) | $P_f$ (%) | $F_1$ | Testing Time |
|---|---|---|---|---|---|---|---|
| Faster RCNN | 1304 | 943 | 361 | 62.78 | 27.68 | 0.672 | 1.019 s |
| CMS-RCNN | 1491 | 1126 | 365 | 74.97 | 24.48 | 0.752 | 1.064 s |
| Proposed method | 1538 | 1327 | 211 | **88.35** | **13.72** | **0.873** | 2.180 s |

By changing the confidence score threshold of detection results on one testing image, different values of $p_d$ and $p_f$ are obtained, which produces the performance curves of different methods in Figure 7. As shown in the figure, the proposed method has the highest detection probability in a given false alarm probability. Similarly, with a specific $p_d$, the proposed method has the lowest false alarm probability. Therefore, the proposed method performs better than Faster RCNN and CMS-RCNN.

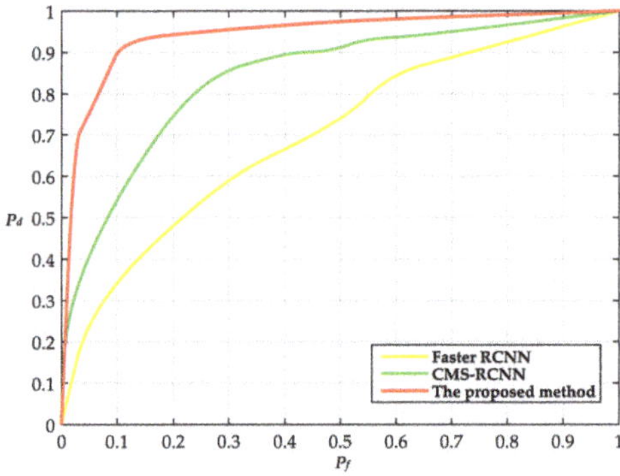


**Figure 7.** The performance curves of different methods. The yellow, green and red curves represent the performance of Faster RCNN, CMS-RCNN and the proposed method respectively. The $x$ label and $y$ label represent $p_f$ and $p_d$ respectively.

## 4. Discussion

Experiments on combination strategies and the influence of context information verify the effectiveness of the proposed method in ship detection, especially in small-sized targets detection. The comparisons with Faster RCNN and CMS-RCNN demonstrate the necessity of resolution improvement and additional context information.

Since the proposed method omits sea–land segmentation which traditional methods required, it provides the possibility for the network to detect ships nearshore, where traditional methods cannot perform well because of the limited accuracy of sea–land segmentation. The equal treatment of land and sea area also brings some undesirable false alarms on the land as shown in Figure 8. The red, yellow and purple boxes represent the detected target, the false alarms and missing targets respectively.

**Figure 8.** The detection results of the proposed method near the harbor area. The red, yellow and purple boxes represent the detected target, the false alarms and missing targets respectively.

Table 4 records the main categories of the false alarms in one test image. It is found that almost 65% of false alarms are building facilities on land, which are able to be ruled out with sea–land segmentation in image preprocessing. Some harbor facilities also are incorrectly detected as ships. While in the open ocean area, some noises, such as azimuth ambiguity and speckle, which have bright intensity will be picked up by the model. Islands, one of most annoying false alarms in the traditional method, are the least common false alarm category.

**Table 4.** The categories of false alarms.

| Categories | Building | Harbor | Island | Noise | Total |
|---|---|---|---|---|---|
| Number | 68 | 17 | 3 | **18** | **106** |

In order to analyze the characteristics of false alarms, some typical patches are displayed in the blue box of Figure 9. Visually, most of them are extremely similar to true positive targets. That is, they are brighter than their context and are shaped similar to ships, which means that the network values the visual features. Those kinds of false alarms are also hard to rule out by some hand-crafted methods. Therefore, some additional discrimination networks need to be trained, aimed at those false alarms and ships.

As shown in the purple box of Figure 9, some missing targets have weak or small intensity, which makes them cause few responses on the shallow layers and go undetected by the network. The missing label of weak and tiny targets on the training dataset is another possible reason for the missing detection, since the performance of the network is driven by the data which is fed into the network. Some of the missing targets are very near to the shore or to some other brighter targets, which makes the network assign them a low confidence score. Additionally, the motion blurring and cross sidelobe of ships also exert adverse effects on classification.

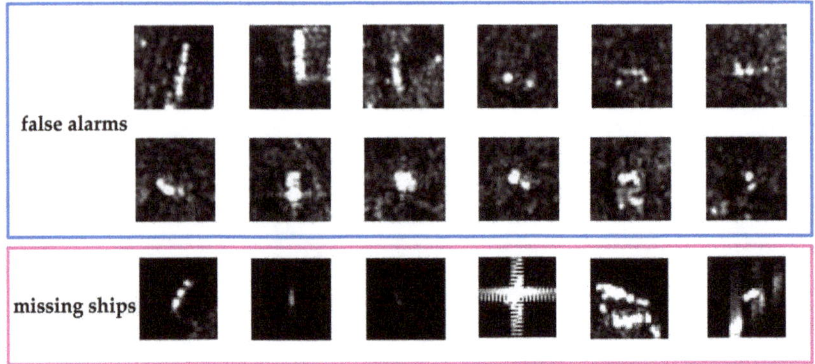


**Figure 9.** Some typical false alarms and missing detection targets among the detection results of the proposed method. The chips in the blue box and the purple box are false alarms and missing ships respectively.

## 5. Conclusions

With the labeled dataset on Sentinel-1, this paper opens up the possibility of utilizing deep neural networks for SAR ship detection. In order to improve the detection of ships on Sentinel-1 SAR imagery, where ships always appear small, layer fusion is employed in a contextual convolutional neural network to obtain semantic and high-resolution feature maps. Additionally, contextual information is added in the object detection network in order to help detectors to rule out false alarms. Experiments conducted in this paper demonstrate the effect of the layer fusion strategy and validate the influence of contextual information. More importantly, experiment results validate that the proposed method improves the detection performance dramatically.

Despite the effectiveness of the proposed method, some weak and tiny targets remain undetected and false alarms on land are hard to rule out. Investigations into the detection of these targets and false alarm discrimination need to be carried out in the future.

**Acknowledgments:** This work is supported partly by the National Natural Science Foundation of China under Grant Nos. 61372163 and 61331015. The authors would like to thank European Space Agency (ESA) for providing free Sentinel-1 data online. The authors would also like to thank the anonymous reviewers for their very competent comments and helpful suggestions.

**Author Contributions:** Miao Kang conceived and designed the experiments; Miao Kang performed the experiments and analyzed the data; Kefeng Ji, Xiangguang Leng and Zhao Lin contributed materials; Miao Kang and Kefeng Ji wrote the paper.

**Conflicts of Interest:** The authors declare no conflict of interest.

## References

1. Brusch, S.; Lehner, S.; Fritz, T.; Soccorsi, M. Ship suveillance with TerraSAR-X. *IEEE Trans. Geosci. Remote Sens.* **2011**, *49*, 1092–1103. [CrossRef]

2. Crisp, D.J. A ship detection system for RADARSAT-2 dual-pol multi-look imagery implemented in the ADSS. In Proceedings of the 2013 IEEE International Conference on Radar, Adelaide, Australia, 9–12 September 2013; pp. 318–323.
3. Torres, R.; Snoeij, P.; Geudtner, D.; Bibby, D.; Davidson, M.; Attema, E.; Potin, P.; Rommen, B.; Floury, N.; Brown, M.; et al. GMES Sentinel-1 mission. *Remote Sens. Environ.* **2012**, *120*, 9–24. [CrossRef]
4. Crisp, D.J. The state-of-the-art in ship detection in Synthetic Aperture Radar imagery. *Org. Lett.* **2004**, *35*, 2165–2168.
5. Marino, A.; Sugimoto, M.; Ouchi, K.; Hajnsek, I. Validating a Notch Filter for Detection of Targets at Sea with ALOS-PALSAR Data: Tokyo Bay. *IEEE J. Sel. Top. Appl. Earth Obs. Remote Sens.* **2014**, *7*, 74907–74918. [CrossRef]
6. Pelich, R.; Longépé, N.; Mercier, G.; Hajduch, G.; Garello, R. AIS-Based Evaluation of Target Detectors and SAR Sensors Characteristics for Maritime Surveillance. *IEEE J. Sel. Top. Appl. Earth Obs. Remote Sens.* **2015**, *8*, 3892–3901. [CrossRef]
7. Wang, C.; Bi, F.; Zhang, W.; Chen, L. An Intensity-Space Domain CFAR Method for Ship Detection in HR SAR Images. *IEEE Geosci. Remote Sens. Lett.* **2017**, *14*, 529–533. [CrossRef]
8. Zhi, Z.; Ji, K.; Xing, X.; Zou, X.; Zhou, H. Ship Surveillance by Integration of Space-borne SAR and AIS—Review of Current Research. *J. Navig.* **2014**, *67*, 177–189. [CrossRef]
9. Fingas, M.F.; Brown, C.E. Review of Ship Detection from Airborne Platforms. *Can. J. Remote Sens.* **2001**, *27*, 379–385. [CrossRef]
10. Guo, Y.; Liu, Y.; Oerlemans, A.; Lao, S.; Wu, S.; Lew, M.S. Deep learning for visual understanding. *Neurocomputing* **2016**, *187*, 27–48. [CrossRef]
11. Druzhkov, P.N.; Kustikova, V.D. A survey of deep learning methods and software tools for image classification and object detection. *Pattern Recognit. Image Anal.* **2016**, *26*, 9–15. [CrossRef]
12. Girshick, R.; Donahue, J.; Darrell, T.; Malik, J. Rich Feature Hierarchies for Accurate Object Detection and Semantic Segmentation. In Proceedings of the 2014 IEEE Conference on Computer Vision and Pattern Recognition (CVPR 2014), Columbus, OH, USA, 23–28 June 2014; pp. 580–587.
13. Redmon, J.; Divvala, S.; Girshick, R.; Farhadi, A. You Only Look Once: Unified, Real-Time Object Detection. In Proceedings of the 2015 IEEE Conference on Computer Vision and Pattern Recognition (CVPR 2015), Boston, MA, USA, 7–12 June 2015; pp. 779–788.
14. Liu, W.; Anguelov, D.; Erhan, D.; Szegedy, C.; Fu, C.; Berg, A.C. Ssd: Single shot multibox detector. In Proceedings of the European Conference on Computer Vision, Amsterdam, The Netherlands, 8–16 October 2016; pp. 21–37.
15. Ren, S.; He, K.; Girshick, R.; Sun, J. Faster R-CNN: Towards Real-Time Object Detection with Region Proposal Networks. *IEEE Trans. Pattern Anal. Mach. Intell.* **2017**, *39*, 1137–1149. [CrossRef] [PubMed]
16. Long, M.; Wang, J.; Jordan, M.I. Deep transfer learning with joint adaptation networks. *arXiv*, **2016**, arXiv:1605.06636.
17. Gupta, U.; Chaudhury, S. Deep transfer learning with ontology for image classification. In Proceedings of the 5th National Conference on Computer Vision, Pattern Recognition, Image Processing and Graphics (NCVPRIPG), Patna, India, 16–19 December 2015; pp. 1–4.
18. Ravishankar, H.; Sudhakar, P.; Venkataramani, R.; Thiruvenkadam, S.; Annangi, P.; Babu, N.; Vaidya, V. Understanding the Mechanisms of Deep Transfer Learning for Medical Images. *arXiv*, **2017**, arXiv:1704.06040.
19. Simonyan, K.; Zisserman, A. Very deep convolutional networks for large-scale image recognition. *arXiv*, **2014**, arXiv:1409.1556.
20. Cai, Z.; Fan, Q.; Feris, R.S.; Vasconcelos, N. A unified multi-scale deep convolutional neural network for fast object detection. In Proceedings of the 14th European Conference on Computer Vision, Amsterdam, The Netherlands, 8–16 October 2016; pp. 354–370.
21. Shelhamer, E.; Long, J.; Darrell, T. Fully convolutional networks for semantic segmentation. *IEEE Trans. Pattern Anal. Mach. Intell.* **2017**, *39*, 640–651. [CrossRef] [PubMed]
22. Kong, T.; Yao, A.; Chen, Y.; Sun, F. Hypernet: Towards accurate region proposal generation and joint object detection. In Proceedings of the 2016 IEEE Conference on Computer Vision and Pattern Recognition (CVPR 2016), Seattle, WA, USA, 27–30 June 2016; pp. 845–853.
23. Zagoruyko, S.; Lerer, A.; Lin, T.Y.; Pinheiro, P.O.; Gross, S.; Chintala, S.; Dollár, P. A multipath network for object detection. *arXiv*, **2016**, arXiv:1604.02135.

24. Ronneberger, O.; Fischer, P.; Brox, T. U-net: Convolutional networks for biomedical image segmentation. In Proceedings of the 18th International Conference on Medical Image Computing and Computer-Assisted Intervention, Munich, Germany, 5–9 October 2015; Springer: Cham, Switzerland, 2015; pp. 234–241.
25. Divvala, S.K.; Hoiem, D.; Hays, J.H.; Efros, A.; Hebert, M. An empirical study of context in object detection. In Proceedings of the 2009 IEEE Conference on Computer Vision and Pattern Recognition (CVPR 2009), Miami, FL, USA, 20–25 June 2009; pp. 1271–1278.
26. Galleguillos, C.; Belongie, S. Context based object categorization: A critical survey. *Comput. Vis. Image Underst.* **2010**, *114*, 712–722. [CrossRef]
27. Liu, W.; Rabinovich, A.; Berg, A.C. Parsenet: Looking wider to see better. *arXiv*, 2015; arXiv:1506.04579.
28. Bell, S.; Lawrence Zitnick, C.; Bala, K.; Girshick, R. Inside-outside net: Detecting objects in context with skip pooling and recurrent neural networks. In Proceedings of the IEEE Conference on Computer Vision and Pattern Recognition (CVPR 2016), Seattle, WA, USA, 27–30 June 2016; pp. 2874–2883.
29. Zhu, C.; Zheng, Y.; Luu, K.; Savvides, M. CMS-RCNN: Contextual Multi-Scale Region-based CNN for Unconstrained Face Detection. In Proceedings of the IEEE Conference on Computer Vision and Pattern Recognition (CVPR 2016), Seattle, WA, USA, 27–30 June 2016.
30. Zeiler, M.D.; Fergus, R. Visualizing and Understanding Convolutional Networks. In Proceedings of the 14th European Conference on Computer Vision, Zurich, Switzerland, 6–12 September 2014.
31. Radford, A.; Metz, L.; Chintala, S. Unsupervised representation learning with deep convolutional generative adversarial networks. *arXiv*, **2015**, arXiv:1511.06434.
32. Dumoulin, V.; Visin, F. A guide to convolution arithmetic for deep learning. *arXiv*, **2016**, arXiv:1603.07285.
33. Hu, P.; Ramanan, D. Finding tiny faces. *arXiv*, **2016**, arXiv:1612.04402.
34. Sentinels Scientific Data Hub. Available online: https://scihub.copernicus.eu/ (accessed on 1 March 2017).
35. Leng, X.; Ji, K.; Zhou, S.; Zou, H. An adaptive ship detection scheme for spaceborne SAR imagery. *Sensors* **2016**, *16*, 1345. [CrossRef] [PubMed]
36. OpenSAR. Available online: http://opensar.sjtu.edu.cn/ (accessed on 29 March 2017).
37. Abadi, M.; Agarwal, A.; Barham, P.; Brevdo, E.; Chen, Z.; Citro, C.; Corrado, G.; Jeffrey, D.; Devin, M.; et al. Tensorflow: Large-scale machine learning on heterogeneous distributed systems. *arXiv*, **2016**, arXiv:1603.04467.
38. Faster-RCNN_TF. Available online: https://github.com/smallcorgi/Faster-RCNN_TF (accessed on 24 May 2017).

*Article*

# Refocusing of Moving Targets in SAR Images via Parametric Sparse Representation

**Yichang Chen [1,2], Gang Li [2,*], Qun Zhang [1,3] and Jinping Sun [4]**

[1] Institute of Information and Navigation, Air Force Engineering University, Xi'an 710077, China; cyc_2007@163.com (Y.C.); zhangqunnus@gmail.com (Q.Z.)
[2] Department of Electronic Engineering, Tsinghua University, Beijing 100084, China
[3] Key Laboratory for Information Science of Electromagnetic Wave (Ministry of Education), Fudan University, Shanghai 200433, China
[4] School of Electronic and Information Engineering, Beijing University of Aeronautics and Astronautics, Beijing 100191, China; sunjinping@buaa.edu.cn
* Correspondence: gangli@tsinghua.edu.cn; Tel.: +86-10-6279-4095

Academic Editors: Xiaofeng Yang, Xiaofeng Li, Ferdinando Nunziata and Alexis Mouche
Received: 12 June 2017; Accepted: 31 July 2017; Published: 2 August 2017

**Abstract:** In this paper, a parametric sparse representation (PSR) method is proposed for refocusing of moving targets in synthetic aperture radar (SAR) images. In regular SAR images, moving targets are defocused due to unknown motion parameters. Refocusing of moving targets requires accurate phase compensation of echo data. In the proposed method, the region of interest (ROI) data containing the moving targets are extracted from the complex SAR image and represented in a sparse fashion through a parametric transform, which is related to the phase compensation parameter. By updating the reflectivities of moving target scatterers and the parametric transform in an iterative fashion, the phase compensation parameter can be accurately estimated and the SAR images of moving targets can be refocused well. The proposed method directly operates on small-size defocused ROI data, which helps to reduce the computational burden and suppress the clutter. Compared to other existing ROI-based methods, the proposed method can suppress asymmetric side-lobes and improve the image quality. Both simulated data and real SAR data collected by GF-3 satellite are used to validate the effectiveness of the proposed method.

**Keywords:** moving target imaging; parametric sparse representation (PSR); region of interest (ROI); synthetic aperture radar (SAR)

---

## 1. Introduction

Moving target imaging is an important task of synthetic aperture radar (SAR) [1–6]. Since SAR imaging algorithms were originally designed for stationary targets, the main challenge for SAR imaging of moving targets is to compensate the phase error caused by non-cooperative motion of targets. It is well known that the Doppler frequency shift caused by target movement results in the imaging position offset along the azimuth direction. Further, the change of the azimuth frequency modulation rate, which is related to the acceleration in range and the velocity in azimuth of the target, introduces defocus of moving targets in regular SAR images [7]. From the viewpoint of process flow, the existing methods of SAR moving target imaging can be generally divided into two types. (1) Methods based on the raw radar data [8–12]. These methods deal with the entire echo data reflected from both stationary background and moving targets. In fact, only a small part of the entire data is related to the moving targets, and therefore, processing on entire data will induce a large amount of clutter. The displaced phase center antenna (DPCA) and along track interferometry (ATI) can effectively suppress the clutter [13,14], which requires multiple channels and therefore is not suitable for single-channel SAR systems. (2) Methods based on the region of interest (ROI) data [15–17]. This type of methods can effectively remove clutter and easily detect moving targets. Since only a small image contacting the ROI data is extracted from regular SAR imaging result, the amount of data to be processed is significantly reduced. Zhang et al. [16] proposed a high-resolution SAR imaging method of ground moving targets with defocused ROI data. The authors derive an exact analytic expression of the ROI data without approximation of the slant range. Ref. [16] assumed that the phase error arises from the unknown azimuth and range velocities in a specific model. Then, the azimuth and range equivalent velocities are estimated by a 2-D search, such that the maximum contrast of moving target image is reached. The image of target is obtained by Stolt interpolation with each pair of equivalent velocity parameters. However, the equivalent velocity cannot reflect the actual motion parameters of target in acceleration state, which results in the difficulty of determining the search interval of equivalent velocities. Even if the accurate equivalent velocities are estimated, the model mismatch problem would introduce high asymmetric side-lobes in the imaging result when the target is in an acceleration state.

Since the radar echo reflected from man-made moving targets are usually stronger than the background, in recent years, many sparsity-aware methods have been applied to SAR moving target parameter estimation and imaging [18–25]. Ref. [23,25] summarize the latest application of sparse processing in SAR systems. An approach to motion parameter estimation with low pulse repetition frequency based on compressed sensing (CS) theory is proposed in [19]. Ref. [20] also proposes a method for motion parameter estimation of moving target based on the raw radar data. However, complicated clutter suppression is required before motion parameter estimation. Ref. [21] presents a method for imaging of moving targets with multi-static SAR using an overcomplete dictionary. In Ref. [22], the authors divide a phase error into three subcategories and correct them by using a nonquadratic regularization approach. However, due to the heavy computational burden, this method may not be suitable for these targets occupying a large number of resolution cells. Onhon et al. utilize the sparsity-driven autofocus framework to solve the problem of moving target imaging in [18], where the phase error induced by target movement is corrected by the non-quadratic regularization approach. However, this method ignores the relationship between the phase error and the motion parameters. It needs to independently estimate the phase error of each sample data, thus increasing the unnecessary computational burden.

In this paper, we propose a parametric sparse representation method for SAR imaging of moving targets with ROI data. The parametric sparse representation technique has been utilized to ISAR imaging of rotating targets [26–28], SAR motion compensation [29], and moving target motion parameters estimation [12]. Differing from the previous work where the process flow starts from entire raw data, the proposed method in this paper is based on ROI data. Firstly, the complex image in ROI that contains the defocused moving target is extracted from the regular SAR image. Then, the principal components of the ROI data, which correspond to the dominant scatterers of a moving target, are represented by a parametric transform that is uniquely determined by a phase compensation parameter. By updating the reflectivities of the target and the parametric transform in an iterative fashion, the phase compensation parameter can be accurately estimated and the focused image of moving target can be obtained when the iterative process converges. Differing from the method in [16], which realizes the phase compensation by searching 2-D equivalent velocities of moving targets, in this paper, the phase compensation is achieved by estimating the phase compensation parameter in an iterative fashion. Experimental results based on both simulated and space-borne SAR data demonstrate that the proposed method outperforms the method in [16] in terms of the imaging quality of moving targets. Simulations also show that, compared to the method in [16], the proposed algorithm in this paper has a higher tolerance for model mismatch problem and is capable of providing satisfactory refocused image even when the target is moving in acceleration state.

The rest of this paper is organized as follows. The SAR signal model is reviewed in Section 2. The proposed method for moving target imaging is formulated in Section 3. The performance of the proposed method is evaluated, with both simulated and measured data, in Section 4. Conclusions are presented in Section 5.

## 2. Signal Model

The geometry relationship between the platform and a moving target for side-looking SAR is shown in Figure 1. The horizontal axis denotes the azimuth direction, and the vertical axis denotes the slant-range direction. Assume that the SAR platform flies straight at a speed of $V$ along the $x$-axis, and the velocities of a moving target in azimuth and range are $v_x$ and $v_r$, respectively.

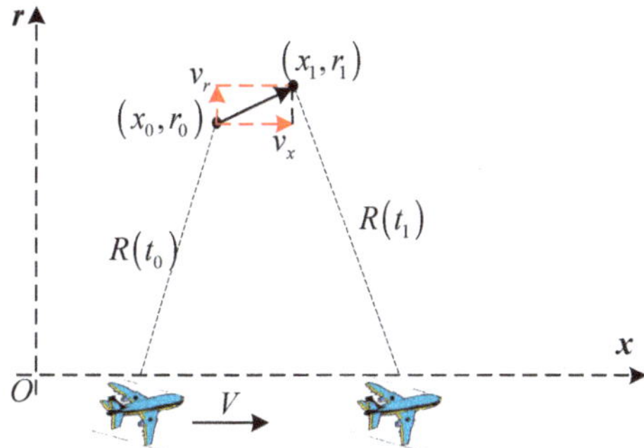


**Figure 1.** Geometry of synthetic aperture radar (SAR) imaging of a moving target.

The slow time is denoted as $t_s = nT$ with an integer $n$, and $T$ is the pulse repetition time. Thus, the radar antenna phase center (APC) position at $t_s$ is $(Vt_s, 0)$, and the position of the moving target is $(x_0 + v_x t_s, r_0 + v_r t_s)$, where $(x_0, r_0)$ is the initial position when $t_s = 0$. The instantaneous distance between the moving target and the radar can be expressed as

$$R(t_s) = \sqrt{(Vt_s - x_0 - v_x t_s)^2 + (r_0 + v_r t_s)^2} \tag{1}$$

Suppose that the radar transmits a linear frequency-modulated (LFM) signal as

$$s(t, t_s) = \mathrm{rect}(t/T_p) \cdot \exp\left(j2\pi f_c t + j\pi\gamma t^2\right), \tag{2}$$

where $t$ is the fast time, $T_p$ is the pulse width, $\mathrm{rect}(\cdot)$ denotes the rectangular function, $f_c$ is the carrier frequency, and $\gamma$ is the chirp rate. The baseband echo of the moving target can be expressed as

$$s_r(t, t_s) = \sigma \cdot \mathrm{rect}\left(\frac{t - 2R(t_s)/c}{T_p}\right) \cdot \mathrm{rect}\left(\frac{t_s}{T_a}\right) \cdot \exp\left(-j4\pi f_c R(t_s)/c + j\pi\gamma(t - 2R(t_s)/c)^2\right), \tag{3}$$

where $T_a$ is the synthetic aperture time, and the target scattering coefficient $\sigma$ is assumed to be constant during the observation time interval. By taking the two-dimensional Fourier transform, the signal of the moving target in 2-D frequency domain can be expressed as

$$\begin{aligned} &S_r(f_r, f_a) \\ &= \sigma \cdot W_r(f_r) \cdot W_a(f_a) \cdot \exp\left\{j \cdot \left[-2\pi f_a \tfrac{\delta}{v_e^2} - \pi \tfrac{f_r^2}{\gamma} - \tfrac{4\pi[x_0 v_r + r_0(V - v_x)]}{c v_e} \sqrt{(f_c + f_r)^2 - \tfrac{c^2 f_a^2}{4 v_e^2}}\right]\right\}, \end{aligned} \tag{4}$$

where $f_r$ and $f_a$ are the range and azimuth frequencies, respectively, $W_r(f_r)$ and $W_a(f_r)$ are the range and azimuth envelope function, respectively, $\delta = x_0(V - v_x) - r_0 v_r$ and $v_e = \sqrt{(V - v_x)^2 + v_r^2}$. After conventional matched filtering and Stolt interpolation, (4) becomes [16]

$$S_{r\_Stolt}(f_r, f_a) = \sigma \cdot W_r(f_r) \cdot W_a(f_a) \cdot \exp\left\{ j \cdot \left[ -2\pi f_a \frac{\delta}{v_e^2} - \frac{4\pi[x_0 v_r + r_0(V - v_x)]}{c v_e} \sqrt{(f_c + f_r)^2 + \frac{c^2 f_a^2}{4}\left(\frac{1}{V^2} - \frac{1}{v_e^2}\right)} \right] \right\}. \quad (5)$$

It can be seen from (5) that the high-order residual items still exist after the conventional SAR imaging processing, which means that the image of the moving target is defocused. To achieve a refocused image of moving target, it is necessary to compensate the residual phase items. The filter for residual phase compensation can be constructed as [16]

$$H_1(f_r, f_a) = \exp\left\{ j \cdot \left[ \frac{4\pi R_{ref}}{c} \sqrt{(f_c + f_r)^2 + \frac{c^2 f_a^2}{4}\left(\frac{1}{V^2} - \alpha\right)} - \frac{4\pi R_{ref}}{c}(f_c + f_r) \right] \right\}, \quad (6)$$

where $R_{ref}$ is the reference distance, and $\alpha$ is the phase compensation parameter, which is the parameter to be estimated in this paper. The phase compensation parameter is related to the motion parameters of target, and it can be expressed as $\alpha = 1/\left[(V - v_x)^2 + v_r^2\right]$ when the target is in uniform motion state. The phase compensation parameter becomes $\alpha = 1/V^2$ when the target is stationary. This results in $H_1(f_r, f_a) = 1$ in (6), which means that no phase compensation is required for the data reflected from stationary targets. In the proposed method, we can initialize the phase compensation parameter as $1/V^2$. By taking the two-dimensional inverse Fourier transform in (5), we can obtain the regular SAR complex image of observed scene, in which the sub-image that containing moving target is the so-called ROI data. The extracted ROI sub-image from the focused stationary background can be expressed as

$$s_{r\_ROI}(\bar{t}, \bar{t}_s) = \sigma \cdot W_r\left(\bar{t} - \frac{2R_{ROI}(\bar{t}_s)}{c}\right) \cdot W_a\left(\bar{t}_s - \frac{\delta}{v_e^2}\right) \cdot \exp\left(-j\frac{4\pi}{\lambda} R_{ROI}(\bar{t}_s)\right), \quad (7)$$

where $\bar{t}$ and $\bar{t}_s$ denote the sampling index of ROI date in range and azimuth time domain, respectively, and

$$R_{ROI}(\bar{t}_s) = \sqrt{\left(\frac{x_0 v_r + r_0(V - v_x)}{v_e}\right)^2 - \left(\bar{t}_s - \frac{\delta}{v_e^2}\right)^2 \cdot \left(\frac{1}{V^2} - \frac{1}{v_e^2}\right)}, \quad (8)$$

In the next section, we will describe how to estimate the phase compensation parameter and refocus target image with the ROI data.

## 3. Parametric Sparse Representation Method for Moving Target Imaging

In this section, we formulate the parametric sparse representation method for imaging of moving targets. We denote the original image size as $N_r \times N_a$, and the ROI data size as $n_r \times n_a$. Note that the data size is significantly reduced, but the signal bandwidth of moving target remains unchanged. We define the range and azimuth frequency of ROI data as $\overline{f}_r$ and $\overline{f}_a$, respectively. By taking two-dimensional Fourier transform in Equation (7), the ROI data in 2-D frequency domain can be expressed as

$$S_{r\_ROI}(\overline{f}_r, \overline{f}_a) = \sigma \cdot W_r(\overline{f}_r) \cdot W_a(\overline{f}_a) \cdot \exp\left\{ j \cdot \left[ -2\pi \overline{f}_a \frac{\delta}{v_e^2} - \frac{4\pi[x_0 v_r + r_0(V - v_x)]}{c v_e} \sqrt{\left(f_c + \overline{f}_r\right)^2 + \frac{c^2 \overline{f}_a^2}{4}\left(\frac{1}{V^2} - \frac{1}{v_e^2}\right)} \right] \right\}. \quad (9)$$

The phase compensation filter of ROI data can be rewritten as

$$H_{2(\alpha)}\left(\overline{f}_r, \overline{f}_a\right) = \exp\left\{ j \cdot \left[ \frac{4\pi R_{ref}}{c} \sqrt{\left(f_c + \overline{f}_r\right)^2 + \frac{c^2 \overline{f}_a^2}{4}\left(\frac{1}{V^2} - \alpha\right)} - \frac{4\pi R_{ref}}{c}\left(f_c + \overline{f}_r\right) \right] \right\}. \quad (10)$$

The refocused image of the moving target can be obtained through a refocusing transform $\Gamma(\cdot)$:

$$\Theta = \Gamma(s_{r\_ROI}) = \mathbf{\Psi}_r^{-1} \cdot \left[ (\mathbf{\Psi}_r \cdot s_{r\_ROI} \cdot \mathbf{\Psi}_a) \circ H_{2(\alpha)} \right] \cdot \mathbf{\Psi}_a^{-1}, \quad (11)$$

where $\Theta$ is the refocusing result of moving target, $\mathbf{\Psi}$ and $\mathbf{\Psi}^{-1}$ are the Fourier matrix and the inverse Fourier matrix, respectively. The subscripts $r$ and $a$ denote the range direction and azimuth direction, respectively, and $\circ$ denotes Hadamard product. As shown in Equation (11), through a series of matrix operations, the refocusing transform $\Gamma(\cdot)$ can achieve the conversion from the defocused ROI data to the refocusing result of moving target. It is obvious that the image quality of $\Theta$ depends on the phase compensation parameter $\alpha$, i.e., different phase compensation parameter produces different filter function $H_{2(\alpha)}$ and, therefore, different refocusing result $\Theta$. Thus, $\Theta$ can be regarded as a function of $\alpha$ and denoted by $\Theta_{(\alpha)}$. With a wrong value of $\alpha$, the refocusing transform $\Gamma(s_{r\_ROI})$ will induce the high-order phase error in azimuth and range, and thus a blurred image is most likely obtained. With accurate estimate of $\alpha$, the phase error of the ROI data can be well compensated, and accordingly, a well-focused moving target image will be obtained. From the above consideration, refocusing of the moving target can be carried out in an iterative fashion, i.e., the sparse imaging result $\Theta$ and the phase compensation parameter $\alpha$ are iteratively updated.

### 3.1. Update the Sparse Solution

At the $p$-th iteration, denotes the estimate of the phase compensation parameter $\alpha^{(p)}$. Accordingly, the filter function $H_{2(\alpha^{(p)})}$ and refocusing transform $\Gamma(s_{r\_ROI})$ can be constructed according to Equations (10) and (11). The refocusing process described in (11) is reversible, and the inverse transform can be written as

$$s_{r\_ROI} = \Gamma^{-1}(\Theta) = \mathbf{\Psi}_r^{-1} \cdot \left[ (\mathbf{\Psi}_r \cdot \Theta \cdot \mathbf{\Psi}_a) \circ H_{2(\alpha)}^* \right] \cdot \mathbf{\Psi}_a^{-1}, \quad (12)$$

where $(\cdot)^*$ denotes the conjugate operation. Since the moving target is usually sparse in 2-D space domain, we can obtain the moving target imaging results by solving the following unconstrained problem [30,31]

$$\min_{\Theta} \|s_{r\_ROI} - \Gamma^{-1}(\Theta)\|_2^2 + \lambda \|\Theta\|_1, \quad (13)$$

where $\|\cdot\|_1$ and $\|\cdot\|_2$ denote $l_1$ and $l_2$ norms, respectively, and $\lambda > 0$ is the regularization parameter that balances the recovery error and the sparsity of the solution. Given the value of $\alpha^{(p)}$, Equation (13) is a standard problem of sparse signal recovery. In this paper, we utilize the soft iterative thresholding algorithm [32] to solve Equation (13). The main steps are summarized below.

**Soft Iterative Thresholding Algorithm**

**Input:** $s_{r\_ROI}$, $\lambda$, $\alpha^{(p)}$, $\Gamma(\cdot)$, and $\Gamma^{-1}(\cdot)$

**Initialization:** Let the iterative counter $k = 1$, residual matrix $R_0 = s_{r\_ROI}$, and $\hat{\Theta}_0 = \mathbf{0}^{n_r \times n_a}$.

**Iteration:** at the $k$-th iteration $(k > 1)$

(1) Update the sparse result by $\hat{\Theta}_k = \text{soft}(\hat{\Theta}_{k-1} + \Gamma(R_{k-1}), \lambda)$, where $\text{soft}(x, \lambda) = \text{sign}(x) \cdot \max(|x| - \lambda, 0)$.

(2) Update the residual matrix by $R_k = s_{r\_ROI} - \Gamma^{-1}(\hat{\Theta}_k)$.

(3) Increment $k$, and return to Step (1) until the stopping criterion is met. Here the stopping criterion is $\|\hat{\Theta}_k - \hat{\Theta}_{k-1}\|_2 / \|\hat{\Theta}_{k-1}\|_2 \leq \varepsilon$. The selection of the threshold value $\varepsilon$ is related to the precision requirement.

**Output:** $\hat{\Theta}^{(p)} = \hat{\Theta}_k$.

### 3.2. Update the Estimate of Phase Compensation Parameter

Given the sparse solution $\hat{\Theta}^{(p)}$, the phase compensation parameter estimate can be updated by

$$\alpha^{(p+1)} = \alpha^{(p)} + \zeta \cdot \rho, \tag{14}$$

where $\zeta$ and $\rho$ are iterative direction and iterative step-length, respectively. Further, the parameters $\{\zeta, \rho\}$ can be estimated by minimizing the recovery error, that is

$$\min \| s_{r\_ROI} - \Gamma^{-1}\left(\hat{\Theta}^{(p)}\right) \|_{\mathrm{F}}, \tag{15}$$

where $\| \cdot \|_{\mathrm{F}}$ denotes F-norm of matrix. Substituting Equations (12) into (15), we have

$$\min \| \mathbf{Y} - \mathbf{U} \circ H^{*}_{2(\alpha)} \mathbf{\Psi}_a^{-1} \|_{\mathrm{F}}, \tag{16}$$

where $\mathbf{Y} = \mathbf{\Psi}_r \cdot s_{r\_ROI}$, and $\mathbf{U} = \mathbf{\Psi}_r \cdot \hat{\Theta}^{(p)} \cdot \mathbf{\Psi}_a$. Define $B_{(\alpha)} = \mathbf{U} \circ H^{*}_{2(\alpha)}$. According to (9) and (10), we have the expression of each element of $B_{(\alpha)}$ as

$$B_{(\alpha)}\left(\overline{f}_r, \overline{f}_a\right) = U\left(\overline{f}_r, \overline{f}_a\right) \cdot \exp\left\{ -j \left[ \frac{4\pi R_{ref}}{c} \sqrt{\left(f_c + \overline{f}_r\right)^2 + \frac{c^2 \overline{f}_a^2}{4}\left(\frac{1}{V^2} - \alpha\right)} - \frac{4\pi R_{ref}}{c}\left(f_c + \overline{f}_r\right) \right] \right\} \tag{17}$$

By taking the first-order Taylor expansion of $B_{(\alpha)}$, we have

$$B_{(\alpha)} = B_{(\alpha^{(p)})} + \left. \frac{dB_{(\alpha)}}{d\alpha} \right|_{\alpha = \alpha^{(p)}} \cdot \Delta\alpha \tag{18}$$

where $\Delta\alpha$ is the first-order increment. According to Equation (17), the derivative of $B_{(\alpha)}$ can be expressed as

$$\frac{dB_{(\alpha)}\left(\overline{f}_r, \overline{f}_a\right)}{d\alpha} = B_{(\alpha)}\left(\overline{f}_r, \overline{f}_a\right) \cdot \left( j \frac{\pi R_{ref} c \overline{f}_a^2}{2\sqrt{\left(f_c + \overline{f}_r\right)^2 + \frac{c^2 \overline{f}_a^2}{4}\left(\frac{1}{V^2} - \alpha\right)}} \right) \tag{19}$$

Then, Equation (16) can be rewritten as

$$\min \| \mathbf{Y} - B_{(\alpha^{(p)})} \cdot \mathbf{\Psi}_a^{-1} - \left. \frac{dB_{(\alpha)}}{d\alpha} \right|_{\alpha = \alpha^{(p)}} \cdot \mathbf{\Psi}_a^{-1} \cdot \Delta\alpha \|_{\mathrm{F}} \tag{20}$$

Define

$$\Xi^{(p)} = \left[ \mathbf{Y} - B_{(\alpha^{(p)})} \cdot \mathbf{\Psi}_a^{-1} \right]_{Vector}, \tag{21a}$$

$$Z^{(P)} = \left[ \left. \frac{dB_{(\alpha)}}{d\alpha} \right|_{\alpha = \alpha^{(p)}} \cdot \mathbf{\Psi}_a^{-1} \right]_{Vector}, \tag{21b}$$

where $[\cdot]_{Vector}$ denotes the operation of stacking vectors one underneath the other sequentially, i.e., the size of $\Xi^{(p)}$ and $Z^{(P)}$ are $n_r n_a \times 1$. Define

$$C^{(p)} = \begin{bmatrix} \mathrm{real}\left(Z^{(P)}\right) \\ \mathrm{imag}\left(Z^{(P)}\right) \end{bmatrix}, \tag{22a}$$

$$D^{(p)} = \begin{bmatrix} \mathrm{real}\left(\Xi^{(p)}\right) \\ \mathrm{imag}\left(\Xi^{(p)}\right) \end{bmatrix}, \tag{22b}$$

where real$(\cdot)$ and imag$(\cdot)$ denote the real and imaginary parts of a complex number, respectively. Then, the real-valued solution of the first-order increment can be directly obtained as

$$\Delta\alpha^{(p)} = \left(\left[C^{(p)}\right]^{\mathrm{T}} C^{(p)}\right)^{-1}\left[C^{(p)}\right]^{\mathrm{T}} D^{(p)}, \tag{23}$$

where $[\cdot]^{\mathrm{T}}$ denotes the transpose operation. Then the iterative direction and iterative step-length can be calculated as

$$\begin{cases} \zeta = \Delta\alpha/|\Delta\alpha| \\ \rho = \kappa \cdot |\Delta\alpha| \end{cases}, \tag{24}$$

where $\kappa$ is the convergence parameter associated with the convergence speed, and it can be determined empirically. In the experiments of this paper, we set $\kappa = 10$. Finally, the phase compensation parameter can be updated by Equation (14).

By iteratively updating $\Theta$ and $\alpha$ as described above, the complete procedure of the parametric sparse representation method for moving target imaging is summarized in Figure 2. The iterative process is terminated until $\left|\alpha^{(p+1)} - \alpha^{(p)}\right| < \eta$, where $\eta$ is the convergence threshold.

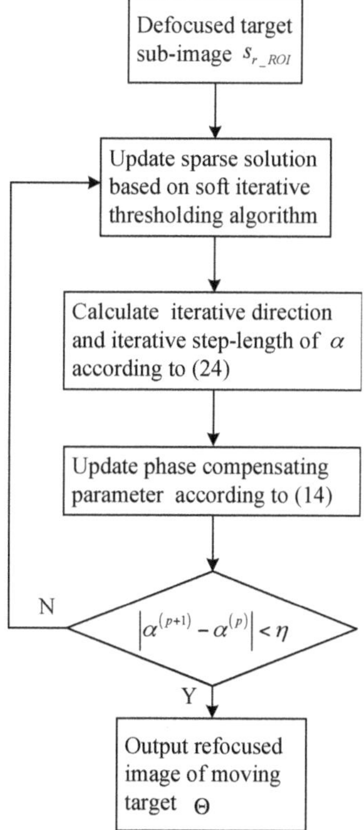


**Figure 2.** Flowchart of the parametric sparse representation method.

## 4. Experimental Results

The experimental results based on simulated data and real SAR data collected by GF-3 satellite are presented in this section to demonstrate the effectiveness of the proposed method.

### 4.1. Simulated Data

First of all, the proposed method is verified by using simulated data. The main system parameters are as follows. The carrier frequency is 10 GHz, the scene center range is 10 km, the platform velocity is 150 m/s, the transmitted bandwidth is 300 MHz, and the pulse duration is 2.2 μs. The simulated scene contains 2 stationary reference scatterers (S1–S2) and a rigid-body moving target consisting of 4 scatterers (M1–M4), as shown in Figure 3. We first assume that the target only has constant-speed components in azimuth and range without acceleration. The actual velocity components of the target are set to be $v_x = 10\,\mathrm{m/s}$ and $v_r = 5\,\mathrm{m/s}$. Figure 4 shows the imaging result obtained by the range migration algorithm (RMA) with the entire data, where the defocused sub-image containing the moving target, i.e., ROI, is indicated by the red dashed box.

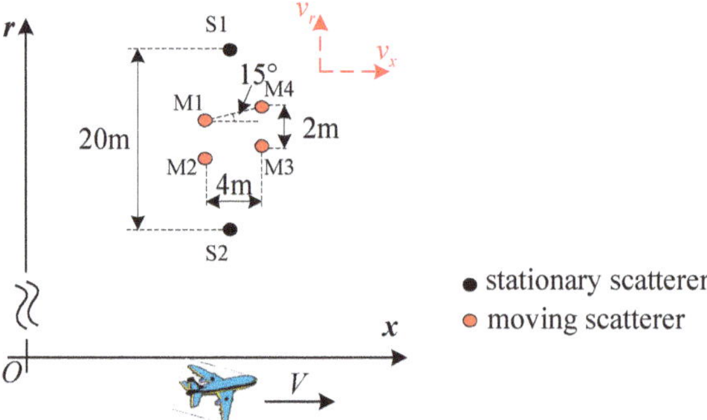


**Figure 3.** Simulation scene of a rigid-body moving target and two stationary reference points.

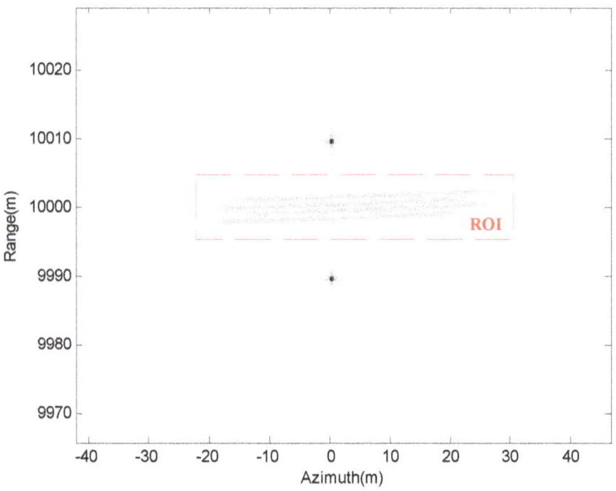


**Figure 4.** Range migration algorithm (RMA) imaging result with entire data.

By extracting the ROI sub-image from the regular SAR image, the data size is reduced to $n_r \times n_a$, i.e., 30 × 1051. In the proposed method, the initialized value of phase compensation parameter is set as $\alpha^{(0)} = 1/150^2$. The convergence threshold in Figure 2 is set as $\eta = \alpha^{(0)}/10^4$. The convergence process of the phase compensation parameter estimation is shown in Figure 5. For comparison, the algorithm proposed in [16] is also employed in this experiment. The defocused ROI data and refocused results are shown in Figure 6. By 2-D searching, the method of [16] can accurately estimate the target equivalent velocity. From Figure 6b,c, we can see that the image produces by the proposed method has a stronger contrast. The reason is that the sparse based method can effectively suppress the side-lobes.

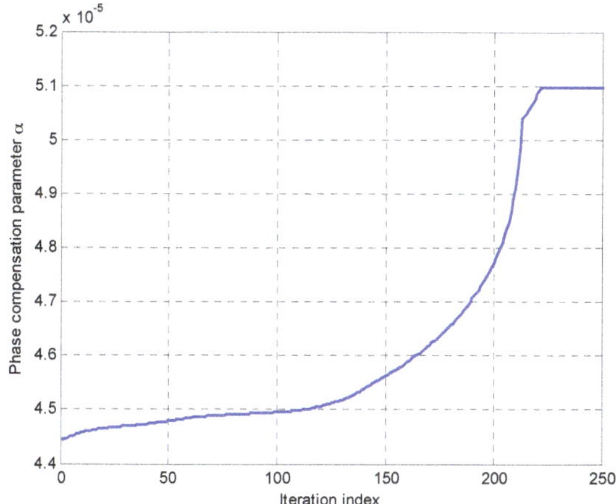


**Figure 5.** Convergence processes of the proposed method.

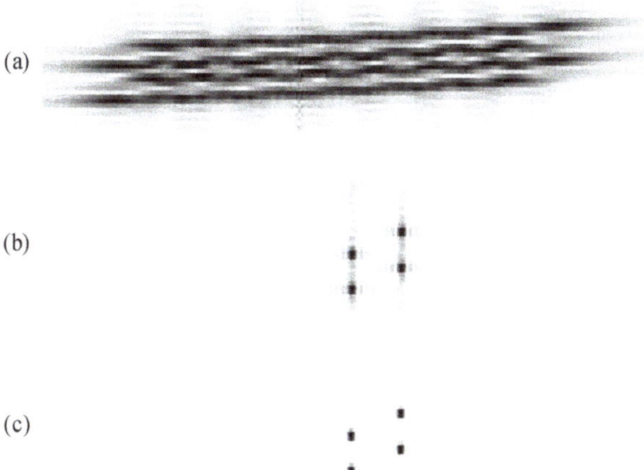


**Figure 6.** (**a**) Defocused region of interest (ROI) data; (**b**) refocused result obtained by the method in [16]; (**c**) refocused result obtained by the proposed method.

In the second simulation, the target in acceleration state is considered. The actual motion parameters of the point target are set to be $v_x = 10\,\mathrm{m/s}$, $v_r = 5\,\mathrm{m/s}$, azimuth acceleration $a_x = 1\,\mathrm{m/s^2}$, and range acceleration $a_r = 1\,\mathrm{m/s^2}$. By extracting the ROI sub-image from the regular SAR image, the data size is $41 \times 2201$. The convergence process of the phase compensation parameter estimation is shown in Figure 7. The refocusing results obtained by the method in [16], and the proposed method are shown in Figure 8a,b, respectively. It is clear that the focusing performance of the method in [16] is seriously deteriorated. The reason is that this method cannot eliminate the high-order phase error caused by accelerative motion of the target. Moreover, we take the scatterer M3 on the target as example to analyze the refocusing quality. The range and azimuth profiles obtained by different methods are shown in Figures 9 and 10, respectively. One can find that the sub-image produced by the proposed method can effectively suppress the asymmetric side-lobe and improve the image quality, which benefits from the superiority of the sparse constraint.

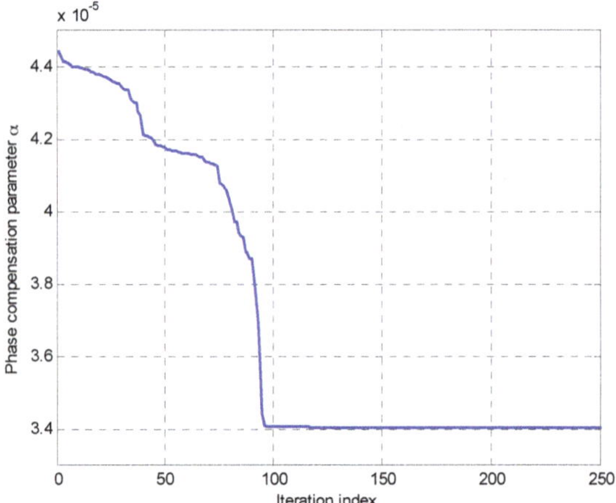


Figure 7. Convergence processes of the proposed method.

Figure 8. Refocusing result obtained by (a) method proposed in [16]; (b) the proposed method.

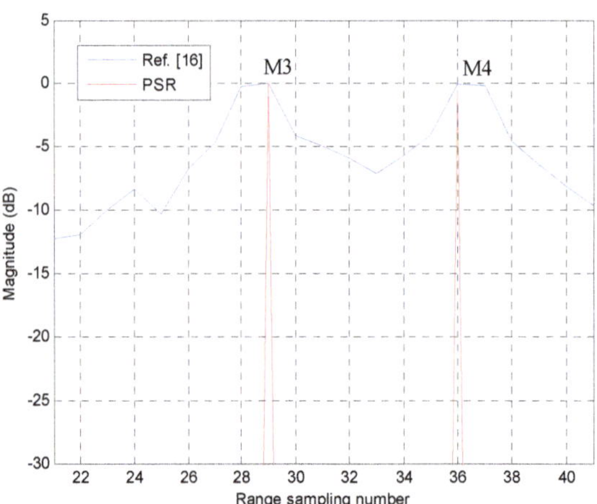


Figure 9. The range profiles of moving scatterer M3.

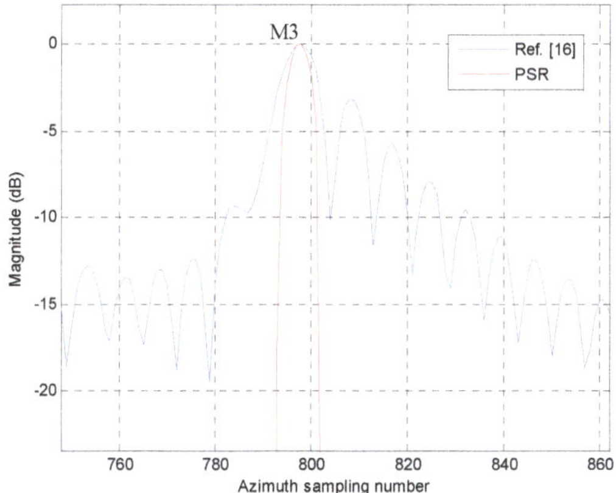


**Figure 10.** The azimuth profiles of moving scatterer M3.

### 4.2. Space-Borne Measured Data

We now present the experimental results by using the GF-3 space-borne SAR data containing moving ships. The regular SAR image of the sea surface is shown in Figure 11. We can see that the whole image is focused well, but some ships are defocused due to their movements. Three ROIs named T1, T2, and T3 containing moving ships are cropped from the original complex image, respectively. These three ships are processed sequentially using the method in [16] and the proposed method, respectively. Figure 12 shows the convergence processes of the phase compensation parameter of these moving ships. The refocused results obtained by the method in [16] and the proposed method are compared in Figures 13–15. It can be observed that the proposed method can successfully reconstruct images of moving ships and significantly suppress the side-lobes. To quantitatively compare different algorithms in terms of the image quality, the image entropy values of the refocused ROI images are listed in Table 1. The smaller value of image entropy means better focusing effect. As shown in Table 1, the proposed method can provide better image quality than the method in [16].

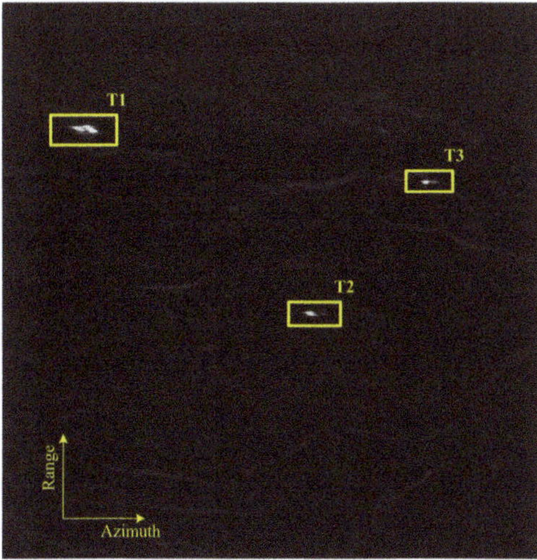


**Figure 11.** Regular imaging result of real data from GF-3 space-borne SAR.

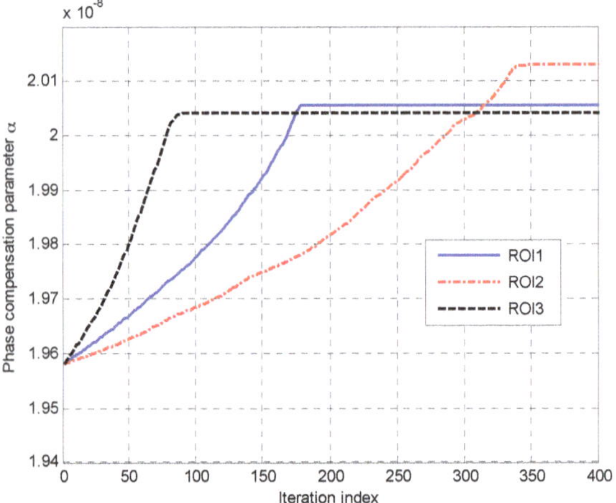


**Figure 12.** Convergence processes of the proposed method.

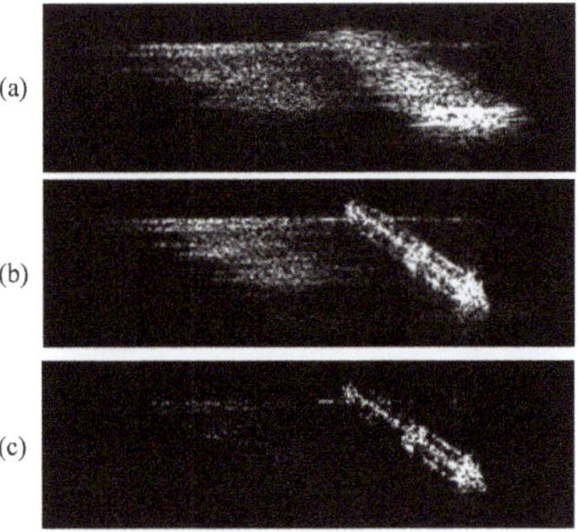


**Figure 13.** (**a**) Defocused ROI data of ship T1; (**b**) refocused result obtained by the method in [16]; (**c**) refocused result obtained by the proposed algorithm.

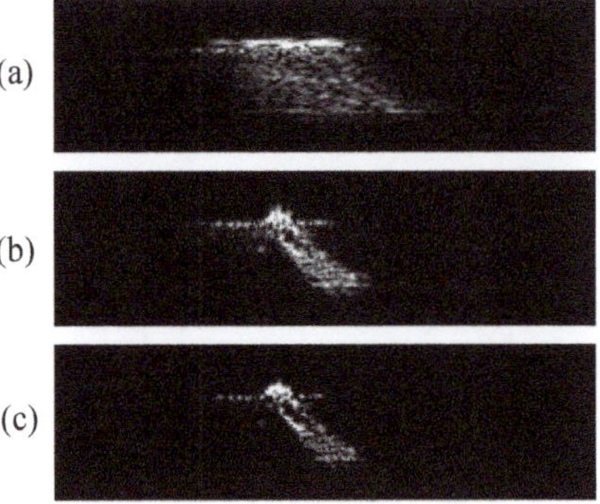


**Figure 14.** (**a**) Defocused ROI data of ship T2; (**b**) refocused result obtained by the method in [16]; (**c**) refocused result obtained by the proposed algorithm.

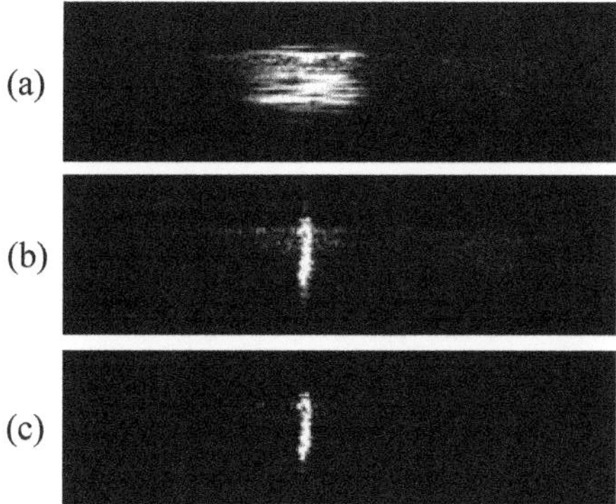


**Figure 15.** (**a**) Defocused ROI data of ship T3; (**b**) refocused result obtained by the method in [16]; (**c**) refocused result obtained by the proposed algorithm.

**Table 1.** The image entropy of the refocused ROI images.

| Targets and Methods | Ship T1 | | Ship T2 | | Ship T3 | |
|---|---|---|---|---|---|---|
| | Ref. [16] | PSR | Ref. [16] | PSR | Ref. [16] | PSR |
| **Image Entropy** | 6.4149 | 3.5389 | 6.3011 | 3.9863 | 5.077 | 4.0953 |

## 5. Conclusions

In this paper, we have presented a parametric sparse representation method for moving target imaging in SAR with ROI data. The ROI data extracted from regular SAR images is represented in a sparse fashion through a parametric refocusing transform. Then, the spare image of the target and the phase compensation parameter are estimated by solving a joint the optimization problem through an iterative process. The proposed method works on the complex ROI data, rather than the raw entire data, which is helpful in reducing the amount of data and alleviating the clutter. Particularly, the proposed method can suppress asymmetric side-lobes and improve the image quality of moving targets, compared to the method in [16]. Experimental results based on both simulated and real space-borne SAR data validate the effectiveness of this method on refocusing the image of moving target using ROI data. Through our experiments, with the initial values as set in Section IV, we have never encountered the case where the proposed algorithm does not converge. A theoretical convergence analysis of the proposed algorithm will be studied in our future work.

**Acknowledgments:** This work was supported in part by the National Natural Science Foundation of China under Grants 61422110, 61631019, 61471019, and 61661130158, and in part by the National Ten Thousand Talent Program of China (Young Top-Notch Talent), and in part by the Tsinghua National Laboratory for Information Science (TNList), and in part by the Tsinghua University Initiative Scientific Research Program.

**Author Contributions:** All authors contributed extensively to the study presented in this manuscript. Gang Li proposed the methodology. Yichang Chen was responsible for data processing. Qun Zhang and Jinping Sun contributed to algorithm review, result discussion and manuscript review.

**Conflicts of Interest:** The authors declare no conflict of interest.

## References

1. Werness, S.A.S.; Carrara, W.G.; Joyce, L.S.; Franczak, D.B. Moving target imaging algorithm for SAR data. *IEEE Trans. Aerosp. Electron. Syst.* **1990**, *26*, 57–67. [CrossRef]
2. Zhou, F.; Wu, R.; Xing, M.; Bao, Z. Approach for single channel SAR ground moving target imaging and motion parameter estimation. *IET Radar Sonar Navig.* **2007**, *1*, 59–66. [CrossRef]

3. Perry, R.P.; DiPietro, R.C.; Fante, R. SAR imaging of moving targets. *IEEE Trans. Aerosp. Electron. Syst.* **1999**, *35*, 188–200. [CrossRef]
4. Huang, P.; Liao, G.; Yang, Z.; Xia, X.G.; Ma, J.; Zheng, J. Ground maneuvering target imaging and high-order motion parameter estimation based on second-order keystone and generalized Hough-HAF transform. *IEEE Trans. Geosci. Remote Sens.* **2014**, *55*, 320–335. [CrossRef]
5. Gao, G.; Shi, G.; Yang, L.; Zhou, S. Moving target detection based on the spreading characteristics of SAR interferograms in the magnitude-phase plane. *Remote Sens.* **2015**, *7*, 1836–1854. [CrossRef]
6. Graziano, M.D.; DErrico, M.; Rufino, G. Wake component detection in X-band SAR images for ship heading and velocity estimation. *Remote Sens.* **2016**, *8*, 498. [CrossRef]
7. Li, G.; Xia, X.G.; Xu, J.; Peng, Y.N. A velocity estimation algorithm of moving targets using single antenna SAR. *IEEE Trans. Aerosp. Electron. Syst.* **2009**, *45*, 1052–1062. [CrossRef]
8. Zhu, S.Q.; Liao, G.S.; Qu, Y.; Zhou, Z.G.; Liu, X.Y. Ground moving targets imaging algorithm for synthetic aperture radar. *IEEE Trans. Geosci. Remote Sens.* **2011**, *49*, 462–477. [CrossRef]
9. Zhu, D.Y.; Li, Y.; Zhu, Z.D. A keystone transform without interpolation for SAR ground moving-target imaging. *IEEE Geosci. Remote Sens. Lett.* **2007**, *4*, 18–22. [CrossRef]
10. Yang, J.G.; Huang, X.T.; Jin, T.; Thompson, J.; Zhou, Z.M. New approach for SAR imaging of ground moving targets based on a keystone transform. *IEEE Geosci. Remote Sens. Lett.* **2011**, *8*, 829–833. [CrossRef]
11. Sun, G.C.; Xing, M.D.; Xia, X.G.; Wu, Y.R.; Bao, Z. Robust ground moving-target imaging using deramp-keystone processing. *IEEE Trans. Geosci. Remote Sens.* **2013**, *51*, 966–982. [CrossRef]
12. Gu, F.F.; Zhang, Q.; Chen, Y.C.; Huo, W.J.; Ni, J.C. Parametric sparse representation method for motion parameter estimation of ground moving target. *IEEE Sens. J.* **2016**, *16*, 7646–7652. [CrossRef]
13. Wang, H.S.C. Mainlobe clutter cancellation by DPCA for space-based radars. In Proceedings of the IEEE Aerospace Applications Conference, Crested Butte, CO, USA, 3–8 February 1991.
14. Pascazio, V.; Schirinzi, G.; Farina, A. Moving target detection by along-track interferometry. In Proceedings of the International Geoscience Remote Sensing Symposium, Sydney, Australia, 9–13 July 2001.
15. Martorella, M.; Giusti, E.; Berizzi, F.; Bacci, A.; Dalle Mese, E. ISAR based technique for refocusing non-cooperative targets in SAR images. *IET Radar Sonar Navig.* **2012**, *6*, 332–340. [CrossRef]
16. Zhang, Y.; Sun, J.; Lei, P.; Li, G.; Hong, W. High-resolution SAR-based ground moving target imaging with defocused ROI data. *IEEE Trans. Geosci. Remote Sens.* **2016**, *54*, 1062–1073. [CrossRef]
17. Sjogren, T.K.; Vu, V.T.; Pettersson, M.I. Moving target refocusing algorithm for synthetic aperture radar images. In Proceedings of the IEEE International Geoscience Remote Sensing Symposium, Honolulu, HI, USA, 25–30 July 2010.
18. Onhon, N.O.; Cetin, M. SAR moving object imaging using sparsity imposing priors. *EURASIP J. Adv. Signal Process.* **2017**. [CrossRef]
19. Wu, Q.; Xing, M.; Qiu, C.; Liu, B.; Bao, Z.; Yeo, T.S. Motion parameter estimation in the SAR system with low PRF sampling. *IEEE Geosci. Remote Sens. Lett.* **2010**, *7*, 450–454. [CrossRef]
20. Khwaja, A.S.; Ma, J. Applications of compressed sensing for SAR moving-target velocity estimation and image compression. *IEEE Trans. Instrum. Meas.* **2011**, *60*, 2848–2860. [CrossRef]
21. Stojanovic, I.; Karl, W.C. Imaging of moving targets with multi-static SAR using an overcomplete dictionary. *IEEE J. Sel. Top. Signal Process.* **2011**, *4*, 164–176. [CrossRef]
22. Onhon, N.Ö.; Cetin, M. A sparsity-driven approach for joint SAR imaging and phase error correction. *IEEE Trans. Image Process.* **2012**, *21*, 2075–2088. [CrossRef] [PubMed]
23. Cetin, M.; Stojanovic, I.; Onhon, N.Ö.; Varshney, K.R.; Samadi, S.; Karl, W.C.; Willsky, A.S. Sparsity-driven synthetic aperture radar imaging: Reconstruction, autofocusing, moving targets, and compressed sensing. *IEEE Signal Process. Mag.* **2014**, *31*, 27–40. [CrossRef]
24. Zhao, L.F.; Wang, L.; Bi, G.A.; Yang, L. An autofocus technique for high-resolution inverse synthetic aperture radar imagery. *IEEE Trans. Geosci. Remote Sens.* **2014**, *52*, 6392–6403. [CrossRef]
25. Zhao, L.F.; Wang, L.; Yang, L.; Zoubir, A.M.; Bi, G. The race to improve radar imagery: An overview of recent progress in statistical sparsity-based techniques. *IEEE Signal Process. Mag.* **2016**, *33*, 85–102. [CrossRef]
26. Rao, W.; Li, G.; Wang, X.; Xia, X.-G. Parametric sparse representation method for ISAR imaging of rotating targets. *IEEE Trans. Aerosp. Electron. Syst.* **2014**, *50*, 910–919. [CrossRef]

27. Rao, W.; Li, G.; Wang, X.; Xia, X.-G. Adaptive sparse recovery by parametric weighted L1 minimization for ISAR imaging of uniformly rotating targets. *IEEE J. Sel. Top. Appl. Earth Obs. Remote Sens.* **2013**, *6*, 942–952. [CrossRef]
28. Li, G.; Zhang, H.; Wang, X.; Xia, X.-G. ISAR 2-D imaging of uniformly rotating targets via matching pursuit. *IEEE Trans. Aerosp. Electron. Syst.* **2012**, *48*, 1838–1846. [CrossRef]
29. Chen, Y.C.; Li, G.; Zhang, Q.; Zhang, Q.J.; Xia, X.G. Motion compensation for airborne SAR via parametric sparse representation. *IEEE Trans. Geosci. Remote Sens.* **2017**, *55*, 551–562. [CrossRef]
30. Donoho, D.L. Compressed sensing. *IEEE Trans. Inf. Theory* **2006**, *52*, 1289–1306. [CrossRef]
31. Candes, E.J.; Romberg, J.; Tao, T. Robust uncertainty principles: Exact signal reconstruction from highly incomplete frequency information. *IEEE Trans. Inf. Theory* **2006**, *52*, 489–509. [CrossRef]
32. Fang, J.; Xu, Z.; Zhang, B.; Hong, W.; Wu, Y. Fast compressed sensing SAR imaging based on approximated observation. *IEEE J. Sel. Top. Appl. Earth Obs. Remote Sens.* **2014**, *7*, 352–363. [CrossRef]

*Article*

# An ML-Based Radial Velocity Estimation Algorithm for Moving Targets in Spaceborne High-Resolution and Wide-Swath SAR Systems

**Tingting Jin [1,2], Xiaolan Qiu [1], Donghui Hu [1] and Chibiao Ding [1,*]**

[1] Key Laboratory of Technology in Geo-Spatial Information Processing and Application System, Institute of Electronics, Chinese Academy of Sciences, Beijing 100190 China; jintingting13@mails.ucas.ac.cn (T.J.); xlqiu@mail.ie.ac.cn (X.Q.); dhhu@mail.ie.ac.cn (D.H.)

[2] University of the Chinese Academy of Sciences, Beijing 100049, China

* Correspondence: cbding@mail.ie.ac.cn; Tel.: +86-138-0107-7025

Academic Editors: Xiaofeng Yang, Xiaofeng Li, Ferdinando Nunziata, Alexis Mouche and Prasad S. Thenkabail
Received: 28 February 2017; Accepted: 21 April 2017; Published: 26 April 2017

**Abstract:** Multichannel synthetic aperture radar (SAR) is a significant breakthrough to the inherent limitation between high-resolution and wide-swath (HRWS) compared with conventional SAR. Moving target indication (MTI) is an important application of spaceborne HRWS SAR systems. In contrast to previous studies of SAR MTI, the HRWS SAR mainly faces the problem of under-sampled data of each channel, causing single-channel imaging and processing to be infeasible. In this study, the estimation of velocity is equivalent to the estimation of the cone angle according to their relationship. The maximum likelihood (ML) based algorithm is proposed to estimate the radial velocity in the existence of Doppler ambiguities. After that, the signal reconstruction and compensation for the phase offset caused by radial velocity are processed for a moving target. Finally, the traditional imaging algorithm is applied to obtain a focused moving target image. Experiments are conducted to evaluate the accuracy and effectiveness of the estimator under different signal-to-noise ratios (SNR). Furthermore, the performance is analyzed with respect to the motion ship that experiences interference due to different distributions of sea clutter. The results verify that the proposed algorithm is accurate and efficient with low computational complexity. This paper aims at providing a solution to the velocity estimation problem in the future HRWS SAR systems with multiple receive channels.

**Keywords:** synthetic aperture radar (SAR); high-resolution and wide-swath (HRWS); velocity estimation; Doppler ambiguities; maximum likelihood (ML)

## 1. Introduction

Remote sensing for civilian and military applications sets a high requirement on both the spatial resolution and swath coverage for synthetic aperture radar (SAR). However, conventional SAR systems can barely achieve high-resolution and wide-swath (HRWS) images simultaneously [1]. Higher pulse repetition frequency (PRF) is needed to obtain higher azimuth resolution, while lower PRF is required to acquire a wider range swath. Multichannel SAR in the azimuth, which can overcome this inherent limitation, has attracted much attention in recent years [2]. The launch of the TerraSAR-X satellite in 2007 [3], the ALOS-2 satellite in 2014 [4], and the Chinese Gaofen-3 satellite in 2016, which all contain a dual-receive channel mode, demonstrated the feasibility of this technique. Spaceborne HRWS SAR with more receive channels is one of the prospects of SAR systems. Moving target indication and imaging is one of the primary applications of spaceborne HRWS SAR systems, especially for ocean remote sensing [2,3]. Estimation of the target's velocity is crucial for target relocation, focused imaging and false target suppression [5–15].

The moving target's velocity can be divided into the radial velocity and the azimuth velocity, which stands for the cross-track and along-track velocities of the moving target, respectively. For a spaceborne HRWS SAR system, the azimuth velocity of a motion target is far smaller than that of the satellite, and thus can be ignored. The effects of the radial velocity are predominant [11], and are listed as follows:

(1) The linear Range Cell Migration (RCM) is caused by the radial velocity after range compression of the moving target;
(2) The azimuth offset of a moving target's location is proportional to its radial velocity; and
(3) The reconstructed echo of a moving target will introduce a frequency-dependent phase mismatch, leading to false targets along the azimuth after imaging.

Thus, estimating the radial velocity is a key procedure to relocation and precise imaging of a moving target.

For a spaceborne multichannel SAR, the low PRF is transmitted to achieve wide swath images with low range ambiguity levels, at the cost of under-sampled data in the azimuth and Doppler spectrum ambiguity for a single channel. Unambiguous imaging of a single channel echo is not feasible. Therefore, a reconstruction algorithm is introduced [1] to suppress the Doppler spectrum ambiguity before obtaining a HRWS image. Most of the previous studies place an emphasis on moving target indication (MTI) with the assumption that there is no Doppler ambiguity for each channel. For example, in the along-track interferometry (ATI) method [6], the Eigen-decomposition method of the covariance matrix [7] processes in the image domain of each channel. However, the main problem of a HRWS SAR MTI system is under-sampled data in the azimuth for signal-channel echo.

In recent years, several methods have been proposed that are focused on moving target indication for HRWS SAR systems, which aim at estimating velocity before imaging. In [8], estimating the radial velocity is transformed to the direction-of-arrival (DOA) estimation of the echoes. By constructing the spatial spectrum of the moving target, the radial velocity can be estimated by maximizing the spectrum. However, without analyzing the efficiency or considering the sea clutter, the analysis is not comprehensive. In [9–11], the imagery quality of the moving target is weighed by some criterion, and the radial velocity is estimated by searching for the value which optimizes the imagery quality. These approaches need iteration, thus guaranteeing the accuracy with the sacrifice of efficiency.

In [12], the Radon transform is applied to estimate the slope of the Doppler spectrum of the single-channel echo, and the radial velocity is proportional to the slope. The computational load is large for the implementation of Radon transform with each searched velocity. Additionally, the redundant information of the multi-channel signal is not taken full advantage of. Yang et al. and Wang et al. [13,14] transform the velocity estimation problem to measuring the azimuth offset, which is proportional to its radial velocity mentioned above. However, these methods need additional processing of the image and determination between false targets and the real one. Furthermore, these methods lack detailed analysis of performance under sea clutter distributions.

In this paper, we propose a novel algorithm for velocity estimation and unambiguous imaging of the moving target in a spaceborne HRWS SAR system. In addition, the estimation accuracy under different sea clutter distributions are discussed. Firstly, we deduce the echo of a moving target for multichannel SAR systems, and obtain the relationship of the radial velocity and the cone angle. Considering that the cone angles are sparse in space for a certain Doppler frequency [16], we apply the maximum likelihood (ML) method to estimate the cone angle as well as the radial velocity. Then the signal reconstruction and compensation of the phase mismatch caused by target motion are processed, followed by focused imaging to suppress false targets. Finally, we discuss the estimation of a moving ship interfered with different sea clutter distributions. The merit of this algorithm is that it does not need iteration or Eigen-decomposition, thus the computational complexity is not large. More importantly, this algorithm can estimate velocities of multiple moving targets as it does not need too many samples of Doppler bins, making estimating of adjacent targets possible.

This paper begins with signal model of the moving target for a HRWS SAR system in Section 2. In Section 3, detailed descriptions of the proposed velocity estimation algorithm and the imaging process of multichannel moving target echoes are given, and the Cramér-Rao lower bound of velocity estimation is deduced. Section 4 presents the experimental results of estimation accuracy under different conditions, followed by the performance analysis in Section 5. Section 6 draws conclusions and discusses future perspectives.

## 2. Echo Model of the Moving Target

### 2.1. Ideal Echo Model

For spaceborne multichannel SAR systems, the echo of the moving target has the following characteristics:

- The velocity of the moving target is treated as constant during the antenna beam scanning as the satellite is moving fast.
- The azimuth velocity of the moving target is negligible as it is much smaller than the satellite velocity.
- The radial velocity can be treated as the same for each receive channel as the radar beam is very narrow.

The geometry of the HRWS SAR system is depicted in Figure 1. The x-axis points to the direction of the platform velocity of the satellite, the z-axis points away from the Earth's center, and the three axes satisfy the orthogonal right-hand rule. The velocity of the platform is $v_s$, $R_0$ is the shortest slant distance of the target, and $R_b$ is the corresponding ground range. The full area of the antenna is used as the transmitter, and is split into $M$ channels in the azimuth as receivers. The antenna transmits chirp signals at the center (Tx), and Rx1-Rx$M$ receive echoes simultaneously. The azimuth resolution of the multichannel SAR system depends on the aperture size of a single receive channel. The distance between two receive channels is $d$, the radial velocity and the azimuth velocity of the moving target are $v_r$ and $v_a$, respectively.

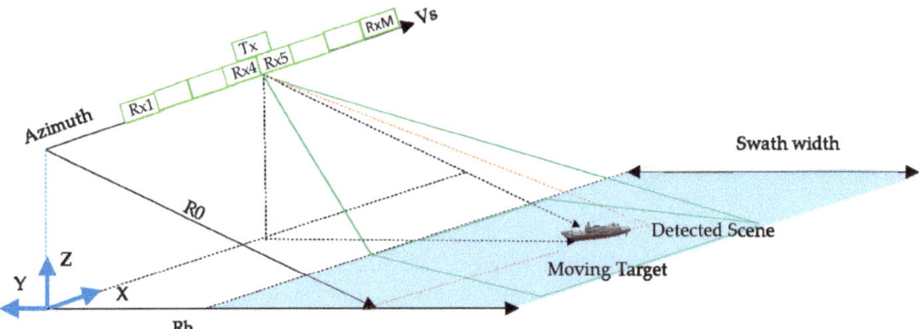


**Figure 1.** The imaging geometry of the high-resolution and wide-swath (HRWS) synthetic aperture radar (SAR) system.

The distance between the transmit center and the moving target is donated as $R_T(\eta)$, and the distance between the $m$-th receiver and the moving target is $R_{Rm}(\eta)$, expressed as:

$$R_T(\eta) = \sqrt{\left((v_s - v_a)\eta\right)^2 + (v_r\eta + R_0)^2} \approx R_0 + v_r\eta + \frac{(v_s\eta)^2}{2R_0} \quad (1)$$

$$R_{Rm}(\eta) \approx \sqrt{\left((v_s - v_a)\eta + x_m\right)^2 + (v_r\eta + R_0)^2} \approx R_0 + v_r\eta + \frac{(v_s\eta + x_m)^2}{2R_0} \quad (2)$$

where $\tau, \eta$ denote the range time and azimuth time, respectively, and:

$$x_m = \left(m - \frac{M+1}{2}\right)d, \quad m = 1, 2..., M \quad (3)$$

For the sake of convenient expression, the moving target is modeled as an ideal point target with constant radar cross section (RCS). Then the received signal of the $m$-th channel can be expressed as:

$$\begin{aligned} s_m(\tau,\eta) = \sigma \cdot rect\left[\frac{\tau - \left(R_T(\eta)+R_{Rm}(\eta)\right)/c}{T_p}\right] \cdot \exp\left\{j\pi K_r\left[\tau - \left(R_T(\eta) + R_{Rm}(\eta)\right)/c\right]^2\right\} \\ \cdot rect\left(\frac{\eta}{T_s}\right) \cdot \exp\left[-j\frac{2\pi}{\lambda} \cdot \left(R_T(\eta) + R_{Rm}(\eta)\right)\right] \end{aligned} \quad (4)$$

where $c$ is the speed of light, $T_s$ is the synthetic aperture time, and $T_p$ and $K_r$ are the pulse width and chirp rate, respectively. $\sigma$ stands for the overall amplitude weighting of the target, containing the target backscatter coefficient, the weighted coefficient of the antenna pattern, and weighting factors of electromagnetic wave propagation in space. Substituting Equations (1) and (2) into Equation (4), the Doppler centroid and the Doppler rate can be written as:

$$f_{dc} = -\left(\frac{2v_r}{\lambda} + \frac{v_s x_m}{\lambda R_0}\right), K_a \approx -\frac{2v_s^2}{\lambda R_0} \quad (5)$$

While for a static target or clutter, $v_r = 0$, the Doppler centroid and the Doppler rate are

$$f_{dc} = -\frac{v_s x_m}{\lambda R_0}, K_a \approx -\frac{2v_s^2}{\lambda R_0} \quad (6)$$

Define $\phi_c$ and $\phi_t$ as the cone angles of the clutter and the moving target, respectively. From Equations (5) and (6), the existence of the target motion result in a certain offset of the Doppler frequency. For a side-looking SAR system, the relationship of the Doppler frequency $f_a$ and the cone angle $\phi_{\{c,t\}}$ can be expressed as [8]:

$$f_a(\phi_c) = \frac{2v_s}{\lambda}\sin\phi_c \quad (7)$$

$$f_{t,a}(\phi_t) = \frac{2v_s}{\lambda}\sin\phi_t + \Delta f_{t,a}(\phi_t) = \frac{2v_s}{\lambda}\sin\phi_t + \frac{2v_r}{\lambda} \quad (8)$$

Figure 2a shows the linear relationship between $f_a$ and $\sin\phi_{\{c,t\}}$, where the dotted line indicates the clutter, and the solid line the ground moving target. For a HRWS SAR system, a low PRF is adopted to eliminate the range ambiguities and enlarge the coverage, at the cost of the Doppler ambiguity. Then the relationship between $f_a$ and $\sin\phi_{\{c,t\}}$ is shown as Figure 2b in practical applications, where the Doppler spectrum of the clutter and the moving target are both folded. The corresponding moving target signal of the $m$-th channel in the Doppler domain is expressed as:

$$S_m(\tau, f_a) = \sum_{l=-L}^{L} S_1(\tau, f_a + l \cdot f_p)\exp\left\{j\frac{4\pi}{\lambda}x_m \sin\phi_t(f_a + l \cdot f_p)\right\} \quad (9)$$

where $S_m(\tau, f_a)$, $m = 1, 2...M$ is echo of the $m$-th channel in the range-Doppler domain, $f_p$ is the PRF, and the number of the main Doppler spectrum ambiguity is $N$, $N = 2L + 1$.

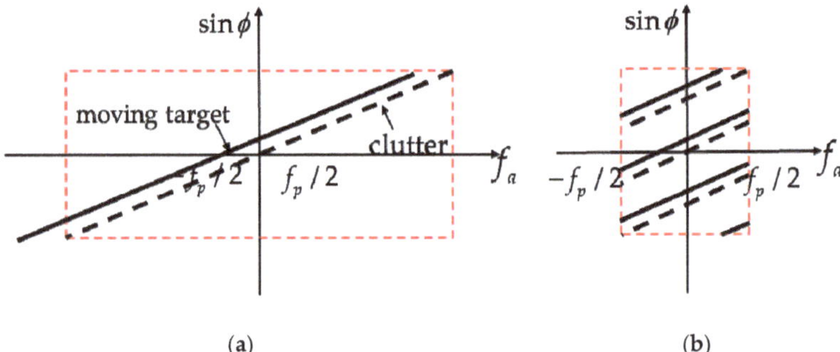


**Figure 2.** Spatial-temporal spectra of echoes: (**a**) unambiguous; and (**b**) Doppler ambiguous.

*2.2. Echo Model with Clutter and Noise*

In reality, the SAR echoes of the moving target are interfered by clutter and noise; the clutter is the background of the detected scene, and the noise comes from the receive chain of each channel. Thus the model of the clutter depends on the detected scene, and the model of the noise is normally white Gaussian noise. Considering the clutter and the noise, the echo in the range-Doppler domain of the $m$-th channel can be expressed as:

$$\begin{aligned} S_m(\tau, f_a) &= \sum_{l=-L}^{L} S_1(\tau, f_a + l \cdot f_p) \exp\left\{j\tfrac{4\pi}{\lambda} x_m \sin\phi_t(f_a + l \cdot f_p)\right\} \\ &+ \sum_{l=-L}^{L} S_{c1}(\tau, f_a + l \cdot f_p) \exp\left\{j\tfrac{4\pi}{\lambda} x_m \sin\phi_t(f_a + l \cdot f_p)\right\} + N_m(\tau, f_a) \end{aligned} \quad (10)$$

where $S_{c1}(\tau, f_a)$ is the clutter signal of the first channel in the range-Doppler domain, $N_m(\tau, f_a)$ is the white Gaussian noise.

For simplification, the echoes of $M$ channels in the vector form can be written as [17]

$$\mathbf{x}(\tau, f_a) = \sigma_t \mathbf{s}(\tau, f_a) + \sum_i \sigma_i \mathbf{c}(\tau, f_a) + \mathbf{n} \quad (11)$$

where bold lowercase letters are used for vectors, $\mathbf{x}(\tau, f_a) = [x_1(\tau, f_a), x_2(\tau, f_a), ..., x_M(\tau, f_a)]^T$ denotes the echo vector in the range-Doppler domain of M channels, and $x_m(\tau, f_a)$ is the echo of the $i$-th channel combined with clutter and noise. $\mathbf{s}(\tau, f_a)$ denotes the echo vector of moving target, $\mathbf{c}(\tau, f_a)$ denotes the echo vector of clutter, and $\mathbf{n}$ denotes the echo vector of thermal noise. $\sigma_t$ and $\sigma_i$ are the complex scattering coefficients of the moving target and the clutter unit, respectively.

## 3. The Proposed Velocity Estimation Algorithm

According to Equation (8), the existence of $v_r$ causes a certain offset of the Doppler frequency; thus, the cone angle of the moving target for a certain Doppler bin is related to $v_r$. In other words, the problem of radial velocity estimation is equivalent to the problem of cone angle estimation. Different from traditional direction of arrival (DOA) estimation, there exists a Doppler ambiguity in the echo for the HRWS SAR system. Fortunately, the steering vector matrix can be constructed considering the Doppler ambiguity for the sparse signal representation.

In this section, we firstly describe the proposed ML-based algorithm. Then the Cramér-Rao lower bound of velocity estimation is deduced. Finally, the HRWS SAR moving target imaging procedure is presented.

### 3.1. Algorithm Discription

To transform the velocity estimation problem to the cone angle estimation, or the steering vector estimation, the echo expressed in Equation (9) can be rewritten in the vector form as the product of the echo of first channel and the steering vector matrix, i.e.,

$$\mathbf{S}(\tau, f_a) = \mathbf{A}_t(f_a)\mathbf{S}_1(\tau, f_a) \tag{12}$$

where,

$$\mathbf{S}(\tau, f_a) = [\mathbf{s}_1(\tau, f_a), \mathbf{s}_2(\tau, f_a), ..., \mathbf{s}_M(\tau, f_a)]^T \tag{13}$$

$$\mathbf{s}_1(\tau, f_a) = [S_1(\tau, f_a - L \cdot f_p), ..., S_1(\tau, f_a + l \cdot f_p), ..., S_1(\tau, f_a + L \cdot f_p)]^T \tag{14}$$

$$\begin{aligned} \mathbf{A}_t(f_a) &= \left[\mathbf{a}_{t,-L}, ..., \mathbf{a}_{t,l}, ..., \mathbf{a}_{t,L}\right] \\ &= \begin{pmatrix} \exp(-j\frac{2\pi}{\lambda}x_1 \cdot \sin\phi_{t,-L}(f_a)) & \cdots & \exp(-j\frac{2\pi}{\lambda}x_1 \cdot \sin\phi_{t,L}(f_a)) \\ \vdots & \ddots & \vdots \\ \exp(-j\frac{2\pi}{\lambda}x_M \cdot \sin\phi_{t,-L}(f_a)) & \cdots & \exp(-j\frac{2\pi}{\lambda}x_M \cdot \sin\phi_{t,L}(f_a)) \end{pmatrix}_{M\times N} \end{aligned} \tag{15}$$

$S_1(\tau, f_a + l \cdot f_p)$ is the $l$-th ambiguous component of the first channel signal, and $\mathbf{a}_{t,l}$ denotes the $l$-th ambiguous steering vector, i.e.,

$$\mathbf{a}_{t,l} = \left[\exp\left(-j\frac{2\pi}{\lambda}x_1 \cdot \sin\phi_{t,l}(f_a)\right), \exp\left(-j\frac{2\pi}{\lambda}x_2 \cdot \sin\phi_{t,l}(f_a)\right), ..., \exp\left(-j\frac{2\pi}{\lambda}x_M \cdot \sin\phi_{t,l}(f_a)\right)\right]^T \tag{16}$$

where

$$\begin{aligned} \sin\phi_{t,l}(f_a) &= \frac{\lambda}{2v_s}\left(f_a + l \cdot f_p + \frac{2v_r}{\lambda}\right) \\ &= \frac{\lambda}{2v_s}\left(f_a + l \cdot f_p\right) + \frac{v_r}{v_s}, \quad l = -L...0...L \end{aligned} \tag{17}$$

From Equation (17), there is a one-to-one correspondence between radial velocity and the cone angle for a moving target.

Then the maximum likelihood (ML) algorithm is applied to estimate $\phi_t$ or $v_r$. The requirements of the ML algorithm are as follows [18]:

1. The signal covariance matrix is positive definite;
2. The number of sampling points is larger than the number of receive channels; and
3. The noises sampled at different Doppler frequencies are uncorrelated, and obey a white Gaussian distribution.

The joint conditional probability density function of the sampled signals from $K$ Doppler frequencies is:

$$f(x_1, x_2, ...x_K) = \prod_{k=1}^{K} \frac{1}{\det\{\pi\sigma_n^2 I\}} \exp\left(-\frac{1}{\sigma_n^2}|x_k - s_k|^2\right) \tag{18}$$

where $\sigma_n^2$ is the average power of $n_k$, $x_k$ is the sampled value of Equation (11), and $s_k$ is the sampled value of moving target signals in range-Doppler domain, i.e., $x_k = s_k + n_k$.

The criterion of the ML estimator is maximizing the following cost function

$$F_{ml}(\phi_t) = tr\left[I - \mathbf{A}\left(\mathbf{A}^{\mathbf{H}}\mathbf{A}\right)^{-1}\mathbf{A}^{\mathbf{H}}\right]\hat{\mathbf{R}}_X \tag{19}$$

where $tr[\cdot]$ represents the trace of a matrix, $\mathbf{A}$ is the steering vector matrix in Equation (15), and $\hat{\mathbf{R}}_X$ is the signal covariance matrix of $x_k$. By searching for $v_r$ in a certain space and computing the steering vector matrix $\mathbf{A}$, the maximum likelihood spectrum can be calculated. The maximum value of the spectrum corresponds to the ML estimation of $v_r$.

The process of the estimator is as follows:

(1) Conduct range compression of the echo of each channel, Equation (4) turns into

$$\begin{aligned} sr_m(\tau,\eta) = \sigma \cdot T_p \cdot rect\left(\tfrac{\eta}{T_s}\right) \cdot \exp\left[-j\tfrac{2\pi}{\lambda} \cdot (R_T(\eta) + R_{Rm}(\eta))\right] \\ \cdot sinc\left\{\pi B\left[\tau - (R_T(\eta) + R_{Rm}(\eta))/c\right]\right\} \end{aligned} \tag{20}$$

where $B = K_r \cdot T_p$ is the bandwidth of the chirp signal. As the *sinc* function varies little around the maximum value, the *sinc* function of Equation (20) for each channel is approximately equal.

(2) Conduct the azimuth Fourier transform, then extract the trajectory of the moving target in the range-Doppler domain. Sample the extracted signal at K Doppler bins to constitute vector $\mathbf{X}$, i.e.,

$$\mathbf{X} = [\mathbf{s}_1+\mathbf{n}_1,\ldots,\mathbf{s}_k+\mathbf{n}_k,\ldots,\mathbf{s}_K+\mathbf{n}_K]^T \tag{21}$$

where

$$\mathbf{s}_k = \left[s_1(\tau, f_{a,k}), s_2(\tau, f_{a,k}), \ldots, s_M(\tau, f_{a,k})\right]^T \tag{22}$$

$$\mathbf{n}_k = [n_1, n_2, \ldots, n_M]^T \tag{23}$$

The signal covariance matrix is acquired by $\hat{\mathbf{R}}_X = \mathbf{X} \cdot \mathbf{X}^H$.

(3) For each searched $v_r$, compute the steering vector matrix $\mathbf{A}$. Find the ML estimation of $v_r$ by substituting $\hat{\mathbf{R}}_X$ and $\mathbf{A}$ into Equation (19). Finally, average the estimated values from K Doppler bins to improve the robustness.

The common point of the proposed algorithm and the Capon spectrum [7] and MUSIC algorithms [8] is that, they all search for the best radial velocity according to some principle by constructing the sampled signal covariance matrix. The difference is the criterion they are based on. The proposed ML-based algorithm has lower complexity than the iterative approaches in [9–11], which need imaging for each possible velocity during the iteration. Moreover, this algorithm can estimate velocities of multiple moving targets as it does not need a large number of Doppler bins, as long as the number is larger than that of the receive channels.

### 3.2. The Cramér-Rao Lower Bound of Velocity Estimation

The Cramér–Rao lower bound (CRLB) is an important evaluation indicator for the effectiveness of parameter estimation. There inevitably exist errors in the ML estimation of $v_r$. The root mean squired error (RMSE) of $v_r$ estimation is compared with the CRLB, and the estimator is viewed as an effective estimate if the RMSE infinitely approaches CRLB with the increase of the signal-to-noise ratio (SNR).

A classical tool for deriving the CRLB is the Fisher information matrix (FIM), and the CRLB is obtained by computing the inverse of the FIM [19]. The FIM is obtained from the second-order derivative of the likelihood function, which is the logarithm of the joint conditional probability density function in Equation (18).

$$F(\phi_t) = -E\left[\frac{\partial^2 \ln f(\mathbf{x};\phi_t)}{\partial \phi_t^2}\right] \tag{24}$$

where $\mathbf{x} = [x_1, x_2, ...x_K]^T$ is the vector constituted by the sampled signals.

From Equation (18), the likelihood function is:

$$\ln f(\mathbf{x};\phi_t) = -\sum_{k=1}^{K}\left\{\frac{1}{\sigma_n^2}|x_k - s_k|^2 - \ln \det\left\{\pi\sigma_n^2 I\right\}\right\} \tag{25}$$

It has been proven in [20] that the CRLB of the ML estimation of the cone angle is expressed as

$$C(\hat{\phi}_{t,ml}) = \frac{\sigma_n}{2K}\left\{\mathrm{Re}\left[\mathbf{H} \odot \hat{\mathbf{R}}_X\right]\right\}^{-1} \tag{26}$$

where:

$$\mathbf{H} = \mathbf{D}^H\left[1 - \mathbf{A}(\mathbf{A}^H\mathbf{A})^{-1}\mathbf{A}^H\right]\mathbf{D} \tag{27}$$

$$\mathbf{D} = [\mathbf{d}_{-L}...\mathbf{d}_0...\mathbf{d}_L] \tag{28}$$

In Equation (26), $\mathbf{d}_l$ is the derivative of the steering vector to the cone angle, i.e.,

$$\mathbf{d}_l = d\mathbf{a}_t(\phi)/d\phi_t|_{\phi_t=\phi_{t,l}} \tag{29}$$

Finally, the CRLB of $v_r$ estimation is obtained from the relationship of $v_r$ and $\phi_t$ in Equation (17), i.e.,

$$C(\hat{v}_{r,ml}) = v_s \cdot \sin\left[C(\hat{\phi}_{t,ml})\right] \tag{30}$$

### *3.3. Processing Flow*

In Section 3.1, the proposed ML-based algorithm is described to estimate the radial velocity. Then the phase errors caused by the moving target of the $M$ channels are compensated before imaging to suppress the false targets. The total processing flow of the moving target's echoes of the multichannel SAR system is illustrated in Figure 3. Detailed descriptions are as follows:

1. Range Compression: Conduct range compression with the echo of each channel.
2. Azimuth Fourier Transform: Perform the Fourier transform in the azimuth to obtain $S_m(\tau, f_a)$, $m = 1, 2...M$
3. Range Bin Selection: Choose the range bins that contain echoes of the moving target. Normally, the range bin of the peak value and its adjacent range bins are selected.
4. ML Estimation of Radial Velocity: The ML method discussed in Section 3.1 is applied to estimate the radial velocity of the moving target.
5. Multichannel Reconstruction: Reconstruct echoes of $M$ channels and compensate for the phase offsets introduced by target motion.
6. Traditional Imaging Algorithm: After reconstruction, the echoes of $M$ channels are combined to equivalent single-channel signal without Doppler ambiguities. The traditional chirp scaling (CS) algorithm can be applied to obtain a focused image of the moving target with suppression of the false targets.

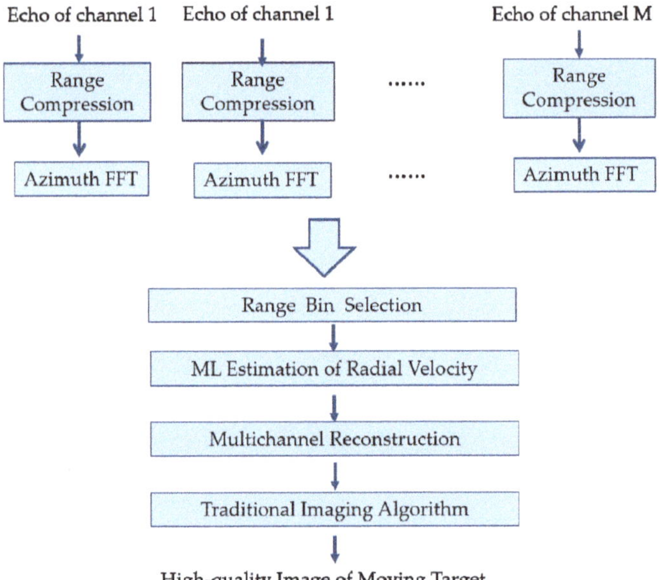


**Figure 3.** Processing flow of the moving target for the HRWS SAR system.

## 4. Experimental Results

In this section, experiments are conducted to evaluate the performance of the proposed ML-based radial velocity estimator. In Section 4.1, we demonstrate the accuracy and the effectiveness of the algorithm, the accuracy is evaluated by the estimation error, and the effectiveness is evaluated by the proximity of the RMSE to the CRLB. After radial velocity estimation, phase offsets among channels caused by target motion are compensated before imaging. The imaging results before and after compensation are compared in Section 4.2. Finally, the estimation accuracy under different distributions of sea clutter is discussed in Section 4.3.

The echo model of the moving target for the HRWS SAR system is shown is Figure 1. Figure 4 shows a diagram of the transmitting and receiving centers. The parameters of the simulated spaceborne multichannel SAR system are listed in Table 1. The number of the Doppler ambiguity $N$ equals 5 according to the parameters.

**Table 1.** Parameters of the simulated spaceborne multichannel SAR system.

| Parameter | Symbol | Value |
|---|---|---|
| Number of Channels | M | 8 |
| Aperture Size | Da | 11.2 m |
| Wavelength | $\lambda$ | 0.05556 m |
| Look Angle | $\theta$ | 53.45 degrees |
| PRF | $f_p$ | 1317.1 Hz |
| Doppler Bandwidth | $B_d$ | 5987.9 Hz |
| Satellite Velocity | $v_s$ | 7586.5 m/s |
| Sample Frequency | $f_s$ | 80 MHz |
| Bandwidth | $B_r$ | 67 MHz |
| Pulsewidth | $T_r$ | 38 μs |

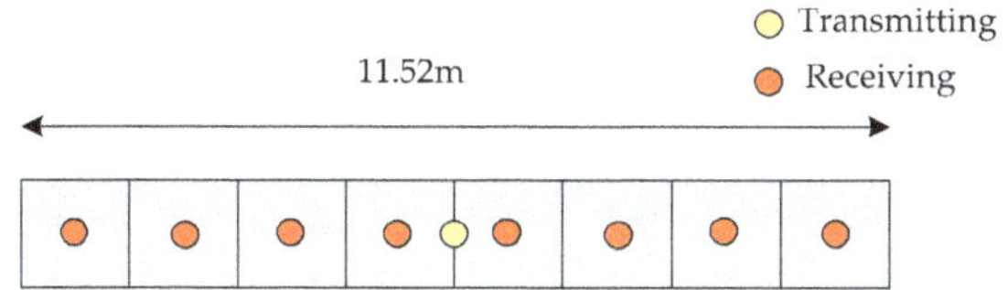


**Figure 4.** Diagram of the transmitting and receiving centers.

Figure 5 shows the comparison of simulated clutter signal and the moving target whose radial velocity is 10 m/s. Figure 5a is the Doppler spectrum of the clutter signal of reconstructed echoes, compared to that of the moving target signal in Figure 5b, where the spectral errors are evident from the frequency-dependent phase mismatch caused by radial velocity. Figure 5c,d compare the trajectory of the clutter signal and the moving target after range compression. We can see an additional linear Range Cell Migration (RCM) after range compression of the moving target. In the following, the trajectory of the moving target is extracted to estimate the radial velocity.

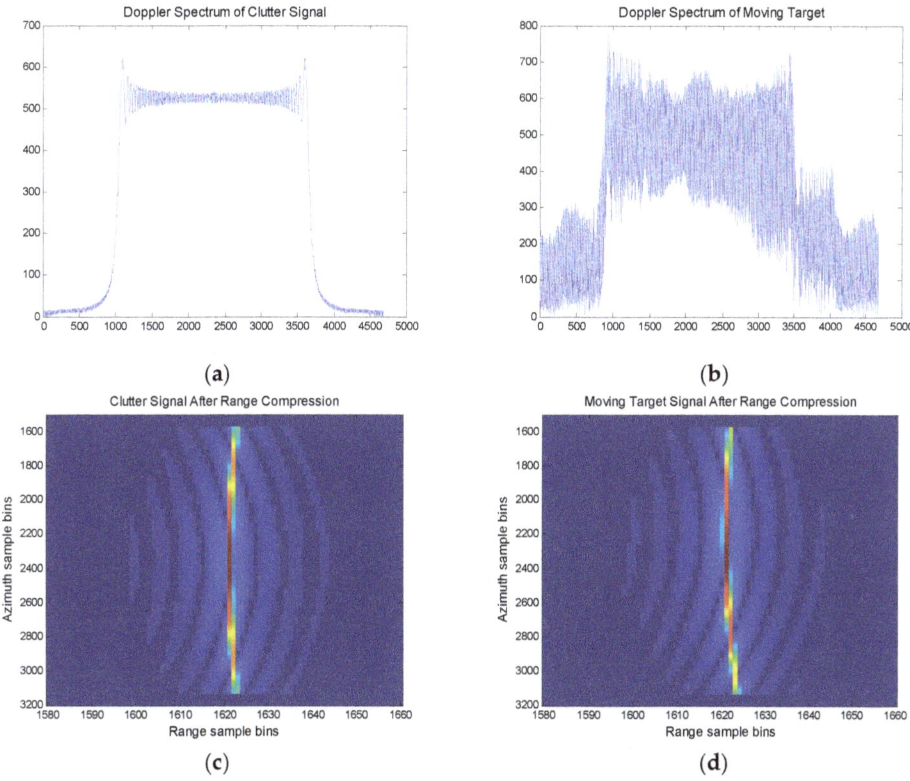


**Figure 5.** Comparison of the Doppler spectra and Range-compressed signals between clutter and the moving target: (**a**) Doppler spectrum of the clutter signal; (**b**) Doppler spectrum of the moving target; (**c**) range-compressed signal of the clutter signal; and (**d**) range-compressed signal of the moving target.

### 4.1. Performance of Radial Velocity Estimaion

In order to verify the performance of the proposed ML-based algorithm, an experiment is conducted to evaluate the estimation accuracy and efficiency under different signal-to-noise ratios (SNR). In the simulation, the radial velocity of the moving target is 10 m/s, and the SNR varies from −5 dB to 20 dB. The searching step is 0.01 m/s and the searching range is 0–20 m/s. The clutter scenario is temporarily not considered in the subsection. In the experiment, the range bin of the peak value from the trajectory of the moving target and its adjacent 10 range bins are selected, and we average the estimated values from 60 Doppler bins to improve the robustness. The estimated radial velocities, the estimation errors and relative errors under different SNRs are listed in Table 2. The maximum likelihood spectrum of radial velocity with the SNR = 0 dB is illustrated in Figure 6.

**Table 2.** Estimated Radial Velocities and Errors under different signal-to-noise ratios (SNRs).

| SNR (dB) | −5 | 0 | 5 | 10 | 15 | 20 |
|---|---|---|---|---|---|---|
| Estimated Value (m/s) | 10.47 | 10.22 | 10.08 | 9.97 | 9.99 | 10.00 |
| Estimation Error (m/s) | 0.47 | 0.22 | 0.08 | 0.03 | 0.01 | 0 |
| Relative Error | 4.7% | 2.2% | 0.8% | 0.3% | 0.1% | 0 |

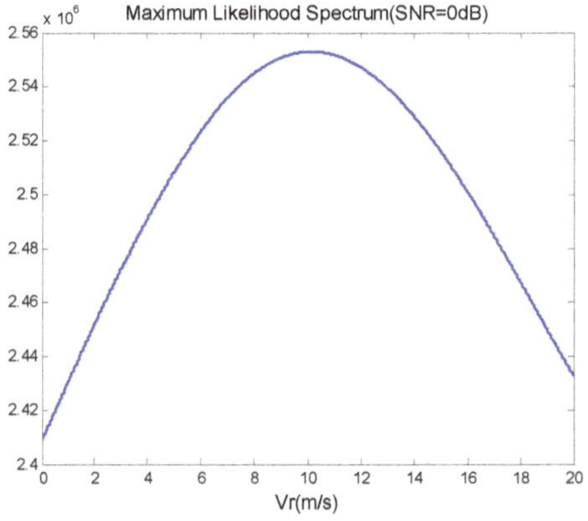


**Figure 6.** Maximum likelihood spectrum of radial velocity.

In the following, the efficiency of the estimator is evaluated by the proximity of RMSE to the CRLB. RMSE is expressed as

$$\sigma_{v_r} = \sqrt{\frac{1}{I}\sum_{i=1}^{I}(\hat{v}_{r,i} - v_r)^2} \tag{31}$$

where $I$ is the number of Monte Carlo experiments. In the comparison, we take the square root of the CRLB in Equation (28). As the computational load is high, 100 Monte Carlo experiments are conducted in the simulation. The RMSE is compared with the CRLB under different SNRs in Figure 7.

From the results in Table 2, the ML-based algorithm can estimate the radial velocity under very strong noise conditions. As a general rule, the SNR is larger than 0 dB, the estimation error is smaller than 0.22 m/s, with the relative error smaller than 2.2%. From the comparison of the RMSE and the CRLB in Figure 7, the RMSE infinitely approaches the CRLB with the increase of the SNR. Thus, the proposed ML-based radial velocity estimator is proven to be effective.

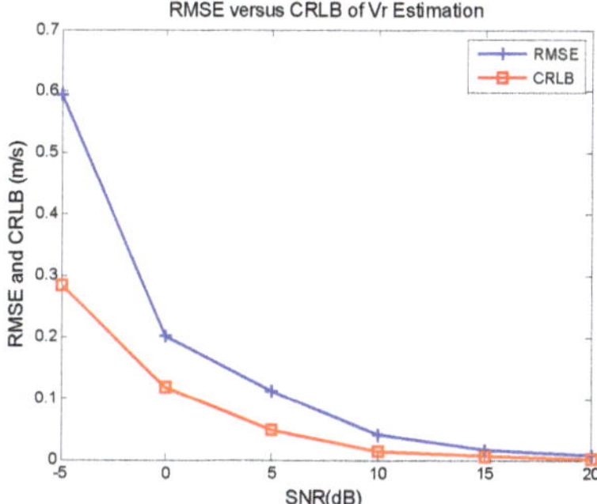


**Figure 7.** Root mean squired error (RMSE) versus Cramér–Rao lower bound (CRLB) of radial velocity estimation.

### 4.2. Imaging Results Before and after Estimation and Compensation

From the received signal of the $m$-th channel expressed in Equation (4), there exists a certain offset in the Doppler frequency compared to the signal of a static target. The frequency offset of each channel results in a frequency-dependent phase mismatch after the multichannel reconstruction introduced in Section 3.3. As a result, the phase mismatches among the channels will cause false targets along the azimuth when imaging. The phase error can be compensated after radial velocity estimation. After compensation for the phase errors, the traditional CS algorithm is applied to the imaging process. Figure 8a gives the trajectory of the moving target after range compression, where the linear RCM in Figure 5d is well corrected with the estimated radial velocity. Finally, the well-focused image of the moving target is obtained as shown in Figure 8b.

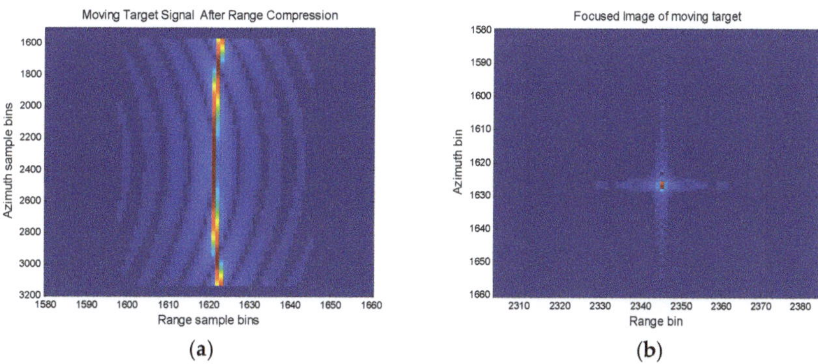


**Figure 8.** Imaging process after compensation for the errors caused by target motion: (**a**) Trajectory of the moving target after range compression; and (**b**) focused image of the moving target.

To demonstrate the impact of radial velocity on imaging quality for the multichannel SAR, we compare the imaging results before and after phase error compensation when the SNR = 0 dB in Figure 9. Figure 9a is the imagery of the moving target with traditional imaging process for multichannel SAR, where false targets are uniformly distributed along the azimuth around the real target. This is the impact of radial velocity on the multichannel SAR imaging. Figure 9b is the imagery after compensation for the phase offsets with the estimated velocity, where false targets are much suppressed and invisible. In Figure 10, the azimuth profiles of imaging results of the moving target are shown, Figure 10a corresponds to the azimuth profile of Figures 9a and 10b corresponds to the azimuth profile of Figure 9b.

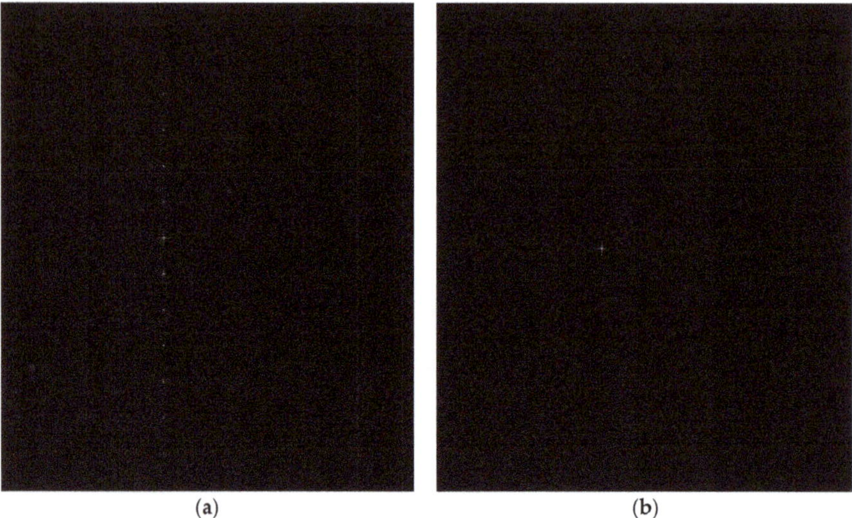

Figure 9. Imaging results before and after compensation for the errors caused by target motion: (a) before compensation; and (b) after compensation.

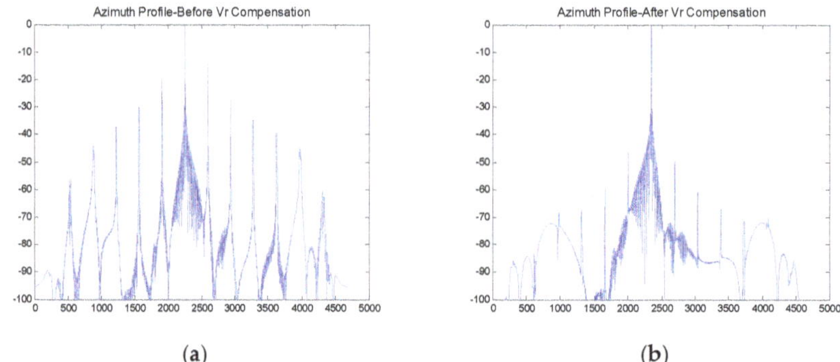


Figure 10. Azimuth profile of imaging result of the moving target: (a) before compensation; (b) after compensation.

To quantifiably describe the impact of radial velocity, we compute the maximum power of false targets relative to the real one. The maximum power of false targets corresponding to Figure 9a is −13.49 dB. Table 3 summarizes the maximum powers after compensation for phase errors under different SNRs, which are smaller than −46.38 dB when the SNR is larger than 0 dB.

Table 3. The maximum power of false targets.

| SNR (dB) | −5 | 0 | 5 | 10 | 15 | 20 |
|---|---|---|---|---|---|---|
| Maximum Power (dB) | −40.48 | −46.38 | −52.92 | −55.08 | −55.23 | −58.59 |

### 4.3. Estimation Accuracy under Clutter Interference

One of the primary applications of the spaceborne multichannel SAR system is the remote sensing of the sea surface, and sea clutter should be considered when detecting the moving ships. In the simulation, four commonly-used clutter models are simulated as the background of the detected scene. Rayleigh distribution is normally viewed as the magnitude probability function of the homogeneous scene. Weibull distribution, Log-normal distribution, and K-distribution are main magnitude models of the heterogeneous sea surfaces. The simulated clutter-interfered moving target signals are processed with the procedure in Section 3.3. In the simulation, the radial velocity of the moving target is 10 m/s, the signal-to-clutter ratio (SCR) is 0 dB and the SNR is 5 dB. The estimated radial velocities, the estimation errors and relative errors under different clutter distributions are listed in Table 4. Table 5 summarizes the maximum powers of false targets before and after compensation for estimated phase mismatches. We compare the imaging results before and after compensation under Weibull distribution clutter in Figure 11. In Figure 12, the azimuth profiles of imaging results are shown, Figure 12a corresponds to the azimuth profile of Figure 11a, and Figure 12b corresponds to the azimuth profile of Figure 11b.

**Table 4.** Estimated radial velocities and errors under different clutter distributions.

| Clutter Distribution | Rayleigh Distribution | Weibull Distribution | Log-Normal Distribution | K-Distribution |
|---|---|---|---|---|
| Estimated Value (m/s) | 10.33 | 10.57 | 10.38 | 10.37 |
| Estimation Error (m/s) | 0.33 | 0.57 | 0.38 | 0.37 |
| Relative Error | 3.3% | 5.7% | 3.8% | 3.7% |

**Table 5.** The maximum power of false targets under different clutter distributions.

| Clutter Distribution | Rayleigh Distribution | Weibull Distribution | Log-Normal Distribution | K-Distribution |
|---|---|---|---|---|
| Before Compensation (dB) | −13.49 | −13.50 | −13.47 | −13.50 |
| After Compensation (dB) | −46.46 | −42.30 | −43.57 | −45.35 |

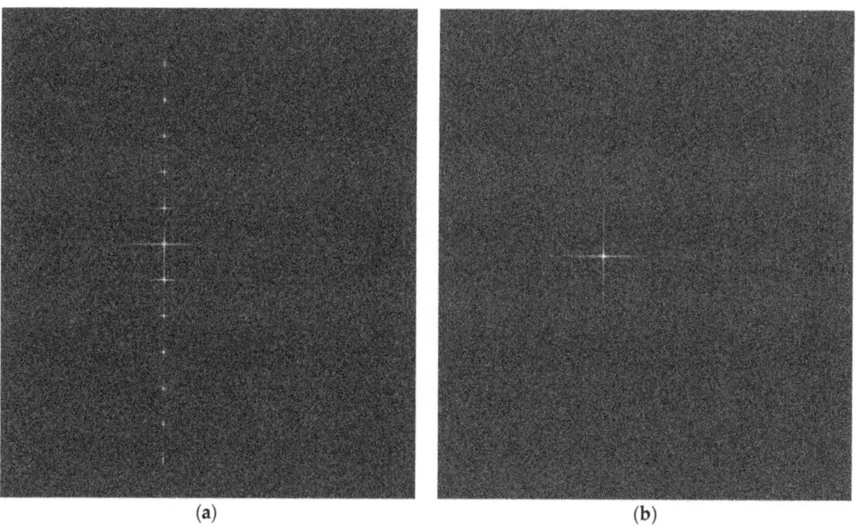

**Figure 11.** Imaging results before and after compensation under Weibull clutter: (**a**) before compensation; and (**b**) after compensation.

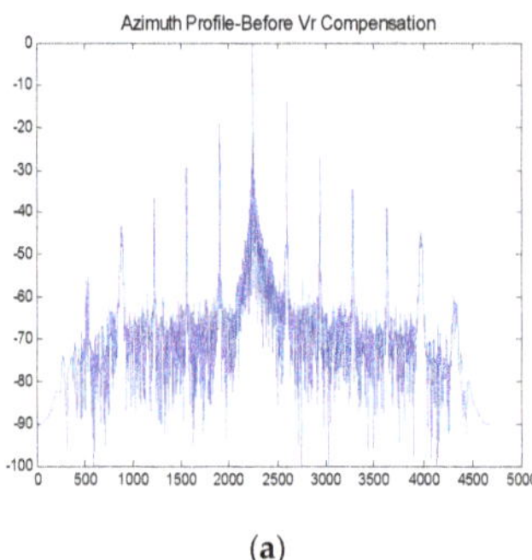


(a)

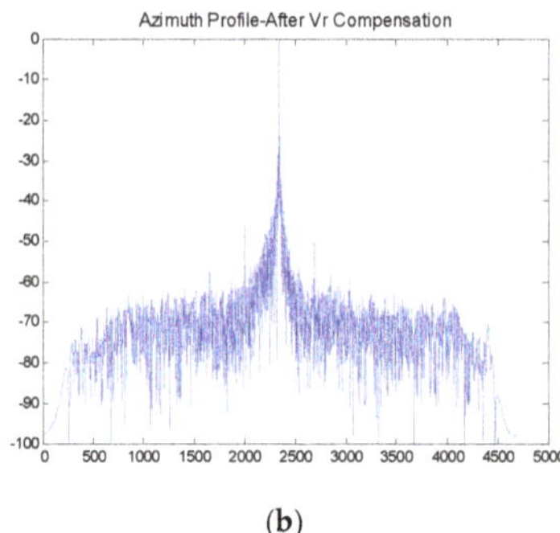


(b)

**Figure 12.** Azimuth profile of imaging result of the moving target under Weibull clutter: (**a**) before compensation; and (**b**) after compensation.

## 5. Discussion

In Section 4, this paper conducted a comprehensive experiment to analyze the performance of the proposed ML-based radial velocity estimation algorithm. From the estimated value under different SNRs in Section 4.1, the relative estimation error is smaller than 2.2% when the SNR is larger than 0dB. Thus this algorithm is accurate enough for applications. From the imaging results of the moving target in the clutter-free situation in Figure 9, the false targets are rather obvious along the azimuth with the traditional multichannel SAR imaging algorithm, and are visually unapparent after compensation for the azimuth offsets with the estimated radial velocity. The maximum power of false targets after compensation has been suppressed by more than 30 dB.

When the moving target is interfered by the sea clutter, the proposed algorithm can still estimate the radial velocity. However, the performance is poorer than the ideal condition. Considering different sea clutter distributions with SCR = 0 dB and SNR = 5 dB, the relative estimation error is at least 3.3%, compared to that of 0.8% without clutter. Figure 11 demonstrates that the ML-based algorithm can also estimate velocity and suppress false targets even in the interference of strong sea clutters. Despite the maximum power of false targets with sea clutter interference being larger than that without clutter, the deterioration of the performance is tolerable. Combined with Figures 9 and 11, the applicability of the proposed estimation algorithm under different practical conditions is verified.

In terms of the estimation effectiveness, we have verified it in Figure 7, where the RMSE infinitely approaches the CRLB with the increase of the SNR. As for the computational complexity, this algorithm does not require iteration or matrix Eigen-decomposition, thus the computational load only lies in the searching process of radial velocity. Finally, we have demonstrated that the proposed algorithm can obtain accurate, efficient, and real-time estimation of the velocity of moving targets for HRWS SAR systems.

## 6. Conclusions

A novel algorithm is proposed to estimate the velocity of the moving target for the spaceborne HRWS SAR system. The main impact of the radial velocity is an additional Doppler spectrum shift in the echo of each channel compared to that of the static target, leading to false targets along the azimuth. According to the characteristics of a moving target signal, a maximum likelihood based algorithm is proposed to estimate the sparse cone angle of the target, obtaining the radial velocity indirectly. Moreover, for the peculiarity of the multichannel SAR system, the Doppler ambiguity is considered in the estimation. After velocity estimation, the multichannel echoes are reconstructed and phase mismatches are compensated to obtain the fine and unambiguous SAR image. The experimental results show high accuracy of the proposed method even under different sea clutter distributions. The effectiveness of the algorithm is also verified by comparing the RMSE and the CRLB. The proposed algorithm can precisely estimate the moving target's velocity for the special mode of HRWS SAR system, providing a reference for applications in remote sensing of the sea surface in future spaceborne multichannel SAR systems.

**Acknowledgments:** This work was jointly supported by the National Science Foundation of China under Grant No. 61331017.

**Author Contributions:** Tingting Jin, Xiaolan Qiu, Donghui Hu and Chibiao Ding initiated the research. Under supervision of Xiaolan Qiu, Tingting Jin performed the experiments and analysis. Tingting Jin wrote the manuscript, and Xiaolan Qiu and Donghui Hu revised the manuscript. All authors read and approved the final version of the manuscript.

**Conflicts of Interest:** The authors declare no conflict of interest.

## References

1. Gebert, N.; Krieger, G.; Moreira, A. Digital beamforming on receive: Techniques and optimization strategies for high-resolution wideswath SAR imaging. *IEEE Trans. Aerosp. Electron. Syst.* **2009**, *45*, 564–592. [CrossRef]
2. Baumgartner, S.V.; Krieger, G. Simultaneous high-resolution wide-swath SAR imaging and ground moving target indication: Processing approaches and system concepts. *IEEE J. Sel. Top. Appl. Earth Obs. Remote Sens.* **2016**, *8*, 1–13. [CrossRef]
3. Suchandt, S.; Runge, H.; Steinbrecher, U. Ship detection and measurement using the TerraSAR-X dual-receive antenna mode. In Proceedings of the 2010 IEEE International Geoscience and Remote Sensing Symposium (IGARSS), Honolulu, HI, USA, 25–30 July 2010; pp. 2860–2863.
4. Rosenqvist, A.; Shimada, M.; Suzuki, S.; Ohgushi, F.; Tadono, T.; Watanabe, M.; Tsuzuku, K.; Watanabe, T.; Kamijo, S.; Aoki, E. Operational performance of the ALOS global systematic acquisition strategy and observation plans for ALOS-2 PALSAR-2. *Remote Sens. Environ.* **2014**, *155*, 3–12. [CrossRef]
5. Dragosevic, M.V.; Burwash, W.; Shen, C. Detection and estimation with RADARSAT-2 moving-object detection experiment modes. *IEEE Trans. Geosci. Remote Sens.* **2012**, *50*, 3527–3543. [CrossRef]
6. Gao, R.; Shi, G.T.; Yang, L.; Zhou, S.L. Moving target detection based on the spreading characteristics of SAR interferograms in the magnitude-phase plane. *Remote Sens.* **2015**, *7*, 1836–1854. [CrossRef]
7. Sikaneta, I.C.; Chouinard, J.Y. Eigen-decomposition of the multi-channel covariance matrix with applications to SAR-GMTI. *Signal Process.* **2004**, *84*, 1501–1535. [CrossRef]
8. Yang, T.L.; Wang, Y. A novel algorithm to estimate moving target velocity for a spaceborne HRWS SAR/GMTI system. In Proceedings of the 2010 IEEE International Geoscience and Remote Sensing Symposium (IGARSS), Beijing, China, 10–15 July 2016; pp. 3242–3245.
9. Baumgartner, S.V.; Krieger, G. Experimental verification of high-resolution wide-swath moving target indication. In Proceedings of the 11th European Conference on Synthetic Aperture Radar (EUSAR), Hamburg, Germany, 6–9 June 2016; pp. 1265–1270.
10. Wang, X.Y.; Wang, R.; Li, N.; Zhou, C. A velocity estimation method of moving target for HRWS SAR. In Proceedings of the 2010 IEEE International Geoscience and Remote Sensing Symposium (IGARSS), Beijing, China, 10–15 July 2016; pp. 6819–6822.
11. Zhang, S.X.; Xing, M.D.; Xia, X.G.; Guo, R.; Liu, Y.Y.; Bao, Z. A novel moving target imaging algorithm for HRWS SAR based on local maximum-likelihood minimum entropy. *IEEE Trans. Geosci. Remote Sens.* **2014**, *52*, 5333–5348. [CrossRef]
12. Wu, Q.; Xing, M.; Qiu, C.; Liu, H.W. Motion parameter estimation in the SAR system with low PRF sampling. *IEEE Geosci. Remote Sens. Lett.* **2010**, *7*, 540–544. [CrossRef]
13. Yang, T.; Li, Z.; Suo, Z.; Bao, Z. Ground moving target indication for high-resolution wide-swath synthetic aperture radar systems. *IET Radar Sonar Navig.* **2014**, *8*, 227–232. [CrossRef]
14. Wang, L.B.; Wang, D.W.; Li, J.J.; Xu, J.; Xie, C.; Wang, L. Ground moving target detection and imaging using a virtual multichannel scheme in HRWS mode. *IEEE Trans. Geosci. Remote Sens.* **2016**, *54*, 1–16. [CrossRef]
15. Zhang, S.X.; Xing, M.D.; Xia, X.G.; Guo, R.; Liu, Y.Y.; Bao, Z. Robust clutter suppression and moving target imaging approach for multichannel in azimuth high-resolution and wide-swath synthetic aperture radar. *IEEE Trans. Geosci. Remote Sens.* **2015**, *53*, 687–709. [CrossRef]

16. Malioutov, D.C.; Etin, M.; Willsky, A.S. A sparse signal reconstruction perspective for source localization with sensor arrays. *IEEE Trans. Signal Process.* **2005**, *53*, 3010–3022. [CrossRef]
17. Gabele, M.; Krieger, G. Moving target signals in high resolution wide swath SAR. In Proceedings of the 9th European Conference on Synthetic Aperture Radar (EUSAR), Friedrichshafen, Germany, 2–5 June 2008; pp. 1–4.
18. Satish, A.; Kashyap, R.L. Maximum likelihood estimation and cramer-rao bounds for direction of arrival parameters with a large sensor array. *IEEE Trans. Geosci. Remote Sens.* **1996**, *44*, 478–491. [CrossRef]
19. Tam, P.K.; Wong, K.T.; Song, Y. An hybrid Cramér-Rao bound in closed form for direction-of-arrival estimation by an "acoustic vector sensor" with gain-phase uncertainties. *IEEE Trans. Signal Process.* **2014**, *62*, 2504–2516. [CrossRef]
20. Stoica, P.; Nehorai, A. Performance study of conditional and unconditional direction-of-arrival estimation. *IEEE Trans. Acoust. Speech Signal Process.* **1990**, *38*, 1783–1795. [CrossRef]

Article

# An Improved Shape Contexts Based Ship Classification in SAR Images

**Ji-Wei Zhu [1,2,*], Xiao-Lan Qiu [1], Zong-Xu Pan [1], Yue-Ting Zhang [1] and Bin Lei [1]**

1 Key Laboratory of Technology in Geo-spatial Information Processing and Application System, Institute of Electronics, Chinese Academy of Sciences, 100190 Beijing, China; xlqiu@mail.ie.ac.cn (X.-L.Q.), zxpan@mail.ie.ac.cn (Z.-X.P.); zhangyueting06@mails.gucas.ac.cn (Y.-T.Z.); leibin@mail.ie.ac.cn (B.L.)

2 University of the Chinese Academy of Sciences, Beijing 100039, China

* Correspondence: zhujiwei1234@126.com; Tel.: +86-10-5888-7208-8932

Academic Editors: Xiaofeng Yang, Xiaofeng Li, Ferdinando Nunziata, Alexis Mouche and Prasad S. Thenkabail
Received: 8 December 2016; Accepted: 4 February 2017; Published: 10 February 2017

**Abstract:** In synthetic aperture radar (SAR) imagery, relating to maritime surveillance studies, the ship has always been the main focus of study. In this letter, a method of ship classification in SAR images is proposed to enhance classification accuracy. In the proposed method, to fully exploit the distinguishing characters of the ship targets, both topology and intensity of the scattering points of the ship are considered. The results of testing the proposed method on a data set of three types of ships, collected via a space-borne SAR sensor designed by the Institute of Electronics, Chinese Academy of Sciences (IECAS), establish that the proposed method is superior to several existing methods, including the original shape contexts method, traditional invariant moments and the recent approach.

**Keywords:** ship classification; improved shape contexts; scattering centers; synthetic aperture radar (SAR) image

## 1. Introduction

Synthetic aperture radar (SAR) has become a valuable tool for maritime surveillance studies. Satellite SAR images are used for ship detection by virtue of their wide coverage and the all-day and any-weather acquisition capability. At present, there have been many literatures on ship detection in SAR imagery. The early researches are generally based on intensity images [1–4]. As early as 1996, an automatic ship and ship wake detection system for spaceborne SAR image is proposed by Eldhuset [1]. In [2], Crisp presents a review of the available literatures on algorithms for ship detection based on intensity images. The information provides by the backscattered intensity collected by a single-polarization SAR is not sufficient to observe metallic targets. To obtain more effective observation techniques, great attention to polarimetric approaches has been paid [5,6]. More recently, many researchers [7,8] propose different approaches relying on the information kept in the spectrum of a single-look complex (SLC) SAR images.

With improvement in the resolution of SAR sensor, as another interesting research field in the domain of marine surveillance, ship classification in SAR images has been receiving greater attention in recent years. Restricted by the amount of samples, many early studies were carried out on simulated images. Graphical electromagnetic computation SAR (GRECOSAR), an orbital polarimetric SAR simulator, is employed by Margarit et al. [9] for ship classification. The usefulness of the simulator on vessel classification studies is assessed. Osman et al. [10] present a back propagation neural network to classify simulated ship images with a correct classification rate of 98%. A ship classifier based on the principal components analysis (PCA) are proposed [11] and the recognition performance of the classifier are measured using various sets of simulated ship images under various aspect angles. However, the above methods require a large amount of data samples. Since it is difficult to get enough real samples, the algorithms can only be carried on simulated images. With the launching of several commercial satellites since the 1990s, such as Envisat and RadarSat-1, the polarimetric SAR is exploited for ship classification [12,13]. Symmetric scattering characterization method (SSCM) [12], a new method based on Poincare sphere representation, is introduced by Touzi [13] for ship characterization. Dominant scatters which are useful for ship characterization are extracted by SSCM. However, the method is sensitive to the system focus setting and Doppler centroid shift. In [14], a new ship classification methodology that takes advantage of single-pol SAR images based on a fuzzy logic (FL) decision rule is presented. Experiments conducted with the ENVISAT images and AIS polls have achieved a classification ratio to 70%. These researches are based on low-resolution images, leading to poor classification accuracy. In recent years, the successful launching of high-resolution and multi-polarization SAR sensors, such as TerraSAR-X and RadarSat-2, provided better observation conditions for ship targets. Researchers paid much more attention to ship classification. Margarit et al. [14] propose an approach combining Pauli data analysis with height-retrieval techniques to classify ships, based on their 3-D geometry. The experimental results prove good performance of algorithm for realistic system, TanDEM-X. By extracting geometric and local Radar Cross Section (RCS) density features and applying the sparse-representation classification (SRC), Xing et al. [15] present a method to recognize ships in the feature space. Zhang et al. [16] consider the scattering components that can represent the structure, materials, orientation, and other information relating to merchant ships. However, the results of their research are mainly limited to civil ships, such as oil tank, bulk and container, which has large difference in shape and appearance. The SRC proposed in Xing's literature [15] requires a large number of ship samples, which is not satisfied by many categories of ships. Besides, the principal axis direction extraction and Zhang's classification criteria [16] are not applicable for warships whose shape difference is not obvious. The key to ship recognition is discriminative feature extraction. Many researchers investigate the effective shape features extraction using the public moving and stationary target acquisition and recognition (MSTAR) dataset which includes several types of armored vehicles with azimuth angles that range from 0 to 360. Zhou et al. [17] put forward scattering center features at different target poses using a global scattering center model. Park et al. [18] propose 12 new features based on the projection length curve. Park and Kim [19] extract the 1-D projection on radius and compressed coefficients by principal component analysis (PCA) based on polar-mapped image. Zernike moments have been introduced by Amoon [20] for target recognition in SAR images. The targets shadow is considered by Papson and Narayanan [21] for classification. However, since it is difficult to obtain enough samples, the above feature extraction methods cannot be applied to warship classification. To sum up, there has not been enough research on ship classification in SAR imagery, especially the warships with a low degree of distinction and those with few samples.

This paper focuses on devising a precise classification of warship targets, without the need for a large number of samples. Besides, compared with the civilian ships, the warships are much closer in shape and structure. In SAR images, the ship targets are composed of scattering points, whose distribution patterns vary from ship to ship. Shape contexts feature [22] is a kind of feature extraction method using log-polar histogram to describe the distribution of contour sampling points in natural images. It has good recognition accuracy and has applied in many fields [23,24]. But so far no literature has applied this method to target classification in SAR imagery. The matching procedure of shape contexts does not require large numbers of samples which are necessary in many machine learning approaches. Due to the characteristics of shape contexts, it would be an effective method for warship classification in SAR images. However, this feature is not rotationally invariant and cannot utilize the intensity of the scattering points. In this work, by modifying the original shape contexts (OSC) method, we propose a novel classification method called improved shape contexts (ISC). Through principal axis extraction of ship by PCA and weighted matching process, both the intensity and the topology of scattering points are utilized in the method. The proposed method is validated by testing on a real ship data set collected via a space-borne SAR sensor designed by the Institute of Electronics, Chinese

Academy of Sciences (IECAS). The result shows that the improved shape contexts method is superior to original method and other traditional moment methods.

The remainder of this letter is organized as follows: Section 2 analyzes the main characteristics of ships in SAR images and presents the authors novel classification method for warships in SAR images; Section 3 validates the proposed method by testing it on real data set collected from a space-borne SAR sensor; the conclusions are presented in Section 4.

## 2. The Proposed Classification Method

### 2.1. Problem Description

According to the mechanism of imaging SAR, the ship targets in SAR imageries are composed of many discrete scattering points. The discrete scattering points are called scattering centers, as shown in Figure 1.

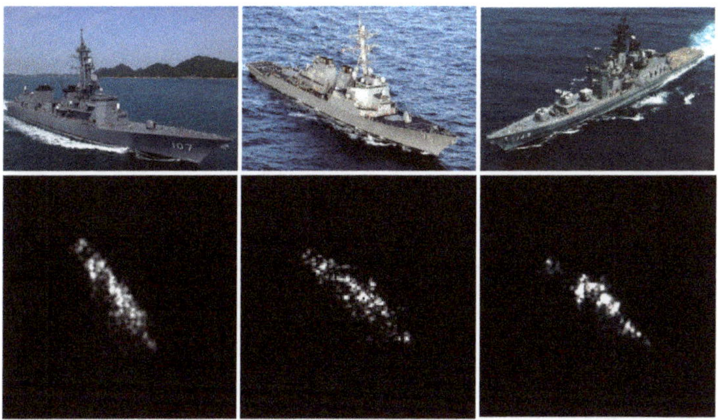

**Figure 1.** The optical and synthetic aperture radar (SAR) images of different types of ships.

The scattering points are generated by the interaction between the radar wave and the superstructure of ships, such as the castle, bridge and mast. Since different types of warships have different types of superstructures, the intensity and the topology of scattering points are also different. They are the best discriminating features to differentiate one type of ship from others in SAR imagery.

However, precise ship classification is always a challenging task for three reasons. First, because of the special imaging process of SAR and the electromagnetic scattering mechanism, the scattering points are sensitive to the azimuth angle of the ship. Second, since the ships in SAR image do not have obvious texture and edge, some local features, such as Scale-Invariant Feature Transform (SIFT) and Histogram of Oriented Gradient (HOG), are no longer applicable. Besides, many types of ships are similar in size, as shown in Figure 1; so, it is not enough to just measure the length and width to distinguish them. Last but not the least, it is quite difficult to collect adequate experimental ship samples; so, it is hard to take advantage of machine learning methods.

In conclusion, the classification strategy should not require many experimental samples. Second, the feature descriptors should have a good description of the ship's global characteristics, specifically the intensity and the topology of scattering points. In addition, since scattering points are sensitive to the azimuth angle of ships, the method should also be robust against changes in scattering intensity.

### 2.2. Proposed Ship Classification Method

In this section, a novel method for classification of warships in SAR images is proposed, based on the improved shape contexts. The proposed method includes two steps: (i) feature extraction and

(ii) feature matching. The feature extraction method takes full advantage of the topology of scattering points, with invariance to image translation and rotation. The feature matching method exploits the intensity of the scattering points.

The proposed method is divided into several stages, as shown in Figure 2. Before attempting to classify the ships, the ships must have been reliably detected by applying the ship detection technique, and separated from the surroundings.

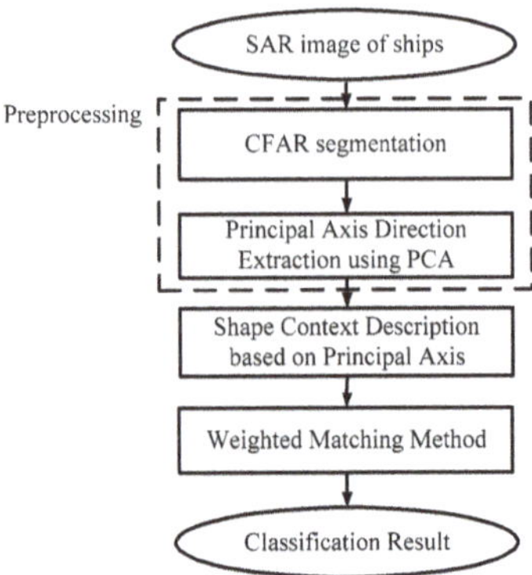


**Figure 2.** The procedure of classification algorithm.

#### 2.2.1. Preprocessing

1. CFAR segmentation

First of all, Constant False Alarm Rate (CFAR) [25] segmentation is used to extract the scattering centers of ships, as shown in Figure 3. In this paper, the 2-parameter CFAR is employed for ship segmentation. Both of the sea clutter and the sidelobes of ships appeared in the segmented image will affect the topology of scattering centers. These interferences will affect the subsequent experimental performance. Besides, the stable scattering centers are more valuable in subsequent experiments. So we set a low false alarm rate ($10^{-6}$) to remove all the sea clutter. Since the sidelobes of warships in our experiments are rather weak, they are also generally removed after segmentation. So, there is no need to suppress the sidelobes. After CFAR segmentation, the stable scattering centers (see Figure 3b) are segmented from the original image.

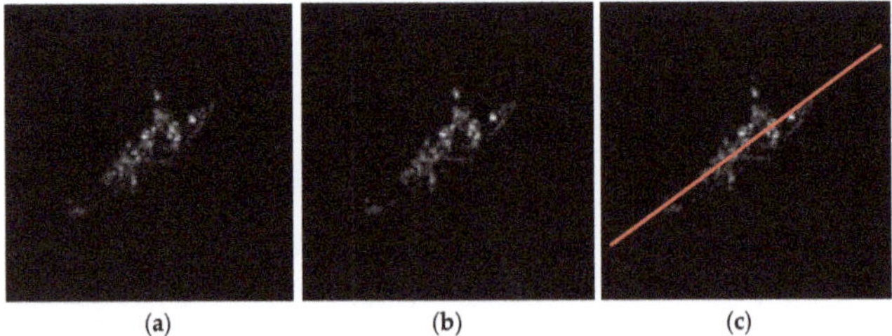

**Figure 3.** The procedure of classification algorithm. Diagram of the principal axis extraction. (**a**) The original image; (**b**) The segmented image; (**c**) The principal axis extraction.

2. Principal axis direction extraction using PCA

Because of the geometrical characteristics of ship targets, the scattering points are distributed along the principal axis of ships. Besides, the scattering points are symmetrically distributed on both sides of ships. So, the principal component analysis (PCA) method [26] can be employed to compute the principal axis of the scattering points of the segmented targets.

2.2.2. The Improved Shape Contexts Method

After extracting the scattering points and the principal axis, the feature extraction and matching methods are implemented for ship classification.

1. Shape contexts description based on principal axis

Shape context (see Figure 4) is a kind of shape feature that uses log-polar histograms to describe the distribution of sampling points of the object contour. The sampling points are defined as $P = \{p_1, p_2, p_3, \ldots p_n\}$, $p_i \in IR^2$, (the meaning of $IR^2$ is two dimensional real number vector space) where $n$ is the total number of points. By selecting the $i$th point $p_i$ as the point of origin, a coarse histogram $h_i$ of the relative coordinates of the remaining $n-1$ points can be computed,

$$h_i(k) = \#\{q \neq p_i : (q - p_i) \in bin(k)\} \quad 1 \leq k \leq K \tag{1}$$

This histogram is defined as the shape context of $p_i$. $q$ denotes the remaining sampling point and $K$ denotes the number of histogram grid partitions. $bin$ are uniform in log-polar space. The polar coordinates of $q$ are defined as $(r, \theta)$. $r$ represents for the distance between $q$ and $p_i$. $\theta$ represents for the angle between the connecting line of the two points and the zero angle axis, measured in anticlockwise direction.

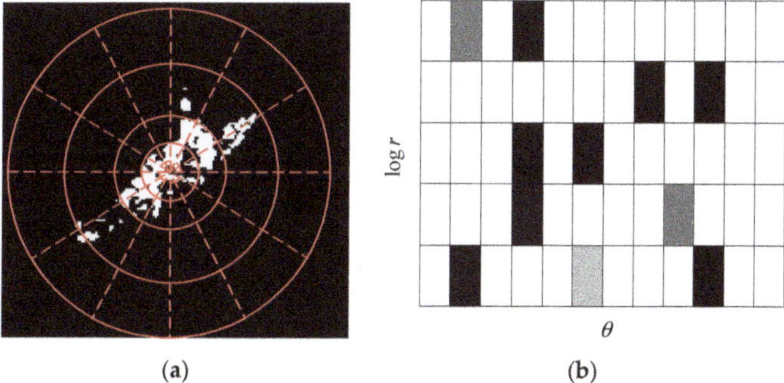


**Figure 4.** The original shape context descriptor. (**a**) The division of sampling points; (**b**) The histogram of sampling points.

Specifically, $\log r$ is divided uniformly into $M$ parts and $\theta$ into $N$ parts; so, the dimensions of histogram $K$ are $M \times N$. The two dimensional histogram is defined as

$$S = \{(m, n) : 1 \leq m \leq M, 1 \leq n \leq N\} \tag{2}$$

Then, the histogram set for all the sampling points is defined as

$$h_i = \{h_i(k) : 1 \leq k \leq K\} = \{h_i(m, n) : (m, n) \in S\} \ , k = (n-1) \times M + m \tag{3}$$

However, the original shape contexts do not satisfy rotation invariance. The histograms of ships of the same type, at different azimuth angles, are vastly different. To enable the feature suitable for recognizing ships with different azimuths, we utilize the extracted principal axis of the ship as the reference angle for shape context. The angle between every scattering point and the reference point is measured by the angle between the connecting line of the two points and the principal axis, measured in anticlockwise direction (see Figure 5). After this measurement, the topology of the scattering points is described by the histograms. Since the angle between every two points depends on the principal axis of the ship, the scattering points of the ship are rotational invariant.

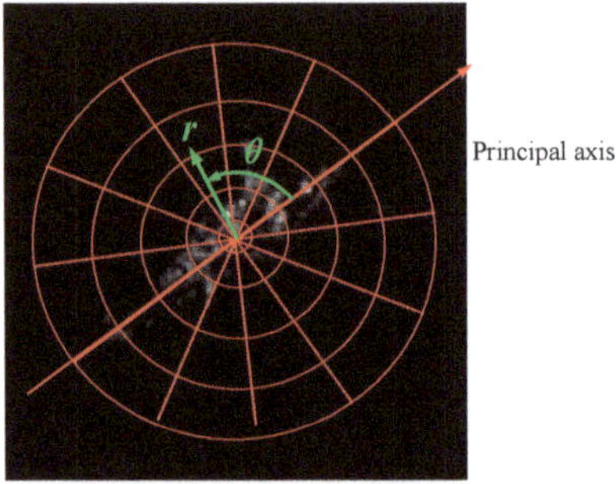


**Figure 5.** Shape contexts description based on principal axis.

2. The weighted feature matching procedure

Because of limited experimental samples, template matching is used to classify different types of ships. Specifically, after calculating the matching costs of the query ship and different candidate ship templates, the candidate ship template with the lowest matching cost would be the final recognition result. The specific shape contexts matching method is as follows.

After obtaining the shape context feature of each scattering point, the distance between the shape contexts of the query ship and the ship template are calculated by $\chi^2$ statistic. The cost of two points is denoted as

$$C_{ij} \equiv C(p_i, q_j) = \frac{1}{2}\sum_{k=1}^{K} \frac{[h_i(k) - h_j(k)]^2}{h_i(k) + h_j(k)} \tag{4}$$

Given the matrix of costs $C_{ij}$ between all pairs of points $p_i$ on the query ship and $q_j$ on the ship template, the next task is minimizing the total matching cost $C = \frac{1}{N}\sum_{i=1}^{N} C(p_i, q_{\pi(i)}) = \frac{1}{N}\sum_{i=1}^{N} C_{i,\pi(i)}$, subject to the constraint that the matching be one-to-one. $q_{\pi(i)}$ is matching point of $p_i$, which is an instance of the square assignment problem that can be solved by the Hungarian method [27]. The matching result of the two ships is shown in Figure 6.

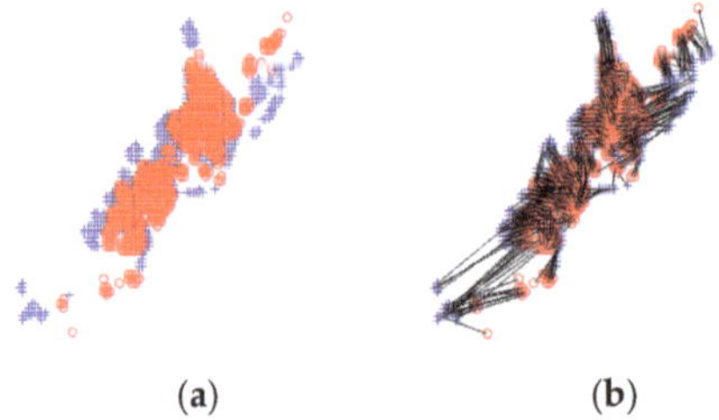


**Figure 6.** An instance of the shape contexts matching. (**a**) The segmented image of the query ship and ship template; (**b**) The shape contexts matching result of the query ship and ship template. Query ship (red), Ship template (blue).

So far, every scattering point of the query ship matches the scattering point of the ship template, with almost similar topology. But we have taken advantage of only the topological relations of the scattering points, without taking into consideration the intensity information of the corresponding scattering points.

The shape context of a scattering point represents its topological relation with other points. Because the same types of ships are similar in structure, the scattering points with similar shape contexts are also similar in intensity. Some scattering points of different types of ships may have similar shape contexts. However, their intensity may be vastly different.

For calculating the matching cost of two ships, the intensities of their scattering points should be taken into account by giving weight to the cost of every two matching points. Specifically, the original matching cost $C_{i,\pi(i)}$ of $p_i$ and $q_{\pi(i)}$ should be multiplied by a weight factor $W_{i,\pi(i)}$, which contains the intensity information of the scattering points.

We believe that the more the similarity between the intensities of two matching points, the smaller would be the matching cost. Assuming that the intensities of the matching points are $I_i(k)$ and $I_{\pi(i)}(k)$, then the weight indicating the degree of similarity in intensity is

$$W^1_{i,\pi(i)} = \left(\frac{I_i(k)}{I_i(k)+I_{\pi(\mathrm{i})}(k)} - \frac{I_{\pi(\mathrm{i})}(k)}{I_i(k)+I_{\pi(\mathrm{i})}(k)}\right)^2 = \left(\frac{I_i(k)-I_{\pi(\mathrm{i})}(k)}{I_i(k)+I_{\pi(\mathrm{i})}(k)}\right)^2 \tag{5}$$

Generally, the scattering points with higher intensity have greater effects on the matching degree of the two ships. So, we multiply the original matching cost $C_{i,\pi(\mathrm{i})}$ with another weight $W^2_{i,\pi(\mathrm{i})} = I_i(k) + I_{\pi(\mathrm{i})}(k)$. Finally, the weighted matching cost is

$$C'_{i,\pi(\mathrm{i})} = W^1_{i,\pi(\mathrm{i})} W^2_{i,\pi(\mathrm{i})} C_{i,\pi(\mathrm{i})} = W^1_{i,\pi(\mathrm{i})} W^2_{i,\pi(\mathrm{i})} C(p_i, q_{\pi(\mathrm{i})}) = \frac{(I_i(k) - I_{\pi(\mathrm{i})}(k))^2}{I_i(k)+I_{\pi(\mathrm{i})}(k)} C(p_i, q_{\pi(\mathrm{i})}) \tag{6}$$

Finally, the total matching cost of the query ship and the ship template is

$$C' = \frac{1}{N}\sum_{i=1}^{N} C'_{i,\pi(\mathrm{i})} = \frac{1}{N}\sum_{i=1}^{N} W^1_{i,\pi(\mathrm{i})} W^2_{i,\pi(\mathrm{i})} C_{i,\pi(\mathrm{i})} = \frac{1}{N}\sum_{i=1}^{N} W^1_{i,\pi(\mathrm{i})} W^2_{i,\pi(\mathrm{i})} C(p_i, q_{\pi(\mathrm{i})}) \tag{7}$$

The type of candidate ship template with the lowest matching cost would be the final classification result of the query ship.

Compared with the original shape contexts method, there are two improvements in our proposed method. Firstly, the principal component analysis (PCA) is employed to compute the principal axis of the scattering points, so the algorithm is suitable for classification of warships under different azimuth angles. Secondly, in the matching stage, the intensities of scattering centers are taken into account by giving weight to the cost of every two matching points. In general, the improved shape contexts method fully describes the distinguishing characteristics of the ship targets. The ship signature is dependent on the azimuth angle of the ship, but given one angle, the algorithm is invariant to rotation and translation.

## 3. Experimental Results

In this section, we implement the experiments on a real ship dataset via a space-borne SAR sensor designed by IECAS to examine the effectiveness of the proposed method. The ships are collected from 32 stripmap-model SAR images with X band, 3m resolution in both azimuth and range directions. The image acquisitions dates vary from 16 May 2008 to 9 September 2015. The categories of the ships are confirmed by the professional interpreters. The data relates to three types of destroyers (total 150 samples). The number of the three kinds of ships is 50, 50 and 50, respectively (See Figure 7). The incidence angle ranges between 40° and 55°. The azimuth angles ranges between 0° and 360°. The ship dataset is extracted from geo-coded images, which have subjected to geometric correction.

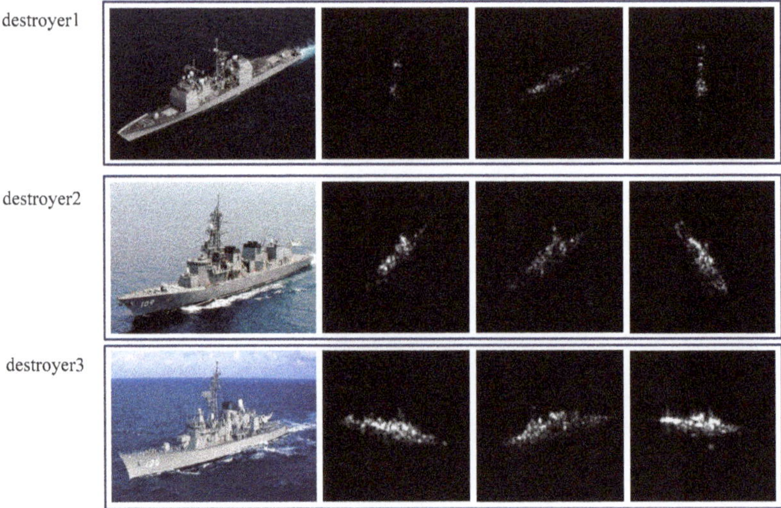


**Figure 7.** The first column represents the optical images of three types of warships. The second column to the fourth column represents the samples corresponding to the three types of warships from different azimuth angles. destroyer1 (**top**), destroyer2 (**middle**), destroyer3 (**bottom**).

Due to the limitation of sample size, the methods we compare are all template matching with features. The features that are commonly used in computer vision for shape matching are Hu moments [28] and Zernike moments [29]. They are invariant to image translation, scaling and rotation, with global characteristic description. Invariant moments are widely used for target recognition and tracking, texture analysis and other related purposes. Besides, we also compare our method with the recent approach [15] proposed by Xing in 2013. The approach proposes to use both geometric features and electromagnetic scattering characteristic for ship classification. The geometric feature includes the ship length (Len), length to width ratio (R), centroid (C), shape complexity (X) and covariance coefficient (V). The electromagnetic scattering characteristic is described by the local RCS density (LRCS), specifically, the pixel intensity ratio of three parts of the ship. The feature vector f denoted as follows.

$$f = [\text{Len}, \text{R}, \text{C}, \text{X}, \text{V}, \text{LRCS}]^T \tag{8}$$

In our study, for the sake of comparison, we conducted classification experiments with the improved shape contexts (ISC) method we propose, the original shape contexts (OSC) method, the feature proposed by the literature [15] with Euclidean distance and the conventional moment features, namely Hu moments and Zernike moments with Euclidean distance. In the ISC and OSC, we use five bins for $\log r$ and 12 equally spaced radial bins for $\theta$. ($K = 60$, $M = 5$ and $N = 12$).

By randomly selecting one of each class as templates and the rest samples for classification test, we carry out the classification experiments. Specifically, we calculate the matching cost of every query ship and the three templates using the five methods cited above, the final classification result is the type of template with the lowest matching cost. As the template samples are selected randomly, the classification experiments are carried out repeatedly for ten times and the results are averaged. The results, presented in Table 1, show that the improved shape contexts method outperforms all the other methods with average classification rate nearly 91%.

Some correctly classified and incorrectly classified ship images of OSC are provided in Figure 8. The third column is misclassified. In this column, the first and third samples are misclassified as destroyer 2 while the second sample is misclassified as destroyer 3. This is mostly because, under some certain azimuth angles, the scattering centers of ships vary greatly due to the mechanism of imaging SAR. Besides, high speed navigation of ships also impacts image quality. These problems are not considered in the proposed algorithm.

**Table 1.** Comparison of Classification Accuracy obtained by Hu moments, Zernike moments, f feature, OSC (Original Shape Contexts) and ISC (Improved Shape Contexts).

| | Hu Moments | Zernike Moments | f Feature | OSC | ISC |
|---|---|---|---|---|---|
| Destoryer1 Classification Accuracy | 70% | 81% | 73% | 88% | 92% |
| Destoryer2 Classification Accuracy | 66% | 78% | 69% | 86% | 90% |
| Destoryer3 Classification Accuracy | 67% | 77% | 66% | 86% | 90% |
| Average Classification Accuracy | 68% | 79% | 69% | 87% | 91% |

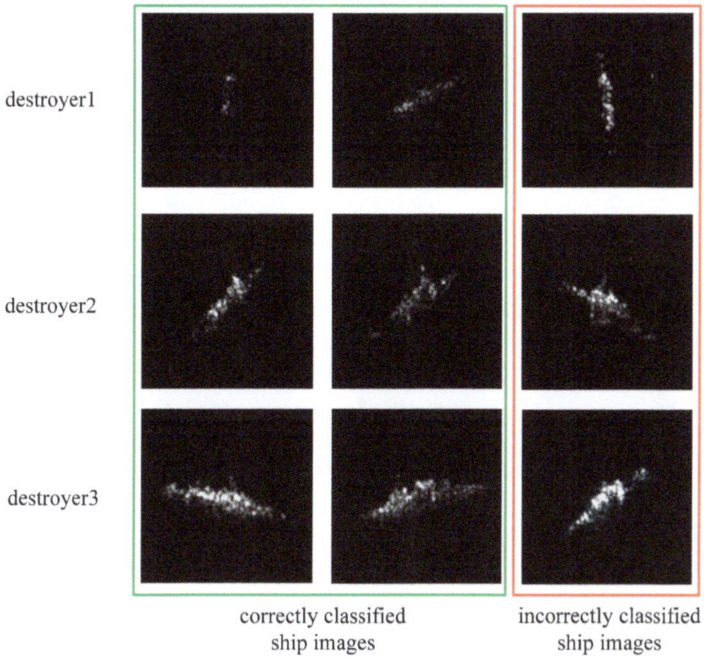


**Figure 8.** The correctly and incorrectly classified ship images of OSC for three types of destroyers.

## 4. Discussion

This paragraph we analyze the possible reasons of the above experimental results. Feature extraction is the key problem of target recognition. Both of Hu and Zernike moments are highly concentrated image features with rotation, translation and scaling invariance. Since Hu moments [28] feature is a kind of low order statistic, it cannot fully describe the details of an image. So the distinguishing ability of Hu moments is limited in many cases. Compared to Hu moments, Zernike moments [29] feature is more robust to noise and small shape changes with less redundant information. However, this paper focuses on the distinction of warships in SAR imagery. Since the above moments are not well designed for the specific targets classification, they cannot fully describe the distinguishing features of warships in SAR imagery. Hu moments and Zernike moments are universal features, so they cannot achieve good experiments results in such a specific dataset. Similarly, despite the effectiveness of f feature vector for civilian ships, the ability of the feature to distinguish warships is also limited. Though local RCS density in f feature vector is designed for civilian ships in SAR imagery, it just divides the ship into several parts and adds up the pixel intensity of each part. As warships are much closer in shape and structure, this simple feature is unable to make a complete use of the distinguishing characteristic of different warships in SAR imagery. As we have analyzed in Section 2.1, the topology and the intensity of the scattering centers are the main distinguishing features of warships in SAR imagery. The improved shape contexts method fully takes advantage of the distinguishing features, thus it can achieve better recognition results.

## 5. Conclusions

In this paper, an effective and precise ship classification method is proposed. The letter first analyzes the main distinguishing features of ships in SAR imagery. In the preprocessing stage of classification, the scattering points of warships are extracted by CFAR, and the principal axis is estimated using PCA. Then we propose the improved shape contexts method, which fully describes the topology and intensity of the scattering points of the ships. It is experimentally demonstrated with the space-borne data that the proposed method outperforms the traditional moments and the recent state-of-art approaches, as also the original shape contexts method, by achieving an average classification accuracy of about 91%. Despite the encouraging results of the novel method, the proposed algorithm needs many improvements. First, more types of ship samples are needed to validate the applicability of the method to a wide ranging variety of ships. Second, since the features of warships change with the azimuth angle in SAR images, more stable features should be looked for to strengthen the robustness of the method. Finally, the scattering mechanism of the ship should be studied in greater depth in future.

**Acknowledgments:** This work was jointly supported by the National Science Foundation of China under Grant No. 61331017.

**Author Contributions:** Jiwei Zhu, Xiaolan Qiu, Zongxu Pan, Yueting Zhang and Bin Lei initiated the research. Under supervision of Xiaolan Qiu, Jiwei Zhu performed the experiments and analysis. Jiwei Zhu, Xiaolan Qiu, Zongxu Pan, Yueting Zhang and Bin Lei wrote and revised the manuscript. All authors read and approved the final version of the manuscript.

**Conflicts of Interest:** The authors declare no conflict of interest.

## References

1. Eldhuset, K. An automatic ship 472 and ship wake detection system for spaceborne SAR images in coastal regions. *IEEE Trans. Geosci. Remote Sens.* **1996**, *34*, 1010–1019. [CrossRef]
2. Crisp, D.J. The State-of-the-Art in Ship Detection in Synthetic Aperture Radar Imagery. Available online: http://www.dsto.defence.gov.au/publications/2660/ (accessed on 5 February 2004).
3. Wang, C.; Liao, M.; Li, X. Ship detection in SAR images based on the Alpha-stable distribution. *Sensors* **2008**, *8*, 4948–4960. [CrossRef] [PubMed]
4. Wackerman, C.C.; Friedman, W.G.; Pichel, W.G.; Clemente-Colon, P.; Li, X. Automatic detection of ships in RADARSAT-1 SAR imagery. *Can. Remote Sens.* **2001**, *27*, 568–577. [CrossRef]
5. Nunziata, F.; Migliaccio, M.; Brown, C.E. Reflection symmetry for polarimetric observation of man-made metallic targets at sea. *IEEE J. Ocean. Eng.* **2012**, *37*, 384–394. [CrossRef]
6. Marino, A. A notch filter for ship detection with polarimetric SAR data. *IEEE J. Sel. Top. Appl. Eearth Obs. Remote Sens.* **2013**, *6*, 1219–1232. [CrossRef]
7. Anfinsen, S.N.; Brekke, C. Statistical models for constant false alarm rate ship detection with the sublook correlation magnitude. *IEEE Trans. Geosci. Remote Sens.* **2012**, *53*, 5626–5629.
8. Marino, A.; Sanjuan-Ferrer, M.J.; Ouchi, K. Ship detection with spectral analysis of Synthetic Aperture Radar: A comparison of new and well-known algorithms. *Remote Sens.* **2015**, *7*, 5416–5439. [CrossRef]
9. Margarit, G.; Mallorqui, J.J.; Rius, J.M.; Sanz-Marcos, J. On the usage of GRECOSAR, an orbital polarimetric SAR simulator of complex targets, for vessel classification studies. *IEEE Trans. Geosci. Remote Sens.* **2006**, *44*, 3517–3526. [CrossRef]

10. Osman, H.; Pan, L.; Blostein, S.D.; Gagnon, L. Classification of ships in airborne SAR imagery using back propagation neural networks. *Proc. SPIE* **1997**, *1997*, 126–136.
11. Gouaillier, V.; Gagnon, L. Ship silhouette recognition using principal components analysis. *Proc. SPIE* **1997**, *1997*, 59–69.
12. Margarit, G.; Tabasco, A. Ship classification in single-Pol SAR images based on fuzzy logic. *IEEE Trans. Geosci. Remote Sens.* **2011**, *49*, 3129–3138. [CrossRef]
13. Touzi, R.; Raney, R.K.; Charbonneau, F. On the use of permanent symmetric scatters for ship characterization. *IEEE Trans. Geosci. Remote Sens.* **2004**, *42*, 2039–2045. [CrossRef]
14. Margarit, G.; Jordi, J.M.; Fabregas, X. Single-pass polarimetric SAR interferometry for vessel classification, *IEEE Trans. Geosci. Remote Sens.* **2007**, *45*, 3494–3502. [CrossRef]
15. Xing, X.; Ji, K.; Zou, H.; Chen, W.; Sun, J. Ship classification in TerraSAR-X images with feature space based sparse representation. *IEEE Geosci. Remote Sens. Lett.* **2013**, *10*, 1562–1566. [CrossRef]
16. Zhang, H.; Tian, X.; Wang, C.; Wu, F.; Zhang, B. Merchant vessel classification based on scattering component analysis for COSMO-SkyMed SAR images. *IEEE Geosci. Remote Sens. Lett.* **2013**, *10*, 1275–1279. [CrossRef]
17. Zhou, J.; Shi, Z.; Cheng, X.; Fu, Q. Automatic target recognition of SAR images based on global scattering center model. *IEEE Geosci. Remote Sens. Lett.* **2011**, *49*, 3713–3729.
18. Park, J.; Park, S.; Kim, K. New discrimination features for SAR automatic target recognition. *IEEE Geosci. Remote Sens. Lett.* **2013**, *10*, 476–480. [CrossRef]
19. Park, J.; Kim, K. Modified polar mapping classifier for SAR automatic target recognition. *IEEE Geosci. Remote Sens. Lett.* **2014**, *36*, 1092–1106. [CrossRef]
20. Amoon, M.; Rezai, G. Automatic target recognition of Synthetic Aperture Radar (SAR) images based on optimal selection of Zernike moments features. *IET Comput. Vis.* **2014**, *8*, 77–85. [CrossRef]
21. Papson, S.; Narayanan, R. Classification via the shadow Region in SAR Imagery. *IEEE Trans. Aerosp. Electron. Syst.* **2012**, *48*, 969–980. [CrossRef]
22. Belongie, S.; Malik, J.; Puzicha, J. Shape matching and object recognition using shape contexts. *IEEE Trans. Pattern Anal. Mach. Intell.* **2002**, *24*, 509–522. [CrossRef]
23. Amores, J.; Sebe, N.; Radeva, P. Context-based object-class recognition and retrieval by generalized correlograms. *IEEE Trans. Pattern Anal. Mach. Intell.* **2007**, *29*, 1818–1833. [CrossRef] [PubMed]
24. Shi, Y.; Thompson, P.M.; Zubicaray, G.I.; Rose, S.E.; Tu, Z.; Dinov, I.; Toga, A.W. Direct mapping of hippocampal surfaces with intrinsic shape context. *NeuroImage* **2007**, *37*, 792–807. [CrossRef] [PubMed]
25. Ji, K.; Kuang, G.; Su, Y.; Yu, W. Methods of target's peak extraction and azimuth estimation from SAR imagery. *J. Astronaut.* **2004**, *25*, 102–108.
26. Jones, G.; Bhanu, B. Recognizing of articulated objects in SAR images. *Pattern Recognition* **2001**, *34*, 469–485.
27. Papadimitriou, C.; Stieglitz, K. *Combinatorial Optimization: Algorithms and Complexity*; Courier Corporation: North Chelmsford, MA, USA, 1982.
28. Hu, M.K. Visual pattern recognition by moment invariants. *IEEE Trans. Inf. Theory* **1962**, *8*, 179–187.
29. Liao, S.X.; Pawlak, M. On the accuracy of Zernike moments for image analysis. *IEEE Trans. Pattern Anal. Mach. Intell.* **1988**, *20*, 1358–1364. [CrossRef]

*Article*

# Exploring the Potential of Active Learning for Automatic Identification of Marine Oil Spills Using 10-Year (2004–2013) RADARSAT Data

**Yongfeng Cao [1,*], Linlin Xu [2,3] and David Clausi [2]**

1 School of Big Data and Computer Science, Guizhou Normal University, Guiyang 550001, China
2 Systems Design Engineering, University of Waterloo, Waterloo, ON N2L 3G1, Canada; linlinxu618@gmail.com (L.X.); dclausi@uwaterloo.ca (D.C.)
3 School of Land Science and Technology, China University of Geosciences, Beijing 100083, China
* Correspondence: yongfeng.cao@outlook.com; Tel.: +86-150-8595-0430

Received: 13 August 2017; Accepted: 9 October 2017; Published: 13 October 2017

**Abstract:** This paper intends to find a more cost-effective way for training oil spill classification systems by introducing active learning (AL) and exploring its potential, so that satisfying classifiers could be learned with reduced number of labeled samples. The dataset used has 143 oil spills and 124 look-alikes from 198 RADARSAT images covering the east and west coasts of Canada from 2004 to 2013. Six uncertainty-based active sample selecting (ACS) methods are designed to choose the most informative samples. A method for reducing information redundancy amongst the selected samples and a method with varying sample preference are considered. Four classifiers (k-nearest neighbor (KNN), support vector machine (SVM), linear discriminant analysis (LDA) and decision tree (DT)) are coupled with ACS methods to explore the interaction and possible preference between classifiers and ACS methods. Three kinds of measures are adopted to highlight different aspect of classification performance of these AL-boosted classifiers. Overall, AL proves its strong potential with 4% to 78% reduction on training samples in different settings. The SVM classifier shows to be the best one for using in the AL frame, with perfect performance evolving curves in different kinds of measures. The exploration and exploitation criterion can further improve the performance of the AL-boosted SVM classifier but not of the other classifiers.

**Keywords:** oil spill detection; SAR; active learning; active sample selecting; support vector machine; k-nearest neighbor; linear discriminant analysis; decision tree

## 1. Introduction

With the increase of maritime traffic, the accidental and deliberate discharge of oil from ships is attracting growing concern. Using satellite-based synthetic aperture radar (SAR) has been proven to be a cost effective way to survey marine pollution over large-scale sea areas [1,2].

Current and future satellites with SAR sensors that can be used for monitoring oil spills include ERS-1/2, RADARSAT-1/2, ENVISAT (ASAR), ALOS1/2 (PALSAR), TerraSAR-X, Cosmos Skymed-1/2, RISAT-1, Sentinel-1, SAOCOM-1 and the RADARSAT constellation mission. Based on the SAR systems, many commercial or governmental agencies have been building SAR oil-spill detection service, such as the multi-mission maritime monitoring services of Kongsberg Satellite Services (KSAT), Airbus defense and space's oil spill detection service, CleanSeaNet [3] and Integrated Satellite Tracking of Pollution (ISTOP). To be more operational, automatic oil spill classification system with real-time, fully operational and wider water coverage capability is needed [1], as Solberg et al. state [4] "The currently manual services is just a first step toward a fully operational system covering wider waters".

Due to their ability to smooth sea surface, oil spills usually appear as dark spots on SAR images. However, other sea features, such as low wind areas and biogenic slicks, also produce smooth sea surface and result in dark formations on SAR imagery. These sea features are usually called "look-alikes". The existence of look-alikes imposes huge challenge on SAR oil spill detection systems. In an automatic or semiautomatic SAR oil spill detection system, three steps are sequentially performed to identify oil spills [1,5–8]: (i) dark-spot detection for identifying all candidates that belong to either oil spills or look-alikes; (ii) feature extraction for collecting object-based features, such as the mean intensity value of dark-spots, for discriminating oil spills and look-alikes; and (iii) classification for separating oil spills and look-alikes using the features extracted.

After identifying all candidates and collecting their features, classification approaches predominantly determines the performance of oil spill detection systems. Many classifiers have been used to detect oil spills including a combination of statistical modeling and rule-based approaches [1,4,5,9,10], artificial neural network (ANN) models [11–16], decision tree (DT) models [3,8,15,17], fisher discrimination or multi regression analysis approaches [18], fuzzy classifiers [3,19], support vector machine (SVM) classifiers [9,20] and K-nearest neighbors (KNN) based classifiers [17,21]. A comparison of SVM, ANN, tree-based ensemble classifiers (bagging, bundling and boosting), generalized additive model (GAM) and penalized linear discriminant analysis on a relatively fair standard has been conducted [22] with the conclusion that the tree-based classifiers, i.e., bagging, bundling and boosting approaches, generally perform better than the other approaches, i.e., SVM, ANN and GAM.

Most classifiers that have been adopted for oil spill detection are supervised classifiers which need training samples to "teach" themselves before performing classification tasks. To get good generalization performance, a large number of training samples are needed to deal with the curse of dimensionality [23]. In the case of oil spill classification, high feature dimensionality are usually needed to cover the complex characteristics of look-alikes and oil spills [1,6].

Although effective classifier learning requires a large number of labeled samples, verifying/labeling and accumulating enough number of samples for training an automatic system with reasonable performance could be very difficult, costly and time-consuming for the following reasons. First, oil spills are rare and fast-changing events, which tend to disappear before being verified by ships or aircrafts, because, after a short time span, mostly within several hours [24], the oil spills will become difficult to distinguish. Second, verifying an oil spill using airplane/vessel is usually very expensive. Third, verified/labeled samples from different SAR platforms may not be sharable, because of the different imaging parameters, such as band, polarization mode, spatial resolution, etc. Even for images from the same SAR platform, the standards of confidence levels, pre and post procedures, etc. must be normalized so that the samples from different institutions can be shared.

Limited by difficulties in verifying oil spills, researchers rely mainly on human experts to manually label the targets. For example, Table 1 indicates that the largest number of verified oil spills is only 29 adopted by Solberg et al. [10], and other researchers predominantly used the expert-labeled samples, although they did not explicitly report the proportion of the verified samples. Nevertheless, using expert-labeled samples is problematic for the following reasons. First, expert-labeling produces inconsistency between the labels (or confidence levels) given by different experts [1,25]. Second, training the system with expert-labeled samples leads to system that can hardly outperform the experts who "teach" the system.

**Table 1.** Summary of the dataset adopted in different SAR oil spill detection researches. Note that in the #samples column; (X, Y) indicates there are X oil spills and Y look-alikes and, in the #Samples verified column; (X, Y) indicates there are X verified oil spills and Y verified look-alikes.

| Literature | SAR Sensor | #Image | #Samples | #Samples Verified | Location (Time Span) |
|---|---|---|---|---|---|
| [22] | RADARSAT1 | 93 | (98, 94) | (21, N/A) | West and East coasts of Canada (2004–2008) |
| [9] | Envisat | 103 | (41, 12, 245) | N/A | Baltic Sea and North See (2003–2005) |
| [10] | Envisat RADARSAT1 | 127 | (37, 12, 110) | (29, N/A) | Baltic Sea and North See (2003–2004) |
| [4] | Envisat | 84 | (71, 6980) | N/A | European waters (N/A) |
| [7,8,12–14] | ERS-2 | 24 | (69, 90) | N/A | Mediterranean Sea (N/A) |
| [19] | ERS-1/2 | 12 | N/A | N/A | Mediterranean Sea (N/A) |
| [24] | ERS-1/2 | 1600 | (1638, N/A) | N/A | Mediterranean Sea (1999) |
| [18] | ERS-1/2 Envisat | N/A | (153, 237) | N/A | Mediterranean Sea (N/A) |
| [26] | ERS-1/2 Envisat | 15,533 | (9299, N/A) | N/A | Mediterranean Sea (1999–2004) |
| [26] | ERS-1/2 Envisat | 3165 | (1227, N/A) | N/A | Black sea (1999–2004) |
| [15] | Envisat | 47 | (80, 155) | N/A | Galicia coast, Spain (2007–2011) |
| [3] | Envisat RADARSAT | 118 | (361, 5728) | N/A | European Waters (2009–2012) |
| [17] | ERS-1 RADARSAT1 | 9 | (41, 896) | N/A | N/A |

Considering the cost and difficulties in verifying oil spill candidates, one key issue in learning an oil spill classification system is to effectively reduce the number of verified samples required for classifier training without compromising the accuracy and robustness of the resulting classifier. Suppose that the current verified samples are insufficient for building an accurate oil spill detection system, and that new samples are required to be verified for increasing the size of the training set. In a conventional supervised classification system, we will not be able to know which samples have higher priority to be verified, because, as indicated in Figure 1, the communication between the conventional classification system and the sample collecting system is one-way directed, where the collected samples are used to train the classifier with no feedback from the classifier on what kind of samples are more informative and urgently needed. Without knowing the values and importance of the samples to the classifiers, the costly verification effort may only lead to training samples that are redundant, useless or even misleading. Although verifying more samples can increase the possibility of obtaining relevant training sample, it will greatly increase the time span and cost for building the system.

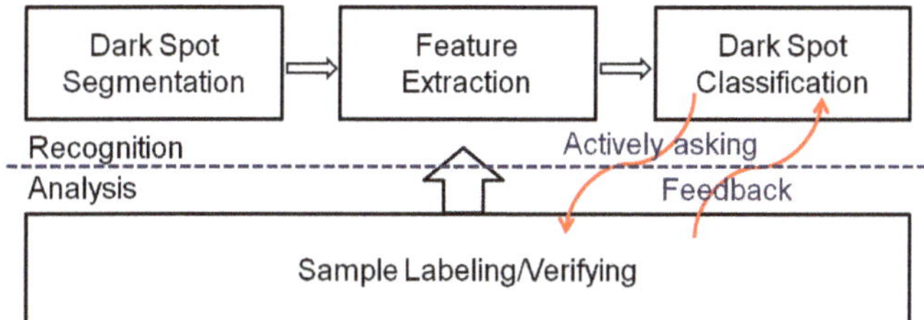


**Figure 1.** The role of active learning in the oil spill detection system. In conventional supervised classification system, the communication between the classification system (upper part) and the training sample collecting system (lower part) is one-way directed (as indicated by the big black arrow), where the collected samples are used to train the classifier with no feedback from the classifier on what kind of samples are most informative and urgently needed. However, in the active learning boosted system, the interaction between the two systems are bi-directional (as indicated by the red arrows), where the classifier will "ask for" the most relevant samples to be verified/labeled in order to learn the classifier in an efficient and effective manner. Considering the cost and difficulty in verifying the oil spill candidate, such an active learning process can greatly reduce the cost and time for building a detection system by reducing the number of candidates that needed to be verified without compromising the robustness and accuracy of the resulting classifier.

The need for reducing training samples without compromising the accuracy of resulting classifiers motivates us to study the potential of introducing AL into the oil spill detection systems. AL is a growing area of research in machine learning [27], and has been widely used in many real-world problems [27–29] and remote sensing classification [30–36]. The insight from AL is that allowing a machine learning algorithm to designate the samples for training could make it achieve higher accuracy with fewer training samples. As indicated by Figure 1, in an AL-boosted system, the interaction between the classification system and the training sample collecting system are bi-directional, where the classifier will "ask for" the most relevant samples to be verified/labeled in order to construct the classifier in an efficient and effective manner. Considering the cost and difficulty to verify the oil spill candidate, such an active learning process can greatly reduce the cost and time for building a detection system by reducing the number of candidates that needed to be verified without compromising the robustness and accuracy of the resulting classifier.

In this paper, we explore the potential of AL in training classifiers for the purpose of oil spill identification using 10 years (2004–2013) of RADARSAT data off the east and west coasts of Canada, which contains 198 RADARSAT-1 and RADARSAT-2 ScanSAR full scene images. Based on these images, we obtain 267 labeled samples, of which there are 143 oil spills and 124 look-alikes. We split these labeled samples into a simulating-set and a test-set, using the simulating-set to simulate an AL process involving a number of AL iterations, and using the test-set to calculate the performance of classifiers in each iteration. We start with a small number of training samples for initializing the classifiers, and with the AL iteration, we progressively select more samples and add them to the training set. Such a process ends when all samples in the simulating-set has been selected. Since the most important issue in AL is how to effectively select the most informative samples, we design six different active sample selection (ACS) methods to choose informative training samples. Moreover, we also explore the ACS approach with varying sample preference and the approach to reduce information redundancy among the selected samples. Four commonly used classifiers (KNN, SVM, LDA and DT) are coupled with ACS methods to explore the interaction between classifiers and ACS methods. Three kinds of measures are adopted to highlight different aspect of classification performance of these AL-boosted classifiers. Finally, to reduce the bias caused by the splitting of simulating set and test set in an effective manner, we adopt a six-fold cross validation approach to randomly split the labeled samples into six folds, using five for simulating and one for testing until all the folds have been used for testing once. To our best knowledge, this work is the first, effort except our very preliminary work [37], to explore the potential of AL for oil spill classification.

## 2. Dataset and Methods

The scheme of exploring the potential of AL for identifying marine oil spill is as Figure 2. In preprocessing step, the labeled samples are split into three parts. One part is put into the training samples pool for initially training classifiers. One part is put into the sample selection pool for sample selection in AL process. The third is for testing the performance of AL-boosted classifiers. The AL-boosted process starts by training classifiers with samples in the training sample pool. Samples in the sample selection pool are then classified by the trained classifiers, whose output are used to help the ACS approach to select some (here we select ten) most "informative" samples from the sample selection pool. These selected samples are added into the training samples pool to train classifiers in next iteration. The process keeps iterating until no sample left in the sample selection pool and all samples have been used to train the classifiers. We adopt four classifiers (i.e., SVM, LDA, KNN and DT), with each one coupled with all of the different ACS approaches. Three complementary numerical measures are calculated for each AL-boosted classifier using the testing set. The six-fold cross validation technique is used to get bias-reduced measures.

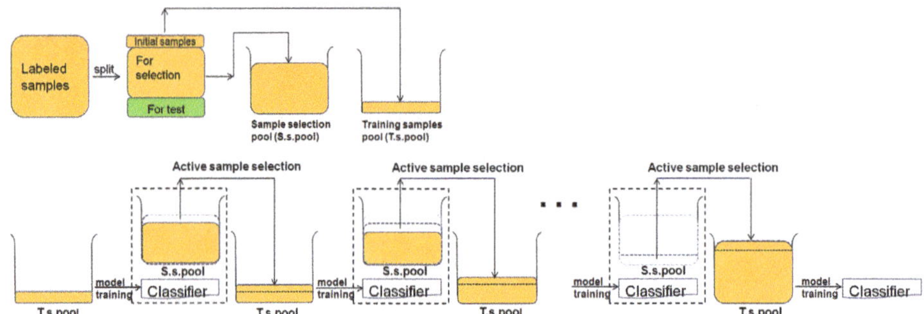


**Figure 2.** The scheme of exploring the potential of AL in building oil spill classification systems using ten-year RADARSAT data.

### 2.1. Dataset

The dataset used in this study contains 198 RADARSAT (188 RADASAT-1, 10 RADASAT-2) images (mode: ScanSAR narrow beam; swath width: 300 km; pixel spacing: 50 m) covering the east and west coasts of Canada from 2004 to 2013. Contained in these images are 143 oil spills and 124 look-alikes, all labeled by human experts in CIS of Environment Canada for a program called Integrated Satellite Tracking of Pollution (ISTOP). Because the boundaries of all labeled dark spots have been drawn by experts in CIS, we will not perform the dark spot detection process. Given the dark-spots in pixel-format, features extracted as input to classifiers can be categorized into four groups: (i) physical properties; (ii) geometric shape; (iii) texture; and (iv) contextual information [1,6,8]. Choosing the most relevant feature set for classification is not easy, because of the fact that feature selection is a complex issue depending on many factors such as the study area, the dataset, the classifiers, and the evaluation measures. Many researchers tried to study the relative importance of features for feature selection, but their conclusions are not in consistency due to their different experiment settings. For example, Karathanassi et al. [19] grouped 13 features into sea state dependent features and sea state independent features; Topouzelis et al. [7,8] examined 25 most commonly used features based on neural networks and decision tree forest, and selected several feature-subsets that are of most importance; Mera et al. [15] applied principal component analysis (PCA) to 17 shape related features and finally selected 5 principal components for their automatic oil spill detection system; and Xu et al. [22] implemented the permutation-based variable accuracy importance (PVAI) technique to evaluate feature's importance relative to different criteria and they found that different types of classifier tended to present different patterns on feature ranking and PVAI values.

Due to the lack of unified criteria for feature selection, we decided to use as many relevant features as could be extracted. We finally obtained 56 features, of which, there are 32 features about physical properties, 19 features about geometric shape, and 5 features about texture characteristics (see Table 2). Wind and ship information that describe the contextual information of the identified objects are not included because these information and techniques needed to detect them are not available to us. Readers could refer to Solberg et al. [4] for manually setting wind information, Espedal et al. [38] for using wind history information and Hwang et al. [39] and Salvatori et al. [40] for automatic wind information detection from SAR images. The features used in this study have varying ranges of values. We normalize these features by linearly mapping the values from their ranges to [0, 1].

**Table 2.** Features extracted based on the dark-spot objects for classifying oil spills from look-alikes.

| No | Type | Features | Code |
|---|---|---|---|
| 1 | Geometric | Target area in number of pixels | A |
| 2 | | Target perimeter in number of pixels | P |
| 3 | | Target Complexity measure C1 = P^2/A | C1 |
| 4 | | Target Complexity measure C2 = P/A | C2 |
| 5 | | Target Complexity measure C3 = P/(2*sqrt(pi*area)) | C3 |
| 6 | | The length of the major axis of the ellipse that has the same normalized second central moments as the object region. | Length |
| 7 | | The length of the minor axis of the ellipse that has the same normalized second central moments as the object region. | Width |
| 8 | | The eccentricity of the ellipse that has the same second-moments as the object region | Ecce. |
| 9 | | Target Spreading measures S = Length/Width | S |
| 10–12 | | The first three of Hu's invariant planar moments [41] | H1–H3 |
| 13–19 | | The first seven of Zernike moments [42] | Z1–Z7 |
| 20 | Physical | Average intensity value of the object | MeO |
| 21 | | Standard deviation of gray-scale intensity values of the object | SDO |
| 22 | | Average intensity value of the background area (a limited area near and outside object) | MeB |
| 23 | | Standard deviation of the intensity value of the background area | SDB |
| 24 | | Maximum intensity value of the object | MaxO |
| 25 | | Minimum intensity value of the object | MinO |
| 26 | | Power-to-Mean Ratio of the Object, SDO/MeO | PMRO |
| 27 | | Power-to-Mean Ratio of the Background area, SDB/MeB | PMRB |
| 28 | | Ratio between MeO and MeB | MeR |
| 29 | | Ratio between SDO and SDB | SDR |
| 30 | | Ratio between PMRO and PMRB | PMRR |
| 31 | | Difference between MeB and MeO | MeD |
| 32 | | Difference between SDB and SDO | SDD |
| 33 | | Difference between PMRB and PMRO | PMRD |
| 34 | | The difference between MeB and MinO | MaxC |
| 35 | | Average gradient value of the object area | MeGO |
| 36 | | Standard deviation of the gradient value of the object area | SDGO |
| 37 | | Average gradient value of the background area. | MeGB |
| 38 | | Standard deviation of the gradient value of the background area | SDGB |

**Table 2.** *Cont.*

| No | Type | Features | Code |
|---|---|---|---|
| 39 | | Average gradient value of the object border area. | MeGBo |
| 40 | | Standard deviation of the gradient value of the object border area | SDGBo |
| 41 | | Maximum gradient value of the object | MaxGO |
| 42 | | Minimum gradient value of the object | MinGO |
| 43 | | Ratio between SDGO and MeGO | PMRGO |
| 44 | | Ratio between SDGB and MeGB | PMRGB |
| 45 | | Ratio between MeGO and meGB | MeGR |
| 46 | | Ratio between SDGO and SDGB | SDGR |
| 47 | | Ratio between PMRGB and PMRGO | PMRGR |
| 48 | | Difference between MeGB and MeGO | MeGD |
| 49 | | Difference between SDGB and SDGO | SDGD |
| 50 | | Difference between PMRGB and PMRGO | PMRGD |
| 51 | | Difference between MeGB and MinGO | MaxGC |
| 52 | Textural | GLCM Contrast | Cont. |
| 53 | | GLCM Correlation | Corr. |
| 54 | | GLCM Energy | Ener. |
| 55 | | GLCM Homogeneity | Homo. |
| 56 | | GLCM Entropy | Entr. |

### 2.2. Classifiers Used

Four commonly used classifiers (SVM, KNN, LDA and DT) are adopted to be integrated into the AL framework. We prefer choosing commonly used classifiers in our study because: (i) we are focusing on exploring the effectiveness of the AL approaches, rather than finding the best classifiers; (ii) complex classifiers introduce more hyper-parameters that may complicate the performance, making it difficult to analyze the role of active learning; and (iii) conclusions drawn from commonly used classifiers may apply on high-level classifiers built on them, while it is not true vice versa. We set the hyper parameters of each classifier fixed across all AL learning iterations.

#### 2.2.1. Support Vector Machine (SVM)

SVM is a "local" classifier whose decision boundary depends on a small number of supporting vectors/samples, which means finding the most relevant set of samples via AL is crucial for the performance. It is a well-known classifier for remote sensing applications [33,43] and particularly for oil spill classification [9,20,22]. Here, we use LIBSVM [44], the radial kernel, C = 1 and gamma = 0.07.

#### 2.2.2. K Nearest Neighbors

KNN classifies a sample by a majority vote of this sample's k nearest neighbors. It is widely used in remote sensing society [45–47] and particularly for oil spill classification [17,21]. We here simply set the only hyper parameter k = 9.

#### 2.2.3. Linear Discriminant Analysis

LDA predicts the class membership based on the posterior probabilities of different classes. It assumes that the densities of predictors conditioned on class membership are Gaussian. Many modifications of LDA exist, such as Penalized LDA [48], null-space LDA [49], Dual-Space LDA [50], Probabilistic LDA [51], Global-local LDA [52], etc. Nirchio et al. [18] used LDA and Xu et al. [22] used the Penalized LDA directly for oil spill classification. Here, we use the basic LDA, hoping that conclusions drawn from LDA could apply to the other variants.

2.2.4. Decision Tree

DTs are flexible classifiers that recursively split the input dataset into subset [53]. The class label of a test sample is predicted by applying the decision criteria from the root to the leaves to determine which leaf it falls in. Because of its capability of easily providing an intelligible model of the data, decision tree is very popular and widely used for classification purpose either directly [3,15] or as the elemental classifier of state-of-the-art ensemble techniques such as bagging, bundling and boosting for achieving better generality performance [8,22,54–56]. Here, we use the DT supported by the classification and regression tree (CART) algorithm [53]. ClassificationTree class in Matlab 2012b was used, with all parameters set as default.

## 2.3. Active Learning

The AL process iteratively helps the classifier to identify and adopt the most informative samples for training the classifier in an efficient and effective manner. In each iteration, a classifier is first trained with current training set, then active sample selecting (ACS) methods choose L (here L = 10) samples that are most informative for current classifier, obtain their labels and add them to the training set. The ACS method responsible for informative sample selection is of key importance to the success of the AL process. To explore the influence of different ACS methods on the classifier performance, here, we choose the uncertainty criterion to define the informativeness of samples and design six basic ACS methods based on it. Two strategies that may further improve the informativeness of the selected samples by adjusting sample preference in iterations and reducing redundancy amongst samples are also considered.

2.3.1. Six Basic ACS Methods

We choose the most widely used uncertainty/certainty criterion for describing the informativeness of samples [27] and propose six ACS methods based on it. Here, the certainty of a sample being an oil-spill is defined by its posterior probability that is usually implemented as the soft-outputs of classifiers. For KNN classifier, the posterior probability of input $x$, is defined as $p(oil|x) = \#(oil\ spills\ in\ N(x))/K$, where $N(x)$ is the $K$ nearest neighbors of $x$. For LDA classifier, it is defined as $p(oil|x) =\propto p(x|oil)$, where $p(x|oil)$ is a multivariate normal density trained by LDA. For SVM classifier, we use libSVM toolkit [44] which obtains the probability according to the work of Wu et al. [57]. Traditionally, DT can only provide piecewise constant estimate of the class posterior probabilities, since all the samples classified by a leaf share the same posterior probabilities. Some improvements have been proposed [58] for getting a more smooth estimation of the class posterior probabilities. For simplicity, we here still use the traditional estimation, i.e., posterior probability of input $x$ classified by a leaf, is defined as $p(oil|x) = k/n$, where $k$ is the number of training samples being classified into the oil spill class by the leaf, and n is the total number of training samples being classified by the leaf.

Sorting all samples in the sample selection pool according to their certainty of being oil spills $p(oil|x)$, their certainty of being look-alikes $1 - p(oil|x)$ and their uncertainty of classification $1 - abs(0.5 - p(oil|x))$ in descending order, we obtain three sequences, which are denoted by q1, q2 and q3, respectively.

Let L be the number of samples selected in each iteration of active learning, and w1, w2 and w3 be the percentages of samples that will be selected from q1, q2 and q3, respectively. Then, our ACS algorithm here can be denoted with ACS(w1,w2,w3), where w1 + w2 + w3 = 1. It means at each iteration, our algorithm selects respectively w1*L, w2*L and w3*L samples from q1, q2 and q3 with no replicates. All the selected samples are then labeled and put into training sample set. We design here our ACS methods according to the different setting of (w1, w2, w3) in Table 3.

**Table 3.** Six basic active sample selecting (ACS) methods.

| Method | Parameters | Description |
|---|---|---|
| ACS-1 | NULL | Randomly select samples |
| ACS-2 | w1 = 1, w2 = 0, w3 = 0 | Prefer samples more like oil spills |
| ACS-3 | w1 = 0, w2 = 1, w3 = 0 | Prefer samples more like look-alikes |
| ACS-4 | w1 = 0.5, w2 = 0.5, w3 = 0 | Take half of samples from ACS-2 and half from ACS-3 |
| ACS-5 | w1 = 0, w2 = 0, w3 = 1 | Prefer samples with high uncertainty of classification |
| ACS-6 | w1 = 0.25, w2 = 0.25, w3 = 0.5 | Take half of the samples from ACS-4 and the other half from ACS-5 |

### 2.3.2. Adjusting Sample Preference in Iterations

One drawback of the six basic ACS methods lies in the fact that the sampling method dictated by w1, w2 and w3 is fixed across all iterations of the AL process, while model learning may prefer different sampling methods in different learning stages. Therefore, a better strategy is to adjust the w1, w2 and w3 values during AL iterations. We proposed a method based on such idea, i.e., in first iterations of AL, model learning needs samples of more certainty; with continuing iterations, samples of more uncertainty are more and more relevant; and, after a certain point, the sample preference of model learning should be fixed to achieve stabilized learning. This idea is inspired by the insight of the criteria of exploitation and exploration [59]. Accordingly, we set w = w1 + w2 with w1 = w2, and set w3 = 1 − w. Let x = i/#iterations be the normalized index of the *i*th iteration. The function that describe how the value $w$ change with $x$ is set as follow (see Figure 3 for the illustration),

$$w(x) = \begin{cases} 1 - c \cdot \exp(a(x-b)), & x \leq b \\ 1-c & x > b \end{cases} \qquad x, c, b \in [0,1] \tag{1}$$

where $c$ determines the minimum value of $w$, $b$ is the index of iteration after which $w$ will keep constant, and $a$ controls the changing rate of $w$.

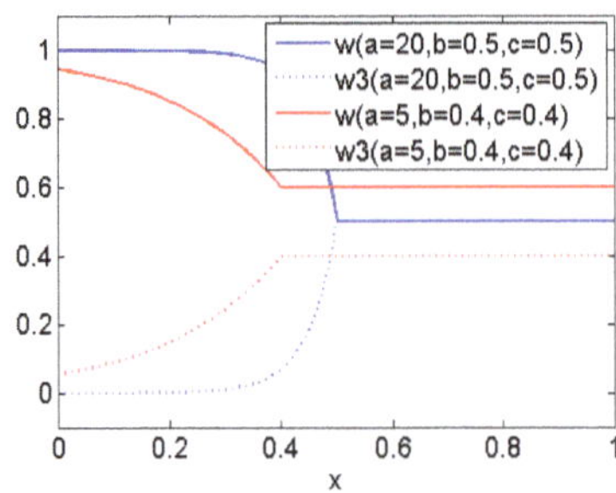


**Figure 3.** Illustration of the function w(x) and w3(x) = 1 − w(x) with different parameter setting of a, b and c.

### 2.3.3. Reducing Redundancy amongst Samples (RRAS)

In each AL iteration, we select L samples to increase the current training set. However, the selected L samples may have overlap of information [27]. To maximize information in selected samples, we adopt a strategy similar to clustering-based diversity criterion [32,33], i.e., the unlabeled samples are divided into clusters with the k-means method in each AL iteration. When selecting samples according to three sequences (q1, q2 and q3), if the candidate sample shares a cluster with any of the existing samples in the same sequence, this candidate will be discarded and keep on considering next sample. We set the k of k-means as min(L, M), where M is the number of samples available for selection at each iteration.

### 2.4. Performance Measures

Three different performance measures that cover demands of different end users are considered here. All these measures are calculated at each AL iteration, and finally shown as trajectories along time (iterations). The mean measures over all iterations are calculated for the convenience of numerical comparison. These measures originate from the confusion matrix (Table 4).

**Table 4.** Confusion matrix (oil spill is the positive class and look-alikes is the negative class).

| | **Predicted as Positive** | **Predicted as Negative** |
|---|---|---|
| **Actually Positive** | True Positives (TP) | False Negatives (FN) |
| **Actually Negative** | False Positive (FP) | True Negatives (TN) |

#### 2.4.1. Overall Performance

The Receiver Operating Characteristics (ROC) curve displays the trade-off between false positive rate (FPR = FP/(TN + FP)) and true positive rate (TPR = TP/(TP + FN)) with the varying of decision points [60]. For the convenience of showing the performance variation over AL iterations, we reduce the ROC curves to single scalar measure by counting the area under the ROC curve (AUC) [60], which stands for the "probability that the classifier will correctly rank a randomly chosen positive instance higher than a randomly chosen negative instance" [61]. We here use AUC to evaluate the overall performance of different methods. We denote this measure as AUC-All.

#### 2.4.2. High TPR Performance

An oil spill classification system should predict correctly as high percentage of true oil spills as possible, it means the classification systems may be tuned to be with fixed high TPRs. To deal with this situation, Xu et al. [22] uses the FPR at a fixed high TPR, i.e., 0.8, to evaluate the performance of different classifiers. To be more general, we here use a measure considering TPR from 0.5 to 1. The area that is under the curve of ROC with TPR from 0.5 to 1 is used for this purpose (see Figure 4). We denote this measure as AUC-H. The higher the value of AUC-H is, the more possible that classifier perform well when it is tuned to with high TPRs.

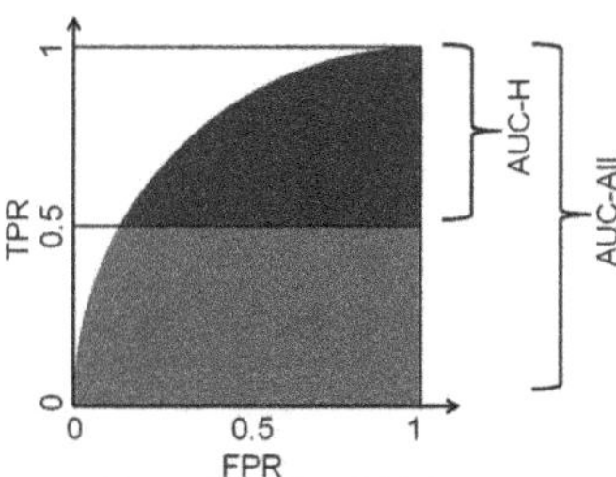


**Figure 4.** Illustration of over-all performance measure (denoted as AUC-All), and of high-TPR performance measure (denoted as AUC-H). TPR stands for true positive rate, and FPR stands for false positive rate.

#### 2.4.3. Sorting Performance

For the oil spill detection system aiming to send alarms to the investigation performing institutions, the performance of sorting the input dark slicks well to make the former part of the sequence has higher accuracy of being true oil spills, seems to be very important.

The precision-recall curve, where precision = TP/(TP + FP), recall = TP/(TP + FN) = TPR, is a good tool for evaluating the sorting performance of the information retrieving systems in which one class is of more importance than others, and it gives more informative details than ROC when the TPR is small [62]. In order to show the trajectories of the performance improvement of different ACS methods, we reduce the precision-recall curves to single scalar measures by counting the mean precision. Considering the fact that only a small percent of alarms will be verified at last (in our study dataset, only 17 percent of alarms were verified after being sent for investigation), we only calculate the mean precision when recall is from 0 to 0.5. We denote this measure as MP-L.

### 2.5. Cost Reduction Measure

A significant benefit of using AL is that AL can reduce the number of training samples required for achieving a reliable system, as such reducing the time and the money for collecting the training samples. We here measure the cost reduction brought by an AL method with 1-R, where R is the ratio between the number of training samples used to achieve a designated performance and the total number of training samples in our study. Here, the designated performances are set as 90%, 92%, 94%, 96%, 98% and 100% of the baseline performance that was achieved by the classier using all training samples without AL. The performance measures are described in Section 2.4.

### 2.6. Initial Training Set

As indicated in Figure 2, some initial training samples are required to train the classifiers, whose output will be used to guide the sample selection and enable the start of AL iterations. Here, we randomly select ten samples, of which there are five look-alikes and five oil spills, to be the initial training set. Considering that AL may be sensitive to initial training set, 100 separate runs are performed and the average performances over them are used.

## 3. Results and Discussion

### 3.1. Performance of ACS Methods

We want ACS methods to help a classifier improve its performance stably and as quickly as possible. Thus, the ideal curve of a classifier's performance over iterations should be stably ascending, steep in the fore part and flat in the back part. Figure 5 shows the graphs of performance evolving over iterations of ACS methods coupled with SVM, KNN, LDA and DT classifier. By averaging the performance values of each curve in Figure 5, we get the mean performance values in Figure 6. The classifiers show different characteristics.

One observation is that in the case of high-PTR performance, KNN, LDA and DT which show bad (flat, unstable or descending) trends in Figure 5 have much better performance number in Figure 6 than SVM which shows good trend (stably ascending, steep in the fore part and flat in the back part) in curves. This might cause confusion when we choose a better classifier to work with ACS methods, because any variance of factors such as pre-processing, feature selection, parameter setting for classifiers, might dramatically change the mean performance value of a classifier and comparing classifiers in mean performance is very difficult to be on a fair base, we here suggest that more trust should be put on the trends of curves which more likely present the intrinsic features and less trust on the performance values that could be affected by too many factors. Based on this principle, Figure 5 shows five bad situations (KNN in high-PRT performance, LDA in all three kinds of performance and DT in high-PRT performance), in which ACS methods work badly and three good situations (SVM in all three kinds of performance), in which ACS methods work well.

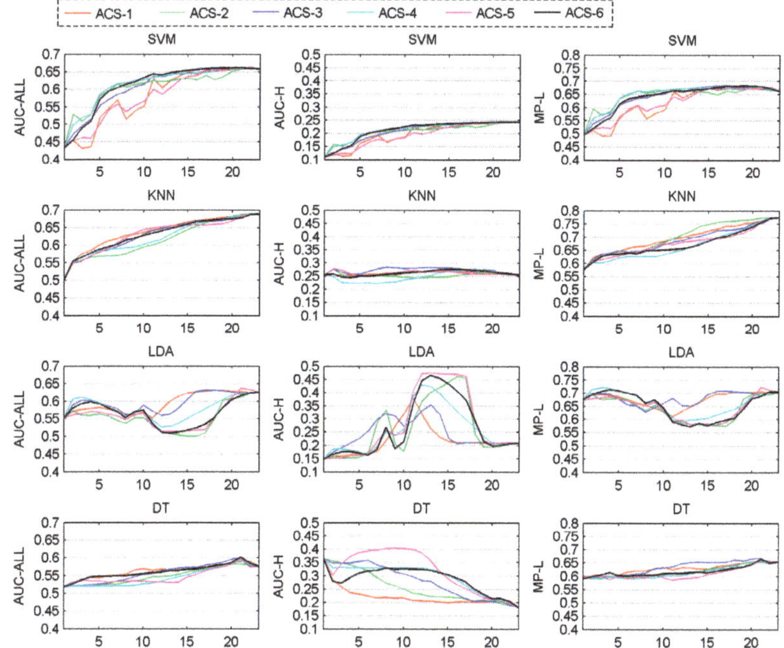


**Figure 5.** Graphs of performance evolving over iterations (horizontal axis) of ACS methods with SVM, KNN, LDA and DT classifier. AUC-ALL, AUC-H and MP-L are the measures of overall performance, high-TPR performance and sorting performance, respectively.

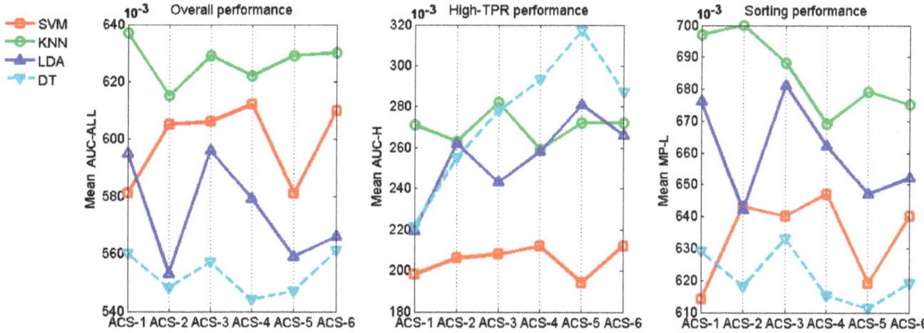


**Figure 6.** The mean performance values of ACS methods coupled with SVM, KNN, LDA and DT classifier. (These mean values are calculated by averaging performance values of each curve in Figure 5.)

KNN-based ACS methods show almost horizontal curves in the high-PTR performance, which means the increasing of training samples from any ACS method will not bring obvious improvement in high-PTR performance o KNN. DT-based ACS methods show almost descending curves in the high-PTR performance, which means the increasing of training samples causes the drop of performance. These phenomena might arise partly from the fact that the distributions of oil spill and look-alikes are heavily overlapping. When improving a system pursuing high-PTR performance, KNN and DT classifier should be considered carefully.

LDA-based ACS methods show dramatic fluctuations in their performance curves. LDA classifier takes the Gaussian assumption for the underlying distribution of oil spills and look-alikes. The big fluctuations means newly added samples change dramatically the shape of distributions learned previously. To deal with this problem, keeping a smooth change of distribution shape should also be considered in future when adding samples from ACS methods.

SVM-based ACS methods show good (stable and ascending, especially ascending quickly in the fore part and becoming flat in the back part of iterations) patterns in all three kinds of performance. That would be a merit when more than one services (each asks for a different performance) are demanded. In this case, only one (not the number of services) system needed to be built.

In the case of overall performance and sorting performance, randomly selecting samples (ACS-1) is always the best or second best when it is coupled with KNN, LDA or DT classifier, but not with SVM classifier. That may be because that KNN, LDA and DT are classifiers that try to make a decision based on some statistics from all samples in a certain region, while SVM makes the decision only on a few key samples (support vectors). Obviously, randomly selected samples (as from ACS-1) are more likely to maintain the statistics of the underlying distribution from which our training and test dataset come, but less likely to contain some key samples for SVM classifier.

The results in Figures 5 and 6 also show that choosing a good ACS method for a specified classifier should be based on considering at least two important factors: the kind of performance chosen for optimization and the learning stages. A classifier may prefer different ACS methods in different performance measures. For example, the DT classifier favors ACS-1 in the overall performance and sorting performance but favors ACS-5 in the high-PTR performance. A classifier may also prefer different ACS methods in different stages of learning process. For example, SVM classifier favors ACS-2 and ACS-4 (methods that prefer choosing samples of more certainty) at the first half of the iterations in our study but ACS-5 and ACS-6 (methods that prefer choosing samples of more uncertainty and that choose samples half of more certainty and half of more uncertainty) at the second half.

### 3.2. Cost Reduction Using ACS Methods

Figure 7 shows the cost reduction of using ACS methods to boost SVM, KNN, LDA and DT classifiers for achieving different destination performance. By selecting the maximum cost reduction for each classifier to achieve each destination performance, we obtain Table 5.

It should be noted that a big cost reduction in Figure 5 and Table 5 does not always means that ACS methods work successfully in that situation. The gray elements in Table 5 show five bad situations (also mentioned in Section 3.1), in which the cost reductions are very big but the ACS methods actually work so poorly that it is not necessary to analyze the cost reduction of these situations.

It can be seen that a considerable cost reduction can be achieved using ACS to boost classifiers. Taking the SVM, for instance, to get D5 destination performance, the maximum 43%, 30% and 70% reductions of cost can be obtained in overall performance, high-PTR performance and sorting performance, respectively. For D6 destination performance, the maximum 26%, 4% and 61% reductions of cost can be obtained in overall performance, high-PTR performance and sorting performance, respectively.

**Table 5.** The maximum cost reduction of ACS methods for achieving designated performance D1 to D6 with SVM, KNN, LDA and DT classifiers. The numbers marked with gray show situations in which there are big cost reductions but ACS methods actually work poorly.

| | **Overall Performance (%)** | | | | | | **High-TPR Performance (%)** | | | | | | **Sorting Performance (%)** | | | | | |
|---|---|---|---|---|---|---|---|---|---|---|---|---|---|---|---|---|---|---|
| | **D1** | **D2** | **D3** | **D4** | **D5** | **D6** | **D1** | **D2** | **D3** | **D4** | **D5** | **D6** | **D1** | **D2** | **D3** | **D4** | **D5** | **D6** |
| **SVM** | 74 | 70 | 61 | 57 | 43 | 26 | 61 | 57 | 52 | 43 | 30 | 4 | 78 | 78 | 78 | 74 | 70 | 61 |
| **KNN** | 70 | 61 | 57 | 39 | 26 | 9 | 91 | 91 | 91 | 91 | 91 | 91 | 57 | 48 | 43 | 39 | 35 | 26 |
| **LDA** | 96 | 91 | 91 | 91 | 91 | 87 | 87 | 83 | 83 | 83 | 83 | 78 | 96 | 96 | 96 | 96 | 96 | 91 |
| **DT** | 91 | 87 | 83 | 61 | 57 | 26 | 96 | 96 | 96 | 96 | 96 | 96 | 96 | 96 | 83 | 70 | 57 | 48 |

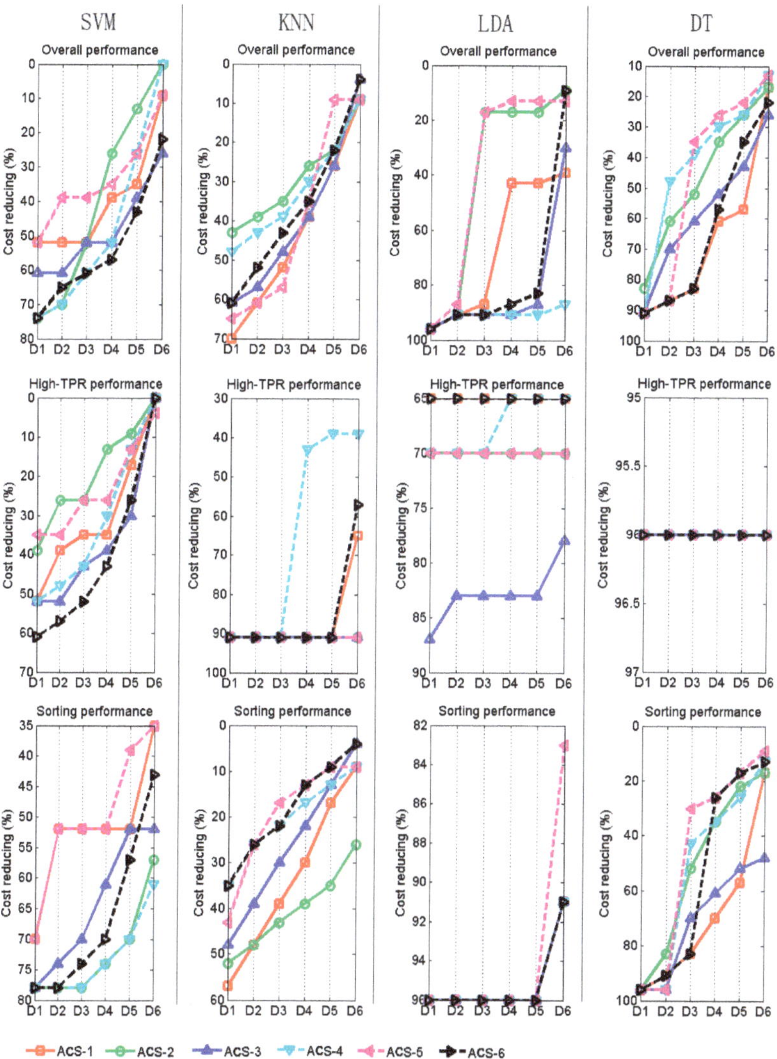


**Figure 7.** The cost reduction for achieving different destination performances with ACS methods coupled with SVM, KNN, LDA and DT classifier. Designated performances D1 to D6 stand for 90% to 100% of the baseline performance achieved by the classier using all training samples without AL.

It can be seen in Figure 7 that, for SVM classifier, ACS-6 shows the best performance curve in overall performance and high-PTR performance, while ACS-2 and ACS-4 show the best curves in sorting performance. For DT classifier, ACS-1 works best in overall performance and sorting performance. For KNN classifier, ACS-1 work best in overall performance and ACS-2 works best in sorting performance.

There is still great potential to improve the cost reduction performance by improving the data preparation and tuning of classifiers, such as feature selecting, preprocessing, optimizing the parameters of model and using other definition of the informativeness preferred by the classifier.

Another method that may further improve the reduction of cost is to carry out the system serving and system training at the same time after the system has already had a sound performance. It can be seen in Table 5 that, using a sound designated performance (such as one of D1–D5), leads to a bigger cost reduction than selecting the perfect designated performance D6. Moreover, the income of service might cover the cost of continuing the system training once the system has been put into working.

### 3.3. Reducing Redundancy Amongst Samples (RRAS)

With SVM classifier, it can be seen in Table 6 that RRAS reduces the mean performance of most ACS methods except ACS-1. Because the key samples (support vectors) only exist in a relatively small region near the decision boundary for SVM classifier, increasing the diversity of samples by RRAS would not help choose samples from the key region, which is very small compared with the whole feature space. Compared to not using RRBS (Figure 5), performance fluctuations in Figure 8 are increased at the former part of iterations and weakened at the end part of iterations, and ACS-2, ACS-4 change from the first and second best methods to the first and second worst at the first few iterations of all graphs. These phenomena could also be caused by the increased randomness of selected samples by RRAS. It seems that RRAS is not suitable for SVM-based ACS methods in our case.

For KNN classifier, the RRAS does not bring significant improvement in the shape of curves in Figure 8 compared to curves in Figure 5, but does slightly increase the mean performance values of more than half of ACS methods (see Table 6). For KNN classifier, the more thorough the training sample dataset can represent the true underlying distribution, the higher performance it can obtain. The RRAS can improve the representative ability of selected samples by first grouping all samples into clusters and then selecting one from each cluster as the representative sample and therefore seems suitable for some KNN-based ACS methods.

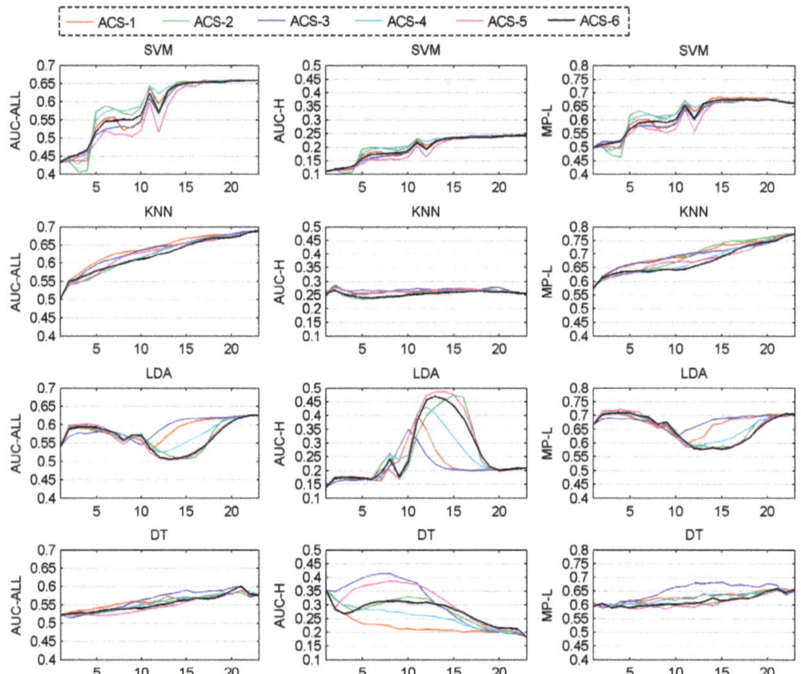


**Figure 8.** Graphs of performance evolving over iterations (horizontal axis) for ACS methods coupled with SVM, KNN, LDA and DT classifier (RRAS are used here). AUC-ALL, AUC-H and MP-L are the measures of overall performance, high-TPR performance and sorting performance, respectively.

In Table 6, there are some performance improvements and drops for ACS methods with LDA and DT classifier, but it is hard to analyze the reasons.

**Table 6.** The difference of mean performance numbers between using and not using RRAS. The positive differences are marked with gray. A1 to A6 stand for ACS-1 to ACS-6.

| | **Overall Performance ($10^{-3}$)** | | | | | | **High-TPR Performance ($10^{-3}$)** | | | | | | **Sorting Performance ($10^{-3}$)** | | | | | |
|---|---|---|---|---|---|---|---|---|---|---|---|---|---|---|---|---|---|---|
| | **A1** | **A2** | **A3** | **A4** | **A5** | **A6** | **A1** | **A2** | **A3** | **A4** | **A5** | **A6** | **A1** | **A2** | **A3** | **A4** | **A5** | **A6** |
| **SVM** | 1.3 | −17 | −28 | −19 | −14 | −26 | 0 | −5 | −14 | −8 | −6 | −15 | 4.2 | −22 | −28 | −23 | −15 | −22 |
| **KNN** | 0 | 5.9 | 2.1 | 1.8 | −5 | −9 | 1.1 | 8.3 | −3 | 5.3 | 1.7 | −7 | 1.5 | −6 | 6.8 | 10 | 1 | 1 |
| **LDA** | −7 | 15 | −4 | −2 | 6.3 | 1.8 | −1 | 0 | −34 | −15 | −12 | −3 | −2 | 14 | −1 | 1 | 6.6 | 2 |
| **DT** | −3 | 4.4 | 5 | 11 | 0 | −7 | 0.7 | 17 | 29 | −39 | −12 | −11 | −1 | −3 | 14 | 11 | −2.2 | −4 |

## 3.4. Adjusting Sample Preference in Iterations

Figure 9 shows the performance-evolution graph of the ACS method with varying parameter for sample selection, coupled with SVM. It can be seen that the ACS method with parameter w varying according to Function (1), at most iterations, matches or outperforms any of the other fixed-w methods (ACS-4 is with w = 1, ACS-5 with w = 0, and ACS-6 with w = 0.5). Here, we set the parameters of Function (1) as a = 20, b = 0.5, c = 0.5. The varying-w method obtained mean performance values (by averaging performance numbers on its curves in Figure 9) 0.614, 0.213 and 0.650 for overall performance, high-TPR performance and sorting performance, respectively. These mean values are all better than the best ones of SVM classifier in Figure 6, i.e., 0.612, 0.212 and 0.647.

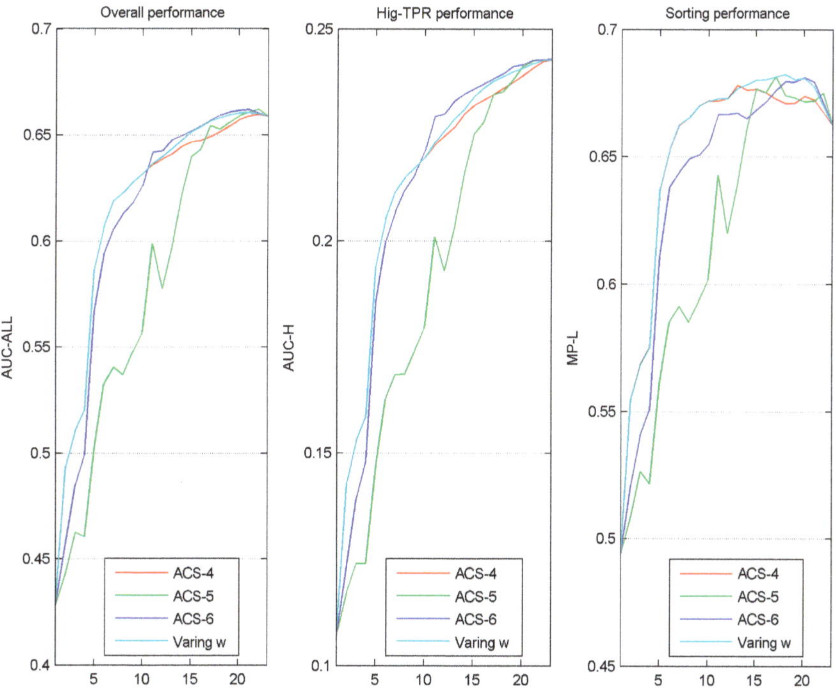


**Figure 9.** Performance-evolution graph of the ACS method with varying parameter w, coupled with SVM.

Although adjusting parameter w in an iteration-based manner seems to be a good strategy for improving system performance in our study, the use of such a strategy must have a theoretical or empirical basis. Otherwise, there is no guarantee that such good results might also be achieved in other datasets. Although the exploration-and-exploitation criterion [59] is a commonly accepted one (our method here is derived from it), it is only suitable for SVM classifier in this study. The unsuitability for KNN, LDA and DT classifier can be seen clearly in Figure 5, in which, when coupled with these classifiers, ACS methods preferring samples of certainty does not perform well at the former part of iterations and so do the ACS methods preferring samples of uncertainty at the end part. Criteria similar to exploration-and-exploitation suitable for SVM need to be found and verified for other classifiers in the future.

## 4. Conclusions

In this study based on a ten-year RADARSAT dataset covering west and east coasts of Canada, AL has shown its great potential of training sample reduction (for example, a 4% to 78% reduction on training samples can be achieved in different settings when using AL to boost SVM classifier) in constructing oil spill classifiers. That means the real-world projects of constructing oil spill classification systems (especially when it is hard to accumulate a large number of training samples, such as when supervising new water area) or improving existed systems may benefit from using AL methods. In the cases where AL are used for classifier training, we boldly suggest that the expensive, time-sensitive, and difficult field verification work should be conduct only for those targets that are identified by AL as the "important" targets to significantly improve the classifier training efficiency. AL could reduce training data, whether they are obtained by expert-labeling or by field-verifying. Generally, field-verified data are better than expert-labeled data for training due to the higher verification accuracy of the field investigation approach. Nevertheless, when it is hard or impossible to do fieldwork, asking experts for labeling is also acceptable. Both labeling approaches benefit from the efficient learning process of the AL method for classifier construction.

Our study shows that not all classifiers can benefit from using AL methods according to all measures. In some cases (in this paper, KNN in high-PRT measure, LDA in all three kinds of performance measures and DT in high-PRT measure), the AL methods may not help improve performance, or even reduce it.

Of the four classifiers tested in this paper, the SVM is the best for using AL methods for the following reasons. First, it can benefit greatly from some basic ACS methods (in our case, ACS-2, ACS-4 and ACS-6), showing perfect performance evolving curves with steeply ascending fore parts and flat back parts. Second, the good ACS methods for SVM in one kind of performance measure will also be good in other kinds of performance measures. That would be a merit when more than one services asking for different kinds of performance are demanded. In this case, only one system needed to be built and that surely will greatly reduce costs. Third, its performance could be further improved by ACS method using exploration-and-exploitation criterion, which considers different sample preference in different learning stages.

The exploration-and-exploitation criterion is suitable for SVM but not for KNN, LDA and DT classifiers. The criteria considering different sample preference in different learning stages and being suitable for other classifiers may also exist and should be found and studied in the future.

The thorough knowledge of a classifier's preference on training samples is the key to achieve efficient AL-based classification system, because all AL operations, such as choosing AL strategy and ACS methods, adjusting sample selection preference in iterations, depend on knowing a classifier's sample favoritism to identify the best samples that satisfy the demands of the classifier. Thus, further study should also focus on investigating the sample preference mechanism of a classifier to build a high-performance AL-based frame on it.

**Acknowledgments:** This work was partially supported by NSF of China (41161065 and 41501410), Canadian Space Agency, ArcticNet, NSF of Guizhou ([2017]1128, [2017]5653) and Educational Commission of Guizhou (KY[2016]027).

**Author Contributions:** Yongfeng Cao conceived, designed and performed the experiments; and wrote the paper. Linlin Xu helped conceive and design the experiments and revised the paper. David Clausi guided the direction and revised the paper.

**Conflicts of Interest:** The authors declare no conflict of interest.

## References

1. Brekke, C.; Solberg, A.H.S. Oil spill detection by satellite remote sensing. *Remote Sens. Environ.* **2005**, *95*, 1–13. [CrossRef]
2. Salberg, A.B.; Rudjord, Ø.; Solberg, A.H.S. Oil spill detection in hybrid-polarimetric SAR images. *IEEE Trans. Geosci. Remote Sens.* **2014**, *52*, 6521–6533. [CrossRef]
3. Singha, S.; Vespe, M.; Trieschmann, O. Automatic Synthetic Aperture Radar based oil spill detection and performance estimation via a semi-automatic operational service benchmark. *Mar. Pollut. Bull.* **2013**, *73*, 199–209. [CrossRef] [PubMed]
4. Solberg, A.H.S.; Storvik, G.; Solberg, R.; Volden, E. Automatic detection of oil spills in ERS SAR images. *IEEE Trans. Geosci. Remote Sens.* **1999**, *37*, 1916–1924. [CrossRef]
5. Solberg, A.H.S. Remote Sensing of Ocean Oil-Spill Pollution. *Proc. IEEE* **2012**, *100*, 2931–2945. [CrossRef]
6. Topouzelis, K. Oil Spill Detection by SAR Images: Dark Formation Detection, Feature Extraction and Classification Algorithms. *Sensors* **2008**, *6*, 6642–6659. [CrossRef] [PubMed]
7. Topouzelis, K.; Stathakis, D.; Karathanassi, V. Investigation of genetic algorithms contribution to feature selection for oil spill detection. *Int. J. Remote Sens.* **2009**, *30*, 611–625. [CrossRef]
8. Topouzelis, K.; Psyllos, A. Oil spill feature selection and classification using decision tree forest on SAR image data. *ISPRS J. Photogramm. Remote Sens.* **2012**, *68*, 135–143. [CrossRef]
9. Brekke, C.; Solberg, A.H.S. Classifiers and confidence estimation for oil spill detection in ENVISAT ASAR images. *IEEE Geosci. Remote Sens. Lett.* **2008**, *5*, 65–69. [CrossRef]
10. Solberg, A.H.S.; Brekke, C.; Husøy, P.O. Oil spill detection in Radarsat and Envisat SAR images. *IEEE Trans. Geosci. Remote Sens.* **2007**, *45*, 746–754. [CrossRef]
11. Del Frate, F.; Petrocchi, A.; Lichtenegger, J.; Calabresi, G. Neural networks for oil spill detection using ERS-SAR data. *IEEE Trans. Geosci. Remote Sens.* **2000**, *38*, 2282–2287. [CrossRef]
12. Topouzelis, K.; Karathanassi, V.; Pavlakis, P.; Rokos, D. Detection and discrimination between oil spills and look-alike phenomena through neural networks. *ISPRS J. Photogramm. Remote Sens.* **2007**, *62*, 264–270. [CrossRef]
13. Topouzelis, K.; Karathanassi, V.; Pavlakis, P.; Rokos, D. Dark formation detection using neural networks. *Int. J. Remote Sens.* **2008**, *29*, 4705–4720. [CrossRef]
14. Topouzelis, K.; Karathanassi, V.; Pavlakis, P.; Rokos, D. Potentiality of feed-forward neural networks for classifying dark formations to oil spills and look-alikes. *Geocarto Int.* **2009**, *24*, 179–191. [CrossRef]
15. Mera, D.; Cotos, J.M.; Varela-Pet, J.; Rodríguez, P.G.; Caro, A. Automatic decision support system based on sar data for oil spill detection. *Comput. Geosci.* **2014**, *72*, 184–191. [CrossRef]
16. Singha, S.; Velotto, D.; Lehner, S. Near real time monitoring of platform sourced pollution using TerraSAR-X over the North Sea. *Mar. Pollut. Bull.* **2014**, *86*, 379–390. [CrossRef] [PubMed]
17. Kubat, M.; Holte, R.; Matwin, S. Machine learning for the detection of oil spills in satellite radar images. *Mach. Learn.* **1998**, *23*, 1–23. [CrossRef]
18. Nirchio, F.; Sorgente, M.; Giancaspro, A.; Biamino, W.; Parisato, E.; Ravera, R.; Trivero, P. Automatic detection of oil spills from SAR images. *Int. J. Remote Sens.* **2005**, *26*, 1157–1174. [CrossRef]
19. Karathanassi, V.; Topouzelis, K.; Pavlakis, P.; Rokos, D. An object-oriented methodology to detect oil spills. *Int. J. Remote Sens.* **2006**, *27*, 5235–5251. [CrossRef]
20. Mercier, G.; Girard-Ardhuin, F. Partially supervised oil-slick detection by SAR imagery using kernel expansion. *IEEE Trans. Geosci. Remote Sens.* **2006**, *44*, 2839–2846. [CrossRef]
21. Ramalho, G.L.B.; Medeiros, F.N.S. Oil Spill Detection in SAR Images using Neural Networks. In Proceedings of the 18th International Conference on Pattern Recognition (ICPR'06), Hong Kong, China, 20–24 August 2006.

22. Xu, L.; Li, J.; Brenning, A. A comparative study of different classification techniques for marine oil spill identification using RADARSAT-1 imagery. *Remote Sens. Environ.* **2014**, *141*, 14–23. [CrossRef]
23. Hastie, T.; Tibshirani, R.; Friedman, J. *Elements of Statistical Learning: Data Mining, Inference, and Prediction*, 2nd ed.; Springer: New York, NY, USA, 2009; pp. 22–26. ISBN 978-0387848570.
24. Pavlakis, P.; Tarchi, D.; Sieber, A.J. On the monitoring of illicit vessel discharges using spaceborne SAR remote sensing—A reconnaissance study in the Mediterranean sea. *Ann. Des. Telecommun. Telecommun.* **2001**, *56*, 700–718.
25. Indregard, M.; Solberg, A.H.S.; Clayton, P. *D2-Report on Benchmarking Oil Spill Recognition Approaches and Best Practice*; Technical Report, Eur. Comm. 2004, Archive No. 04-10225-A-Doc, Contract No:EVK2-CT-2003-00177; European Commission: Brussels, Belgium, 2004.
26. Ferraro, G.; Meyer-Roux, S.; Muellenhoff, O.; Pavliha, M.; Svetak, J.; Tarchi, D.; Topouzelis, K. Long term monitoring of oil spills in European seas. *Int. J. Remote Sens.* **2009**, *30*, 627–645. [CrossRef]
27. Settles, B. Active Learning Literature Survey. *Mach. Learn.* **2010**, *15*, 201–221.
28. Wang, M.; Hua, X.-S. Active learning in multimedia annotation and retrieval. *ACM Trans. Intell. Syst. Technol.* **2011**, *2*, 1–21. [CrossRef]
29. Fu, Y.; Zhu, X.; Li, B. A survey on instance selection for active learning. *Knowl. Inf. Syst.* **2013**, *35*, 249–283. [CrossRef]
30. Ferecatu, M.; Boujemaa, N. Interactive remote-sensing image retrieval using active relevance feedback. *IEEE Trans. Geosci. Remote Sens.* **2007**, *45*, 818–826. [CrossRef]
31. Rajan, S.; Ghosh, J.; Crawford, M.M. An Active Learning Approach to Hyperspectral Data Classification. *IEEE Trans. Geosci. Remote Sens.* **2008**, *46*, 1231–1242. [CrossRef]
32. Demir, B.; Persello, C.; Bruzzone, L. Batch-mode active-learning methods for the interactive classification of remote sensing images. *IEEE Trans. Geosci. Remote Sens.* **2011**, *49*, 1014–1031. [CrossRef]
33. Persello, C.; Bruzzone, L. Active and Semisupervised Learning for the Classification of Remote Sensing Images. *IEEE Trans. Geosci. Remote Sens.* **2014**, *52*, 6937–6956. [CrossRef]
34. Chen, R.; Cao, Y.F.; Sun, H. Active sample-selecting and manifold learning-based relevance feedback method for synthetic aperture radar image retrieval. *IET Radar Sonar Navig.* **2011**, *5*, 118. [CrossRef]
35. Cui, S.; Dumitru, C.O.; Datcu, M. Semantic annotation in earth observation based on active learning. *Int. J. Image Data Fusion* **2014**, *5*, 152–174. [CrossRef]
36. Samat, A.; Gamba, P.; Du, P.; Luo, J. Active extreme learning machines for quad-polarimetric SAR imagery classification. *Int. J. Appl. Earth Obs. Geoinf.* **2015**, *35*, 305–319. [CrossRef]
37. Cao, Y.F.; Xu, L.; Clausi, D. Active learning for identifying marine oil spills using 10-year radarsat data. In Proceedings of the 2016 IEEE International Geoscience and Remote Sensing Symposium (IGARSS), Beijing, China, 10–15 July 2016; pp. 7722–7725.
38. Espedal, H.A.; Wahl, T. Satellite SAR oil spill detection using wind history information. *Int. J. Remote Sens.* **1999**, *20*, 49–65. [CrossRef]
39. Hwang, P.A.; Stoffelen, A.; van Zadelhoff, G.-J.; Perrie, W.; Zhang, B.; Li, H.; Shen, H. Cross-polarization geophysical model function for C-band radar backscattering from the ocean surface and wind speed retrieval. *J. Geophys. Res. Ocean.* **2015**, *120*, 893–909. [CrossRef]
40. Salvatori, L.; Bouchaib, S.; DelFrate, F.; Lichtenneger, J.; Smara, Y. Estimating the Wind Vector from Radar Sar Images When Applied To the Detection of Oil Spill Pollution. In Proceedings of the Fifth International Symposium on GIS and Computer Catography for Coastal Zone Management, CoastGIS'03, Genoa, Italy, 16–18 October 2003.
41. Hu, M.K. Visual pattern recognition by moment invariants. *IEEE Trans. Inf. Theory* **1962**, *IT-8*, 179–187.
42. Khotanzad, A.; Hong, Y.H. Invariant image recognition by Zernike moments. *IEEE Trans. Pattern Anal. Mach. Intell.* **1990**, *12*, 489–497. [CrossRef]
43. Mountrakis, G.; Im, J.; Ogole, C. Support vector machines in remote sensing: A review. *ISPRS J. Photogramm. Remote Sens.* **2011**, *66*, 247–259. [CrossRef]
44. Chang, C.C.; Lin, C.J. *LIBSVM: A Library for Support Vector Machines*. Available online: http://www.csie.ntu.edu.tw/~cjlin/libsvm/ (accessed on 1 September 2015).
45. Zhu, H.; Basir, O. An adaptive fuzzy evidential nearest neighbor formulation for classifying remote sensing images. *IEEE Trans. Geosci. Remote Sens.* **2005**, *43*, 1874–1889. [CrossRef]

46. Blanzieri, E.; Melgani, F. Nearest neighbor classification of remote sensing images with the maximal margin principle. *IEEE Trans. Geosci. Remote Sens.* **2008**, *46*, 1804–1811. [CrossRef]
47. Yang, J.-M.; Yu, P.-T.; Kuo, B.-C. A Nonparametric Feature Extraction and Its Application to Nearest Neighbor Classification for Hyperspectral Image Data. *IEEE Trans. Geosci. Remote Sens.* **2010**, *48*, 1279–1293. [CrossRef]
48. Hastie, T.; Buja, A.; Tibshirani, R. Penalized Discriminant Analysis. *Ann. Stat.* **1995**, *23*, 73–102. [CrossRef]
49. Chen, L.F.; Liao, H.Y.M.; Ko, M.T.; Lin, J.C.; Yu, G.J. New LDA-based face recognition system which can solve the small sample size problem. *Pattern Recognit.* **2000**, *33*, 1713–1726. [CrossRef]
50. Wang, X.; Tang, X. Dual-space linear discriminant analysis for face recognition. In Proceedings of the 2004 IEEE Computer Society Conference on Computer Vision and Pattern Recognition, Washington, DC, USA, 27 June–2 July 2004; Volume 2, pp. 564–569.
51. Li, P.; Fu, Y.; Mohammed, U.; Elder, J.H.; Prince, S.S.J.D. Probabilistic models for Inference about Identity. *IEEE Trans. Pattern Anal. Mach. Intell.* **2012**, *34*, 144–157. [CrossRef]
52. Zhang, D.; He, J.; Zhao, Y.; Luo, Z.; Du, M. Global plus local: A complete framework for feature extraction and recognition. *Pattern Recognit.* **2014**, *47*, 1433–1442. [CrossRef]
53. Breiman, L.; Friedman, J.H.; Olshen, R.A.; Stone, C.J. *Classification and Regression Trees*; Taylor Francis Ltd.: Burlington, MA, USA, 1984; Volume 5.
54. Breiman, L. Random forests. *Mach. Learn.* **2001**, *45*, 5–32. [CrossRef]
55. Hothorn, T.; Lausen, B. Bundling classifiers by bagging trees. *Comput. Stat. Data Anal.* **2005**, *49*, 1068–1078. [CrossRef]
56. Friedman, J.H. Greedy Function Approximation: A Gradient Boosting Machine. *Ann. Stat.* **2001**, *29*, 1189–1232. [CrossRef]
57. Wu, T.F.; Lin, C.J.; Weng, R.C. Probability estimates for multi-class classification by pairwise coupling. *J. Mach. Learn. Res.* **2004**, *5*, 975–1005.
58. Alvarez, I.; Bernard, S.; Deffuant, G. Keep the decision tree and estimate the class probabilities using its decision boundary. In Proceedings of the 20th International Joint Conference on Artificial Intelligence, Hyderabad, India, 6–12 January 2007; pp. 654–659.
59. Cebron, N.; Berthold, M.R. Active learning for object classification: From exploration to exploitation. *Data Min. Knowl. Discov.* **2009**, *18*, 283–299. [CrossRef]
60. Swets, J.A. Measuring the accuracy of diagnostic systems. *Science* **1988**, *240*, 1285–1293. [CrossRef] [PubMed]
61. Fawcett, T. An introduction to ROC analysis. *Pattern Recognit. Lett.* **2006**, *27*, 861–874. [CrossRef]
62. Davis, J.; Goadrich, M. The Relationship between Precision-Recall and ROC Curves. In Proceedings of the 23rd International Conference Machine Learning ICML'06, Pittsburgh, PA, USA, 25–29 June 2006; pp. 233–240.

*Article*

# Mechanisms of SAR Imaging of Shallow Water Topography of the Subei Bank

**Shuangshang Zhang [1], Qing Xu [1,*], Quanan Zheng [2] and Xiaofeng Li [3]**

[1] College of Oceanography, Hohai University, Nanjing 210098, China; ang104@yeah.net
[2] Department of Atmospheric and Oceanic Science, University of Maryland, College Park, MD 20742, USA; qzheng2@umd.edu
[3] GST, NESDIS/NOAA, College Park, MD 20740, USA; xiaofeng.li@noaa.gov
* Correspondence: maggiexu@hhu.edu.cn; Tel.: +86-025-83787340

Received: 27 September 2017; Accepted: 20 November 2017; Published: 22 November 2017

**Abstract:** In this study, the C-band radar backscatter features of the shallow water topography of Subei Bank in the Southern Yellow Sea are statistically investigated using 25 ENVISAT (Environmental Satellite) ASAR (advanced synthetic aperture radar) and ERS-2 (European Remote-Sensing Satellite-2) SAR images acquired between 2006 and 2010. Different bathymetric features are found on SAR imagery under different sea states. Under low to moderate wind speeds (3.1~6.3 m/s), the wide bright patterns with an average width of 6 km are shown and correspond to sea surface imprints of tidal channels formed by two adjacent sand ridges, while the sand ridges appear as narrower (only 1 km wide), fingerlike, quasi-linear features on SAR imagery in high winds (5.4~13.9 m/s). Two possible SAR imaging mechanisms of coastal bathymetry are proposed in the case where the flow is parallel to the major axes of tidal channels or sand ridges. When the surface Ekman current is opposite to the mean tidal flow, two vortexes will converge at the central line of the tidal channel in the upper layer and form a convergent zone over the sea surface. Thus, the tidal channels are shown as wide and bright stripes on SAR imagery. For the SAR imaging of sand ridges, all the SAR images were acquired at low tidal levels. In this case, the ocean surface waves are possibly broken up under strong winds when propagating from deep water to the shallower water, which leads to an increase of surface roughness over the sand ridges.

**Keywords:** SAR; Subei Bank; shallow water topography

---

## 1. Introduction

A bathymetric measurement of shallow water is of fundamental importance to coastal environment research and resource management. The traditional bathymetric survey uses a shipboard sonar, single-beam, or multi-beam sounding system, which can provide high-precision data but is costly and inefficient. With the development of remote sensing techniques, the shallow water depth can be measured with high efficiency [1–4]. A spaceborne synthetic aperture radar (SAR), in particular, provides valuable information of shallow water topography in all-weather and day-night conditions with a high spatial resolution (a few to tens of meters). Although the SAR signal does not penetrate through sea water, the bathymetric features of shallow water (water depth < 50 m) or even deep water (water depth of about 600 m) can still be observed indirectly through the interaction between the ocean current and the underwater topography [5–12]. Shallow water bathymetric features were first discovered on radar images in 1969 [6,13]. Since then, many researchers have investigated the radar imaging mechanism of underwater topography in shallow waters. In 1984, Alpers and Hennings [7] developed a one-dimensional (1-D) SAR imaging model under the assumption that the current velocity is primarily normal to the direction of the major axis of topographic corrugation in the un-stratified ocean. The model was further enhanced by Van der Kooij et al. [14], Vogelzang et al. [15], and Romeiser and Alpers [16]. For a stratified ocean, Zheng et al. [8] obtained dynamical solutions for the vertical propagation of disturbance signals induced by underwater topography from the ocean bottom to the surface. All of these studies have shown that under the condition of a tidal current perpendicular to topographic features, the underwater topography can be imaged by SAR. However, recent satellite observations show that when the tidal current is parallel to topographic corrugations such as underwater sand ridges, sand bars, or tidal channels, the shallow water topography can also appear on SAR imagery [9,17–19]. These observations cannot be explained using the existing 1-D radar imaging model. Considering the tidal convergence, Li et al. [9] developed a two-dimensional (2-D) analytical model for the interpretation of SAR imaging of underwater sand ridges parallel to the tidal current. Recently, Zheng et al. [17] analyzed the secondary circulation induced by the flow parallel to the topographic corrugation by solving the three-dimensional (3-D) disturbance governing equations of the shear-flow. The theoretical results were applied to interpret SAR imaging of tidal channels. The above studies show that different bathymetric features might appear on SAR imagery. Then, under what dynamic conditions can shallow water bathymetry be observed by SAR in the case of the current being parallel to underwater topographic corrugations? Particularly, when will the sand ridges or tidal channels be shown on SAR imagery? The answers to these questions are still unclear.

With large amounts of sediment input from the river runoff, the radial sand ridges offshore from the middle Jiangsu coast in the Southern Yellow Sea (also called Subei Bank) were formed as a sediment physiognomy and represent an ideal region for harbor construction, agricultural development, and fishery production [20] (see Figure 1a). The distinguished characteristic of the topography in this area is the unique distribution of a group of tidal channels and shallow sand ridges radiating from Jianggang city [21] (see Figure 1b), which encompass an area larger than 200 km long and 140 km wide [22]. The major axes of the topographic corrugations are roughly parallel to the semidiurnal tidal currents.

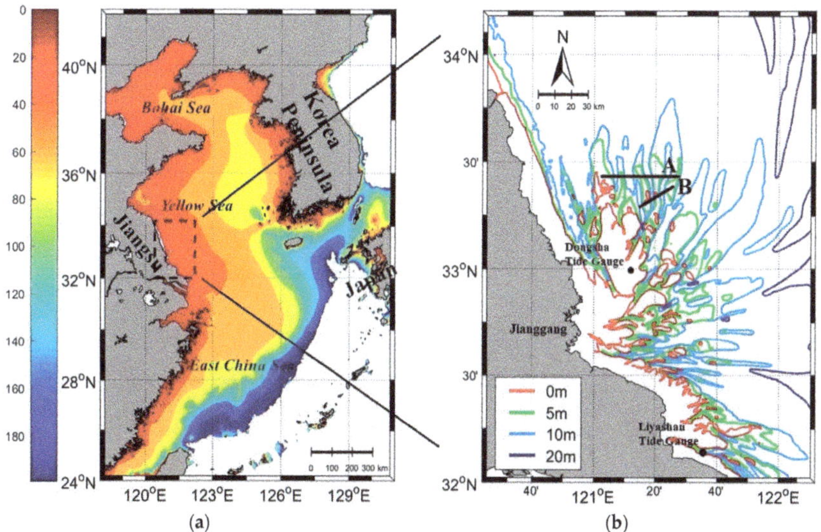


**Figure 1.** (**a**) Bathymetry (m) of the Yellow Sea and (**b**) Subei Bank boarded by dashed lines in panel (**a**) The bathymetry data are from ETOPO2 (National Centers for Environmental Information, 2006) for (**a**) and Sea Chart (published by China Navy Hydrographic Office, 2013) for (**b**). The cross sections A and B in (**b**) (black lines) are primarily perpendicular to the paralleled bright stripes on SAR imagery in Figure 2. The black dots denote the locations of the Dongsha and Liyashan tide gauges.

In this study, the radar backscatter features of the shallow water topography of Subei Bank are investigated using ENVISAT (Environmental Satellite), ASAR (advanced synthetic aperture radar), and European Remote-Sensing Satellite-2 (ERS-2) SAR images. We analyze the influences of wind, current, and tide on the capability of C-band SAR in observing the underwater topography in this region, and try to find out the possible radar imaging mechanisms.

## 2. Data and Methods

The SAR data used in this study include 16 ENVISAT ASAR images and nine ERS-2 SAR images over Subei Bank in the Southern Yellow Sea acquired between 2006 and 2010. All these C-band SAR images are VV-polarized with a nominal spatial resolution of 30 × 30 m [23]. The ASAR system has been designed to provide continuity with ERS SAR by the European Space Agency (ESA). Compared with the ERS-1/2 SAR launched in 1991/1995, ASAR, launched in 2002, features extended observational capabilities, three new modes of operation, and improved performances [24]. Figure 2 presents examples of three typical types of SAR images over Subei Bank.

SAR not only observes oceanic or atmospheric phenomena, but also provides direct measurements of sea surface roughness that is related to sea surface wind speed (e.g., [25–28]). In this study, the sea surface wind speed (at 10-m height) is derived from SAR using the C-band geophysical model function CMOD5 [29] with wind direction interpolated from the six-hourly blended sea surface wind data from NOAA/National Climatic Data Center (NCDC). The NOAA/NCDC blended sea winds with a spatial resolution of 0.25° × 0.25° are generated by blending observations from multiple satellites, which fills in the data gaps (in both time and space) of the individual satellite samplings and reduces the subsampling aliases and random errors [30].

In order to investigate the contribution of the tidal current and height to the SAR imaging of shallow water topography, we use the Tidal Model Driver (TMD) tide data to demonstrate the tide condition when SAR images were acquired. TMD is a package for accessing the harmonic constituents and making predictions of tidal height and currents [31,32]. As shown in Figure 3, compared with the data from two tide gauges and the Tide Table, the TMD results perform well in the tidal phase but present a systematic underestimation of tidal amplitude, which is possibly caused by the input of inaccurate water depth data in the tidal model in this region. By fitting the TMD results with in situ observations of the tidal height at Dongsha tide gauge in the lunar month of July 2014 (Figure 3a), we obtain a relationship between the observed tidal height $\zeta$ (m) and the TMD output $\zeta_{TMD}$ (m):

$$\zeta = 1.4612 * \zeta_{TMD} + 0.0016 \tag{1}$$

To validate the relationship, the TMD outputs of the tidal height in lunar August 2014, were corrected by Equation (1) and compared with in situ measurements from the same tide gauge (Figure 3b). One can see that the root mean square error (RMSE) between the TMD results and tide gauge observations is decreased significantly from 0.64 to 0.39 m after the correction. Similar results can be obtained for Liyashan tide gauge data collected in lunar September 2016, with the RMSE decreasing from 0.75 to 0.40 m (Figure 3c). Therefore, in the following study, the TMD output of the tidal height at SAR imaging time is corrected using Equation (1) for further analysis.

The bathymetry data of the whole study area are generated from the Sea Chart published by the China Navy Hydrographic Office in 2013 [33]. However, most Sea Chart data were measured in 1979. In addition, the data are relatively sparse and antiquated because of the evolution of sand ridges induced by the action of tidal current year after year [34]. For more accurate and higher-resolution water depth data, we carried out a field survey along the two cross sections A and B (Figure 1b) in December, 2016. The measured water depth data are used to interpret the bathymetric features of Subei Bank on SAR imagery.

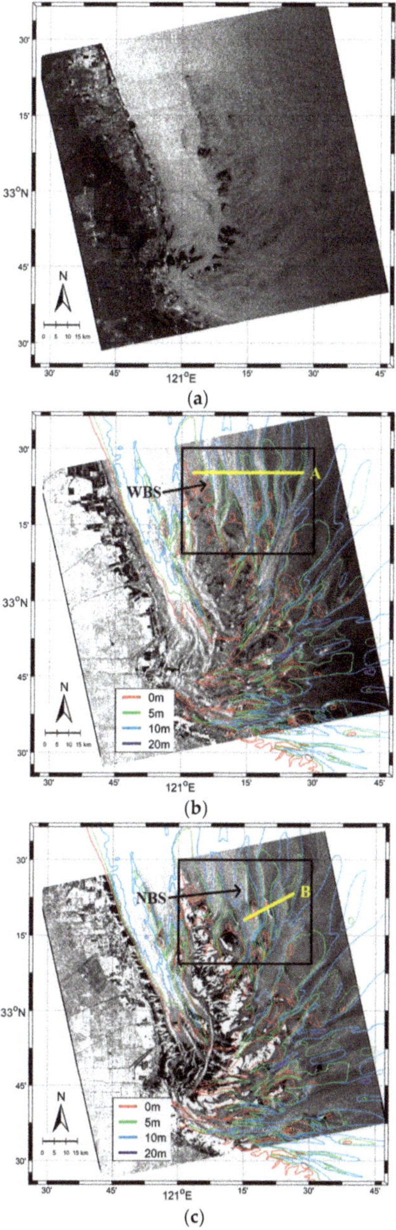


**Figure 2.** Examples of three typical ENVISAT ASAR images over Subei Bank: (**a**) image without any bathymetric features acquired at 13:45:32 UTC on 22 December 2008; (**b**) image with bathymetric features shown as wide bright stripes (WBS) in the small region denoted by the black rectangle, acquired at 13:45:29 UTC on 13 October 2008; (**c**) image with bathymetric features shown as narrow bright stripes (NBS) in the same region as (**b**), acquired at 13:45:28 UTC on 11 February 2008. The contours are water depth (m). The cross sections A and B (yellow lines, also shown as black lines in Figure 1b) are perpendicular to the paralleled bright stripes on SAR images.

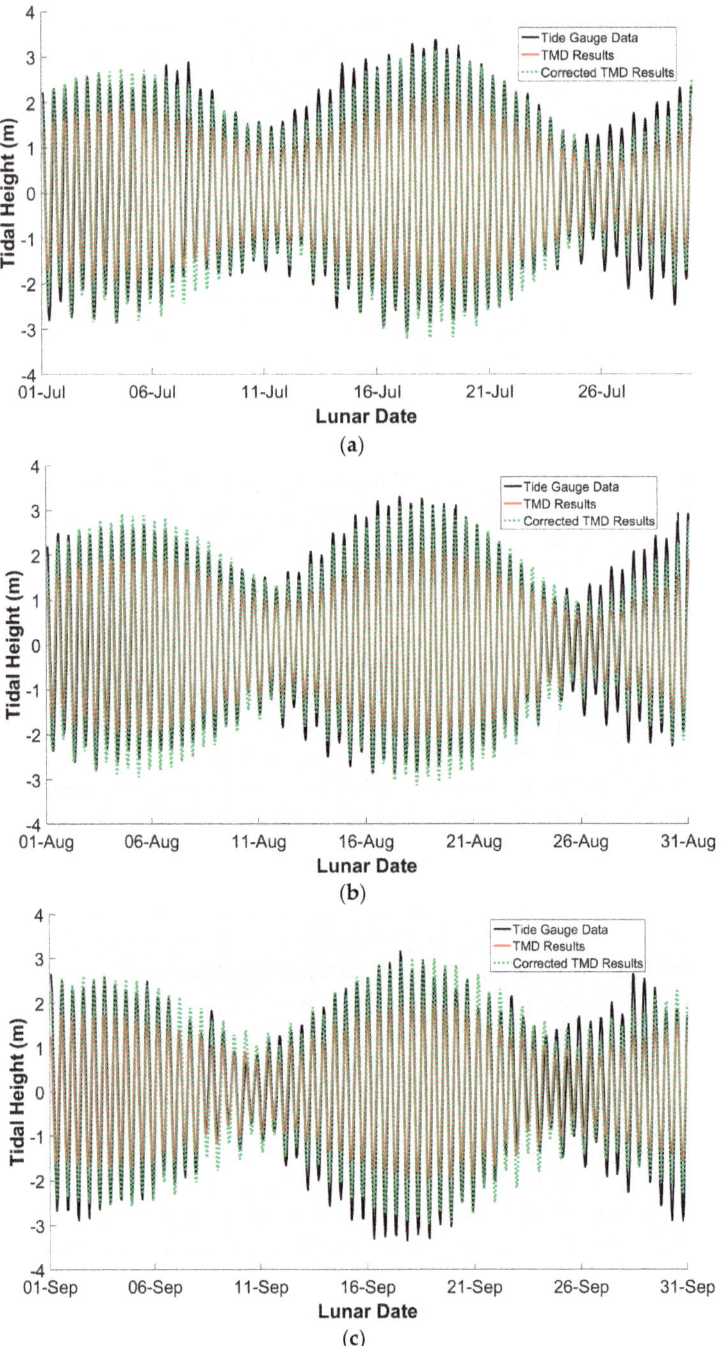


**Figure 3.** Comparison of the TMD results of the tidal height with in situ observations in lunar July (**a**) and August (**b**), 2014 at Dongsha tide gauge, and lunar September, 2016 at Liyashan tide gauge (**c**).

## 3. Bathymetric Features of Subei Bank on SAR Imagery

From Figure 2, one can see the tidal channels or sand ridges are not always clearly shown on SAR imagery (see Figure 2a). Under certain sea states and wind conditions, the shallow water topography appears as fingerlike features (see Figure 2b,c). What is interesting is that distinct bathymetric features of the same region are shown on SAR images acquired at different times. In particular, an apparent difference occurs in the northeastern area (see the black boxes in Figure 2b,c). As shown in Figure 2b, there are some paralleled wide bright patterns in this region, and the average width of the stripes is about 6 km. However, the locations of the bright stripes change and are much narrower in Figure 2c with an average width of only 1 km.

Among the 25 SAR images over Subei Bank, there are a total of eight SAR images without any bathymetric features (e.g., Figure 2a) and 17 images showing obvious bathymetric features. The paralleled wide bright patterns appear on five SAR images (Figure 4). By examining the Sea Chart bathymetric data, we find the locations of the wide bright stripes mainly coincide with the deep water area (>10 m) in this region. The relationship can be seen more clearly in Figure 5, which shows the variation of the SAR derived normalized radar backscatter cross section (NRCS) and water depth along the cross section A. Apparently, the wide bright stripes on this type of SAR image correspond to the deep water region, i.e., the tidal channels. The other 12 SAR images show obviously much narrower bright stripes at different locations (Figure 6). Comparing the variation of the NRCS with water depth along the cross section B (Figure 7), one can clearly see that these narrow bright stripes are sea surface imprints of underwater sand ridges. One may also notice that the SAR signal enhancement in Figure 7 does not take place exactly over the crest of the sand ridge measured in 2016, but with an offset of about 0.5 km westward (see the dashed blue line in Figure 7). The possible reason for this is that the topography of Subei Bank changes with time under the action of strong tidal currents [34]. To investigate the evolution of the sand ridges, we collect two optical images from Landsat_7 Enhanced Thematic Mapper Plus (ETM+) in 2008 (Figure 8a) and Landsat_8 Operational Land Imager (OLI) in 2016 (Figure 8b), respectively. The spatial resolution of the images is 30 m. Figure 8c shows the edge of the sand ridge in the study area extracted from the Landsat images. The deviation of the edge lines indicates that the sand ridges moved a little to the northeast from 2008 to 2016. This may partly explain why there is a small deviation between the locations of the peak of SAR observed NRCS and the crest of the sand ridge.

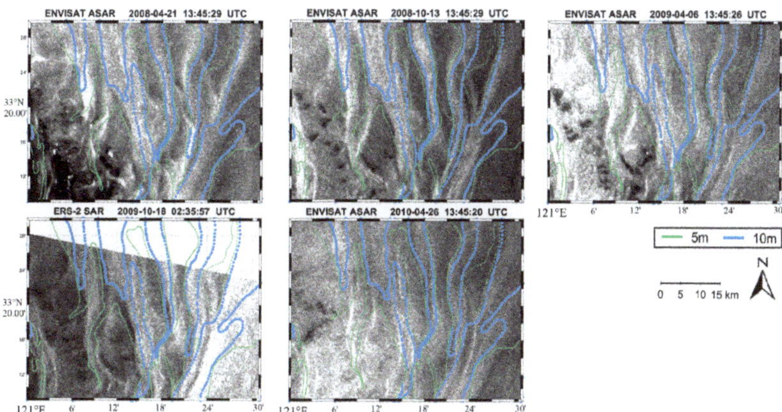


**Figure 4.** SAR sub-images over Subei Bank with bathymetric features shown as wide bright stripes. Green and blue lines are water depth contours of 5 and 10 m, respectively.

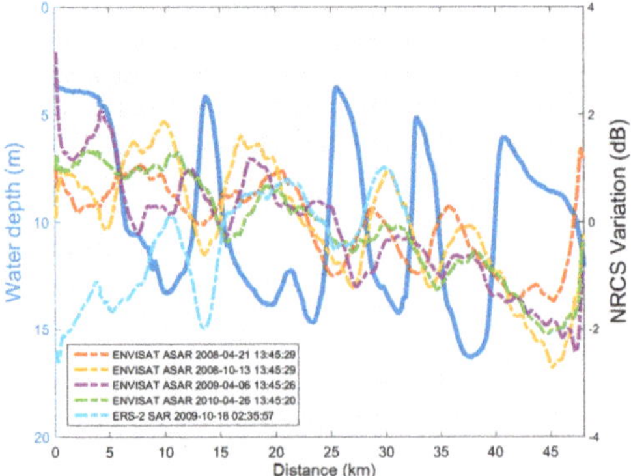


**Figure 5.** The water depth measured in December, 2016 (m) in blue solid line and NRCS Variation (dB) calculated from SAR data along the cross section A. The distance is measured from the left to the right for each cross section.

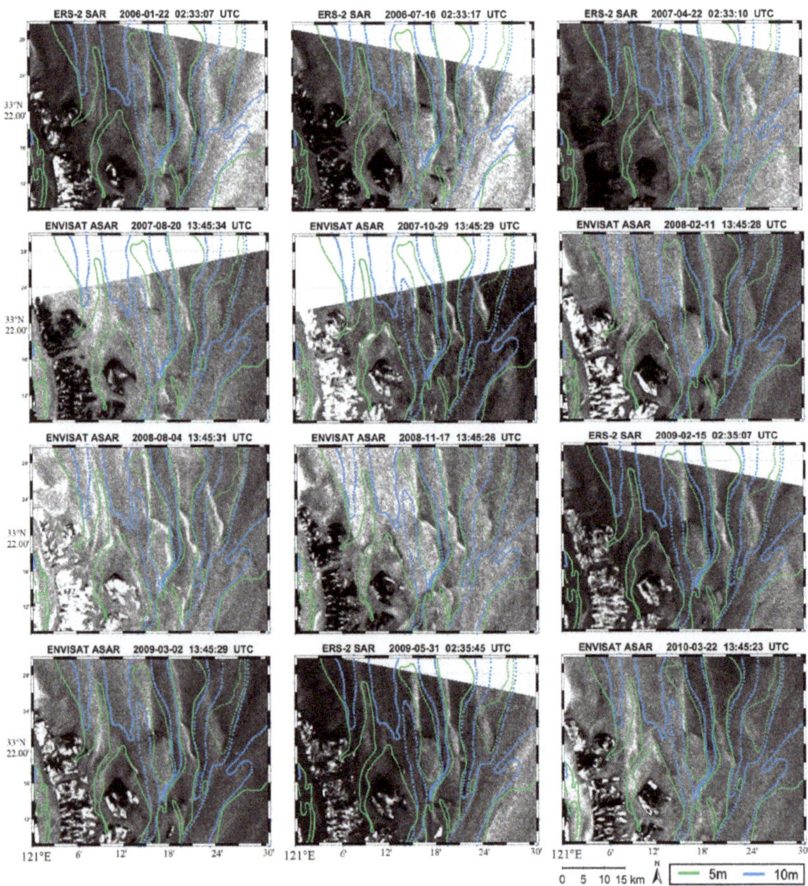


**Figure 6.** Same as Figure 4 but for SAR sub-images with bathymetric features shown as narrow bright stripes.

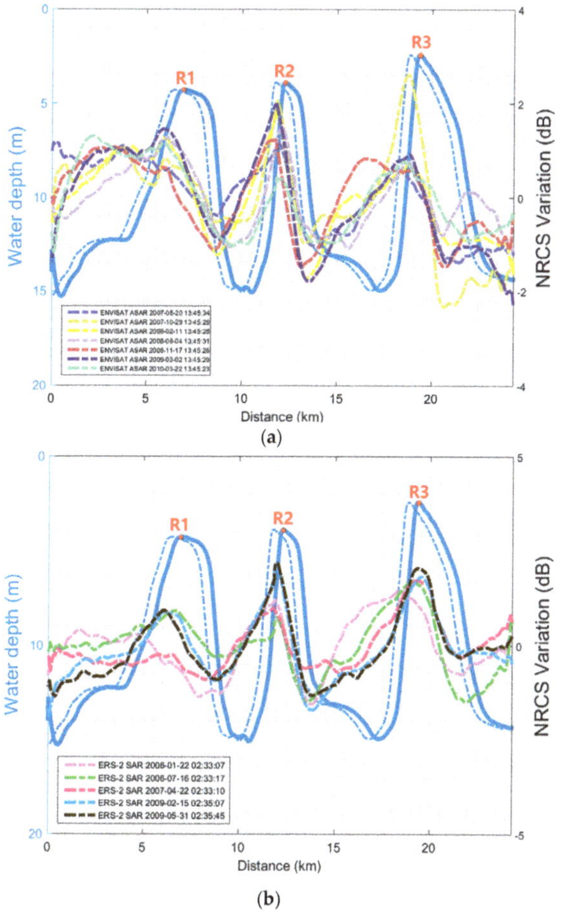


**Figure 7.** Same as Figure 5 but along the cross section B: (**a**) for ENVISAT ASAR images and (**b**) for ERS-2 SAR images. The locations of the crests of the three sand ridges along cross section B are marked as R1, R2, and R3, respectively. The blue dashed line is the same as the blue solid line but has a deviation of 0.5 km westward.

Table 1 shows ambient wind, current, and tide conditions at the acquisition time of 25 SAR images with or without obvious bathymetric features over Subei Bank. Most of the images with obvious underwater topographic features (13/17) were acquired during the flood tide, while most of those without any features (7/8) were acquired during the ebb tide. Comparing Figure 4 with Figure 6 and judging from the extent of the shoal exposed to the sea surface, we find that the water level at the time when the sand ridges were observed by SAR should be much lower than that when the tidal channels were imaged. This is further validated by the corrected TMD results. The values of the tidal heights when the SAR images with sand ridge features were acquired are all negative, and the water levels are below the mean sea level by over 1.3 m. For the images with tidal channel features, however, the tidal height is much larger and the water levels are all above the mean sea level. Another interesting thing to note is that the tidal channels were observed by SAR under low to moderate winds (3.1~6.3 m/s), while the sand ridges were detected at much higher wind speeds (5.4~13.9 m/s). This means that both the tidal height and wind may play a significant role in the SAR imaging of shallow water topography in this region.

**Table 1.** Wind, current, and tide conditions at an imaging time of 25 SAR images.

| | Satellite | Date | Bathymetric Features [1] | Tidal Phase | Wind Direction [2] | Tidal Height (m) | Wind Speed (m/s) | Wave Height (m) | Wave Breaking Depth (m) | Instantaneous Water Depth (m) |
|---|---|---|---|---|---|---|---|---|---|---|
| 1 | ENVISAT | 2008-04-21 | WBS | flood | 105 | 0.29 | 6.30 | 1.22 | 1.90 | 4.61 |
| 2 | ENVISAT | 2008-10-13 | WBS | flood | 174 | 1.61 | 4.80 | 0.71 | 1.10 | 5.93 |
| 3 | ENVISAT | 2009-04-06 | WBS | flood | 172 | 1.83 | 3.10 | 0.29 | 0.46 | 6.15 |
| 4 | ERS-2 | 2009-10-18 | WBS | flood | 195 | 1.90 | 5.60 | 0.96 | 1.50 | 6.22 |
| 5 | ENVISAT | 2010-04-26 | WBS | flood | 210 | 1.76 | 4.60 | 0.65 | 1.01 | 6.08 |
| 6 | ERS-2 | 2006-01-22 | NBS | ebb | 351 | −1.61 | 6.10 | 1.14 | 1.90 | 1.86 |
| 7 | ERS-2 | 2006-07-16 | NBS | flood | 158 | −1.75 | 10.50 | 3.38 | 5.63 | 1.72 |
| 8 | ERS-2 | 2007-04-22 | NBS | flood | 20 | −1.46 | 9.80 | 2.94 | 4.90 | 2.01 |
| 9 | ENVISAT | 2007-08-20 | NBS | ebb | 125 | −1.75 | 10.10 | 3.12 | 5.20 | 1.72 |
| 10 | ENVISAT | 2007-10-29 | NBS | flood | 24 | −1.46 | 6.60 | 1.33 | 2.22 | 2.01 |
| 11 | ENVISAT | 2008-02-11 | NBS | flood | 355 | −2.34 | 9.20 | 2.59 | 4.32 | 1.13 |
| 12 | ENVISAT | 2008-08-04 | NBS | flood | 132 | −1.46 | 6.30 | 1.22 | 2.03 | 2.01 |
| 13 | ENVISAT | 2008-11-17 | NBS | flood | 351 | −2.34 | 13.90 | 5.91 | 9.86 | 1.13 |
| 14 | ERS-2 | 2009-02-15 | NBS | flood | 17 | −1.46 | 6.50 | 1.29 | 2.16 | 2.01 |
| 15 | ENVISAT | 2009-03-02 | NBS | flood | 28 | −1.90 | 5.80 | 1.03 | 1.72 | 1.57 |
| 16 | ERS-2 | 2009-05-31 | NBS | ebb | 200 | −2.04 | 5.40 | 0.89 | 1.49 | 1.43 |
| 17 | ENVISAT | 2010-03-22 | NBS | ebb | 149 | −1.31 | 6.60 | 1.33 | 2.22 | 2.16 |
| 18 | ERS-2 | 2006-02-26 | none | flood | 355 | 1.61 | 8.62 | 2.27 | 3.67 | 5.51 |
| 19 | ENVISAT | 2006-09-04 | none | ebb | 45 | 1.02 | 8.00 | 1.96 | 3.16 | 4.92 |
| 20 | ENVISAT | 2008-03-17 | none | ebb | 124 | 1.32 | 5.90 | 1.07 | 1.72 | 5.22 |
| 21 | ENVISAT | 2008-12-22 | none | ebb | 300 | 0.59 | 7.60 | 1.77 | 2.85 | 4.49 |
| 22 | ERS-2 | 2009-03-22 | none | ebb | 342 | 0.44 | 9.20 | 2.59 | 4.18 | 4.34 |
| 23 | ERS-2 | 2009-09-13 | none | ebb | 162 | −0.58 | 3.30 | 0.33 | 0.54 | 3.32 |
| 24 | ENVISAT | 2009-09-28 | none | ebb | 156 | −0.29 | 6.70 | 1.37 | 2.22 | 3.61 |
| 25 | ENVISAT | 2010-01-11 | none | ebb | 346 | 1.32 | 10.20 | 3.18 | 5.14 | 5.22 |

[1] 'WBS' and 'NBS' denote wide and narrow bright stripes, respectively. [2] The wind direction is measured in degrees clockwise from due north. A wind coming from the north (i.e., the northerly wind) has a wind direction of 0° and the southerly wind has a direction of 180°.

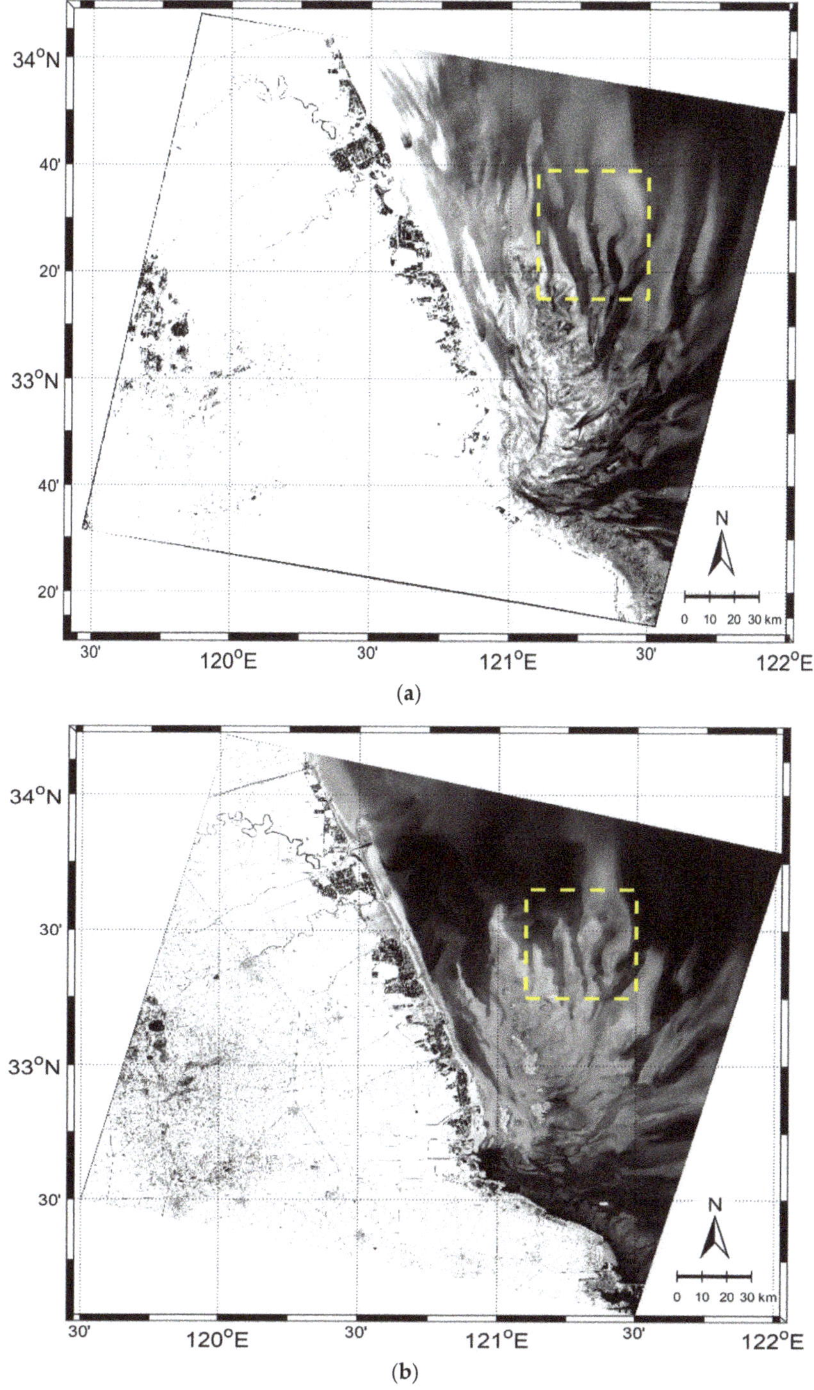


**Figure 8.** *Cont.*

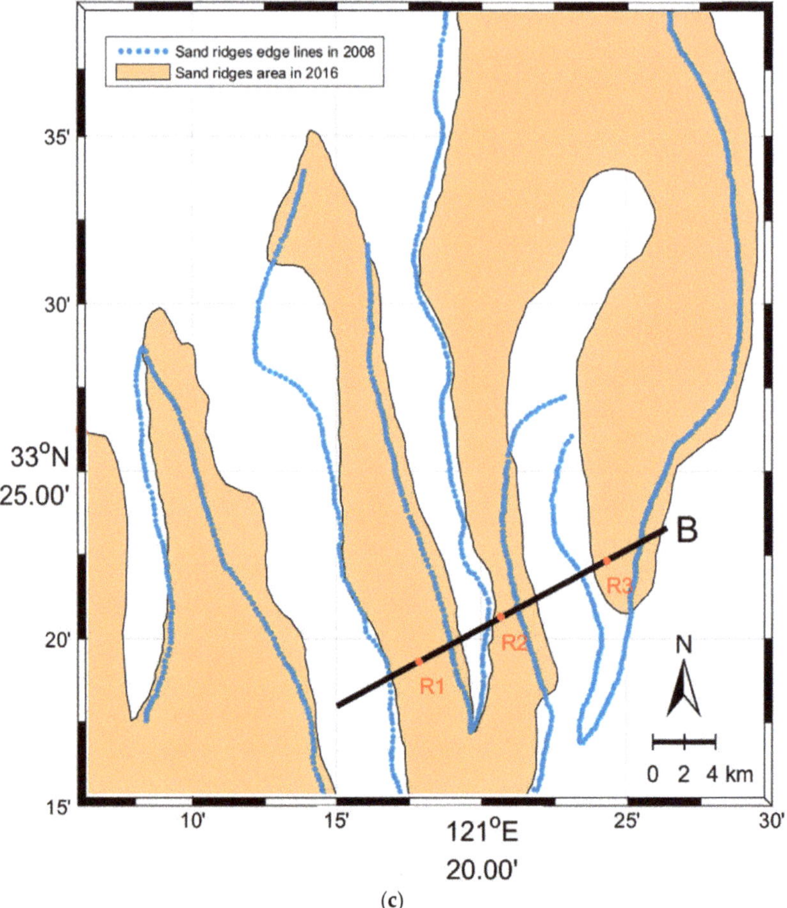


(c)

**Figure 8.** Landsat images over Subei Bank: (**a**) Landsat_7 ETM+ image acquired at 02:20:53 UTC on 24 April 2008; (**b**) Landsat_8 OLI image acquired at 02:30:41 UTC on 18 February 2016; and (**c**) the edge of the sand ridges in the study area (yellow dashed box in (**a**,**b**)) extracted from Landsat images. The dotted blue line denotes the edge extracted from (**a**) in 2008, and the light brown patch denotes the sand ridge area extracted from (**b**) in 2016. The field survey along cross section B is the same as that in Figure 1b with R1, R2, and R3 representing the locations of the sand ridge crests measured in 2016.

## 4. SAR Imaging Mechanisms

Why does the underwater topography in the same region have distinctive radar backscatter features on SAR imagery? In this section, we discuss the possible imaging mechanisms of SAR imaging of shallow water topography over Subei Bank.

### 4.1. SAR Imaging of Tidal Channels

The existing SAR imaging theories of underwater topography are based on the following three processes: (1) the current and topography interaction generates sea surface current divergence or convergence zones; (2) the divergence and convergence of the current modulate the wind-generated sea surface wave spectrum; and (3) the variation of short surface wave height induces the backscatter variations seen in the SAR image [9]. Considering the sidewall friction, the surface, and the bottom

Ekman layers, the authors proposed a physics model to analyze the secondary circulation induced by the flow parallel to underwater topographic corrugation (see Figure 9) [17]. The analytical solutions show that in the case where the direction of the surface Ekman current is opposite to the mean flow, there is a surface current convergent zone along the central line of a canal. Using this model, we tried to find the possible factors causing the sea surface imprints of tidal channels on SAR images over Subei Bank.

For the small area in this study, the wind and tidal current conditions over the tidal channels are nearly the same. Therefore, we can take one tidal channel as a representation to analyze the SAR imaging mechanism of the tidal channels. As sketched in Figure 9, we consider the flow in a long canal with a free surface and rectangular cross section with two flat sidewalls. The sidewalls have a height $D$ and the bottom has a width $2b$. A Cartesian coordinate system is set up with its origin located at the bottom. The vertical axis $z$ is positive upward. The horizontal axis $y$ is perpendicular to the central line and the vertical walls and positive leftward. The horizontal axis $x$ is parallel to the walls and positive downstream. The 3-D scales of the canal, $L1$, $L2$ (=$2b$), and $L3$ (=$D$) satisfy $L1 >> L2 >> L3$. The mean flow $(\overline{u}, \overline{v}, \overline{w})$ is driven by a pressure gradient externally imposed by a large-scale process, such as the tidal waves or the ocean circulation, and is thus considered a stable process. Due to the confinement of sidewalls, the mean flow is 1-D and parallel to the $x$-axis, i.e., $(\overline{v}, \overline{w}) = 0$, and has horizontal and vertical velocity shears. The horizontal shear can be described by a parabolic profile as a plane Poiseuille flow [35]

$$\overline{u}(y) = -\frac{y}{\mu}\frac{dP_0}{dx}(b - \frac{y}{2}), \tag{2}$$

where $\mu$ is the dynamic viscosity, and $dP_0/dx$ is the externally imposed pressure gradient. On the other hand, considering the existence of surface and bottom Ekman layers, we suppose that the vertical shear has a sinusoidal profile with an apex at $H$, as follows:

$$\overline{u}(z) = sin\frac{\pi z}{2H},\ 0 \ll z \ll D. \tag{3}$$

Thus, we have:

$$\overline{u}(y,z) = -\frac{y}{\mu}\frac{dP_0}{dx}\left(b - \frac{y}{2}\right)sin\frac{\pi z}{2H},\ 0 \ll y \ll 2b,\ 0 \ll z \ll D. \tag{4}$$

After solving the governing equations and taking some approximations (see Appendix A), we obtain the analytical solutions:

$$w(y,z) = w_0(sin\frac{\pi z}{2H})^{[1+\frac{y(2b-y)}{2(b-y)^2}]}, \tag{5}$$

$$v(y,z) = -v_0\left[\frac{y(2b-y)}{2(b-y)}\right](sin\frac{\pi z}{2H})^{\frac{y(2b-y)}{2(b-y)^2}}cos\frac{\pi z}{2H}, \tag{6}$$

where $v_0 = \pi(2H)^{-1}w_0$.

Solutions of Equations (5) and (6) are graphically shown in Figure 10. For the study area, we take $D$ = 15 m, $2b$ = 6 km, and $w_0$ = 0.01 m/s. One can see the secondary circulation consists of a pair of current vortexes with opposite signs distributed symmetrically on the two sides of the central line of the channel, a cyclonic vortex on the right and an anti-cyclonic vortex on the left. The mean flow ($\vec{V}$) shear drives upwelling along two sidewalls, and the stronger it is, the closer it is to the sidewalls. In the case of the presence of a surface Ekman layer where the direction of the Ekman current ($\vec{V_E}$) component is opposite to that of the mean flow ($H < D$ and $\vec{V_E}\cdot\vec{V} < 0$), the two vortexes converge at the central line of the canal in the upper layer. Thus, there is a surface current convergent zone along the central line of the canal. In addition, the convergence gets stronger with the increase of $H$ in the case of $H < D$ (see Figure 10a–c), which may imply that the strong tidal current and weak wind are favorable for the SAR imaging of the tidal channels. We also calculate the convergence value ($\partial v/\partial y$) at the sea surface when $H = 5/6D$. The value is about $10^{-3}$ s$^{-1}$ and increases with the increase of $H$ ($H < D$). Alpers (1985) [36] pointed out that $10^{-3}$ s$^{-1}$ is the typical convergence value for the internal wave imaged by SAR, which is also sufficient to explain the bright stripes over the tidal channels on the SAR images in our study. In the case of the absence of a surface Ekman layer ($H = D$), there is no current convergent zone to be formed at any depth, as shown in Figure 10d. In the case of the presence of a surface Ekman layer with the direction identical to the mean flow ($H > D$, and $\vec{V_E} \cdot \vec{V} > 0$), the two vortexes diverge at the central line of the canal in all the layers, as shown in Figure 10e.

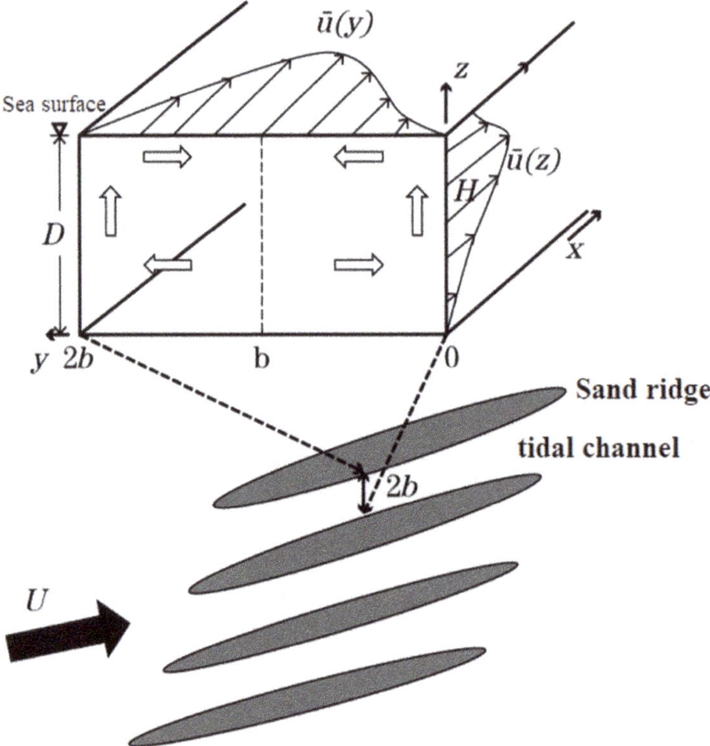


**Figure 9.** The physics model for secondary circulation (large hollow arrows) induced by a shear flow over parallel underwater topographic corrugation.

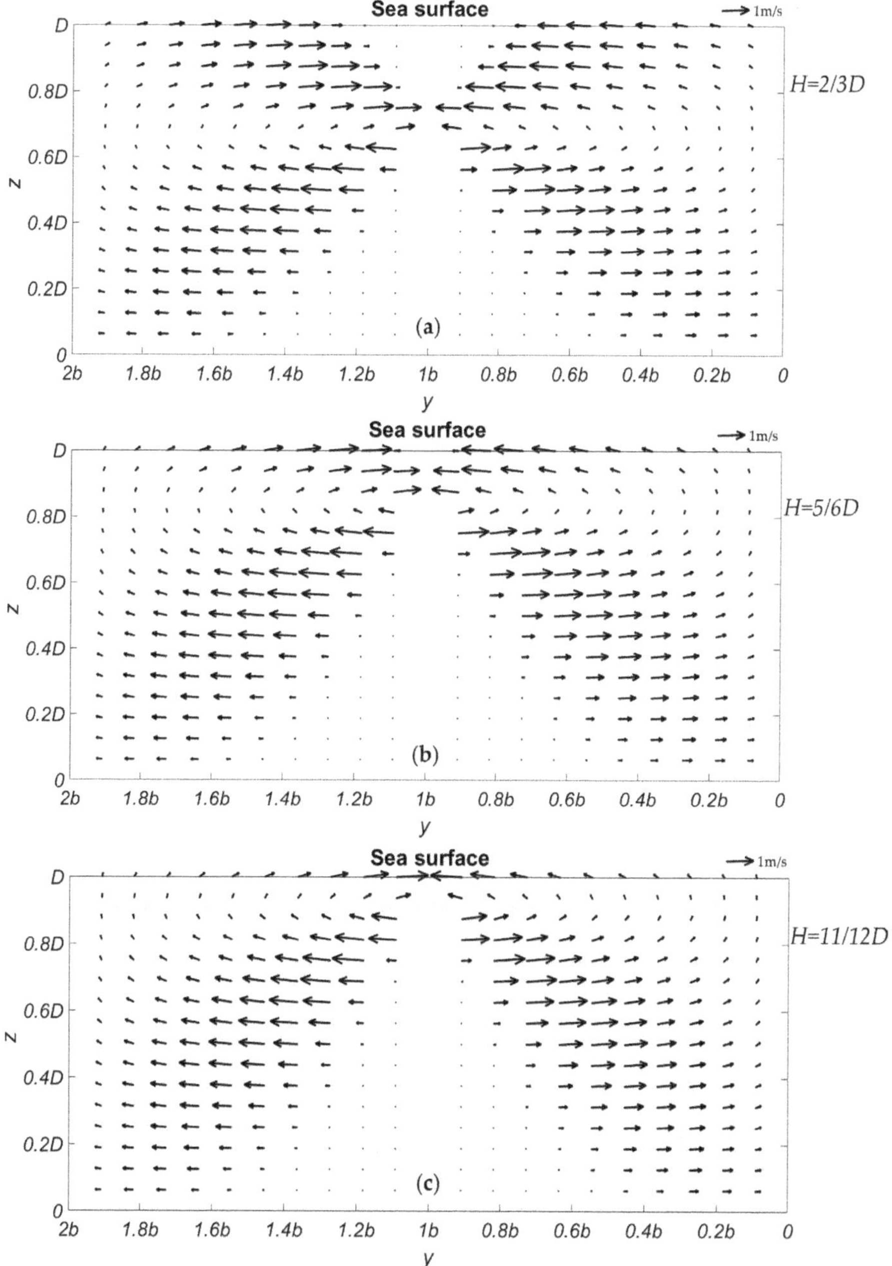


**Figure 10.** *Cont.*

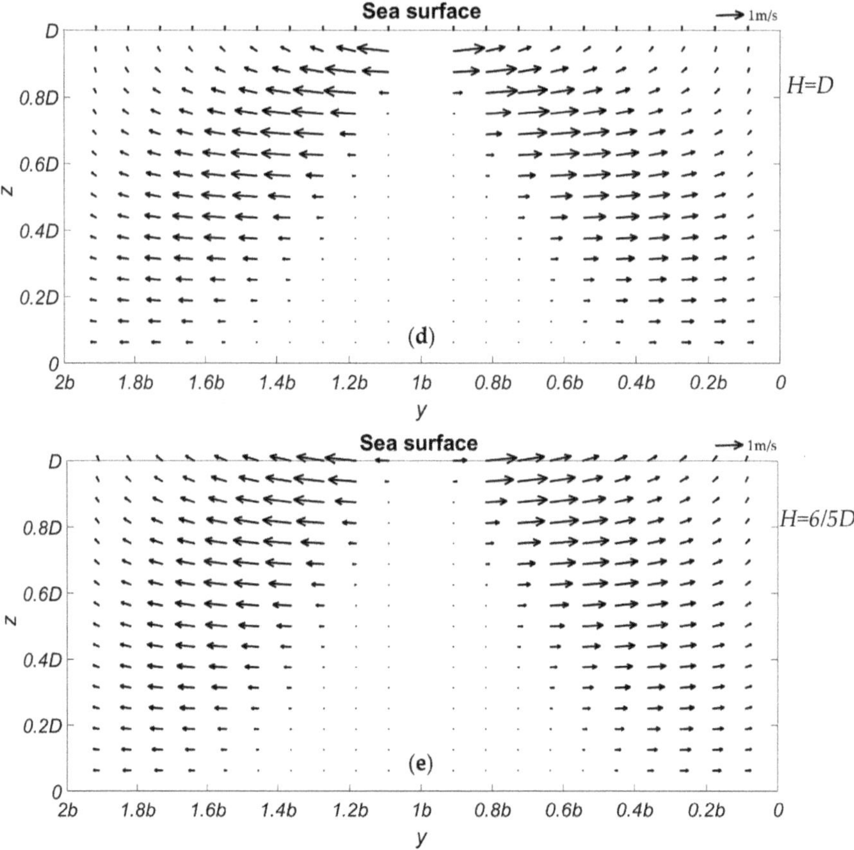


**Figure 10.** Analytical solutions of the secondary circulation induced by a shear flow in a long, rectangular canal. (**a**) There is an upper Ekman layer, in which the Ekman current has a negative component in the mean flow direction ($\vec{V_E} \cdot \vec{V} < 0$) and $H = 2/3D$; (**b**) The same as (**a**) but for $H = 5/6D$; (**c**) The same as (**a**) but for $H = 11/12D$; (**d**) No upper Ekman layer; (**e**) There is an upper Ekman layer, in which the Ekman current has a positive component in the mean flow direction ($\vec{V_E} \cdot \vec{V} > 0$). The vertical velocity is 10 times larger for plotting the solutions.

As shown in Table 1, all five SAR images with sea surface imprints of tidal channels in Subei Bank were acquired during flood tide, implying that the tidal current was mainly flowing southward and was parallel to the submerged sand ridges or tidal channels [37]. Meanwhile, according to the Ekman theory [38], the wind-driven surface Ekman current flows at an angle to the right of the prevailing wind direction. The wind direction of the five wide bright stripes images in Table 1 indicates that the Ekman velocity has a northward component. According to the physic model, when the tidal current and the surface Ekman current have opposite directions ($\vec{V_E} \cdot \vec{V} < 0$), surface current convergence zones occur in the middle of two adjacent sand ridges, i.e., over the tidal channel region in this study. Therefore, the tidal channels appear as wide and bright stripes on the five SAR images.

### 4.2. SAR Imaging of Sand Ridges

For SAR imaging of underwater sand ridges, in most cases (nine out of 12), the secondary circulation theory is not applicable because the relationship between the tidal current and wind direction does not satisfy the necessary dynamic condition. However, as pointed out in the last section, the water levels at the imaging time are far below the mean sea level by over 1.3 m. In this case, the sea surface waves are most likely to break when propagating to shallower waters. Additionally, the ocean wave breaking has been proved to be one of the most frequent oceanic processes in Subei Bank [39]. In the following section, we will determine whether if this is true for the cases when sand ridges were observed by SAR.

The wave breaking generally occurs where the wave height reaches the point that the crest of the wave actually overturns [40]. Nelson and Gonsalvas [41] studied the laboratory and field wave data and developed a wave breaking relationship applicable to the regular and irregular waves:

$$\gamma_b = 0.55 + \exp(-0.012\cot(m)), \qquad m \leq 0.01 \tag{7}$$

where $m$ is the sea floor slope, and $\gamma_b$ is the ratio of the wave height ($H_w$) to wave breaking water depth ($h_b$), i.e.,

$$h_b = \frac{H_w}{\gamma_b}, \tag{8}$$

Here, the wave breaking depth $h_b$ means that a wave will start to break when it reaches an area where the instantaneous water depth is smaller than $h_b$.

For the fully developed ocean waves, the wave height $H_w$ can be expressed as [42]:

$$H_w = \frac{A_0}{g}{U_{10}}^2, \tag{9}$$

where $A_0$ is a non-dimensional constant taken to be 0.3, $g$ is the gravitational acceleration, and $U_{10}$ is the wind speed at 10 m from the sea surface.

The mean seafloor slope of the sand ridges in the study region (see Figure 7) is close to 0.004. Hence we have $\gamma_b = 0.6$. Then, using Equations (8) and (9) and the SAR-derived wind speed, the wave height and the corresponding breaking depth at SAR imaging time are calculated and listed in Table 1. Considering the tidal height, all the instantaneous water depths at the sand ridge locations are smaller than the breaking depth, indicating that the surface waves under the relatively strong winds are quite likely to break when propagating over the extremely shallow sand ridges. The increase of surface roughness induced by breaking waves over the sand ridges will make the sea surface appear as narrow bright stripes on SAR imagery.

### 4.3. Discussion

Note that for some cases where the sand ridges are observed by SAR (cases 6, 7, and 12), or topographic features are not shown on SAR imagery (cases 19, 21, 22, and 25), the tidal current was also opposite to the wind direction. According to the secondary circulation theory proposed above, the tidal channels might also be observed by SAR in these cases. However, the wide bright stripes corresponding to the tidal channels are not shown on these images. Why? If we look at the wind and current conditions in more detail, we find the images were all acquired under high winds (6.1~10.5 m/s), implying relatively high NRCS values throughout the study area. On the other hand, as the output from the TMD model shows, the time differences between the acquisition times of these SAR images and the local high or low tide times are within 1.5 h, indicating that the tidal current velocity might be so weak (and even close to 0) that the convergence does not occur at the surface over the tidal channels, or the signal enhancement generated by the weak convergence is not strong enough to be observed by SAR compared to the ambient high NRCS induced by the winds.

From Figure 7, one can see that the peak NRCS positions exhibit very little movement. One possible reason for this is that the topography of Subei Bank changes slowly with time under the action of strong tidal currents and this change may fluctuate if the sea state changes severely (e.g., typhoon, storm current, etc.) in some years [34,37]. From another perspective, we may be able to use SAR to observe the short-term change and long-term evolution of the sand ridges. For some few cases under a relatively high wind speed where the instantaneous water depth is smaller than the wave breaking depth, since the slope of the sand ridge in the study area is very steep, the relatively strong wind also impelled the breaking wave to quickly propagate to the peak of the sand ridges. Therefore, the breaking wave induced increase in surface roughness is larger over the shallower sand ridge and is observed by SAR.

## 5. Conclusions

In this study, 25 ENVISAT ASAR and ERS-2 SAR images are analyzed to investigate the C-band radar backscatter features of the shallow water topography over Subei Bank in the Southern Yellow Sea of China, where the flow is primarily parallel to the major axes of tidal channels or sand ridges. Based on the statistical analysis, we find the bathymetric features are not always shown on SAR imagery. For SAR images with obvious topographic features, paralleled fingerlike bright stripes appear at different locations and have distinct widths. The tidal channels appear as wide bright stripes with an average width of 6 km on SAR images under low to moderate wind speeds, while the sea surface imprints of underwater sand ridges on SAR imagery are narrow (~1 km wide), quasi-linear, bright stripes at high winds.

Theoretical analysis suggests that the reason why tidal channels are observed by C-band SAR under low to moderate winds is that the tidal current and the wind-driven surface Ekman current have opposite directions. In this case, a convergent zone at the sea surface forms at the central line of the tidal channel due to the convergence of two vortexes in the upper layer. Therefore, the tidal channels are shown as relatively wide bright stripes on SAR imagery. However, the tidal channels might not be able to be detected by SAR at high winds due to the high NRCS value of background seawaters, even if the above dynamic condition is fulfilled. For SAR imaging of the sand ridges in the study area, both the low water level and strong winds provide favorable conditions for the breaking of ocean surface waves when propagating to the shallow waters, thus leading to an increase of SAR observed NRCS over the shallow sand ridges.

**Acknowledgments:** The ENVISAT ASAR data used in this study were provided by the Institute of Space and Earth Information Science of the Chinese University of Hong Kong. The ERS-2 SAR data are from the ESA and the Open Spatial Data Sharing Project of Institute of Remote Sensing and Digital Earth Chinese Academy. This work was supported by the National Key Project of Research and Development Plan of China (Grant No. 2016YFC1401905), the Fundamental Research Funds for the Central Universities (Hohai University) (Grants No. 2015B15914 and 2017B714X14), the Postgraduate Research & Practice Innovation Program of Jiangsu Province (Grant No. KYCX17_0507), and the project of the Priority Academic Program Development of Jiangsu Higher Education Institutions (Ocean Science). The views, opinions, and findings contained in this report are those of the authors and should not be construed as an official NOAA or U.S. Government position, policy, or decision.

**Author Contributions:** Qing Xu and Shuangshang Zhang initiated the research. Under the supervision of Qing Xu, Quanan Zheng and Xiaofeng Li, Shuangshang Zhang performed the experiments and analysis. Shuangshang Zhang and Qing Xu drafted the manuscript. Quanan Zheng and Xiaofeng Li revised the paper. All authors read and approved the final version of the manuscript.

**Conflicts of Interest:** The authors declare no conflict of interest.

## Appendix A. Derivation of Secondary Circulation Solutions

Consider the governing equations for a flow consisting of mean flow and disturbance:

$$\frac{\partial U}{\partial t} + U\frac{\partial U}{\partial x} + V\frac{\partial U}{\partial y} + W\frac{\partial U}{\partial z} - fV = \frac{1}{\rho}\frac{\partial P}{\partial x} + A^2 U + F_x, \quad (A1)$$

$$\frac{\partial V}{\partial t} + U\frac{\partial V}{\partial x} + V\frac{\partial V}{\partial y} + W\frac{\partial V}{\partial z} + fU = \frac{1}{\rho}\frac{\partial P}{\partial y} + A^2 V + F_y, \quad (A2)$$

$$\frac{\partial W}{\partial t} + U\frac{\partial W}{\partial x} + V\frac{\partial W}{\partial y} + W\frac{\partial W}{\partial z} = \frac{1}{\rho}\frac{\partial P}{\partial z} - g + A^2 W + F_z, \quad (A3)$$

$$\frac{\partial U}{\partial x}+\frac{\partial V}{\partial y}+\frac{\partial W}{\partial z}=0, \tag{A4}$$

where $f$ is the Coriolis parameter; $P$ is the pressure; $\rho$ is the water density; $A$ is the kinetic viscosity; $F_x$, $F_y$, and $F_z$ are the components of external forcing; g is the gravitational acceleration; and:

$$U=\overline{u}+u,\ V=\overline{v}+v,\ W=\overline{w}+w,\ P=\overline{p}+p. \tag{A5}$$

The boundary conditions are:

$$U=V=W=0, \qquad for\ z=0, \tag{A6}$$

and

$$U=V=0,\ for\ y=0,\ and\ y=2b. \tag{A7}$$

Substituting (A5) into (A1)–(A4) yields the disturbance governing equations:

$$\frac{\partial u}{\partial t}+\overline{u}\frac{\partial u}{\partial x}+v\frac{\partial \overline{u}}{\partial y}+w\frac{\partial \overline{u}}{\partial z}-fv=\frac{1}{\rho}\frac{\partial p}{\partial x}+A^2u, \tag{A8}$$

$$\frac{\partial v}{\partial t}+\overline{u}\frac{\partial v}{\partial x}+fu=\frac{1}{\rho}\frac{\partial p}{\partial y}+A^2v, \tag{A9}$$

$$\frac{\partial w}{\partial t}+\overline{u}\frac{\partial w}{\partial x}=\frac{1}{\rho}\frac{\partial p}{\partial y}+A^2w, \tag{A10}$$

$$\frac{\partial u}{\partial x}+\frac{\partial v}{\partial y}+\frac{\partial w}{\partial z}=0. \tag{A11}$$

The boundary conditions are:

$$u=v=w=0, \quad for\ z=0, \tag{A12}$$

and

$$u=v=0,\ for\ y=0,\ and\ y=2b. \tag{A13}$$

In order to examine the role of velocity shear in generating the secondary circulation, we further take the following approximations: (1) ignoring the viscous terms; (2) assuming the $x$-coordinate scale of mean flow $\overline{u}$, $\overline{L}$, is much larger than that of the disturbance, $L$, thus resulting in $\partial/\partial t \ll \overline{u}\partial/\partial x$; (3) the $x$-coordinate scale of disturbance is much larger than that of the $y$-coordinate scale, thus resulting in $\partial/\partial x \ll \partial/\partial y$; (4) in Equations (A8)–(A10), the velocity shear terms are much larger than other terms. Thus, we have the simplified disturbance equations:

$$v\frac{\partial \overline{u}}{\partial y}+w\frac{\partial \overline{u}}{\partial z}=0, \tag{A14}$$

$$\overline{u}\frac{\partial v}{\partial x}=0, \tag{A15}$$

$$\overline{u}\frac{\partial w}{\partial x}=0, \tag{A16}$$

$$\frac{\partial v}{\partial y}+\frac{\partial w}{\partial z}=0, \tag{A17}$$

From (A15) and (A16), we have:

$$\frac{\partial v}{\partial x}=\frac{\partial w}{\partial x}=0, \tag{A18}$$

i.e., $v$ and $w$ are independent of $x$. From (A14) and (A17), we derive a secondary circulation equation of $w$:

$$\frac{\partial w}{\partial y}F - \frac{\partial w}{\partial z} = -w\frac{\partial F}{\partial y}, \tag{A19}$$

where $F$ is defined as

$$F \equiv \frac{\frac{\partial \overline{u}}{\partial z}}{\frac{\partial \overline{u}}{\partial y}}. \tag{A20}$$

Equation (A19) has an analytical solution of

$$w(y,z) = w_0(sin\frac{\pi z}{2H})^{[1+\frac{y(2b-y)}{2(b-y)^2}]}, \tag{A21}$$

where $w_0$ is a constant to be determined. From (A14) we have:

$$v = -w(\frac{\partial \overline{u}}{\partial z}/\frac{\partial \overline{u}}{\partial y}), \tag{A22}$$

$$v(y,z) = -v_0\left[\frac{y(2b-y)}{2(b-y)}\right](sin\frac{\pi z}{2H})^{\frac{y(2b-y)}{2(b-y)^2}} cos\frac{\pi z}{2H}, \tag{A23}$$

where $v_0 = \pi(2H)^{-1}w_0$.

## References


1. He, X.; Chen, N.; Zhang, H.; Fu, B.; Wang, X. Reconstruction of sand wave bathymetry using both satellite imagery and multi-beam bathymetric data: A case study of the Taiwan Banks. *Int. J. Remote Sens.* **2014**, *35*, 3286–3299. [CrossRef]
2. Ma, S.; Tao, Z.; Yang, X.; Yu, Y.; Zhou, X.; Li, Z. Bathymetry retrieval from hyperspectral remote sensing data in optical-shallow water. *IEEE Trans. Geosci. Remote Sens.* **2014**, *52*, 1205–1212. [CrossRef]
3. Pacheco, A.; Horta, J.; Loureiro, C.; Ferreira, Ó. Retrieval of nearshore bathymetry from Landsat 8 images: A tool for coastal monitoring in shallow waters. *Remote Sens. Environ.* **2015**, *159*, 102–116. [CrossRef]
4. Wozencraft, J.M.; Millar, D. Airborne LIDAR and integrated technologies for coastal mapping and nautical charting. *Mar. Technol. Soc. J.* **2005**, *39*, 27–35. [CrossRef]
5. Shi, W.; Wang, M.; Li, X.; Pichel, W.G. Ocean sand ridge signatures in the Bohai Sea observed by satellite ocean color and synthetic aperture radar measurements. *Remote Sens. Environ.* **2011**, *115*, 1926–1934. [CrossRef]
6. De Loor, G.P. The observation of tidal patterns, currents, and bathymetry with SLAR imagery of the sea. *IEEE J. Ocean. Eng.* **1981**, *6*, 124–129. [CrossRef]
7. Alpers, W.; Hennings, I. A theory of the imaging mechanism of underwater bottom topography by real and synthetic aperture radar. *J. Geophys. Res.* **1984**, *89*, 1029–1054. [CrossRef]
8. Zheng, Q.; Li, L.; Guo, X.; Ge, Y.; Zhu, D.; Li, C. SAR Imaging and hydrodynamic analysis of ocean bottom topographic waves. *J. Geophys. Res.* **2006**, *111*. [CrossRef]
9. Li, X.; Li, C.; Xu, Q.; Pichel, W.G. Sea surface manifestation of along-tidal-channel underwater ridges imaged by SAR. *IEEE Trans. Geosci. Remote Sens.* **2009**, *47*, 2467–2477. [CrossRef]
10. Li, X.; Yang, X.; Zheng, Q.; Pietrafesa, L.J.; Pichel, W.G.; Li, Z.; Li, X. Deep-water bathymetry feature imaged by spaceborne SAR in the Gulf Stream region. *Geophys. Res. Lett.* **2010**, *37*. [CrossRef]
11. Hennings, I. An historical overview of radar imagery of sea bottom topography. *Int. J. Remote Sens.* **1998**, *19*, 1447–1454. [CrossRef]
12. Zheng, Q.; Holt, B.; Li, X.; Liu, X.; Zhao, Q.; Yuan, Y.; Yang, X. Deep-water seamount wakes on SEASAT SAR image in the Gulf Stream region. *Geophys. Res. Lett.* **2012**, *39*. [CrossRef]
13. De Loor, G.P.; Hulten, B.V. Microwave Measurements over the North Sea. *Bound. Layer Meteorol.* **1978**, *13*, 119–131. [CrossRef]
14. Van der Kooij, M.W.A.; Vogelzang, J.; Calkoen, C.J. A simple analytical model for brightness modulations caused by submarine sand waves in radar imagery. *J. Geophys. Res.* **1995**, *100*, 7069–7082. [CrossRef]

15. Vogelzang, J.; Wensink, G.J.; Calkoen, C.J.; Van der Kooij, M.W.A. Mapping Submarine Sand Waves with Multiband Imaging Radar: Experimental Results and Model Comparison. *J. Geophys. Res.* **1997**, *102*, 1183–1192. [CrossRef]
16. Romeiser, R.; Alpers, W. An improved composite surface model for the radar backscattering cross section of the ocean surface, 2, Model response to surface roughness variations and the radar imaging of underwater bottom topography. *J. Geophys. Res.* **1997**, *102*, 25251–25267. [CrossRef]
17. Zheng, Q.; Zhao, Q.; Yuan, Y.; Xian, L.; Hu, J.; Liu, X.; Yin, L.; Ye, Y. Shear-flow induced secondary circulation in parallel underwater topographic corrugation and its application to satellite image interpretation. *J. Ocean. Univ. China* **2012**, *11*, 427–435. [CrossRef]
18. Wang, X.; Zhang, H.; Li, X.; Fu, B.; Guan, W. SAR imaging of a topography-induced current front in a tidal channel. *Int. J. Remote Sens.* **2015**, *36*, 3563–3574. [CrossRef]
19. Zhang, S.; Xu, Q.; Cheng, Y.; Li, Y.; Huang, Q. Bathymetric features of Subei Bank on ENVISAT ASAR images. In Proceedings of the IEEE Progress in Electromagnetic Research Symposium (PIERS), Shanghai, China, 8–11 August 2016.
20. Liu, Y.; Li, M.; Cheng, L.; Li, F.; Chen, K. Topographic mapping of offshore sandbank tidal flats using the waterline detection method: A case study on the Dongsha Sandbank of Jiangsu Radial Tidal Sand Ridges, China. *Mar. Geod.* **2012**, *35*, 362–378. [CrossRef]
21. Iwenfei, N.; Wang, Y.; Zou, X.; Zhang, J.; Gao, J. Sediment dynamics in an offshore tidal channel in the southern Yellow Sea. *Int. J. Remote Sens.* **2014**, *29*, 246–259. [CrossRef]
22. Song, Y.; Zhang, J. Study on Evolutions of Jiangsu Radiating Sandbanks Based on SAR Images. In *Advances in SAR Oceanography from Envisat and ERS Missions, Proceedings of the SEASAR 2006, Frascati, Italy, 23–26 January 2006*; Lacoste, H., Ouwehand, L., Eds.; ESA Publications Division: Noordwijk, The Netherlands, 2006; Available online: https://earth.esa.int/workshops/seasar2006/proceedings/papers/p11_song.pdf (accessed on 1 November 2017).
23. Desnos, Y.L.; Buck, C.; Guijarro, J.; Suchail, J.L.; Torres, R.; Attema, E. ASAR-ENVISAT's advanced synthetic aperture radar. *ESA Bull.* **2000**, *102*, 91–100.
24. Moreira, A.; Prats-Iraola, P.; Younis, M.; Krieger, G.; Hajnsek, I.; Papathanassiou, K.P. A tutorial on synthetic aperture radar. *IEEE Geosci. Remote Sens. Mag.* **2013**, *1*, 6–43. [CrossRef]
25. Lin, H.; Xu, Q.; Zheng, Q. An overview on SAR measurements of sea surface wind. *Prog. Nat. Sci.* **2008**, *18*, 913–919. [CrossRef]
26. Xu, Q.; Lin, H.; Li, X.; Zuo, J.; Zheng, Q.; Pichel, W.G.; Liu, Y. Assessment of an analytical model for sea surface wind speed retrieval from spaceborne SAR. *Int. J. Remote Sens.* **2010**, *31*, 993–1008. [CrossRef]
27. Yang, X.; Li, X.; Zheng, Q.; Gu, X.; Pichel, W.G.; Li, Z. Comparison of ocean surface winds retrieved from QuikSCAT scatterometer and Radarsat-1 SAR in offshore waters of the U.S. West Coast. *IEEE Geosci. Remote Sens. Lett.* **2010**, *8*, 163–167. [CrossRef]
28. Yang, X.; Li, X.; Pichel, W.G.; Li, Z. Comparison of ocean surface winds from ENVISAT ASAR, MetOp ASCAT scatterometer, buoy measurements, and NOGAPS model. *IEEE Trans. Geosci. Remote Sens.* **2011**, *49*, 4743–4750. [CrossRef]
29. Hersbach, H.; Stoffelen, A.; De, S.H. An improved c-band scatterometer ocean geophysical model function: CMOD5. *J. Geophys. Res.* **2007**, *112*, 225–237. [CrossRef]
30. Zhang, H.; Bates, J.J.; Reynolds, R.W. Assessment of composite global sampling: Sea surface wind speed. *Geophys. Res. Lett.* **2006**, *33*, L17714. [CrossRef]
31. Egbert, G.D.; Erofeeva, S.Y. Efficient inverse modeling of barotropic ocean tides. *J. Atmos. Ocean. Technol.* **2002**, *19*, 183–204. [CrossRef]
32. Padman, L.; Erofeeva, S. *Tide Model Driver (TMD) Manual*; Earth & Space Research: Seattle, WA, USA, 2005.
33. The Navigation Guarantee Department of the Chinese Navy Headquarters. *Sea Chart of Yellow Sea: Sheyanghe Kou to Lvsi Gang*; China Navigation Publications Press: Tianjin, China, 2013.
34. Chen, J.; Wang, Y.; Zhang, R.; Lin, X. Stability study on the Dongsha Sandbanks in submarine radial sand ridges field off Jiangsu coast. *Ocean Eng.* **2007**, *25*, 105–113.
35. Kundu, P.K. *Fluid Mechanics*; Academic Press: New York, NY, USA, 1990; pp. 263–298.
36. Alpers, W. Theory of radar imaging of internal waves. *Nature* **1985**, *314*, 245–247. [CrossRef]
37. Zhang, C.; Zhang, D.; Zhang, J.; Wang, Z. Tidal current-induced formation—Storm-induced change—Tidal current-induced recovery. *Sci. China Ser. D Earth Sci.* **1999**, *42*, 1–12. [CrossRef]

38. Ekman, V.W. On the influence of the earth's rotation on ocean currents. *Matem. Astr. Fysik.* **1905**, *2*, 1–52.
39. Yang, Y.; Feng, W. Numerical simulation of wave fields in radial sand ridge filed of southern yellow sea. *J. Hohai Univ.* **2010**, *38*, 457–461.
40. Evander, L. *Breaking Wave*; Alphascript Publishing: Beau Bassin, Mauritius, 2010.
41. Nelson, R.C.; Gonsalves, J. Surf zone transformation of wave height to water depth ratios. *Coast. Eng.* **1992**, *17*, 49–70. [CrossRef]
42. Hubert, W.E. A preliminary report on numerical sea condition forecasts. *Mon. Weather Rev.* **1957**, *85*, 200–204. [CrossRef]

*Article*

# Underwater Topography Detection in Coastal Areas Using Fully Polarimetric SAR Data

**Xiaolin Bian [1,2], Yun Shao [1], Wei Tian [1], Shiang Wang [3], Chunyan Zhang [1,*], Xiaochen Wang [1,2] and Zhixin Zhang [1,2]**

1 Institute of Remote Sensing and Digital Earth, Chinese Academy of Sciences, Beijing 100101, China; bianxl@radi.ac.cn (X.B.); shaoyun@radi.ac.cn (Y.S.); tianwei@radi.ac.cn (W.T.); wsadw@sina.com (S.W.); wangxc@radi.ac.cn (C.W.); zhangzx2014@radi.ac.cn (Z.Z.)
2 University of Chinese Academy of Sciences, Beijing 100049, China
3 National Ocean Technology Center, Tianjin 300112, China
* Correspondence: zhangcy@radi.ac.cn; Tel.: +86-10-6488-9214

Academic Editors: Xiaofeng Yang, Xiaofeng Li, Ferdinando Nunziata, Alexis Mouche and Prasad S. Thenkabail
Received: 27 February 2017; Accepted: 31 May 2017; Published: 4 June 2017

**Abstract:** Fully polarimetric synthetic aperture radar (SAR) can provide detailed information on scattering mechanisms that could enable the target or structure to be identified. This paper presents a method to detect underwater topography in coastal areas using high resolution fully polarimetric SAR data, while less prior information is required. The method is based on the shoaling and refraction of long surface gravity waves as they propagate shoreward. First, the surface scattering component is obtained by polarization decomposition. Then, wave fields are retrieved from the two-dimensional (2D) spectra by the Fast Fourier Transformation (FFT). Finally, shallow water depths are estimated from the dispersion relation. Applicability and effectiveness of the proposed methodology are tested by using C-band fine quad-polarization mode RADARSAT-2 SAR data over the near-shore area of the Hainan province, China. By comparing with the values from an official electronic navigational chart (ENC), the estimated water depths are in good agreement with them. The average relative error of the detected results from the scattering mechanisms based method and single polarization SAR data are 9.73% and 11.53% respectively. The validation results indicate that the scattering mechanisms based methodology is more effective than only using the single polarization SAR data for underwater topography detection, and will inspire further research on underwater topography detection with fully polarimetric SAR data.

**Keywords:** shallow water; swell waves; water depth; dispersion relationship; quad-polarization; Bragg scattering

## 1. Introduction

Underwater features and ocean bathymetry is an indispensable information for coastal engineering and management and coastal resources exploitation and protection [1,2]. For example, safely navigating, offshore fishery and aquaculture, research on tide and biodiversity, planning for seawalls and wharf and other human activities are carried out in these areas where water depths less than 100 m. Detailed knowledge of water depth is very useful for them. Conventionally, water depth surveying are carried out by sonar measurements from dedicated vessels, which are accurate for point-measurement, but are not only expensive and time consuming but also difficult in shallow water areas, especially in some special water areas where routine surveying cannot be achieved [2,3].

Synthetic aperture radar (SAR) is an active microwave remote sensor that has the ability to image targets on the earth in both day and night, and for almost all weather conditions [1,4]. Therefore, SAR is widely applied in earth observation, especially in ocean observation. SAR polarimetry is the science of acquiring, processing and analyzing the polarization state of an electromagnetic field that will improve both the ability to characterize physical properties of objects and the retrieval of bio—or geophysical properties of the earth's surface and subsurface [5,6]. Fully polarimetric (HH, HV, VH and VV) SAR can provide detailed information on scattering mechanisms (single bounce scattering, double bounce scattering and triple—and higher-order bouncing scattering) that could enable the target or the structure to be identified [5]. Under favorable conditions, sea floor topographic features in shallow water areas can be detected from SAR image intensity directly or other features in SAR image indirectly [3,4,7–14]. Therefore, SAR based method that needs less cost and labor and will enrich the means for underwater topographic surveying and mapping.

Underwater topography is conventionally detected by variations of the SAR image intensity which relies on the presence of a current, along with the presence of small-scale waves on the sea surface to provide the radar back scatter [1,3,8,10,14,15]. The typical of this method is Bathymetry Assessment System (BAS) which was based on the theory of imaging processes of underwater bottom topography by SAR proposed by Alpers and Hennings [3,14]. Li et al. shown that sand ridges parallel to tidal currents can also be imaged by SAR [16]. Bi et al. applied polarization information for underwater topography detection [17]. Under special conditions, deep-water (>500 m) bathymetry features can also be imaged by SAR [11]. However, the intensity variations based method depends on some real-time or near-real-time data (such as sea surface wind field and current field) and the calculation procedure of the algorithm is very complicated.

In contrast, swell patterns has been suggested to be applied in the detection of underwater topography. It relies on refraction of long surface gravity waves as they propagate shoreward and requires the assumption that the same wave train exists throughout the image [18]. Although this method has been used since World War II to estimate water depths based on changes in observable wave characteristics, approaches to detect bathymetry from swell patterns in SAR images have been proposed in the past two decades [4,7,9,12,13,18–22]. Pleskachevsky et al. explored bathymetry by the synergetic use of multiple remote sensed data resources (radar data from TerraSAR-X and optical data from QuickBird satellite) for coastal areas. Water depth between 20 m and 60 m is obtained with accuracy in order of 15% from SAR-based method. Water depth less than 20 m is detected by the optical-based method using sunlight reflection analysis. The depths in the range of 20 m up to 10 m represent the domain where the synergy of data from both sources arises [12]. Bian et al. estimated shallow water depth in coastal region where water depth is less than 10 m using HJ-1C (Huan Jing-1C, Huan Jing means environment in Chinese) SAR in S band [7]. In addition, this methodology had been developed by data processing [22] and using modulation transfer function (MTF) to account for the dependence of the strength of radar signatures of ocean waves on their wavenumber and propagation direction relative to the radar look direction [18]. Applicability of linear dispersion relationship and detected algorithm has been applied and discussed [7,21]. Comparing to the intensity variations based method, the swell patterns based method is simple and requires little prior information, but sharp water depths cannot be detected because they are filtered by the algorithm. Although the swell patterns based applications has been proposed and developed for bathymetry estimation using different band (L-band, S-band, C-band and X-band) spaceborne SAR data, they used single polarization data within the allowable error range.

The aim of this study is to develop a detection method based on swell patterns and scattering mechanism using fully polarimetric SAR data. First, different scattering components are obtained by polarization decomposition. Subsequently, shoaling waves are tracked by fixed grid from open sea to shoreline. Wavelength and wave direction can be retrieved by means of the Fast Fourier Transformation (FFT) computation from the dominant scattering component image. Finally, shallow water depth can be estimated using the linear dispersion relationship of ocean waves [1,21]. Moreover, this paper analyzes viability and potential performance for the case of application.

This paper is organized as follows. Section 2 describes materials and the swell patterns based method. Section 3 presents the detection results with one case study in Sanya coastal region, Hainan province, China. Section 4 discusses the proposed method for the application. In the end, conclusions are presented in Section 5.

## 2. Materials and Methods

### 2.1. SAR Data and Auxiliary Data

The fully polarimetric datasets of single look complex (SLC) fine quad-polarization C-band full-resolution RADARSAT-2 SAR data are collected for underwater topography detection. They are acquired on 18 September 2009 at 10:50:44 UTC over Sanya coastal region with coverage of 25 km by 25 km, Hainan province, China. Figure 1a–d shows the amplitude images of HH-polarized, HV-polarized, VV-polarized and color-coded image of the Freeman-Durden target decomposition: red, Ps; green, Pd; and blue, Pv with pixel spacing of 5 m respectively. In the monostatic backscattering case, the reciprocity constrains the scattering matrix to be symmetrical, that is, HV = VH. Therefore, VH-polarized image is not given. RADARSAT-2 satellite operates from a sun-synchronous orbit at a height of 798 km, and its platform velocity ($V$) is about 7.46 km/s. Acquisition parameters for the collected SAR datasets are given in Table 1.

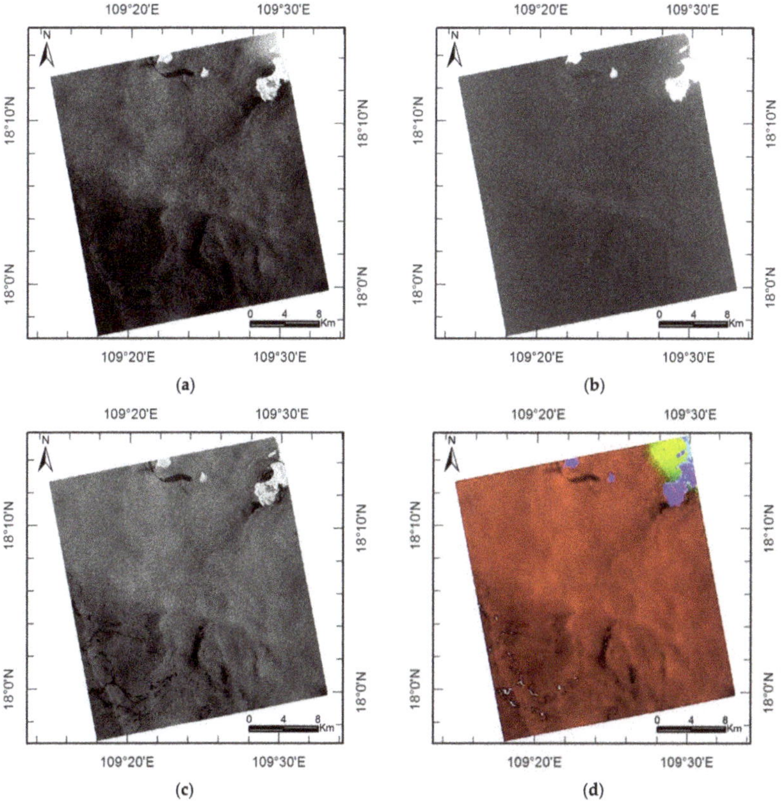


**Figure 1.** The C-band fine quad-polarization RADARSAT-2 images and the polarization decomposition result in Sanya coastal region, Hainan province, China, acquired on September 18, 2009 at 10:50:44 UTC. RADARSAT-2 Data and Product © MacDonald, Dettwiler and Associates Ltd., All Rights Reserved. (**a**) HH-polarized image; (**b**) HV-polarized image; (**c**) VV-polarized image; (**d**) Color-coded image of the Freeman-Durden decomposition red, Ps; green, Pd; and blue, Pv.

**Table 1.** RADARSAT-2 quad-polarization SAR data acquisition parameters.

| Name | Satellite Height | Slant Range Near Edge | Incidence Angle Near Range | Incidence Angle Far Range |
|---|---|---|---|---|
| Value | 793.93 km | 919.8 km | 32.353 deg | 34.013 deg |

Two different auxiliary data are used in this paper. For water depth information, the ENC (http://map.enclive.cn/) with different scale from the Navigation Guarantee Department (NGD) of the Chinese Navy Headquarter (CNH) is used for comparing and analyzing the detected results. Figure 2 shows low scale of water depth map from the ENC. For sea state, a wind vectors map from microwave scatterometer QuikSCAT and a photo was taken before the start time of the collected RADARSAT-2 data sets as shown in Figure 3 are used for estimating cut-off wavelength.

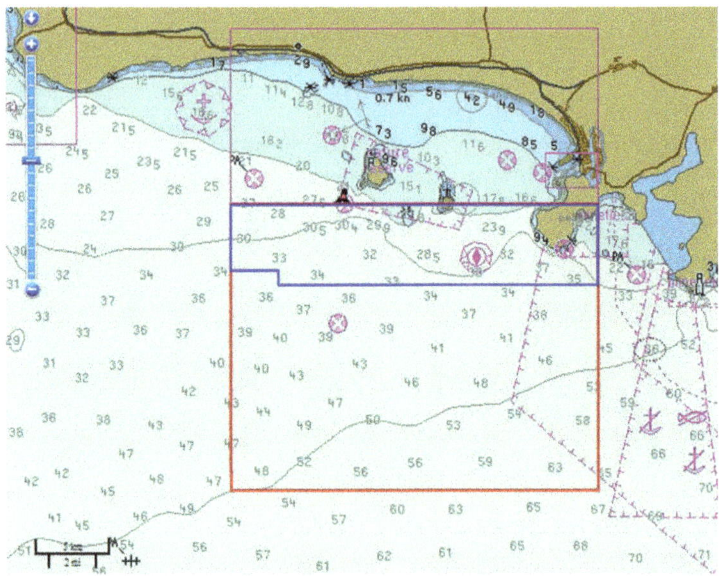

**Figure 2.** Water depth map from electronic navigational chart via map.enclive.cn. The unit of water depth in this image is the meter. Low scale of water depths (46 pairs) in the red box are collected for comparison. High scale of water depths (305 pairs) in the blue polygon are also collected for comparison, but they are not presented in this figure. Water depth map © www.enclive.cn, All Rights Reserved.

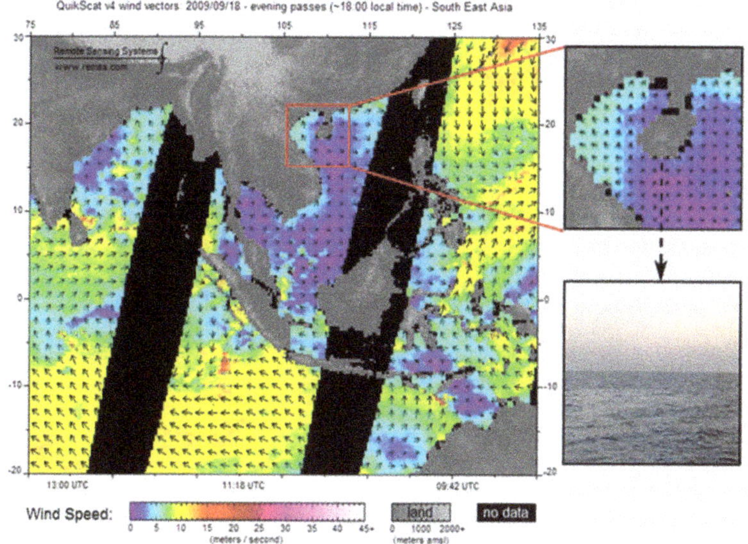


**Figure 3.** A wind vectors map from microwave scatterometer QuikSCAT at 18:00 (local time) on 18 September 2009. The study area is outlined in red box. The lower-right portion is a photo of sea state over the study area before the start time of the collected RADARSAT-2 datasets.

## 2.2. Methods

### 2.2.1. SAR Imaging of Waves

SAR transmits radar signals and receives backscattered returns whose levels are directly dependent on ocean surface roughness. Ocean roughness is comprised of the mean surface slope, dominated by the long-wavelength field, and the short-scale waves that range between capillary and short-gravity waves. With incidence angle between 20° and 60°, Bragg scattering is the dominant backscatter mechanism, but not under all wind and wave states [1,4,23–25]. Under Bragg scattering, the normalized radar cross section (NRCS) of ocean is proportional to the amplitude of the small-scale waves called Bragg waves. Bragg waves are ocean surface waves which have wavelength equal to the projection of the SAR electromagnetic wavelength onto the local ocean surface and which are propagating either directly toward or away from the look direction of SAR [1]. It is expressed by the follow relation:

$$\lambda_B = 0.5\lambda_r \sin\theta_i \tag{1}$$

where $\lambda_B$ is the wavelength of the Bragg wave, $\lambda_r$ is the SAR electromagnetic wavelength, and $\theta_i$ is the local incidence angle. For incidence angle less than 15°, the specular scattering takes a significant contribution in the radar backscatter, while for incidence angle larger than 70°,the wedge scattering is most prevalent [1,26].

The combination of Bragg scattering with the two-scale approximation is used to describe the interaction of short and long waves, but not under all wind and wave states and radar system configurations. When imaging ocean surface waves, the main modulation mechanisms have been identified as three modulation mechanisms: title modulation, hydrodynamic modulation and velocity bunching modulation [1,6,27]. Title modulation means the longer waves change the local slope of the shorter wave fields. It is strongest for waves travelling in the range direction. Hydrodynamic modulation means the longer waves change the distribution of the shorter wave fields.

Velocity bunching modulation means the motion of the long waves alters the SAR imaging process that is unique to SAR imaging systems. If the azimuth shift is less than a wavelength, azimuth-traveling waves are linearly mapped in the SAR imagery. On the contrary, when the displacements are greater than a wavelength, the mapping of waves on the imagery will be nonlinear and distorted. It is greatest for waves travelling in the azimuth direction. When long waves are propagating in a direction perpendicular to the SAR platform, title modulation and hydrodynamic modulation have the strongest effects on the SAR returns [1,28].

Under specific conditions, the imaging mechanism of sea surface can be assumed linear. In general, the imaging mechanism of sea surface by SAR is not linear [1,23]. Therefore, a SAR image cannot be interpreted as a picture of the surface because the orbital motion of longer ocean waves with azimuth component allows the waves to be imaged through the velocity bunching mechanism, the random motions of the ocean surface caused by the short scale waves introduce random position shifts in the azimuth direction that in turn degrade the azimuth resolution and thus limits the detectable ocean wavelengths [1,4,23]. Even though the SAR image contains swell patterns, it may not be suitable for underwater topography detection. However, linear imaging can be assumed under specific conditions, such as not extreme wind speed and sea state, absence of currents, and swell patterns characterized by wavelengths that are greater than cut-off conditions. In other words, the image modulation of azimuth-travelling waves by velocity bunching is linear if the ratio of wave height to wavelength is small. The cut-off wavelength *Lmin* can be estimated by a simple empirical relationship:

$$Lmin = \frac{R}{V}\sqrt{H} \tag{2}$$

where $R$ is the slant range, $V$ is the platform velocity, $H$ is the significant wave height. The lower the satellite orbit and the smaller the wave height, the shorter shoaling waves will be detected. When these conditions are satisfied, the swell wavelengths imaged by SAR represent an unbiased estimate of the true swell wavelengths [21].

### 2.2.2. Freeman-Durden Decomposition

The Freeman-Durden decomposition is a technique for fitting a physically based, three-component mechanism model to the polarimetric SAR observations, without utilizing any ground truth measurements [5,29]. The mechanisms are a canopy scatter from a cloud of randomly oriented dipoles, even—or double-bounce scatter from a pair of orthogonal surfaces with different dielectric constants, and Bragg scattering from a moderately rough surface [5]. This model can be used to determine to first order, what are the dominant scattering mechanisms that give rise to observed backscatter in polarimetric SAR data. The contribution of each scattering mechanism from Freeman-Durden three-component decomposition can be estimated to the span, following:

$$\mathrm{Span} = |\mathrm{S_{HH}}|^2 + 2|\mathrm{S_{HV}}|^2 + |\mathrm{S_{VV}}|^2 = P_S + P_D + P_V \tag{3}$$

where $P_S$ is the power of surface scattering component, $P_D$ is the power of double-bounce scattering component, and $P_V$ is the power of volume scattering component.

Unlike ground covers, backscattering from ocean surface is considerably homogeneous in scattering mechanism, and can be, in most cases, characterized by a two-scale titled Bragg scattering model. The contribution of Bragg surface scattering is given by

$$P_S = f_S\left(1 + |\beta|^2\right) \tag{4}$$

where $f_S$ corresponds to the contribution of the single-bounce scattering to the $|\mathrm{S_{VV}}|^2$ components, with

$$f_S = |\mathrm{S_{VV}}|^2 \text{ and } \beta = \frac{\mathrm{S_{HH}}}{\mathrm{S_{VV}}} \tag{5}$$

### 2.2.3. Wave Shoaling and Refraction

Shoaling and refraction of waves occur when the waves are in shallow water. If the water depth is less than half of the wavelength, then the wave is considered to be in shallow water. As waves travel from deep to shallow water, their shape alters, for example, wave height increase, speed decrease, and length decrease as wave orbits become asymmetrical. This process is called shoaling. Wave refraction is the process by which wave crests realign themselves as a result of decreasing water depths. Varying depths along a wave crest cause the crest to travel at different phase speeds, with those parts of the wave in deeper water moving faster than those in shallow water [1]. This process continues until the crests become parallel to depth contours or the wave breaks regardless of the original direction in deeper water.

Phenomena of swell wave shoaling and refraction due to underwater morphology start appearing in intermediate water depth because surface waves begin to "feel" the bottom when sea water depth is shorter than about half of the swell wavelength. It means that bathymetry through swell wave modulation can be performed only for water depth values lower than half the swell wavelength, that is, the limit water depth value [21].

### 2.2.4. Linear Dispersion Relationship

The linear dispersion relationship is based on the Airy wave theory. For freely propagating ocean waves, the wave number, wave frequency and water depth are not independent, but are linked by the wave dispersion relation [2,30]. If the effect of current can be neglected, the linear dispersion relation is given by:

$$\omega^2 = gktanh(kd) \tag{6}$$

where $\omega$ is the angular wave frequency ($\omega = 2\pi/T$, $T$ is the wave period), $g$ is the gravitational acceleration (9.8 $\mathrm{m/s^2}$), $k$ is the wave number ($k = 2\pi/L$, $L$ is wavelength ), $d$ is the water depth in the range between $L/20$ and $L/2$.

Using Equation (6), water depth can be written as Equation (7).

$$d = \frac{L}{4} ln\left(\frac{2\pi g + \omega^2 L}{2\pi g - \omega^2 L}\right) \tag{7}$$

### 2.2.5. Description of the Algorithm for Underwater Topography Detection

In this paper, an algorithm based on swell patterns and scattering mechanism using fully polarimetric SAR data is presented for underwater topography detection. Because microwave signals emitted by SAR are able to penetrate into just a few centimeters, they are unable to reach the seabed [1,21]. Therefore, underwater topography is detected indirectly. In the near-shore region, surface waves can "feel" the bottom, causing changes in the wave's length and direction of wave propagation. Shoaling waves are tracked by FFT computation. Shallow water depth can be estimated due to the fact that SAR images are able to image swell waves in the ocean and their wavelength can be connected to the local depth. There are four steps in underwater topography detection from fully polarimetric SAR data:

(1) Obtaining dominant scattering component. Under moderate wind conditions and at intermediate incidence angle, sea surface scattering which calls for Bragg scattering is a single-reflection and dominant scattering mechanism [5]. Therefore, the Bragg scattering component from polarimetric decompositions can be used for underwater topography detection. In this paper, Freeman-Durden three-component decomposition is used.

(2) Estimation of angular wave frequency. The angular wave frequency can be calculated from other sources or first guesses of initial water depth or wave period. Firstly, the collected SAR images are analyzed. Only those have obvious swell wave features over the shallow water may be used to estimate water depth. Then wavelength and direction of the swell waves are calculated and analyzed. How to

estimate swell wave parameters will be given in the next step in detail. Sub images over the isobaths are chosen for the initial angular wave frequency estimation, whereas sub images over different depth are chosen for swell patterns analyzing. Finally, the angular wave frequency is calculated by the linear dispersion relation (Equation (6)) from first guesses of initial water depth or water depth from other resources [4,12]. It can also be calculated by wave period from other resources (buoys, weather services).

If swell wave belongs to shallow water domain (the water depth is less than half of the wavelength), where it can be influenced by the underwater topography. The threshold minimum peak period ($T_{\text{min}}$) for the peak wavelength ($L_{\text{max}}$) is used to distinguish the specified swell wavelength belongs to deep water domain or shallow water domain. It is obtained from deep water relation:

$$T_{\text{min}} = \sqrt{2\pi L_{\text{max}}/g} \tag{8}$$

where $g$ is the gravitational acceleration. For water depth estimation, the wave period should be greater than peak period. If the wave period is smaller than peak period, the wave belongs to deep water domain.

(3) Retrieval of wave fields. The swell wave parameters for shallow water depth estimation are calculated by FFT computation from the dominant scattering component image. By means of the FFT for the selected sub image, a two-dimensional (2D) image spectrum in wave number space is obtained. The peak in the 2D spectrum marks wavelength and wave direction of all waves visible in the sub image. The wavelength and direction of swell wave can be estimated from the following formula [4,12,31]:

$$L = \frac{2\pi}{\sqrt{k_{px}^2 + k_{py}^2}} \text{ and } \theta = arctan\left(\frac{k_{py}}{k_{px}}\right) \tag{9}$$

where $L$ is the wavelength, $\theta$ is the wave direction with respect to the sub image, $k_{px}$ and $k_{py}$ are the peak coordinates in the wave number space. The retrieved directions have an ambiguity of 180° due to the static nature of a SAR image [4]. This ambiguity can be solved with information from the cross spectrum or first guess information from other sources. In shallow water where wave shoaling and refraction appear, the ambiguity problem can be solved by manual inspection. In most case, waves can be considered propagating toward the costal line or propagating along the direction that the water becomes shallower.

Swell wave fields map can be retrieved by ray tracing mode, fixed grid mode or the integrated mode [7]. Ray tracing mode tracks the long wave in the wave direction with the distance related to the wavelength. The distance between the neighboring wave rays is fixed. In this way, the wave can be tracked from the open sea to the shoreline and the change in wavelength and direction can be measured. Fixed grid mode tracks the long wave not in the wave direction but at a constant specified shift. In this way, the wave can also be tracked from the open sea to the shoreline uniformly and the change in wavelength can be measured. Integrated mode tracks the long wave by two modes mentioned above. In this paper, wavelength and wave direction are retrieved by fixed grid using FFT.

Starting from the open ocean, FFT-box (e.g., 1024 × 1024) is moved with a constant specified shift (dx and dy, e.g., dx = dy = 30), then a new FFT is computed. The same process repeats until the FFT box reaches both row edge and column edge of the image for the fixed grid mode. After swell wave tracking, a uniform wave field map is produced.

(4) Detection of underwater topography. By inserting the known wave frequency from step 2 and the wave fields map from step 3, shallow water depths are estimated from the linear dispersion relation for ocean gravity waves. In the end, all estimated shallow water depths are analyzed and abnormal data (e.g., water depth over the land and island, water depth from the FFT-box image has invisibility swell patterns) will be removed.

## 3. Results

Based on the assumption of the presence of a single wave system in the SAR image of the study area, the shallow water depths map from the quad-polarization SAR data is presented. Results are compared with the values from an official ENC to evaluate the quality of the proposed method.

### *3.1. Underwater Topography Results*

Figure 4 shows the obtained Bragg scattering component image and image spectra. Before shallow water depth estimation, the fully polarimetric datasets of SLC fine quad-polarization RADARSAT-2 SAR data are processed by PolSARpro version 5.0.4. Data processing includes SLC data import, coherency matrix T3 extraction and geometric rectification. Then dominant scattering component is obtained by the Freeman-Durden decomposition which adopts $5 \times 5$ average window size. There are 12 different subset areas ($1024 \times 1024$ pixels, a to l) are used for image spectrum analyzing and swell wave parameters estimation.

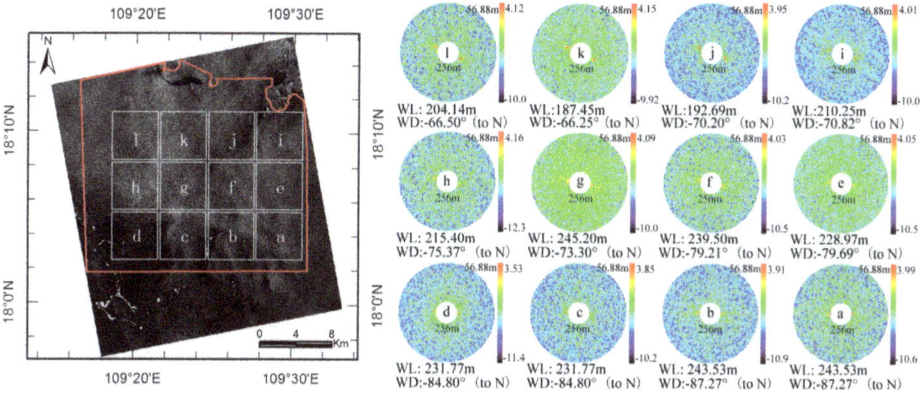


**Figure 4.** Image spectra (to be scaled and converted to bytes) for the subscenes of Bragg scattering component image from Freeman-Durden decomposition divided by three rows (in azimuth direction ) and four columns (in opposite range direction) labeled a to l. The red polygon is used to create a buffer polygon with distance 1.28 km. Results outside the inner line are removed. Each spectrum corresponds to a subscene ($1024 \times 1024$ pixels) of 5.12 km by 5.12 km. In order to display back ground as white value, some features in the down left corner of the image are changed to display as white color. Actually, some of them are biogenic film that should display as black color. The corresponding wavelength (WL) and wave direction (WD) are on the right for wave parameters estimation and analyzing.

By computing the FFT, wavelength and propagation direction of these areas are estimated from the 2D spectrum (wavelength between 56.88 m and 256 m are kept and wind streaks are removed based on analysis of image spectra). The wavelength and the relative deep water depth of the region labeled a (Figure 4) are about 243.53 m and 65.0 m (Figure 2). Accordingly, the angular wave frequency ($\omega$) computed by the dispersion relationship (from Equation (6)) is about 0.48556 rad/s and wave period ($T$) is about 12.94 s. The ratio of water depth (65 m) and wavelength (243.53 m) is less than 0.3 that means these areas belong to shallow water domain. The island and coastal line is on the top of the study area where topography is relatively flat. In general, obtained wavelengths both from east to west and south to north direction are decreasing and the corresponding water depth is becoming lower. Therefore, problem of the 180° ambiguity of the swell wave direction is resolved and swell waves can be considered propagating toward the costal line.

Figure 5 shows wave tracking results. According to the surface swell patterns in the study area, wave field map is tracked by the fixed grid mode. Every arrow means a FFT-box. Its length means three times of wavelength and direction means the propagating direction of wave. Waves over or near the island are removed from the map by buffer polygon with distance 1.28 km from the red polygon as shown in Figure 4. The distance between the tracked FFT-boxes is set to 1600 m. The change in both wavelength and direction caused by wave shoaling and refraction are clearly visible in Figure 5.

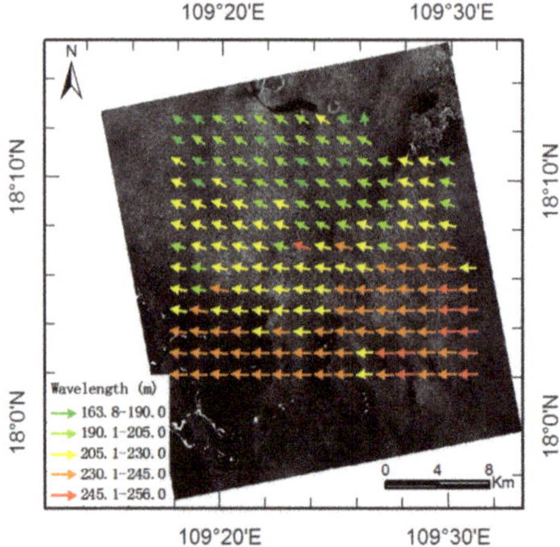


**Figure 5.** Wave fields map by fixed grid mode. Each arrow corresponds to a subscene of 1024 × 1024 pixels. The distance between two arrows is 1.6 km (320 pixels). Its length means three times of wavelength and direction means the propagating direction of swell waves.

The longest tracked wavelength ($L_{max}$) is 256 m as shown in Figure 5. Accordingly, the threshold minimum peak period ($T_{min}$ from Equation (8)) is 12.81 s. The obtained wave period (12.94 s) is greater than minimum peak period (12.81 s) that is also show the study area belongs to shallow water domain.

Figure 6 shows underwater topography results from the Freeman-Durden decomposition component of Bragg scattering image. Swell wave fields are tracked by the fixed grid mode which the distance between the tracked FFT-boxes with size of 1024 × 1024 is set to 150 m. The obtained swell wave fields of each FFT-box are sorted from smallest to largest and the last three are averaged as final wavelength. Shallow water depths are estimated from the averaged wavelengths and the estimated wave period (12.90 s) by the linear dispersion relationship (Equation (7)) based on the assumption of the presence of a single wave system in the SAR image. During the calculation, if water depth is greater than 65 m, it will be replaced by 65 m. The estimated water depths where absolute error of the estimated depth is more than 10 m (nearly 25% of the average water depth) are replaced with the value from the nearest location of the ENC as shown in Figure 2. By comparing Figure 2 with Figure 6, the estimated results have accordance well with water depth map from the ENC and have higher spatial resolution, that is, have more detailed underwater topography.

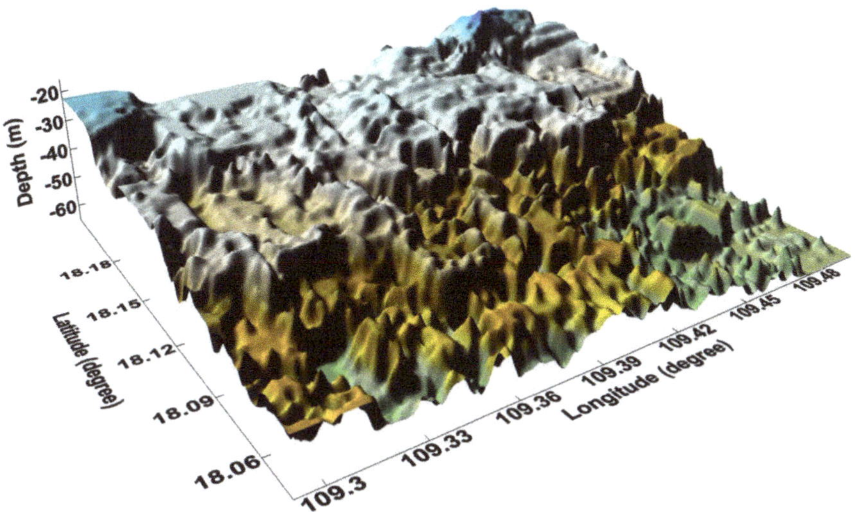


**Figure 6.** Three-dimensional of underwater topography with resolution of 150 m by 150 m. The estimated water depths where absolute error of the estimated depth is more than 10 m are replaced with the value from the nearest location of the ENC as shown in Figure 2.

## 3.2. Comparing with ENC

In order to validate the detected results, there are 46 pairs of the depth estimated from the Bragg scattering component image are selected according to the nearest neighbor distance of the low scale of ENC as shown in Figure 2. When wave period is set to 12.9 s, the average absolute error is about 3.93 m and the average relative error is about 9.73%. The scatter plot as shown in Figure 7 is presented to show the general agreement of the detected results and the water depths from low scale of ENC.

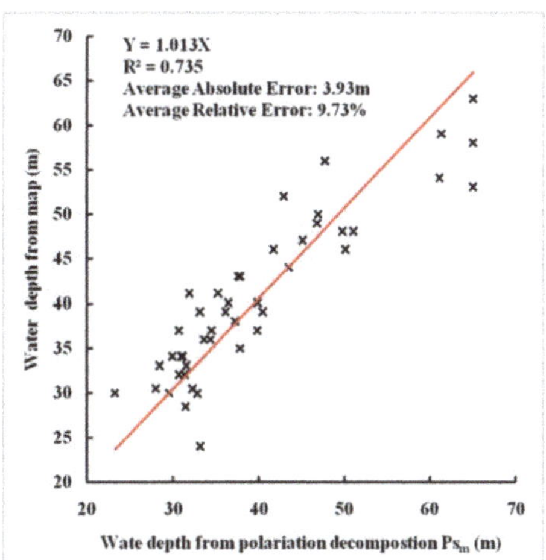


**Figure 7.** Scatter plot of the estimated results and water depths from low scale of ENC, for a wave period of 12.90 s.

For further evaluation, there are 305 pairs of water depth from local high scale of ENC in the blue polygon are also collected for comparison. For local comparison, the average absolute error is about 2.57 m and the average relative error is about 8.76%. Results show that the proposed algorithm is feasible and keep more detailed underwater topography. If water depths from ENC are not obsolete, the estimated water depth can be adjusted that will improve the underwater topography detection accuracy.

In order to evaluate whether some of SAR data pre-processing procedures and the introduced detected algorithm will affect the accuracy of the detected accuracy, speckle filtering, different pixel spacing, different subscene size of FFT box and different input images are used for comparing. In this section, wave period is set to 12.9 s and the distance between two FFT-boxes is set to 150 m unless they are specially explained. For smooth filtering, results of comparison between estimated water depths from SAR image and ENC are shown in Table 2. VVsm and HHsm means smoothed VV-polarized and HH-polarized SAR data with pixel spacing of 5 m respectively. Table 3 presents errors of estimated water depths with different pixel spacing (12.5 m, 10 m, 8 m, 5 m). For example, $VV_5$ means VV-polarized image with pixel spacing of 5 m. Table 4 presents errors of estimated water depths with different subscene size of VV-polarized image (256, 512, 1024) for FFT-box. Take for example, $VV_{256}$ means the size of FFT-box is 256. Table 5 shows comparison results from different input image. $VV_m$, $Ps_m$ and $Ys_m$ means wavelengths from VV-polarized image, surface scattering component image from Freeman-Durden decomposition and Yamaguchi decomposition are the average of the top three wavelengths respectively.

**Table 2.** Errors of estimated water depths from single polarization SAR image.

| Input Data | VV | VVsm | HH | HHsm |
|---|---|---|---|---|
| Low scale average absolute error | 4.37 m | 4.22 m | 5.03 m | 4.30 m |
| Low scale average relative error | 11.30% | 10.30% | 12.42% | 10.92% |
| Local high scale average absolute error | 3.41 m | 3.43 m | 3.95 m | 3.55 m |
| Local high scale average relative error | 11.43% | 11.53% | 13.32% | 12.63% |

**Table 3.** Errors of estimated water depths with different pixel spacing of VV-polarized image.

| Input Data | $VV_{12.5}$ | $VV_{10}$ | $VV_8$ | $VV_5$ |
|---|---|---|---|---|
| Low scale average absolute error | 5.43 m | 4.37 m | 4.54 m | 4.37 m |
| Low scale average relative error | 14.17% | 11.29% | 12.20% | 11.30% |
| Local high scale average absolute error | 3.79 m | 3.24 m | 3.43 m | 3.41 m |
| Local high scale average relative error | 12.32% | 10.99% | 11.65% | 11.43% |

**Table 4.** Errors of estimated water depths with different subscene size of VV-polarized image.

| Input Data | $VV_{256}$ | $VV_{512}$ | $VV_{1024}$ |
|---|---|---|---|
| Low scale average absolute error | 18.2 m | 9.80 m | 4.37 m |
| Low scale average relative error | 46.33% | 26.18% | 11.30% |
| Local high scale average absolute error | 17.80 m | 9.26 m | 3.41 m |
| Local high scale average relative error | 59.18% | 31.36% | 11.43% |

**Table 5.** Errors of estimated water depths with different input image.

| Input Data | HH | VV | Ps | $VV_m$ | $Ps_m$ | $Ys_m$ |
|---|---|---|---|---|---|---|
| Low scale average absolute error | 5.03 m | 4.37 m | 4.05 m | 4.45 m | 3.93 m | 3.95 m |
| Low scale average relative error | 12.42% | 11.30% | 9.84% | 11.53% | 9.73% | 9.82% |
| Local high scale average absolute error | 3.95 m | 3.41 m | 3.2 m | 2.98 m | 2.57 m | 2.46 m |
| Local high scale average relative error | 13.32% | 11.43% | 10.8% | 10.6% | 8.76% | 8.42% |

## 4. Discussion

A method is proposed which is based on swell patterns and dominant scattering component from polarization decomposition. The new method uses fully polarimetric SAR data for underwater topography detection indicates that it will improve the accuracy of detection. For underwater topography detection, the proposed method and SAR image intensity variations based method are not independent but complementary. Although both of them may be not precisely accurate, they can provide general information about depth changes, especially in coastal areas where underwater topography could not easily be detected by other means.

Comparing to conventional amplitude/intensity images, the proposed method is based on scattering component from polarization decomposition that aims at providing more information for applications. Although advances in the high-resolution SAR technology provide the capability of obtaining detailed target signatures, interpreting SAR images of some special targets or structures or estimating parameters are always being a challenge, especially for single polarization SAR. Experimental results as shown in Table 5 indicate that the dominant scattering mechanism based methodology is more effective than only using the single polarization SAR data for underwater topography detection. They also show single polarization SAR image can be used for shallow water depth estimation using linear dispersion relationship with some accuracy that confirm former studies on the swell patterns based method [4,7,18,19,21,22]. For single polarization SAR data, errors from VV-polarized (VV) SAR data are better than from HH-polarized SAR data, but all relative errors are above 10%. Cross-polarization (HV and VH) SAR data do not discuss here because the investigations have demonstrated that the cross-polarization radar backscatter is not sensitive to incidence angle, but is dependent on wind speed especially under high sea states [32,33]. The Freeman-Durden decomposition results as shown in Figure 1 shows that color red is dominant. It confirms that the surface scattering is the dominant backscatter mechanism with incidence angle between 20–60°, but not under all wind and wave states. Errors from the Bragg scattering component (Ps) by Freeman-Duren decomposition are better than from single polarization SAR data. For fully polarimetric SAR data, errors from the Bragg scattering component by processed wavelengths ($Ps_m$) is better than unprocessed wavelengths, and is also better than processed wavelength from VV-polarized SAR data ($VV_m$). The dominant scattering mechanism based method produces good results and reduces the average relative errors to less than 10% that indicates fully polarimetric SAR data can be used for underwater topography detection with good accuracy. In addition, the proposed method is simple but model-based and needs less prior information (even nothing) than the variations of the SAR image intensity based method, especially for relative flat underwater topography. In the test case, we only used one referenced water depth as input parameter, but large covered underwater topography with acceptable errors is obtained. Since ocean wave parameters (e.g., wave height) and other environmental parameters (e.g., wind speed) can be estimated from SAR image [6,31,33–35], the proposed algorithm can be developed in the future.

Results as shown in Table 2 imply that speckle filtering will improve the detected accuracy. Although errors of local high resolution from the filtered VV-polarized data is a little higher than other results, in general, errors from smoothed data (VVsm and HHsm) are better than initial data (HH and VV) that supports view of speckle filtering will improve the detected accuracy. For a given SAR data, different pixel spacing can affect the accuracy of wave detection. Nominal resolution of fine beam modes of RADARSAT-2 is 8 m and sampled pixel spacing of the collected SLC data is 4.73 m. Table 3 presents errors of estimated water depths with different pixel spacing. It shows that pixel spacing larger than 10 m will reduce the accuracy of water depth detection. Consequently, pixel spacing should be set near the nominal resolution. The scale of sea surface waves is usually between 100 m and 600 m [1]. If pixel spacing is set too small, it will take up more processing time and not improve the accuracy. Generally, the size of FFT-box is set to 512 or multiples of 512. Results as shown in Table 4 indicate that the larger the size, the higher the accuracy will get. When the size of FFT-box is changed from 1024 to 512, errors are more than doubled. Although errors from 512 are also better than from

256 that may keep some detailed underwater topography, but the averaged relative error larger than 20% is unacceptable. Meanwhile, the size of FFT-box should not be set too large that will take up processing time and reduce some detailed underwater topography.

The sea state over the study area as shown in Figure 3 is not high [36]. It is supposed that the significant wave height ($H$) is set to 1.0 m and wind speed (10 m over sea) is set to 5.0 m/s. For a given incidence angle (the near range, 32.353 deg), the cut-off wavelength $Lmin$ can be estimated from Equation (2) with slant range ($R$) of 919.8 km as 123.3 m. It is less than the minimum of the detectable wavelengths (163.8 m) that means the estimated results are plausible.

For ocean waves, under the Bragg scattering conditions, backscattered signals in general are from single bounce. The single bounce returns from the ocean surface possess the typical Bragg resonant scattering characteristic. Although single polarization amplitude/intensity images keep some swell patterns, the surface scattering component image keeps swell patterns more visible than them and the other two components (the double-bounce and volume scattering components). Results from Freeman-Durden three-component scattering decomposition indicate that the surface scattering component is the dominant scattering over the sea in our test area. However, Freeman-Durden decomposition contains two important assumptions (the assumed three-component scattering model is not always applicable and reflection symmetry) which limit its applicability [5]. Yamaguchi decomposition is used to check the applicability of the surface scattering component from Freeman-Durden decomposition [37]. The results confirm that the surface scattering is the dominant scattering and the surface scattering component image keeps the detailed swell patterns, and as shown in Table 5, indicate that the dominant scattering component based method is applicable in our test area, and both absolute error and relative error are nearly in the same error range. It should be noted that this study has examined only one scene fully polarimetric SAR data. In the future, the dominant scattering component based method will be tested and developed, and further research on underwater topography detection with fully polarimetric SAR data may be inspired.

## 5. Conclusions

In this paper, an underwater topography detection algorithm based on swell patterns is developed. In comparison to methods published previously, the proposed method depends on the Bragg scattering component which is the dominant scattering of sea surface in most common cases. One scene of fully polarimetric RADARSAT-2 SAR data covering the near shore water of Hainan Island is used in this investigation, and the ENC are used for comparing and analyzing. The average absolute error is within 4.0 m with the average relative error less than 10.0%. For local comparison, the average relative error is less than 9.0%. Due to full polarimetric system provided the most extensive multi-parameter ocean data, fully polarimetric SAR data can be used for feature extraction and information retrieval. Furthermore, the conventional SAR image intensity variations based method is of limited use under less current condition. Therefore, the proposed method can be used for underwater topography detection with less initial input parameters and better accuracy than single polarization SAR data. Although fully polarimetric SAR data have limited swath coverage, they provide unique information on sea surface scattering mechanisms that can be used for underwater topography detection and other applications. As GF-3 Satellite was launched successfully, we may acquire more fully polarimetric SAR data that can be used to verify and develop the proposed method in the future. The dominant scattering component based method can, potentially, be developed for further into fully operational algorithms. The improved method may be used for further underwater topographic mapping with higher temporal resolution and spatial resolution.

**Acknowledgments:** This study is jointly supported by the National Science Foundation of China (Grant No. 41431174), the National Key Research and Development Program of China (Grant No. 2016YFB0502504) and the National Science Foundation of China (Grant No. 61471358). The authors would like to thank the three anonymous reviewers and the editors for their comments and suggestions that improved the manuscript.

**Author Contributions:** Xiaolin Bian, Yun Shao, Wei Tian and Chunyan Zhang conceived and designed the research. Xiaolin Bian and Chunyan Zhang performed the experiments. Xiaolin Bian and Yun Shao analyzed the data and results. Wei Tian and Chunyan Zhang conducted validation. Shiang Wang, Xiaochen Wang and Zhixin Zhang contributed materials. Xiaolin Bian wrote the paper. All authors read and approved the final version of the manuscript.

**Conflicts of Interest:** The authors declare no conflict of interest.

## References

1. Jackson, C.R.; Apel, J.R. *Synthetic Aperture Radar Marine User's Manual*; NOAA: Washington, DC, USA, 2004.
2. Holland, T.K. Application of the linear dispersion relation with respect to depth inversion and remotely sensed imagery. *IEEE Trans. Geosci. Remote Sens.* **2001**, *39*, 2319. [CrossRef]
3. Calkoen, C.J.; Hesselmans, G.H.F.M.; Wensink, G.J.; Vogelzang, J. The bathymetry assessment system: Efficient depth mapping in shallow seas using radar images. *Int. J. Remote Sens.* **2001**, *22*, 2973–2998. [CrossRef]
4. Brusch, S.; Held, P.; Lehner, S.; Rosenthal, W.; Pleskachevsky, A. Underwater bottom topography in coastal areas from TerraSAR-X data. *Int. J. Remote Sens.* **2011**, *32*, 4527–4543. [CrossRef]
5. Lee, J.S.; Pottier, E. *Polarimetric Radar Imaging: From Basics to Applications*; CRC Press: London, UK, 2009.
6. Schuler, D.L.; Lee, J.S.; Kasilingam, D.; Pottier, E. Measurement of ocean surface slopes and wave spectra using polarimetric SAR image data. *Remote Sens. Environ.* **2004**, *91*, 198–211. [CrossRef]
7. Bian, X.; Shao, Y.; Tian, W.; Zhang, C. Estimation of shallow water depth using HJ-1C S-band SAR data. *J. Navig.* **2016**, *69*, 113–126. [CrossRef]
8. Fan, K.; Huang, W.; Lin, H.; Pan, J.; Fu, B.; Gu, Y. Shallow water depth retrieval from space-borne SAR imagery. *J. Oceanogr.* **2011**, *67*, 405–413. [CrossRef]
9. Fan, K.; Huang, W.; He, M.; Fu, B.; Zhang, B.; Chen, X. Depth inversion in coastal water based on SAR image of waves. *Chin. J. Oceanol. Limn.* **2008**, *26*, 434–439. [CrossRef]
10. Yang, J.; Zhang, J.; Meng, J. Underwater topography detection of Taiwan Shoal with SAR images. *Chin. J. Oceanol. Limn.* **2010**, *28*, 636–642. [CrossRef]
11. Li, X.; Yang, X.; Zheng, Q.; Pietrafesa, L.J.; Pichel, W.G.; Li, Z.; Li, X. Deep-water bathymetric features imaged by spaceborne SAR in the Gulf Stream region. *Geophys. Res. Lett.* **2010**, *37*, 96–104. [CrossRef]
12. Pleskachevsky, A.; Lehner, S.; Heege, T.; Mott, C. Synergy and fusion of optical and synthetic aperture radar satellite data for underwater topography estimation in coastal areas. *Ocean Dynam.* **2011**, *61*, 2099–2120. [CrossRef]
13. Wackerman, C.; Lyzenga, D.; Ericson, E.; Walker, D. Estimating Near-Shore Bathymetry Using SAR. In Proceedings of the 1998 IEEE International Geoscience and Remote Sensing Symposium, Seattle, WA, USA, 6–10 July 1998; pp. 1668–1670.
14. Alpers, W.; Ingo, H. A theory of the imaging mechanism of underwater bottom topography by real and synthetic aperture radar. *J. Geophys. Res. Oceans* **1984**, *89*, 10529–10546. [CrossRef]
15. Yang, J.; Zhang, J.; Meng, J. A detection model of underwater topography with a series of SAR images acquired at different time. *Acta Oceanol. Sin.* **2010**, *29*, 28–37. [CrossRef]
16. Li, X.; Li, C.; Xu, Q.; Pichel, W.G. Sea surface manifestation of along-tidal-channel underwater ridges imaged by SAR. *IEEE Trans. Geosci. Remote Sens.* **2009**, *47*, 2467–2477.
17. Bi, X.; Meng, J.; Yang, J.; Liu, Q. Application of SAR polarization information for detection of underwater topography. *J. Remot. Sens.* **2013**, *17*, 34–45. (In Chinese).
18. Monteiro, F. Advanced Bathymetry Retrieval from Swell Patterns in High-Resolution SAR Images. Master Thesis, University of Miami, Coral Gables, FL, USA, December 2013.
19. Mishra, M.K.; Ganguly, D.; Chauhan, P.; Ajai. Estimation of coastal bathymetry using RESAT-1 C-band microwave SAR data. *IEEE Geosci. Remote Sens. Lett.* **2014**, *11*, 671–675. [CrossRef]
20. Li, X.; Lehner, S.; Rosenthal, W. Investigation of ocean surface wave refraction using TerraSAR-X data. *IEEE Trans. Geosci. Remote Sens.* **2010**, *48*, 830–840.
21. Boccia, V. Linear dispersion relation and depth sensitivity to swell parameters: Application to synthetic aperture radar imaging and bathymetry. *Sci. World J.* **2015**, *2015*, 1–10. [CrossRef] [PubMed]

22. Boccia, V.; Renga, A.; Moccia, A.; Zoffoli, S. Tracking of coastal swell fields in SAR images for sea depth retrieval: Application ALOS L-band data. *IEEE J. Sele. Top. Appl. Earth Obs. Remote Sens.* **2015**, *8*, 3532–3540. [CrossRef]
23. Hasselmann, K.; Raney, R.K.; Plant, W.J.; Alpers, W.; Shuchman, R.A.; Lyzenga, D.R.; Rufenach, C.L.; Tucker, M.J. Theory of synthetic aperture radar ocean imaging: A MARSEN view. *J. Geophys. Res. Atmos.* **1985**, *90*, 4659–4686. [CrossRef]
24. Valenzuela, G.R. Theories for the interaction of electromagnetic and oceanic waves—A review. *Bound. Lay Meteorol.* **1978**, *13*, 61–85. [CrossRef]
25. Romeiser, R.; Alpers, W. An improved composite surface model for the radar backscattering cross section of the ocean surface: 2. Model response to surface roughness variations and the radar imaging of underwater bottom topography. *J. Geophys. Res. Atmos.* **1997**, *102*, 25251–25267. [CrossRef]
26. Ouchi, K. A theory on the distribution function of backscatter radar cross section from ocean waves of individual wavelength. *IEEE Trans. Geosci. Remote Sens.* **2000**, *38*, 811–822. [CrossRef]
27. Alpers, W.; Ross, D.B.; Rufenach, C.L. On the detectability of ocean surface waves by real and synthetic aperture radar. *J. Geophys. Res. Atmos.* **1981**, *86*, 6481–6498. [CrossRef]
28. Zheng, Q.; Li, L.; Guo, X.; Yong, G.; Zhu, D.; Li, C. SAR imaging and hydrodynamic analysis of ocean bottom topographic waves. *J.Geophys. Res. Oceans* **2006**, *111*. [CrossRef]
29. Freeman, A.; Durden, S.L. A three-component scattering model for polarimetric SAR data. *IEEE Trans. Geosci. Remote Sens.* **1998**, *36*, 963–973. [CrossRef]
30. Kirbya, J.T.; Dalrympleb, R.A. An approximate model for nonlinear dispersion in monochromatic wave propagation models. *Coast. Eng.* **1986**, *9*, 545–561. [CrossRef]
31. Collard, F.; Ardhuin, F.; Chapron, B. Extraction of coastal ocean wave fields from SAR images. *IEEE J. Oceanic Eng.* **2005**, *30*, 526–533. [CrossRef]
32. Tian, W.; Shao, Y.; Yuan, J.; Wang, S. An experiment for oil spill recognition using RADARSAT-2 image. In Proceedings of the 2010 IEEE International Geoscience and Remote Sensing Symposium, Honolulu, HA, USA, 25–30 July 2010; pp. 2761–2764.
33. Vachon, P.W.; Wolfe, J. C-band cross-polarization wind speed retrieval. *IEEE Geosci. Remote Sens. Lett.* **2011**, *8*, 456–459. [CrossRef]
34. Zhang, B.; Li, X.; Perrie, W.; He, Y. Synergistic measurements of ocean winds and waves from SAR. *J. Geophys. Res. Oceans* **2015**, *120*, 6164–6184. [CrossRef]
35. Shao, W.; Zhang, Z.; Li, X.; Li, H. Ocean wave parameters retrieval from Sentinel-1 SAR imagery. *Remote Sens.* **2016**, *8*. [CrossRef]
36. Zhang, B.; Perrie, W.; Vachon, P.W.; Li, X.; Pichel, W.G.; Guo, J.; He, Y. Ocean vector winds retrieval from c-band fully polarimetric SAR measurements. *IEEE Trans. Geosci. Remote Sens.* **2012**, *50*, 4252–4261. [CrossRef]
37. Yamaguchi, Y.; Sato, A.; Boerner, W.M.; Sato, R.; Yamada, H. Four-component scattering power decomposition with rotation of coherency matrix. *IEEE Trans. Geosci. Remote Sens.* **2011**, *49*, 2251–2258. [CrossRef]

Article

# Detection of Bivalve Beds on Exposed Intertidal Flats Using Polarimetric SAR Indicators

**Wensheng Wang [1,2,3], Martin Gade [3] and Xiaofeng Yang [1,*]**

[1] State Key Laboratory of Remote Sensing Science, Institute of Remote Sensing and Digital Earth, Chinese Academy of Sciences, Beijing 100101, China; wangws@radi.ac.cn

[2] University of Chinese Academy of Sciences, Beijing 100049, China

[3] Institut für Meereskunde, Universität Hamburg, 20146 Hamburg, Germany; martin.gade@uni-hamburg.de

* Correspondence: yangxf@radi.ac.cn; Tel.: +86-10-64806215

Received: 24 June 2017; Accepted: 10 October 2017; Published: 13 October 2017

**Abstract:** We propose new indicators for bivalve (oyster and mussel) beds on exposed intertidal flats, derived from dual-copolarization (HH + VV) TerraSAR-X, Radarsat-2, and ALOS-2 images of the German North Sea coast. Our analyses are based upon the Kennaugh element framework, and we show that different targets on exposed intertidal flats exhibit different radar backscattering characteristics, which manifest in different magnitudes of the Kennaugh elements. Namely, the inter-channel correlation's real ($K_3$) and imaginary ($K_7$) part can be used to distinguish bivalve beds from surrounding sandy sediments, and together with the polarimetric coefficient (i.e., the normalized differential polarization ratio, $K_0/K_4$) they can be used as indicators for bivalve beds using multi-frequency dual-copolarization SAR data. Our results show that continuous bivalve bed monitoring is possible using dual-copolarimetric SAR acquisitions at all radar wavelengths.

**Keywords:** bivalve beds; intertidal flats; Kennaugh elements; SAR; dual-polarization; multi-frequency

---

## 1. Introduction

Intertidal flats are coastal areas between the land and the open sea that are at risk due to climate change, sea level rise, marine pollution, and invasive species [1–3]. The German Wadden Sea is such a dynamic ecosystem, with a high economic and ecological value, and has been a UNESCO World Natural Heritage since 2009 [4]. Pacific oysters are an invasive species that has been rapidly spreading over large parts of the German Wadden Sea during the past decades, thereby having an impact on nutrient cycling, water filtration, and the entire coastal ecosystem [5]. Therefore, a frequent environmental monitoring of this entire area (approx. 4700 km$^2$) is not only important, but also mandatory. However, in-situ measurements in tidal flats face many constraints, because of the repetitive flooding and the shallow water depth [6], and the development of reliable and automated techniques for an accurate classification (especially for bivalve beds) and for an assessment of their changes are strongly required.

Remote sensing is already regarded as a useful tool for the monitoring of intertidal flats. There have been many studies utilizing optical remote sensing techniques to detect oyster/mussel habitats, but the use of those sensors is limited by daytime and cloud coverage [7–9]. Moreover, bivalve beds are often covered by brown algae, which may cause classification errors when using optical data. In addition, there exist difficulties in describing the spectral signatures of bivalves due to a considerable spectral variability.

Radar sensors such as Synthetic Aperture Radar (SAR) can overcome those restrictions because of their all-weather capabilities and independence of daylight and also due to their ability to extract detailed structural information. It was found that bivalve beds and salt marshes have specific SAR signatures and can thus be monitored using space-borne SAR sensors [10–13]. Those bivalves are exposed above the sediments and form a rough reef or bed structure (with sharp and jagged surfaces) and increase the surface roughness locally, which causes stronger radar backscattering signals, making them visible in SAR imagery [11,14,15]. Therefore, surface roughness parameters retrieved from radar backscattering models can be used to distinguish bivalves from surrounding areas [10,16,17]. However, this approach may work only for a limited range of smooth bare soils, while mudflats with intense benthic fauna can be misinterpreted [13,14]. In addition, sandy sediments with sharp rims and rippled surfaces may also cause strong signals in SAR imagery, and hence SAR intensity channels cannot be simply used to get satisfactory classification accuracies.

With the advent of high-performance polarimetric SAR sensors, studies have been conducted to distinguish bivalves from sediments using multiple polarization SAR data. Gade et al. [10] provided both multi-frequency and multi-temporal analyses to improve the identification of bivalve beds in the German Wadden Sea. Similarly, Dehouck et al. [18] used combined TerraSAR-X (TSX) and optical imagery of the French Arcachon Bay to detect mussels, salt marshes, and sandy sediments. Choe et al. [14] demonstrated that quad polarimetric (quad-pol) multi-frequency (C-band and L-band) SAR data can be used to detect oyster reefs, but also demonstrated that indicators derived from L-band data show no significant differences between oyster beds and surrounding areas. Further analyses of indicators based on L-band data are rare, and therefore, the present paper is to propose polarimetric SAR indicators for the detection of bivalve beds that work at all wavelengths.

Commonly applied polarimetric decompositions such as the Cloude-Pottier and Freeman-Durden decomposition have been used for intertidal flats studies [19–22]. However, these decompositions can only be applied to quad-pol SAR data, while operational SAR sensors currently in orbit (such as the German TerraSAR-X, the Italian COSMO-SkyMed, and the European Sentinel-1A/B) provide data in dual-copolarization or dual-polarization mode. Radarsat-2 and ALOS-2 may acquire quad-pol data. However, this is only at the cost of limited spatial resolution and areal coverage. Moreover, bivalve beds on exposed intertidal flats can only be monitored at low tide. The limited time-window requires larger areal coverage with fine resolution. Therefore, further research is needed on the capabilities of existing or newly acquired dual-copolarization (dual-copol; HH + VV) SAR data [23] for monitoring intertidal flats, especially for the detection and classification of sediments and bivalve beds.

The Normalized Kennaugh element framework was developed by Schmitt et al. [24] and provides a method that can be applied on multi-scale, multi-temporal, multi-polarized, multi-frequency, and multi-sensor SAR data in a consistent mathematical manner. The Kennaugh elements allow the interpretation of physical scattering mechanisms (even- and odd-bounce scattering) for dual-polarization data [25], and this decomposition framework may allow a successful inclusion of those data into the Wadden Sea monitoring. Common polarimetric decompositions are either coherent or incoherent. The coherent decompositions based on the Sinclair matrix prohibit multi-looking and require descriptions of distributed targets, while the incoherent decompositions based on either the covariance or the coherency matrix often require large look numbers to keep radiometric stability [23–25]. Therefore, the Kennaugh formulation appears to be the right balance due to the basic description of the backscattering process [24]. Schmitt et al. and Moser et al. [25,26] first applied Kennaugh elements into wetland monitoring using dual-copol SAR data of the Upper Rhine in Germany and Lac Bam in Burkina Faso, respectively. Gade et al. [11] showed that the polarization coefficient (i.e., the normalized differential polarization ratio, referred to as *PC*) can be used to infer indicators for bivalve beds and demonstrated with X-band data. In this paper, we follow on those works using, among others, TerraSAR-X (TSX), Radarsat-2 (RS2), and ALOS-2 (AL2) data, with the specific aim to propose indicators for bivalve beds that were derived from dual-copol SAR data acquired at all wavelengths (X-, C-, and L-band).

The test site in the German Wadden Sea and the dual-copol multi-frequency (X-, C-, and L-band) SAR datasets are introduced in the following Section. The Kennaugh element framework and polarization coefficient are described thereafter, along with a presentation of statistical parameters that were used for the detection of oyster and mussel beds. We then compare and discuss the results of three space-borne sensors (working at X-, C-, and L-band) using the available dual-copol data and finally draw some conclusions.

## 2. Materials and Methods

### 2.1. Test Site and SAR Data

The test site "Amrum" on the northern part of the German Wadden Sea (Figure 1) is located between the islands of Amrum and Föhr and was already subject to previous studies [11,12,27]. This region contains sandy and muddy sediments, vegetated areas, bivalve beds (mainly Pacific oysters and cockles, but also blue mussels) and seagrass meadows; hence it represents an area of a typical mixture of bivalve beds and sediments on exposed intertidal flats. In addition, the bivalve beds in the test area "Amrum" are quite stable and chosen for demonstrating the effectiveness of the following indicators. Blue mussels and Pacific oysters may form extensive bivalve beds on elevated intertidal flats that may also contain areas of bare sediments and water puddles.

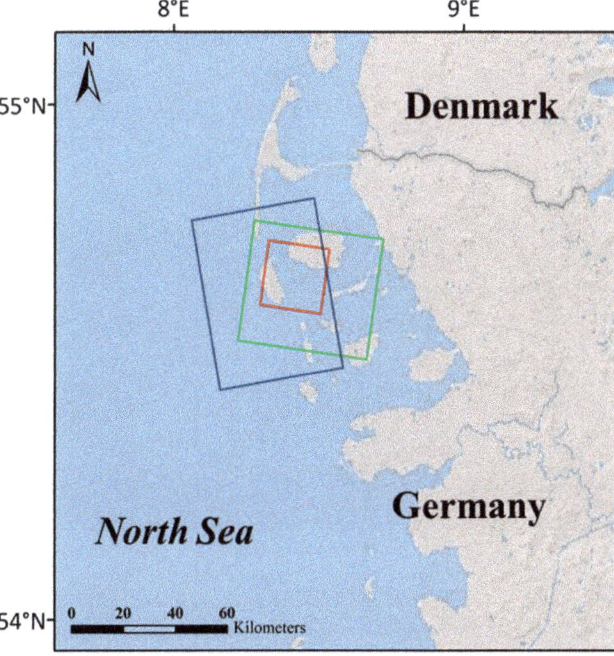


**Figure 1.** The test site "Amrum" in the German Wadden Sea. The solid rectangles delineate the spatial coverage of the TerraSAR-X (TSX) (red), Radarsat-2 (RS2) (green), and ALOS-2 (AL2) (blue) data (Figure 2).

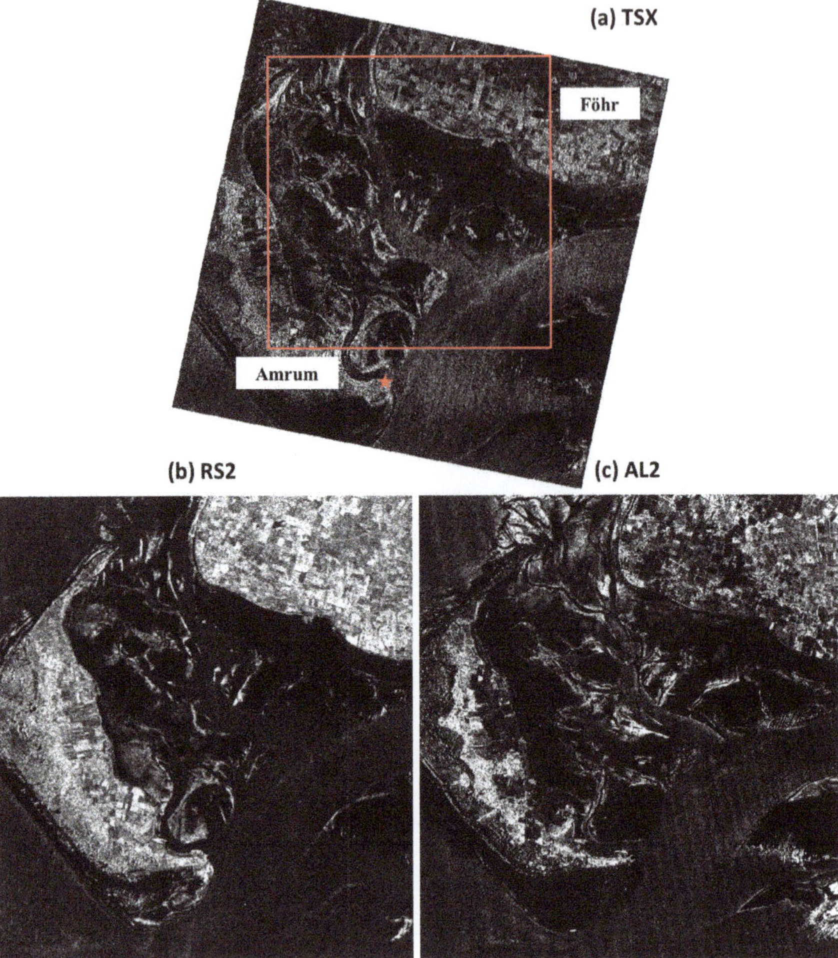


**Figure 2.** Synthetic Aperture Radar (SAR) images of the test site "Amrum" in VV-polarization. (**a**) TSX data acquired on 20 June 2016, at 05:50 UTC © DLR; (**b**) RS2 data on 24 December 2015, at 05:43 UTC, Radarsat-2 data and products © MacDonald, Dettwiler and Associates Ltd. 2015 – All Rights Reserved; (**c**) AL2 data on 29 February 2016, at 23:10 UTC © JAXA. The red rectangle marks the area of interest (8.97 km × 8.31 km); the red star marks the location of tide gauge "Wittdün".

Frequent monitoring of bivalve beds in the test site "Amrum" is conducted during field campaigns as well as based on optical and (single-polarization) SAR imagery [12]. Therefore, a larger number of SAR images was available for the present investigation. The solid rectangles in Figure 1 delineate the locations of the X-, C-, and L-band SAR data used in this study. VV-polarization SAR scenes acquired around low tide by TSX, RS2, and AL2 are shown in Figure 2 as examples. Herein we focused on an 8.97 km × 8.31 km area of interest within the "Amrum" test site, marked by the red rectangle in the upper panel of Figure 2. In all panels, bivalve beds on exposed sediments show up as bright patches. That is because the bivalves stick out of the sediments, increasing the surface roughness locally with higher radar backscattering. Some of the exposed sediments in the inner parts of the intertidal flats appear in dark, likely because of remnant water that effectively flattens the surface. Tidal creeks can be delineated because of their rather smooth surface causing lower radar backscatter and also due to the enhanced surface roughness of the sandy sediments at their rims. As seen in Figure 2, it is difficult to infer the spatial extent of bivalve beds simply using intensity channels. This motivated us to use polarimetric information available at all wavelengths (X-, C-, and L-band).

Three single-look complex (SLC) SAR images of the test site "Amrum" acquired around low tide by TSX, RS2, and AL2 were used for the analyses presented in this paper. The SAR images' pixel sizes range from 1 m × 1 m to 5 m × 5 m. Details of the used SAR images are summarized in Table 1, including the acquisition dates and times, the sensors and frequencies, polarizations, incidence angles, and water levels measured by the tide gauge "Wittdün" on the southern tip of Amrum. The TSX image was acquired in High-Resolution Spotlight mode (dual-copol: HH + VV), the RS2 image in Fine Quad-Pol mode (quad-pol: HH + HV + VH +VV), and the AL2 image in Strip-Map Ultra-Fine mode (quad-pol: HH + HV + VH + VV). However, as our analyses were carried out with the specific aim to demonstrate the potential of dual-copol SAR images for bivalve bed detection, we only used the co-polarization (HH and VV) channels of the RS2 and AL2 data. All SAR images were acquired close to the low tide (around 30 min) and at comparable incidence angles (between 30° and 40°). Validation data was obtained during field excursions in 2015.

**Table 1.** SAR acquisition dates, times, sensors, modes, and environmental conditions during image acquisitions. Polarizations: dual-copol HH + VV (D), quad-pol HH + HV + VH + VV (Q).

| Date/Time [UTC] | Sensor/Band | Polarization/Incidence Angle | Water Level [cm] | Time (UTC)/Water Level [cm] at Low Tide |
|---|---|---|---|---|
| 20 June 2016/05:50 | TSX/X Band | D/31.4° | −160 | 06:22/−171 |
| 24 December 2015/05:43 | RS2/C Band | Q/36.3° | −94 | 05:25/−103 |
| 29 February 2016/23:10 | AL2/L Band | Q/35.3° | −171 | 23:46/−176 |

For the validation of our results, we used in-situ data from monitoring campaigns in 2015 provided by the local National Park Agency. The left panel of Figure 3 shows the VV-pol TSX SAR image, with both Amrum on the left and Föhr on the upper right masked out for better orientation. Figure 3 also shows that many parts of the exposed intertidal flats appear brighter, although not all of them are bivalve beds.

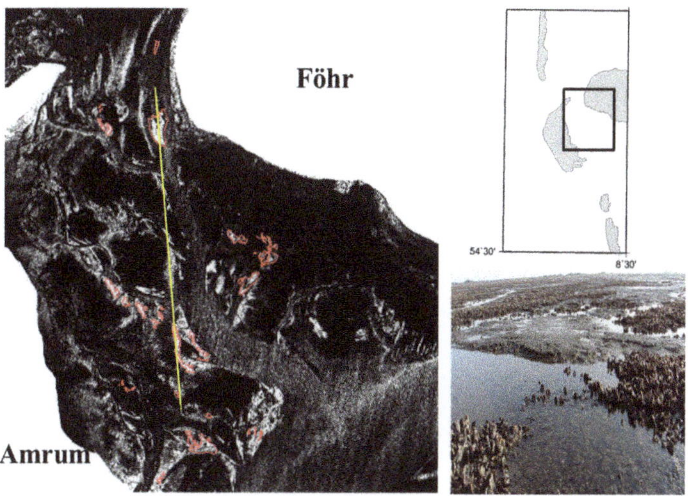


**Figure 3. Left**: TSX VV-pol SAR image of 20 June 2016. The red lines denote the spatial extent of bivalve beds, as obtained from field excursions, the yellow line is a transect through bare flats and tidal creeks, but also through bivalve beds; upper right: map showing the TSX SAR image location; **lower right**: photograph (S. Melchionna) of oyster beds showing the typical structures with elevated patches of oysters, water puddles, and sediments.

Figure 4 shows transects of the normalized radar cross section (NRCS) of the three SAR image along the yellow line in the right panel of Figure 3 (solid curves; blue: TSX; red: RS2; green: AL2). Also added as dashed-dotted lines are the respective noise floors at vertical polarization, the noise equivalent sigma zero (NESZ). Hereafter, we used only those data whose signal-to-noise ratio (SNR) was at least 2 dB. From Figure 4, we can see that at most places, the NRCS was well above the noise floor; however, we also note that at some places the backscattered signal dropped below the noise floor, particularly at L-band (green curve/line).

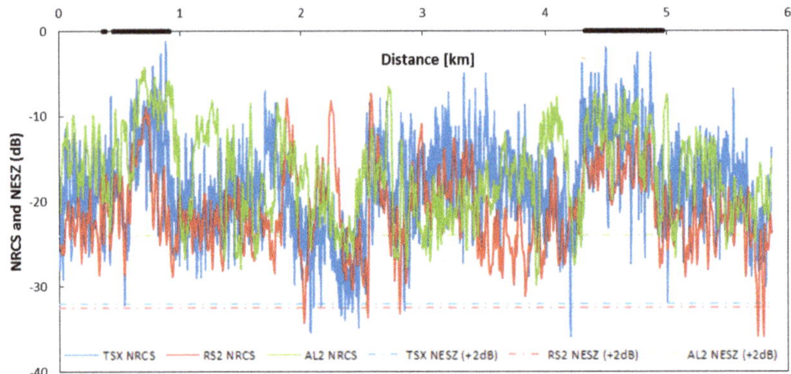


**Figure 4.** Comparisons of the normalized radar cross section (NRCS) (solid lines) and noise equivalent sigma zero (NESZ) (dashed-dotted lines) along the yellow transect line marked in Figure 3. Blue: TSX; red: RS2; green: AL2. The dashed-dotted lines were raised by 2 dB to better demonstrate where the SAR data's SNR was better than 2 dB. The thick horizontal bars on the abscissae indicate the locations of bivalve beds found in field excursions in 2015.

### 2.2. Kennaugh Element Framework

The single-look complex (SLC) TSX, RS2, and AL2 products were processed following a general approach that is based upon the elements extracted from the 4 × 4 Kennaugh matrix, [K], which is computed by linearly transforming the four-dimensional Stokes vector [24]:

$$[K] = \begin{bmatrix} K_0 & K_4 & K_5 & K_6 \\ K_4 & K_1 & K_9 & K_8 \\ K_5 & K_9 & K_2 & K_7 \\ K_6 & K_8 & K_7 & K_3 \end{bmatrix} \quad (1)$$

The Kennaugh element framework has been demonstrated to be applicable to dual- as well as quad-pol data of any wavelengths. In this study, we used dual-copol (HH + VV) TSX data, from which only the Kennaugh elements $K_0$, $K_3$, $K_4$, and $K_7$, can be inferred as follows [24]:

$$K_0 = \frac{1}{2}\{|S_{HH}|^2 + |S_{VV}|^2\} \quad (2)$$

$$K_3 = -Re\{S_{HH}S_{VV}^*\} \quad (3)$$

$$K_4 = \frac{1}{2}\{|S_{HH}|^2 - |S_{VV}|^2\} \quad (4)$$

$$K_7 = Im\{S_{HH}S_{VV}^*\} \quad (5)$$

where $|S_{pp}|^2$ are the image intensities at polarization *pp*, the asterisk (*) denotes the complex conjugate, and the factor $\frac{1}{2}$ appears for symmetry reasons. $K_0$ reflects the total intensity of both HH and VV layers;

$K_3$ is the difference between even- and odd-bounce scattering, where large values indicate a stronger even-bounce scattering than odd-bounce scattering, and vice versa [24]; $K_4$ is the difference between the HH and VV intensities; $K_7$ is the phase shift between even- and odd-bounce scattering. In the case of a dual-pol system (one co-pol and one cross-pol channel) Equations (2) to (5) correspond exactly to the well-known definition of a Stokes vector.

Following [24], the Kennaugh element layers were subsequently multi-looked, calibrated, and ground-range projected (UTM Zone 32, 1 m pixel spacing). These four Kennaugh elements at all wavelengths are shown in Figure 5. In this paper, we use the normalized Kennaugh elements, divided by the total intensity $K_0$, in order to derive nearly normally distributed values per Kennaugh element and physically comparable measurements. Here, we choose the normalized $K_3$ and $K_7$, for detecting oysters and mussels.

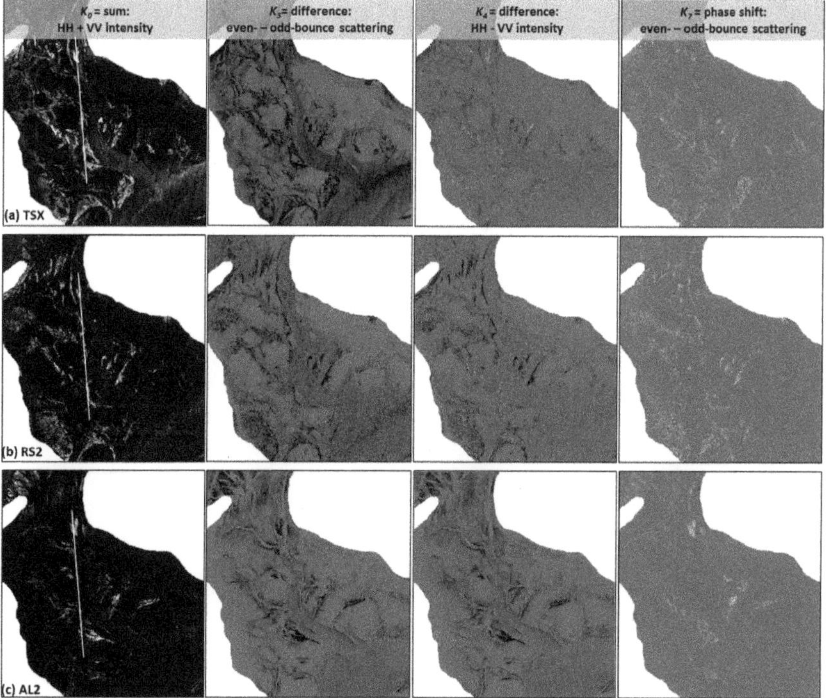


**Figure 5.** The four Kennaugh elements derived from dual-copol SAR data. Columns (from left to right): $K_0$ (sum of HH and VV intensities), $K_3$ (difference between even- and odd-bounce scattering), $K_4$ (difference of HH and VV intensities), $K_7$ (phase shift between even- and odd-bounce scattering); rows (from top to bottom): (**a**) TSX (20 June 2016), (**b**) RS2 (24 December 2015), (**c**) AL2 (29 February 2016).

Moreover, we use the polarization coefficient (*PC*), which was recently used by Gade et al. [11] as a promising parameter to infer indicators for bivalve beds from X-band dual-copol SAR images. *PC* is defined as

$$PC = \frac{|S_{HH}|^2 - |S_{VV}|^2}{|S_{HH}|^2 + |S_{VV}|^2} = \frac{K_4}{K_0} \tag{6}$$

and is identical to the normalized Kennaugh element of $K_4$.

Where the radar backscattering at horizontal polarization strongly dominates over that at vertical polarization *PC* approaches +1, and $-1$ in the opposite case. *PC* is close to 0 where the radar backscatter at both polarizations is similar. In this paper, we further verified the use of *PC* with respect to its effectiveness in detecting bivalve beds not only using X-band data as previous works, but also using data at longer radar wavelengths (C- and L-band).

## 3. Results

The new SAR decomposition framework described above was applied to the three SAR data sets of the test site "Amrum" in two case studies, the first of which were analyses of the real ($K_3$) and imaginary ($K_7$) parts of the inter-channel correlations, and the second analyses of *PC* using longer radar wavelengths (C- and L-band). In both case studies, we focused on potential indicators of bivalve beds.

### 3.1. New Indicators from Single Kennaugh Elements

The dielectric constant of bivalve shells (with lower moisture contents) is lower than that of the surrounding sediments, which would result in a lower radar backscatter from bivalve beds. However, since the bivalves are sticking out of sediments, they increase the surface roughness strongly, which in turn causes an overall stronger radar backscatter. And since they are oriented heterogeneously, this increase in radar backscatter depends on the radar polarization, which is the main factor that causes different polarimetric scattering signatures between bivalves and mudflats.

Along the transect line added to Figures 3 and 5 (from north to south) we calculated profiles of the normalized Kennaugh elements $K_3$ and $K_7$. Figure 6 shows in blue the running mean, $\mu$, of a moving 11-pixel window, in red the corresponding running standard deviation, $\sigma$, for $K_3$ (left column) and $K_7$ (right column) and for the TSX, RS2, and AL2 data (from top to bottom). In addition, we added in green the difference $D$ of both,

$$D_i = \mu_i - \sigma_i \, ; \, i = 3;7 \quad (7)$$

Thick horizontal bars on the abscissae in Figure 6 denote the locations of bivalve beds, as found during field campaigns in 2015.

The panels in the left column ($K_3$) of Figure 6 clearly show that in all radar bands both the (running) mean and (running) standard deviation are increased in the bivalve beds (between 0 km and 1 km, and between 4 km and 5 km), but also in the exposed sand flats in between. This effect is most pronounced in X-band (TSX, upper row) and weakens with increasing radar wavelength (middle row: RS2/C-band; lower row: AL2/L-band). However, we also note that the bivalve beds are the only areas where the standard deviation exceeds the mean value, which results in a negative difference $D$ (green curves). This effect can be used to infer indicators for bivalve beds.

The peaks of the mean values indicate that both oyster/mussel beds and the sandy rims of the intertidal flats show stronger even-bounce backscatter components than odd-bounce, because of their rougher surfaces. This increase in $K_3$ can be caused by either an increase in the even-bounce scattering or by a decrease in the odd-bounce scattering. In contrast, lower $K_3$ values in the surrounding areas, consisting of mudflats and tidal channels, indicate much smoother surfaces. Apparently, the heterogeneous surface structure of bivalves causes a strong spatial variability of even- and odd-bounce backscattering and therefore higher standard deviations, which in turn results in negative differences, $D_3$.

The Kennaugh element $K_7$ holds complementary correlation information, the phase shift between even- and odd-bounce backscattering. The corresponding profiles in Figure 6 (right column) show that the phase differences inside the bivalve beds are always larger, and always show strong variations, so that the running standard deviation always exceeds the running mean. Therefore, the difference of both, $D_7$, is always negative, but extreme values are only found in the bivalve beds. The rougher surfaces of bivalves cause diffuse radar backscattering, resulting in strong variations of the phase differences and, therefore, in standard deviations exceeding the mean values. Similar to the profiles in the left column of Figure 6, this effect weakens with increasing wavelength, i.e., from top to bottom. However, since areas of extremely negative "running differences" $D_7$ correspond well with the validation data, we conclude that this difference can also be used as indicator for bivalve beds.

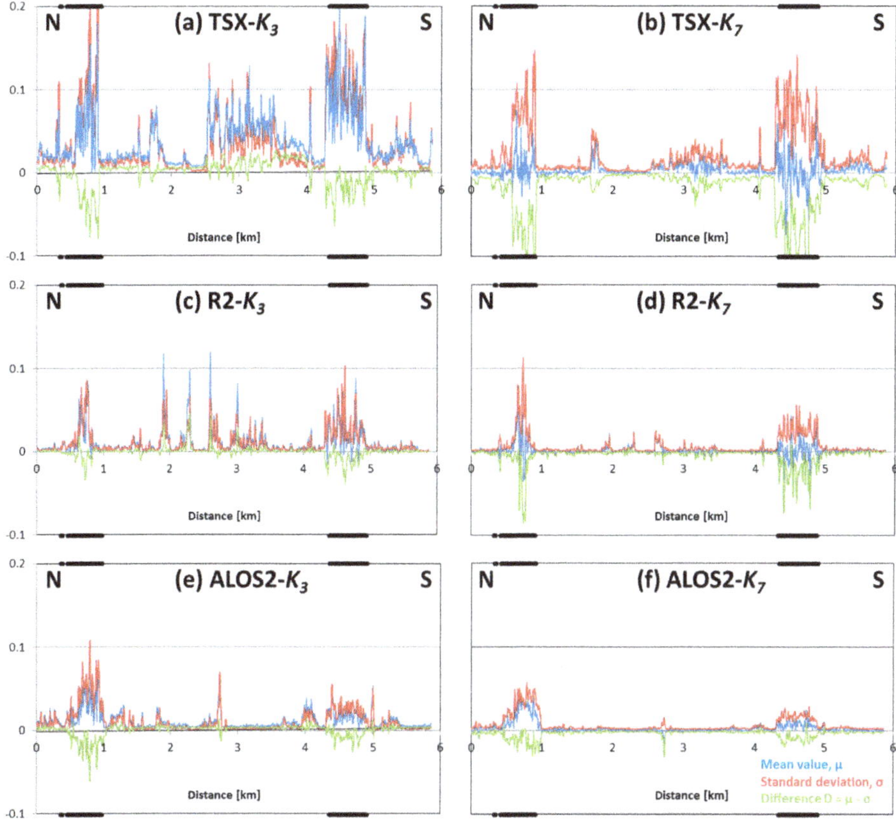


**Figure 6.** Profiles along the transect line included in Figures 3 and 5 from north (N) to south (S). Blue: running mean; red: running standard deviation; green: difference of both; each calculated for $K_3$ (left column) and $K_7$ (right column). Upper row: X-band; middle row: C-band; lower row: L-band. The thick horizontal bars on the top and bottom frames indicate the locations of bivalve beds found in field excursions in 2015.

The profiles for both Kennaugh elements, $K_3$ and $K_7$ (left and right column of Figure 6, respectively), show a strong dependence on radar wavelength: with increasing wavelength (i.e., from top to bottom in Figure 6), the curves become flatter, i.e., the difference between even- and odd-bounce backscattering becomes smaller. One possible reason lies in the size of oysters, which are the prevailing species in those beds: depending on their age, the oysters may become 30 cm long, with the majority being below 20 cm in length. The roughness scale of those beds, therefore, is on the order of a decimeter, and the beds' surfaces are much rougher for an X-band sensor (wavelength 3 cm) than for an L-band sensor (25 cm). This also results in smaller differences in the polarimetric backscattering, i.e., in smaller differences between even- and odd-bounce backscattering. Another, though less important, reason may lie in the season, in which the SAR data were acquired: the TSX data were acquired in early summer (June), during the vegetation period and growing season of the bivalves, while both the C- and L-band data were acquired in winter (December and February, respectively), when storm and high water events are more frequent.

In order to generate maps of bivalve bed indicators we calculated the running mean and standard deviation of a moving 11 pixels × 11 pixels window, and consequently the differences $D_3$ (Figure 7) and $D_7$ (Figure 8). For $D_3$ we set 0 as threshold for the bivalve indicators (i.e., we marked all negative differences $D_3$), and we arbitrarily chose 0.01 to further discriminate between sediments (between 0 and 0.01) and tidal channels and creeks (above 0.01). The results for the three datasets, TSX, RS2, and AL2, are shown in Figure 7, where the blue, orange, and green colors denote bivalve beds, exposed sediments, and tidal channels and creeks, respectively. A comparison with the validation data (red lines in Figure 3) immediately reveals that the difference $D_3$ provides an indicator for bivalve beds on exposed intertidal flats. Moreover, previous studies have also demonstrated that the use of multi-polarization SAR images have some additional potential for the monitoring of sediment surfaces on intertidal flats using SAR sensors [6,11,18,22]. With the given color coding, areas of water coverage, mainly the tidal channels, appears in orange colors, whereas the open sediment flats appear in green and yellow. More research is still required to analyze to what extent this parameter can be used in this respect.

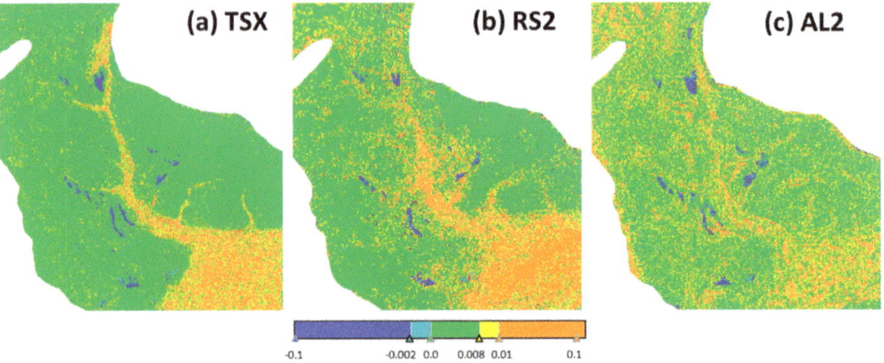


**Figure 7.** Difference $D_3$ of the running mean and standard deviation of Kennaugh element $K_3$ derived from (**a**) TSX X-band data, (**b**) RS2 C-band data, (**c**) AL2 L-band data, each calculated for a moving window of size 11 pixels × 11 pixels.

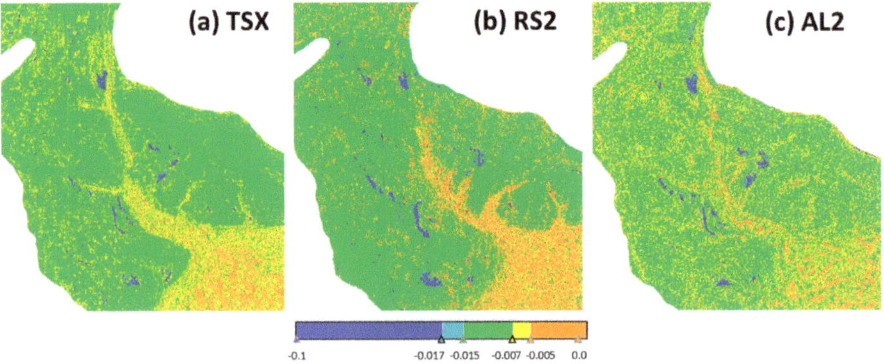


**Figure 8.** Same as Figure 7, but for the Difference $D_7$.

The same analyses were also performed with the difference $D_7$ (Figure 8), with the thresholds set to $-0.015$ (bivalves–sediment) and $-0.005$ (sediment–tidal creeks and channels). The results look very similar to those obtained for $D_3$ (Figure 7).

Figures 7 and 8 demonstrate that the best correlation with the in-situ data is obtained when X-band data are used, and the indicators for bivalves are derived using the Difference $D_3$ (left Panel of Figure 7). This is in line with our earlier observation, that the highest $K_3$ values of the running mean and standard deviation were found at X-band (upper left panel of Figure 6). The detection accuracy was derived as the ratio of the number of pixels correctly assigned to bivalve beds and the total number of pixels in bivalve beds, as given in the validation data from monitoring campaigns. The accuracies ranging from 84.70% to 88.87% for $D_3$ and 81.02% to 85.13% for $D_7$, are given in Table 2. Moreover, a discrimination of exposed sediments (green) and tidal creeks and channels (orange) also appears to be possible using our approach, particularly using X-band data and the difference $D_3$. The most homogeneous backscattering conditions are found in water-covered areas, i.e., in the tidal creeks and channels, and as a result, the difference of tidal channels and exposed sediment flats between the running mean and standard deviation is maximum in those areas. Apparently, this discrimination works well at short radar wavelengths (X-band), although we also note that different water levels (Table 1) may have caused the observed differences.

**Table 2.** Detection accuracies of indicators for bivalve beds at all wavelengths (X-, C-, and L-band).

| Indicator | Detection Accuracy (%) | | |
|---|---|---|---|
| | **TSX** | **RS2** | **AL2** |
| $D_3$ | 88.87 | 86.35 | 84.70 |
| $D_7$ | 85.13 | 83.72 | 81.02 |
| $P$ | 87.72 | 85.59 | 84.72 |

### *3.2. Indicators from Polarization Coefficients*

The polarization coefficient (*PC*) was used by Gade et al. [11,28] as indicator for bivalve beds on exposed intertidal flats. As the abrupt end of the ALOS-1 mission in 2011, we were only able to demonstrate *PC* with TSX X-band data. Here, we build up on those previous works using SAR data acquired at longer wavelengths (C- and L-band). Following [11] the product *P* is defined as

$$P = |\mu_{PC}| \cdot \sigma_{PC} \tag{8}$$

where $|\mu_{PC}|$ and $\sigma_{PC}$ are the absolute value of the running mean and standard deviation, respectively, of the polarization coefficient, *PC*. Because of their increased surface roughness, bivalve beds always cause an increased radar backscatter at both HH and VV polarizations, resulting in both low mean values and low standard deviations, and eventually in low values of the product *P* [11]. Similar to the above analyses, we used a running window of size 11 pixels $\times$ 11 pixels to calculate the product *P*, and the results are shown in Figure 9.

Figure 9 shows that the spatial range of the bivalve beds coincides well with the validation data of the field campaigns (marked in red). We note that the high detection accuracies for the Product P (84.72% to 87.72%) indicate that the use of dual co-pol SAR images has good potential for the monitoring of bivalves even when SAR data acquired at longer wavelengths (L-band) are used. Similar to the above findings, we found that the ability to distinguish sediments from tidal creeks and channels decreases with increasing radar wavelength. It should also be noted that surface scattering is the dominant scattering mechanism in most areas of exposed intertidal flats and the adjacent tidal creeks and channels. As the radar wavelength increases, these surface scatters tend to represent a relatively smoother effect, which can also explain this phenomenon. Moreover, images acquired at incidence angles between 31° and 37° are well suited for detection of bivalve beds because of the stronger radar contrast between the bivalve beds and their surroundings.

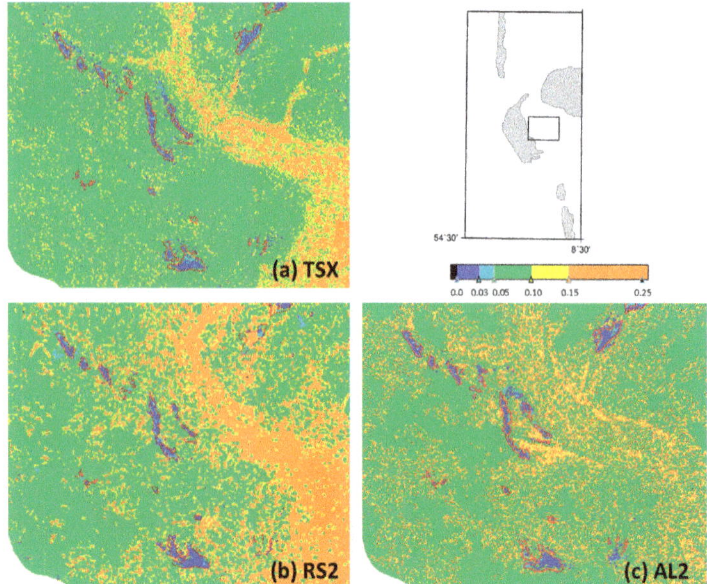


**Figure 9.** Product $P$ of the absolute mean and standard deviation, each calculated for a running 11 pixels × 11 pixels window using (**a**) TSX X-band data [11]; (**b**) RS2 C-band data; (**c**) AL2 L-band data. The borders of bivalve beds found during field campaigns are inserted as red lines, and the map on the upper right shows the location of the area of interest.

## 4. Discussion

Our studies indicate that the use of the Kennaugh element framework has great potential for the detection of bivalve beds on exposed intertidal flats. The spatially complex and rough surface structures of bivalve (oyster/mussel) beds on intertidal flats can be clearly defined through their unique polarimetric signatures, so that they can be discriminated from the surrounding bare sediments. A major advantage of the proposed indicators is that a series of single-acquisition SAR data can be used for a frequent monitoring of intertidal flats on the German North Sea coast, thereby providing valuable input for existing classification schemes that is independent of daylight and weather conditions. The detection of bivalve beds based on the Kennaugh element framework works best using X-band SAR data, but even long radar wavelengths (L-band) can be used to gain reliable results, and thus complement the L-band's inability to distinguish bivalves from surrounding areas in previous studies.

The Kennaugh element framework appears to be superior over other decomposition schemes on quad-pol SAR data and thus, improves the monitoring capabilities in the German Wadden Sea, particularly during the short periods of low tide. E.g., decompositions based on eigenvectors are susceptible to mixed results in very noisy decomposition parameters and thus, require very high look numbers to overcome this problem. Decompositions applied to interpret the components with the help of physical mechanisms based on radar backscattering models potentially also impair the description of natural targets, and are always limited to certain wavelength regimes. The numerical models, in turn, can rare suffice our requirements in wide range of smooth bare soils in the German Wadden Sea, especially targets with vegetation coverage, as well as our existing or newly acquired dual-copol SAR datasets. In this study, the relationship between even- and odd-bounce scattering events helps the detection of bivalves as expected, especially the complementary phase shift between even- and odd-bounce scattering ($K_7$). It was physically interpreted herein and demonstrated the ability to aid the discriminability between bivalve beds and surrounding sediments. Therefore, the Kennaugh decomposition theory has great advantages in describing the scattering features of sediments and habitats on exposed intertidal flats and may help in fully understanding the radar backscattering mechanisms on those flats. This Kennaugh-based approach should be transferable to any other regions as long as the data acquired at low tide, likewise to any other available polarimetric SAR sensors in orbit (such as COSMO-SkyMed and Sentinel-1A/B). This transferable to our available SAR datasets in terms of different wavelengths and characteristics is subject to ongoing research. This is in line with the idea exploited in this paper and could be further extended to other intertidal flats, such as the test sites around the islands of Spiekeroog and Norderney in the German Wadden Sea.

We also note that our results may help in improving the results from field campaigns: not only were our thresholds for the discrimination of bivalve beds, sediments, and tidal creeks and channels chosen arbitrarily, but the borders of the bivalve beds measured in-situ are always subjectively identified and may not reflect the very density threshold, from which bivalves may be detectable from space. In addition, our results also provide predictive guidance for the future field campaigns, based on multi-frequency SAR data and on multi-polarization modes. In this way, frequent monitoring of targets on exposed intertidal flats can be conducted combining the rapid imaging of SAR instruments with field investigations. It should be mentioned that dedicated field campaigns can be conducted only once or twice per year because of the reduced accessibility of these areas. In addition, muddy sediments and tidal channels may impede field surveys of larger areas or their access from land. Thus, such manpower- and time-consuming field campaigns cannot be conducted frequently, and this is also one of our motivations to monitor habitats using remote sensing techniques.

Any discrimination between different bivalve species (e.g., between blue mussels and oysters, or between cockles and oysters) based on polarimetric SAR images is not (yet) possible and requires further analyses. Therefore, further research is still needed to determine to what extent our new parameters can be used for such purposes. In addition, different water levels and seasonal changes can influence the backscattering characteristics of bare sediments [29] and bivalve beds. Therefore, future work will also include the sensitivity of Kennaugh elements to water levels and seasonal changes, e.g., the coverage by brown algae or Fucus. Future studies will be performed using dual-polarization SAR data (with one co-polarization and one cross-polarization channel), as usually acquired in current (Sentinel-1A/B) and future (RCM and SAOCOM) space borne SAR missions.

## 5. Conclusions

We have introduced new polarimetric SAR indicators for bivalve beds on exposed intertidal flats that are based on dual-copol SAR data. The proposed indicators for bivalve beds consist of the normalized real ($K_3$) and imaginary ($K_7$) parts of inter-channel correlations, as well as on the polarization coefficient, i.e., the normalized differential polarization ratio, $PC = K_4/K_0$. Comparing with the in-situ data from the field campaigns, we conclude that the Difference $D_3$ and $D_7$, as well as Product P, are useful indicators for extracting backscattering characteristics of bivalve beds, and can provide frequent and accurate mapping of habitats on intertidal flats.

Using SAR data from different sensors (working at different radar bands) the locations of the newly applied indicators always coincided with those where bivalve beds were encountered during field campaigns. Therefore, we also conclude that bivalve (oyster/mussel) beds can be detected at all radar wavelengths (X-, C-, and L-band), although we note that best results were obtained using X-band (TSX) data. In addition, the proposed indicators work well using images acquired at incidence angles between 31° and 37°.

Our proposed indicators allow the dual-copol SAR datasets for the detection of bivalve beds, exploiting the values of the existing or newly acquired datasets from the SAR sensors in orbit to the largest extent and keeping fine spatial resolution and large areal coverage. Moreover, the fact that these indicators are based on single-acquisition SAR data makes it superior to others based on series of SAR images, particularly for the monitoring of intertidal flats, a highly dynamic environment. Therefore, the results of our polarimetric analyses, as presented herein, have great potential to act as valuable input for existing classification systems.

**Acknowledgments:** The authors are grateful to Jörn Kohlus and Kai Eskildsen of the National Park Agency of the Schleswig-Holsteinean Wadden Sea National Park, who provided a-priori knowledge and in-situ data. One of the authors (W.W.) performed this research is in the frame of a fellowship of the German Academic Exchange Service (DAAD). Tide gauge data were provided by the Waterways and Shipping Board (Wasser- und Schifffahrtsamt, WSA) Tönning. ALOS-2, Radarsat-2, and TerraSAR-X data were provided by JAXA, CSA, and DLR, respectively, under contract RA6-3200 and 5077/OCE0994. RADARSAT is an official mark of the Canadian Space Agency. This work was also supported in part by the National Key R & D Program of China under Grant 2017YFB0502803 and in part by the Student President Foundation of the Institute of Remote Sensing and Digital Earth under Grant Y6SY1400CX.

**Author Contributions:** W. Wang, M. Gade, and X. Yang conceived and designed the data analyses; W. Wang mainly performed the analyses and interpretation of the results, and wrote most sections of this paper; M. Gade acquired the data, participated in analyses and interpretation, and contributed in writing and revising the paper. X. Yang aided with revising the manuscript. The sequence of authors reflects their level of contribution.

**Conflicts of Interest:** The authors declare no conflict of interest.

## References

1. Troost, K. Causes and effects of a highly successful marine invasion: Case-study of the introduced Pacific oyster Crassostrea gigas in continental NW European estuaries. *J. Sea Res.* **2010**, *64*, 145–165. [CrossRef]
2. Lee, Y.K.; Park, J.W.; Choi, J.K.; Oh, Y.; Won, J.S. Potential uses of TerraSAR-X for mapping herbaceous halophytes over salt marsh and tidal flats. *Estuar. Coast. Shelf Sci.* **2012**, *115*, 366–376. [CrossRef]
3. Klemas, V. Remote sensing of emergent and submerged wetlands: An overview. *Int. J. Remote Sens.* **2013**, *34*, 6286–6320. [CrossRef]
4. Van der Wal, D.; Herman, P.M.J. Regression-based synergy of optical, shortwave infrared and microwave remote sensing for monitoring the grain-size of intertidal sediments. *Remote Sens. Environ.* **2007**, *111*, 89–106. [CrossRef]
5. Regniers, O.; Bombrun, L.; Ilea, I.; Lafon, V.; Germain, C. Classification of oyster habitats by combining wavelet-based texture features and polarimetric SAR descriptors. In Proceedings of the IEEE International Geoscience and Remote Sensing Symposium, Milan, Italy, 26–31 July 2015.
6. Gade, M.; Alpers, W.; Melsheimer, C.; Tanck, G. Classification of sediments on exposed tidal flats in the German Bight using multi-frequency radar data. *Remote Sens. Environ.* **2008**, *112*, 1603–1613. [CrossRef]
7. Liu, S.; Li, Z.; Yang, X.; William, G.P.; Yu, Y.; Zheng, Q.; Li, X. Atmospheric frontal gravity waves observed in satellite SAR images of the Bohai Sea and Huanghai Sea. *Acta Oceanol. Sin.* **2010**, *29*, 35–43. [CrossRef]
8. Grizzle, R.E.; Adams, J.R.; Walters, L. Historical changes in intertidal oyster (*Crassostrea Virginica*) reefs in a Florida Lagoon potentially related to boating activities. *J. Shellfish Res.* **2002**, *21*, 749–756.
9. Zheng, Q.; Holt, B.; Li, X.; Liu, X.; Zhao, Q.; Yuan, Y.; Yang, X. Deep-water seamount wakes on SEASAT SAR image in the Gulf Stream region. *Geophys. Res. Lett.* **2012**, *39*, 11604. [CrossRef]
10. Gade, M.; Melchionna, S.; Stelzer, K.; Kohlus, J. Multi-frequency SAR data help improving the monitoring of intertidal flats on the German North Sea coast. *Estuar. Coast. Shelf Sci.* **2014**, *140*, 32–42. [CrossRef]
11. Gade, M.; Melchionna, S. Joint use of multiple Synthetic Aperture Radar imagery for the detection of bivalve beds and morphological changes on intertidal flats. *Estuar. Coast. Shelf Sci.* **2016**, *171*, 1–10. [CrossRef]
12. Müller, G.; Stelzer, K.; Smollich, S.; Gade, M.; Adolph, W.; Melchionna, S.; Kemme, L.; Geißler, J.; Millat, G.; Reimers, H.C. Remotely sensing the German Wadden Sea—A new approach to address national and international environmental legislation. *Environ. Monit. Assess.* **2016**, *188*, 595. [CrossRef] [PubMed]
13. Choe, B.H.; Kim, D.J.; Hwang, J.H.; Oh, Y.; Moon, W.M. Oyster reef signature in tidal flats detected by multi-frequency polarimetric SAR data. In Proceedings of the IEEE International Geoscience and Remote Sensing Symposium, Vancouver, BC, Canada, 24–29 July 2011.
14. Choe, B.H.; Kim, D.J.; Hwang, J.H.; Oh, Y.; Moon, W.M. Detection of oyster habitat in tidal flats using multi-frequency polarimetric SAR data. *Estuar. Coast. Shelf Sci.* **2012**, *97*, 28–37. [CrossRef]
15. Baghdadi, N.; Zribi, M.; Loumagne, C.; Ansart, P.; Anguela, T.P. Analysis of TerraSAR-X data and their sensitivity to soil surface parameters over bare agricultural fields. *Remote Sens. Environ.* **2008**, *112*, 4370–4379. [CrossRef]

16. Park, S.E.; Moon, W.M.; Kim, D.J. Estimation of surface roughness parameter in intertidal mudflat using airborne polarimetric SAR data. *IEEE Trans. Geosci. Remote Sens.* **2009**, *47*, 1022–1031. [CrossRef]
17. Van der Wal, D.; Herman, P.M.J.; van den Dool, A.W. Characterisation of surface roughness and sediment texture of intertidal flats using ERS SAR imagery. *Remote Sens. Environ.* **2005**, *98*, 96–109. [CrossRef]
18. Dehouck, A.; Lafon, V.; Baghdadi, N.; Roubache, A.; Rabaute, T. Potential of TerraSAR-X imagery for mapping intertidal coastal wetlands. In Proceedings of the 4th TerraSAR-X Science Team Meeting, Oberpfaffenhofen, Germany, 14–16 February 2011.
19. Van Beijma, S.; Comber, A.; Lamb, A. Random forest classification of salt marsh vegetation habitats using quad-polarimetric airborne SAR, elevation and optical RS data. *Remote Sens. Environ.* **2014**, *149*, 118–129. [CrossRef]
20. Lee, S.K.; Hong, S.H.; Kim, S.W.; Yamaguchi, Y.; Won, J.S. Polarimetric features of oyster farm observed by AIRSAR and JERS-1. *IEEE Trans. Geosci. Remote Sens.* **2006**, *44*, 2728–2735. [CrossRef]
21. Cheng, T.Y.; Yamaguchi, Y.; Chen, K.S.; Lee, J.S.; Cui, Y. Sandbank and oyster farm monitoring with multi-temporal polarimetric SAR data using four-component scattering power decomposition. *IEICE Trans. Commun.* **2013**, *96*, 2573–2579. [CrossRef]
22. Wang, W.; Yang, X.; Li, X.; Chen, K.; Liu, G.; Li, Z.; Gade, M. A fully polarimetric SAR imagery classification scheme for mud and sand flats in intertidal zones. *IEEE Trans. Geosci. Remote Sens.* **2017**, *55*, 1734–1742. [CrossRef]
23. Ullmann, T.; Schmitt, A.; Jagdhuber, T. Two component decomposition of dual polarimetric HH/VV SAR data: Case study for the tundra environment of the Mackenzie Delta region, Canada. *Remote Sens.* **2016**, *8*, 1027. [CrossRef]
24. Schmitt, A.; Wendleder, A.; Hinz, S. The Kennaugh element framework for multi-scale, multi-polarized, multi-temporal and multi-frequency SAR image preparation. *ISPRS J. Photogramm. Remote Sens.* **2015**, *102*, 122–139. [CrossRef]
25. Moser, L.; Schmitt, A.; Wendleder, A.; Roth, A. Monitoring of the Lac Bam wetland extent using dual-polarized X-band SAR data. *Remote Sens.* **2016**, *8*, 302. [CrossRef]
26. Schmitt, A.; Wendleder, A.; Roth, A.; Brisco, B. Water extent monitoring and water level estimation using multi-frequency, multi-polarized, and multi-temporal SAR data. In Proceedings of the IEEE International Geoscience and Remote Sensing Symposium, Quebec, QC, Canada, 13–18 July 2014.
27. Gade, M.; Melchionna, S.; Kemme, L. Analyses of multi-year Synthetic Aperture Radar imagery of dry-fallen intertidal flats. In Proceedings of the 36th International Symposium of Remote Sensing and Environment, Berlin, Germany, 11–15 May 2015.
28. Gade, M. A polarimetric radar view at exposed intertidal flats. In Proceedings of the IEEE International Geoscience and Remote Sensing Symposium, Beijing, China, 10–15 July 2016.
29. Gade, M.; Wang, W.; Kemme, L. On the imaging of exposed intertidal flats by dual-polarization Synthetic Aperture Radar. *Remote Sens. Environ.* **2017**. under review.

MDPI AG
St. Alban-Anlage 66
4052 Basel, Switzerland
Tel. +41 61 683 77 34
Fax +41 61 302 89 18
http://www.mdpi.com

*Remote Sensing* Editorial Office
E-mail: remotesensing@mdpi.com
http://www.mdpi.com/journal/remotesensing

www.ingramcontent.com/pod-product-compliance
Lightning Source LLC
Chambersburg PA
CBHW051710210326
41597CB00032B/5435

* 9 7 8 3 0 3 8 4 2 7 2 0 9 *